The Philosophy of Physics

Why has philosophy evolved in the way that it has? How have its subdisciplines developed, and what impact has this development exerted on the way the subject is now practiced? Each volume of "The Evolution of Modern Philosophy" will focus on a particular subdiscipline of philosophy and examine how it has evolved into the subject as we now understand it. The volumes will be written from the perspective of a current practitioner in contemporary philosophy, whose point of departure will be the question: How did we get from there to here? Cumulatively, the series will constitute a library of modern conceptions of philosophy and will reveal how philosophy does not in fact comprise a set of timeless questions but has rather been shaped by broader intellectual and scientific developments to produce particular fields of inquiry addressing particular issues.

Roberto Torretti has written a magisterial study of the philosophy of physics that both introduces the subject to the nonspecialist and contains many original and important contributions for professionals in the area. Unlike other fields of endeavor such as art, religion, or politics, all of which preceded philosophical reflection and may well outlive it, modern physics was born as a part of philosophy and has retained to this day a properly philosophical concern for the clarity and coherence of ideas. Any introduction to the philosophy of physics must therefore focus on the conceptual development of physics itself. This book pursues that development from Galileo and Newton through Maxwell and Boltzmann to Einstein and the founders of quantum mechanics. There is also discussion of important philosophers of physics in the eighteenth and nineteenth centuries and of twentieth-century debates. In the interest of appealing to the broadest possible readership the author avoids technicalities and explains both the physics and the philosophical terms.

Roberto Torretti is Professor of Philosophy at the University of Chile.

"... a rich work full of fascinating material on the history of the interaction between physics and philosophy."

Lawrence Sklar, author of *Physics and Chance*

THE EVOLUTION OF MODERN PHILOSOPHY

General Editors:
Paul Guyer and Gary Hatfield (*University of Pennsylvania*)

Roberto Torretti: *The Philosophy of Physics*

Forthcoming:
Paul Guyer: *Aesthetics*
Gary Hatfield: *The Philosophy of Psychology*
Stephen Darwall: *Ethics*
T. R. Harrison: *Political Philosophy*
William Ewald & Michael J. Hallett: *The Philosophy of Mathematics*
Michael Losonsky: *The Philosophy of Language*
David Depew & Marjorie Grene: *The Philosophy of Biology*
Charles Taliaferro: *The Philosophy of Religion*

The Philosophy of Physics

ROBERTO TORRETTI
University of Chile

CAMBRIDGE
UNIVERSITY PRESS

PUBLISHED BY THE PRESS SYNDICATE OF THE UNIVERSITY OF CAMBRIDGE
The Pitt Building, Trumpington Street, Cambridge, United Kingdom

CAMBRIDGE UNIVERSITY PRESS
The Edinburgh Building, Cambridge CB2 2RU, UK   www.cup.cam.ac.uk
40 West 20th Street, New York, NY 10011-4211, USA   www.cup.org
10 Stamford Road, Oakleigh, Melbourne 3166, Australia
Ruiz de Alarcón 13, 28014, Madrid, Spain

First published 1999

Typeface Sabon 10.25/13 pt.   *System* QuarkXPress [BTS]

A catalog record for this book is available from the British Library.

Library of Congress Cataloging-in-Publication Data is available.

0 521 56259 7 hardback
0 521 56571 5 paperback

Transferred to digital printing 2004

For Carla

Contents

Contents   xi

Preface

Like other volumes in "The Evolution of Modern Philosophy" series, this book is meant to introduce the reader to a field of contemporary philosophy – in this case, the philosophy of physics – by exploring its sources from the seventeenth century onward. However, while the modern philosophies of art, language, politics, religion, and so on seek to elucidate manifestations of human life that are much older and probably will last much longer than the philosophical will for lucidity, the modern philosophy of physics has to do with modern physics, an intellectual enterprise that began in the seventeenth century as a central piece of philosophy itself. The theory and practice of physics is firmly rooted in that origin, despite substantial changes in its informational contents, conceptual framework, and explicit aims. A vein of philosophical thinking about the phenomena of nature runs through the four-century-old tradition of physics and holds it together. This philosophy *in* physics carries more weight in the book than the reflections *about* physics conducted by philosophers. Our study of the evolution of the modern philosophy of physics will therefore pay much attention to the conceptual development of physics itself.

The book is divided into seven chapters. The purport and motivation of the first six are summarily described in the short introductions that precede them. The seventh and last chapter – "Perspectives and Reflections" – does not have an introduction, so I shall say something about it here. I had planned to close the book with a survey of current debate on the philosophy of physics in general (beyond the special philosophical problems of relativity and quantum mechanics studied in Chapters Five and Six). But the series editors asked me to give instead my own vision of the subject. Now, my imagination is too weak to encompass a vision of anything so vast, so I sketch instead what I regard as a coherent way of tackling the main issues. I believe that to do this ought to be more fruitful and will agree better with the contemporary spirit of philosophy than to erect some new idol of the forum for others to practice their markmanship on.

It is a welcome feature of contemporary societies that educated people have very different educational backgrounds. However, it makes it difficult to find a common denominator of prerequisites for potential readers of a book like this one. I assume that:

(a) The readers will know the names of great philosophers, such as Descartes, Spinoza, and Kant, and will be vaguely acquainted with some philosophical ideas, such as mind–body dualism, but, for the most part, they will have no professional training in philosophy. I have therefore avoided philosophical jargon and explained all essential philosophical notions.

(b) They are interested in physics and have a good recollection of high-school physics. Some college physics will make many things easier to understand, but it is not indispensable. Previous acquaintance with popular and semipopular books on twentieth-century physics can also be useful.

(c) They enjoyed their high-school mathematics and remember the gist of it; or they have later developed a taste for it and studied it again. I take this to include elementary Euclidean geometry, high-school algebra, and the rudiments of calculus. Mathematics beyond this level is needed only in §4.1.3 on Riemannian geometry; §§5.4 and 5.5 on general relativity and relativistic cosmology; and §§6.2, 6.3, and 6.4 on quantum mechanics. This is supplied in the Supplements at the end of the book and in some footnotes to §4.1.3. They are written in the standard prose style of mathematical textbooks and probably will be inaccessible to someone wholly unacquainted with this form of English. Like all idiolects, this one can only be acquired by practice, for example, by taking a good undergraduate course in modern algebra. Readers who find that they cannot understand the Supplements should just omit the sections listed above; they can also omit §2.5.3, "Analytical Mechanics", which mainly serves as an antecedent to §6.2.

(d) Except in §6.4.3, on quantum logic, readers are not required to know any formal logic. However, philosophy students who have taken a couple of courses in this area will – I expect – be enabled thereby to read and understand the mathematical supplements.

References are usually given by author name and publication year. Multiple works published by the same author in the same year are distinguished by lowercase letters. The choice of letters is arbitrary, except in the case of Einstein, in which, for papers published before 1920, I follow the lettering of the *Collected Papers*. In a few cases – usually "collected writings" – in which the publication year would be uninformative, I use acronyms (mostly standard ones). All coded references are decoded in the Reference list at the end of the book.

Translations from other languages are mine unless otherwise noted. The English translations that I have consulted (e.g., of Kant) are mentioned in the Reference list. In translations from continental languages, I treat Nature and Reason as feminine, when this use of gender contributes to dispel ambiguities.

◆

I warmly thank the friends and colleagues who have assisted me with advice, comments, preprints, offprints, and photocopies while I wrestled with the book: Juan Arana, Harvey Brown, Jeremy Butterfield, Werner Diedrich, John Earman, Bruno Escoubès, Miguel Espinoza, Alfonso Gómez-Lobo, Gary Hatfield, Christian Hermansen, Bernulf Kanitscheider, David Malament, Deborah Mayo, Jesús Mosterín, Ulises Moulines, Michel Paty, Massimo Pauri, Michael Redhead, and Dudley Shapere. I acknowledge with special gratitude the contribution of Francisco Claro, who detected a serious error in the first version of §6.3.2, on the problem of measurement in quantum mechanics. This led me to reformulate that section in a way that corrected the error and placed the whole matter in what I think is a better light. Of course, if new errors turn up in the new version, Dr. Claro bears no responsibility for them.

I also owe thanks to the University of Puerto Rico, where I taught until 1995. Although the book was written after my retirement, it is based on reading done while I was there using its library facilities and the free time that the University generously granted to me for research.

A shorter Spanish version of §3.4 was published in 1996 as "Las analogías de la experiencia de Kant y la filosofía de la física" in *Anales de la Universidad de Chile*. Parts of Chapter Seven were included, in Spanish translation, in my paper "Ruptura y continuidad en la historia de la física", which appeared in 1997 in *Revista de Filosofía* (Universidad de Chile). I thank the editors of these journals for permission to use those materials here.

I am very grateful to Alexis Ruda, Gwen Seznek, and Rebecca Obstler, of Cambridge University Press, for promptly replying to all my queries and requests while the book was been written and produced, and to Elise Oranges for her careful and accurate editing.

As in everything I have done during my adult life, my greatest debt is to Carla Cordua. It is with great joy that I dedicate to her this, my latest book, just as I did the first.

Santiago de Chile, 25 December 1998

✦

CHAPTER ONE

The Transformation of Natural Philosophy in the Seventeenth Century

Physics and philosophy are still known by the Greek names of the Greek intellectual pursuits from which they stem. However, in the seventeenth century they went through deep changes that have conditioned their further development and interaction right to the present day. In this chapter I shall sketch a few of the ideas and methods that were introduced at that time by Galileo, Descartes, and some of their followers, emphasizing those aspects that I believe are most significant for current discussions in the philosophy of physics.

Three reminders are in order before taking up this task.

First, in the Greek tradition, physics was counted as a *part* of philosophy (together with logic and ethics, in one familiar division of it) or even as *the whole* of philosophy (in the actual practice of "the first to philosophize" in Western Asia Minor and Sicily). Philosophy was the grand Greek quest for understanding everything, while physics or "the understanding of nature (*physis*)" was, as Aristotle put it, "about bodies and magnitudes and their affections and changes, and also about the sources of such entities" (*De Caelo*, 268ª1–4). For all their boasts of novelty, the seventeenth-century founders of modern physics did not dream of breaking this connection. While firmly believing that nature, in the stated sense, is not all that there is, their interest in it was motivated, just like Aristotle's, by the philosophical desire to understand. And so Descartes compared philosophy with a tree whose *trunk* is physics; Galileo requested that the title of Philosopher be added to that of Mathematician in his appointment to the Medici court; and Newton's masterpiece was entitled *Mathematical Principles of Natural Philosophy*. The subsequent divorce of physics and philosophy, with a distinct cognitive role for each, although arguably a direct consequence of the transformation they went through together in the

1

seventeenth century, was not consummated until later, achieving its classical formulation and justification in the work of Kant.

Second, some of the new ideas of modern physics are best explained by taking Aristotelian physics as a foil. This does not imply that the Aristotelian system of the world was generally accepted by European philosophers when Galileo and Descartes entered the lists. Far from it. The Aristotelian style of reasoning was often ridiculed as sheer verbiage. And the flourishing movement of Italian natural philosophy was decidedly un-Aristotelian. But the physics and metaphysics of Aristotle, which had been the *dernier cri* in the Latin Quarter of Paris *c.* 1260, although soon eclipsed by the natively Christian philosophies of Scotus and Ockam, achieved in the sixteenth century a surprising comeback. Dominant in European universities from Wittenberg to Salamanca, it was ominously wedded to Roman Catholic theology in the Council of Trent, and it was taught to Galileo at the university in Pisa and to Descartes at the Jesuit college in La Flèche; so it was very much in their minds when they thought out the elements of the new physics.

Finally, much has been written about the medieval background of Galileo and Descartes, either to prove that the novelty of their ideas has been grossly exaggerated – by themselves, among others – or to reassert their originality with regard to several critical issues, on which the medieval views are invariably found wanting. The latter line of inquiry is especially interesting, insofar as it throws light on what was really decisive for the transformation of physics and philosophy (which, after all, was not carried through in the Middle Ages). But here I must refrain from following it.[1]

1.1   Mathematics and Experiment

The most distinctive feature of modern physics is its use of mathematics and experiment, indeed its *joint* use of them.

A physical experiment artificially produces a natural process under carefully controlled conditions and displays it so that its development

---

[1] The medieval antecedents of Galileo fall into three groups: (i) the statics of Jordanus Nemorarius (thirteenth century); (ii) the theory of uniformly accelerated motion developed at Merton College, Oxford (fourteenth century); and (iii) the impetus theory of projectiles and free fall. All three are admirably explained and documented in Claggett (1959a). Descartes's medieval background is the subject of two famous monographs by Koyré (1923) and Gilson (1930).

can be monitored and its outcome recorded. Typically, the experiment can be repeated under essentially the same conditions, or these can be deliberately and selectively modified, to ascertain regularities and correlations. Experimentation naturally comes up in some rough and ready way in every practical art, be it cooking, gardening, or metallurgy, none of which could have developed without it. We also have some evidence of Greek experimentation with purely cognitive aims. However, one of our earliest testimonies, which refers to experiments in acoustics, contains a jibe at those who "torture" things to extract information from them.[2] And the very idea of *artificially* contriving a *natural* process is a contradiction in terms for an Aristotelian. This may help to explain why Aristotle's emphasis on *experience* as the sole source of knowledge did not lead to a flourishing of *experiment*, although some systematic experimentation was undertaken every now and then in late Antiquity and the Middle Ages (though usually not in Aristotelian circles).

Galileo, on the other hand, repeatedly proposes in his polemical writings experiments that, he claims, will decide some point under discussion. Some of them he merely imagined, for if he had performed them, he would have withdrawn his predictions; but there is evidence that he did actually carry out a few very interesting ones, while there are others so obvious that the matter in question gets settled by merely describing them. Here is an experiment that Galileo says he made. Aristotelians maintained that a ship will float better in the deep, open sea than inside a shallow harbor, the much larger amount of water beneath the ship at sea contributing to buoy it up. Galileo, who spurned the Aristotelian concept of lightness as a positive quality, opposed to heaviness, rejected this claim, but he saw that it was not easy to refute it by direct observation, due to the variable, often agitated condition of the high seas. So he proposed the following: Place a floating vessel in a shallow water tank and load it with so many lead pellets that it will sink if one more pellet is added. Then transfer the loaded vessel to another tank, "a hundred times bigger", and check how many more pellets must be added for the vessel to sink.[3] If, as one readily guesses, the difference is 0, the Aristotelians are refuted on this point.

[2] Plato (*Republic*, 537d). The verb βασανίζειν used by Plato means 'to test, put to the question', but was normally used of judicial questioning under torture. The acoustic experiments that Plato had in mind consisted in tweaking strings subjected to varying tensions like a prisoner on a rack.

[3] Benedetto Castelli (*Risposta alle opposizioni*, in Galileo, EN IV, 756).

Turning now to mathematics, I must emphasize that both its scope and our understanding of its nature have changed enormously since Galileo's time. The medieval *quadrivium* grouped together arithmetic, geometry, astronomy, and music, but medieval philosophers defined mathematics as the science of quantity, discrete (arithmetic) and continuous (geometry), presumably because they regarded astronomy and music as mere applications. Even so, the definition was too narrow, for some of the most basic truths of geometry – for example, that a plane that cuts one side of a triangle and contains none of its vertices inevitably cuts one and only one of the other two sides – have precious little to do with quantity. In the centuries since Galileo mathematics has grown broader and deeper, and today no informed person can accept the medieval definition. Indeed, the wealth and variety of mathematical studies have reached a point in which it is not easy to say in what sense they are one. However, for the sake of understanding the use of mathematics in modern physics, it would seem that we need only pay attention to two general traits. (1) Mathematical studies proceed from precisely defined assumptions and figure out their implications, reaching conclusions applicable to whatever happens to meet the assumptions. The business of mathematics has thus to do with the construction and subsequent analysis of concepts, not with the search for real instances of those concepts. (2) A mathematical theory constructs and analyzes a concept that is applicable to any collection of objects, no matter what their intrinsic nature, which are related among themselves in ways that, suitably described, agree with the assumptions of the theory. Mathematical studies do not pay attention to the objects themselves but only to the system of relations embodied in them. In other words, mathematics is about *structure*, and about *types* of structure.[4]

With hindsight we can trace the origin of structuralist mathematics to Descartes's invention of analytic geometry. Descartes was able to solve geometrical problems by translating them into algebraic equations because the system of relations of order, incidence, and congruence between points, lines, and surfaces in space studied by classical geometry can be seen to be embodied – under a suitable interpretation – in the set of ordered triples of real numbers and some of its subsets. The same structure – mathematicians say today – is *instantiated* by geo-

[4] For two recent, mildly different, philosophical elaborations of this idea see Shapiro (1997) and Resnik (1997).

metrical points and by real number triples. The points can be put – in many ways – into one-to-one correspondence with the number triples. Such a correspondence is known as a *coordinate system*, the three numbers assigned to a given point being its *coordinates* within the system. For example, we set up a *Cartesian coordinate system* by arbitrarily choosing three mutually perpendicular planes K, L, M; a given point O is assigned the coordinates $\langle a,b,c \rangle$ if the distances from O to K, L, and M are, respectively, $|a|$, $|b|$, and $|c|$, the choice of positive or negative a (respectively, b, c) being determined conventionally by the side of K (respectively, L, M) on which O lies. The *origin* of the coordinate system is the intersection of K, L, and M, that is, the point with coordinates $\langle 0,0,0 \rangle$. The intersection of L and M is known as the x-axis, because only the first coordinate – usually designated by x – varies along it, while the other two are identically 0 (likewise, the y-axis is the intersection of K and M, and the z-axis is the intersection of K and L). The sphere with center at O and radius r is represented by the set of triples $\langle x,y,z \rangle$ such that $(x - a)^2 + (y - b)^2 + (z - c)^2 = r^2$; thus, this equation adequately expresses the condition that an otherwise arbitrary point – denoted by $\langle x,y,z \rangle$ – lies on the sphere (O,r).

By paying attention to structural patterns rather than to particularities of contents, mathematical physics has been able to find affinities and even identities where common sense could only see disparity, the most remarkable instance of this being perhaps Maxwell's discovery that light is a purely electromagnetic phenomenon (§4.2). A humbler but more pervasive and no less important expression of structuralist thinking is provided by the time charts that nowadays turn up everywhere, in political speeches and business presentations, in scientific books and the daily press. In them some quantity of interest is plotted, say, vertically, while the horizontal axis of the chart is taken to represent a period of time. This representation assumes that time is, at least in some ways, structurally similar to a straight line: The instants of time are made to correspond to the points of the line so that the relations of betweenness and succession among the former are reflected by the relations of betweenness and being-to-the-right-of among the latter, and so that the length of time intervals is measured in some conventional way by the length of line segments.

Such a correspondence between time and a line in space is most naturally set up in the very act of moving steadily along that line, each point of the latter corresponding uniquely to the instant in which the mobile reaches it. This idea is present already in Aristotle's rebuttal of Zeno's "Dichotomy" argument against motion. Zeno of Elea claimed that an athlete could not run across a given distance, because before traversing any part of it, no matter how small, he would have to traverse one half of that part. Aristotle's reply was – roughly paraphrased – that if one has the time t to go through the full distance d one also has the time to go first through 1/2 d, namely, the first half of t (*Phys.* 233ᵃ21 ss.). In fact, Zeno himself had implicitly mapped time into space – that is, he had assigned a unique point of the latter to each instant of the former – in the "Arrow", in which he argues that a flying arrow never moves, for at each instant it lies at a definite place. Zeno's mapping is repeated every minute, hour, and half-day on the dials of our watches by the motion of the hands, and it is so deeply ingrained in our ordinary idea of time that we tend to forget that time, as we actually live it, displays at least one structural feature that is not reflected in the spatial representation, namely, the division between past and future. (Indeed, some philosophers have brazenly proclaimed that this division is "subjective" – by which they mean illusory – so one would do well to forget it . . . if one can.)

There is likewise a structural affinity between all the diverse kinds of continuous quantities that we plot on paper. Descartes was well aware of it. He wrote that "nothing is said of magnitudes in general which cannot also be referred specifically to any one of them," so that there "will be no little profit in transferring that which we understand to hold of magnitudes in general to the species of magnitude which is depicted most easily and distinctly in our imagination, namely, the real extension of body, abstracted from everything else except its shape" (AT X, 441). Once all sorts of quantities are represented in space, it is only natural to combine them in algebraic operations such as those that Descartes defined for line segments.[5] Mathematical physics has been doing it for almost four centuries, but it is important to realize that at one time the idea was revolutionary. The Greeks had a well-developed calculus of proportions, but they would not countenance ratios between heterogeneous quantities, say, between distance and time, or between mass and volume. And yet a universal calculus of ratios would seem to be a fairly easy matter when ratios between homogeneous

---

[5] Everyone knows how to add two segments a and b to form a third segment $a + b$. Descartes showed how to find a segment $a \times b$ that is the *product* of a and b: $a \times b$ must be a segment that stands in the same proportion to a as b stands to the unit segment.

quantities have been formed. For after all, even if you only feel free to compare quantities of the same kind, the *ratios* established by such comparisons can be ordered by size, added and multiplied, and compared with one another as constituting a new species of quantity on their own. Thus, if length b is twice length a and weight w is twice weight v, then the ratio b/a is identical with the ratio w/v and twice the ratio $w/(v + v)$. Euclid explicitly equated, for example, the ratio of two areas to a ratio of volumes and also to a ratio of lengths (Bk. XI, Props. 32, 34), and Archimedes equated a ratio of lengths with a ratio of times (*On Spirals*, Prop. I). Galileo extended this treatment to speeds and accelerations. In the *Discorsi* of 1638 he characterizes uniform motion by means of four "axioms". Let the index i range over $\{1,2\}$. We denote by s_i the space traversed by a moving body in time t_i and by v_i the speed with which the body traverses space s_i in a fixed time. The body moves with *uniform motion* if and only if (i) $s_1 > s_2$ if $t_1 > t_2$, (ii) $t_1 > t_2$ if $s_1 > s_2$, (iii) $s_1 > s_2$ if $v_1 > v_2$, and (iv) $v_1 > v_2$ if $s_1 > s_2$. From these axioms Galileo derives with utmost care a series of relations between spaces, times, and speeds, culminating in the statement that "if two moving bodies are carried in uniform motion, the ratio of their speed will be the product of the ratio of the spaces run through and the inverse ratio of the times", which, if we designate the quantities concerning each body respectively by primed and unprimed letters, we would express as follows:

$$\frac{v}{v'} = \left(\frac{s}{s'}\right)\left(\frac{t'}{t}\right) \qquad (1.1)$$

By taking the reciprocal value of the ratio of times, the ratio of speeds can also be expressed as a ratio of ratios:

$$\frac{v}{v'} = \left(\frac{s}{s'}\right) \Big/ \left(\frac{t}{t'}\right) \qquad (1.2)$$

If we now assume that the body to which the primed quantities refer moves with unit speed, traversing unit distance in unit time, eqn. (1.1.) can be rewritten as:

$$\frac{v}{1} = \left(\frac{s}{1}\right) \Big/ \left(\frac{t}{1}\right) \qquad (1.3)$$

which, except for the pedantry of writing down the 1's, agrees with the familiar schoolbook definition of constant or average speed.

1.2   Aristotelian Principles

The most striking difference between the modern view of nature and
Aristotle's lies in the separation he established between the heavens and
the region beneath the moon. While everything in the latter ultimately
consists of four "simple bodies" – fire, air, water, earth – that change
into one another and into the wonderful variety of continually changing organisms, the heavens consist entirely of aether, a simple body that
is very different from the other four, which is capable of only one sort
of change, viz., circular motion at constant speed around the center of
the world. This mode of change is, of course, minimal, but it is incessant. The circular motion of the heavens acts decisively on the sublunar region through the succession of night and day, the monthly lunar
cycle, and the seasons, but the aether remains immune to reactions
from below, for no body can act on it.

This partition of nature, which was cheerfully embraced by medieval
intellectuals like Aquinas and Dante, ran against the grain of Greek
natural philosophy. The idea of nature as a unitary realm of becoming, in which everything acted and reacted on everything else under
universal constraints and regularities, arose in the sixth century B.C.
among the earliest Greek philosophers. Their tradition was continued
still in the Roman empire by Stoics and Epicureans. Measured against
it, Aristotle's system of the world appears reactionary, a sop to popular
piety, which was deadly opposed to viewing, say, the sun as a fiery rock.
But Aristotle's two-tiered universe was nevertheless unified by deep
principles, which were cleverer and more stimulating than anything put
forward by his rivals (as far as we can judge by the surviving texts),
and they surely deserve no less credit than the affinity between Greek
and Christian folk religion for Aristotle's success in Christendom.
Galileo, Descartes, and other founding fathers of modern physics were
schooled in the Aristotelian principles, but they rejected them with surprising unanimity. It will be useful to cursorily review those principles
to better grasp what replaced them.

Aristotle observes repeatedly that the verb 'to be' has several meanings ("*being* is said in various ways" – *Metaph.* 1003^b5, 1028^a10). The
ambiguity is manifold. We have, first of all, the distinction "according
to the figures of predication" between *being a substance* – a tree, a
horse, a person – and *being an attribute* – a quality, quantity, relation,
posture, disposition, location, time, action, or passion – of substances.[6]

---

[6] 'Substance' translates οὐσία, a noun formed directly from the participle of the verb
εἶναι, 'to be'. So a more accurate translation would be 'being, properly so called'.

Aristotle mentions three other such spectra of meaning, but we need only consider one of them, viz., the distinction between *being actually* and *being potentially*. This is the key to Aristotle's understanding of organic development, which is his paradigm of change (just as organisms are his paradigm of substance – *Metaph.* 1032ª19). Take a corn seed. Actually it is only a small hard yellow grain. Potentially, however, it is a corn plant. While it lies in storage the potentiality is dormant; yet its presence can be judged from the fact that it can be destroyed, for example, if the seed rots, or is cooked, or if an insect gnaws at it. The potentiality is activated when the seed is sown and germinates. From then on the seed is taken over by a process in the course of which the food it contains, plus water and nutrients sucked up from the environment, are organized as the leaves, flowers, and ears of a corn plant. The process is guided by a goal, which our seed inherited from its parents. This is none other than the *morphe* ('form') or *eidos* ('species') of which this plant is an individual realization. The form is that by virtue of which *this* is a corn plant and *that* a crocodile. If a substance is fully and invariably what it is, its form is all that there is to it. Such are the gods. But a substance that is capable of changing in any respect is a compound of form and *matter* (*hyle*, literally 'wood'), a term under which Aristotle gathers everything that is actively or dormantly potential in a substance. Only such substances can be said to have a 'nature' (*physis*) according to Aristotle's definition of this term, that is, an inherent principle of movement and rest.

Although the development of organisms obviously inspired Aristotle's overall conception of change, it is not acknowledged as a distinct type in his classification of changes. This is tailored to his figures of predication. He distinguishes (*a*) the generation and destruction of substances, and (*b*) three types of change in the attributes of a given substance, which he groups under the name *kinesis* – literally, 'movement' –, viz., (*b₁*) alteration or change in quality, (*b₂*) growth and wane or change in quantity, and (*b₃*) motion proper – *phora* in Greek – or change of location. We need consider only (*a*) and (*b₃*), the former because it was believed to involve a sort of matter – in the Aristotelian sense – that eventually came to be conceived as matter in the un-Aristotelian modern sense and the latter because change of location was the only kind of change that this new-fangled matter could really undergo.

But the modern thinkers we are presently interested in were taught to say 'substance' (Lat. *substantia*) for Aristotle's οὐσία, so we better put up with it.

Motion (*phora*) was viewed by Aristotle as one of several kinds of movement (*kinesis*). Organic growth was another, somehow more revealing, kind. He wrote that *kinesis* is "the actuality of potential being as such" (*Phys.* 201ᵃ11), a definition that Descartes dismissed as balderdash (AT X, 426; XI, 39) but that surely makes sense with regard to a corn seed that grows into a plant when its inborn potentialities are actualized. Movement is thus conceived by Aristotle as a way of being. Zeno's arrow surely *is* at one place at any one time, but it *is moving*, not *resting* there, for it is presently exercising its natural potentiality for resting elsewhere, namely, at the center of the universe, where, according to Aristotle, it would naturally come to stand if allowed to fall without impediment.

I mentioned previously Aristotle's doctrine of the four simple bodies from which everything under the moon is compounded. They are characterized by their simple qualities, one from each pair of opposites, hot/cold and wet/dry, and their simple motions, which motivate their classification as light or heavy. Thus fire is *hot*, *dry*, and also *light*, in that it moves naturally in a straight line *away* from the center of the universe until it comes to rest at the boundary of the nethermost celestial sphere; earth is *dry*, *cold*, and also *heavy*, that is, it moves naturally in a straight line *toward* the center of the universe until it comes to rest at it; water is *wet*, *cold*, and *heavy* (though less so than earth); and air is *hot*, *wet*, and *light* (though less so than fire). Aristotle's notion of heaviness and lightness can grossly account for the familiar experience of rising smoke, falling stones, and floating porous timber.[7] But what about the full variety of actual motions? To cope with it, Aristotle employs some additional notions. Although the natural upward or downward straight-line motion of the simple elements is inherent in their compounds, the heaviness of plants and animals – which presumably consist of all four elements, but mostly of earth and water – can be overcome by their supervenient forms. Thus ivy climbs

[7] A reflective mind will find fault with them even at the elementary level. Imagine that a straight tunnel has been dug across the earth from here to the antipodes. Aristotelian physics requires that a stone dropped down this tunnel should stop dead when it reaches the center of the universe (i.e., of the earth), even though at that moment it would be moving faster than ever before. Albert of Saxony, who discussed this thought experiment *c.* 1350, judged the Aristotelian conclusion rather improbable. He expected the stone to go on moving toward the antipodes until it was stopped by the downward pull toward the center of the earth (which, after the stone has passed through it, is exerted, of course, in the opposite direction).

walls and goats climb rocks. The simple bodies and their compounds are also liable to *forced* motion (or rest) *against* their natures, through being pushed/pulled (or stopped/held) by other bodies that move (or rest) naturally. Thus a heavy wagon is forced to move forward by a pair of oxen and a heavy ceiling is stopped from falling by a row of standing pillars. But Aristotelian physics has a hard time with the motion of missiles. This must be forced, for missiles are heavy objects that usually go higher in the first stage of their motion. Yet they are separated from the mover that originally forces them to move against their nature. Aristotle (*Phys.* 266^b27–267^a20) contemplates two ways of dealing with this difficulty. The first way is known as *antiperistasis*: The thrown missile displaces the air in front of it, and the nimble air promptly moves behind the missile and propels it forward and upward; this process is repeated continuously for some time after the missile has separated from the thrower. This harebrained idea is mentioned approvingly in Plato's *Timaeus* (80a1), but Aristotle wisely keeps his distance from it. Nor does he show much enthusiasm for the second solution, which indeed is not substantially better. It assumes that the thrower confers a forward and upward thrust to the neighboring air, which the latter, being naturally light, retains and communicates to further portions of air. This air pushes the missile on and on after it is hurled by the thrower.[8]

Despite its obvious shortcomings, Aristotle's theory of the natural motions of light and heavy bodies is the source of his sole argument for radically separating sublunar from celestial physics. It runs as follows. Simple motions are the natural motions of simple bodies. There are two kinds of simple motions, viz., straight and circular. But all the simple bodies that we know from the sublunar region move naturally in a straight line. Therefore, there must be a simple body whose natural motion is circular. Moreover, just as the four familiar simple bodies move in straight lines to and from the center of the universe, the fifth simple body must move in circles around that center. The nightly spectacle of the rotating firmament lends color to this surpris-

---

[8] John Philoponus, commenting on Aristotle's *Physics* in the sixth century A.D., remarked that if this theory of missile motion were right, one should be able to throw stones most efficiently by setting a large quantity of air in motion behind them. This is exactly what Renaissance Europe achieved with gunpowder. However, the unexpectedly rich and precise experience with missiles provided by modern gunnery has not vindicated Aristotle.

ing argument. Its conclusion agrees well, of course, with fourth-century Greek mathematical astronomy, which analyzed the wanderings of the sun, moon, and planets as resulting from the motion of many nested spheres linked to one another and rotating about the center of the universe with different (constant) angular velocities.[9]

Changes of quality, size, or location, grouped by Aristotle under the name of *kinesis*, suppose a permanent substance with varying attributes. But Aristotelian substances also change into one another in a process by which one substance is generated as another is being destroyed. The generation of plants and animals can be understood as the incorporation of a new form in a suitable combination of simple bodies behaving pliably as matter. But the transmutation of one simple body into another – which, according to Aristotle, occurs incessantly in the sublunar realm – cannot be thus understood. However, the traditional reading of Aristotle assumes that in such cases the change of form is borne by formless matter, an utterly indeterminate being that potentially is anything and yet, despite its complete lack of definition, ensures the numerical identity of what was there and is destroyed with what thereupon comes into being. Recent scholars have questioned this interpretation and the usual understanding of the Aristotelian expression 'prime matter' (*prote hyle*) as referring to the alleged ultimate substratum of radical transformations.[10] Their view makes Aristotle into a better philosopher than he would otherwise be, but this is quite irrelevant to our present study, for the founders of modern science read Aristotle in the traditional way. Indeed, what they call 'matter' appears to have evolved from the 'prime matter' of Aristotelian tradition in the course of late medieval discussions. Ockam, for example, held that prime matter, if it is at all real – as he thought it must be to account for the facts of generation and corruption –, must in some way be actual: "I say that matter is a certain kind of act, for matter exists in the realm of nature, and in this sense, it is not potentially every act for

---

[9] It is important to realize that the celestial physics of Aristotle was deeply at variance with the more accurate system of astronomy that was later developed by Hipparchus and Apollonius and which medieval and Renaissance Europe received through Ptolemy. Each Ptolemaic planet (including the Sun and the Moon) moves in a circle – the *epicycle* – whose center moves in another circle – the *deferent* – whose center is at rest. But not even the deferent's center coincides with the Aristotelian "center of the universe", that is, the point to or from which heavy and light bodies move naturally in straight lines.

[10] King (1956), Charlton (1970, appendix). For a defense of the traditional reading see Solmsen (1958), Robinson (1974), and C. J. F. Williams (1982, appendix).

it is not potentially itself."[11] Such matter "is the same in kind in all things which can be generated and corrupted".[12] Moreover, although the heavenly bodies are incorruptible, Ockam was convinced that they too were formed from that same kind of matter:

> It seems to me then that the matter of the heavens is the same in kind as that of things here below, because as has been frequently said: *one must never assume more than is necessary*. Now there is no reason in this case that warrants the postulation of a different kind of matter here and there, because every thing explained by assuming different matters can be equally accounted for, or better explained by postulating a single kind.[13]

1.3 Modern Matter

'Matter' is just the anglicized form of *materia*, a Latin word meaning 'timber' that Cicero deftly chose for translating Aristotle's *hyle*. 'Matter' may thus in all fairness be seen as a contribution of philosophy to ordinary English. Yet its everyday meaning is a far cry from Aristotle's. In fact, the term could hardly keep its Aristotelian sense in a Christian setting. Christian theologians cherished Plato's myth of the divine artisan who molds matter[14] as potter's clay, but their God did not *encounter* matter as a coeval 'wet-nurse of becoming' (*Timaeus* 52d) but *created* it out of nothing. As an *actual* creature of God's will, Christian matter cannot be purely potential and indeterminate, but comes with all the properties required for God's purpose. Indeed, some seventeenth-century authors thought it most fitting that the world created by an all-knowing, all-powerful God should consist of a single universal stuff that develops automatically into its present splendor from a wisely chosen initial configuration, with no further intervention on His part. Be that as it may, surely the Deity of Christian philosophy knew exactly what He wanted when He created the world and could bring forth a material thoroughly suited to His ends.

Both Plato and Aristotle held that an exact science of nature was

---

[11] Ockam (*Summulae in libros Physicorum*, Pars I, cap. 16, fol. 6ra) quoted in Wolter (1963, p. 134).

[12] Ockam (*Expositio super octo libros Physicorum*, Lib. I, com. 1) quoted in Wolter (1963, p. 130).

[13] Ockam (*Reportatio in Sentent. II*, q. 22, D) quoted in Wolter (1963, p. 146) (my italics).

[14] Plato's word is χώρα, which in ordinary Greek meant 'extension, room, place', and often 'land, country'.

precluded by matter's inherent potentiality for being otherwise. Just as a geometer admires an excellent drawing but does not expect to establish true geometrical relations by studying it, so a "real astronomer" will judge "that the sky and everything in it have been put together by their maker in the most beautiful way in which such works can be put together, but will – don't you think? – hold it absurd to believe that the metrical relation (*symmetrian*) of night to day, of these to month, and month to year, and of the other stars to these and to each other are ever the same and do not deviate at all anywhere, although they are corporeal and visible" (Plato, *Rep.* 530a–b). The predictive success of Eudoxus's planetary models caused Plato to recant, and his spokesman in *Laws* (821b) asserts that "practically all Greeks now slander those great gods, Sun and Moon", for "we say that they and some other stars besides them never go along the same path, and we dub them roamers (*planeta*)." For Aristotle, heavenly motions are exact because they are steered directly by gods, but even gods could not achieve this if the heavens did not consist of aether, which admits no change except rotation on the spot. All other matter is incapable of such unbending regularity, and therefore sublunar events are not liable to mathematical treatment. Therefore, according to Aristotle, physics should not rely on geometrical notions such as 'concave', but rather use concepts like 'snub', which is confined to noses and involves a reference to facial flesh (*Phys.* 194^{a}13; cf. *De an.* 431^{b}13, *Metaph.* 1025^{b}31, 1064^{a}24, 1030^{b}29).

The idea of *created* matter does away with all such limitations. Indeed, the conception of universal matter professed with comparatively minor variations by Galileo, Descartes, and Newton seems expressly designed for mathematical treatment, or, more precisely, for treatment with the resources of seventeenth-century mathematics. The gist of it is concisely stated by Robert Boyle: "I agree with the generality of philosophers, so far as to allow that there is one catholic or universal matter common to all bodies, by which I mean a substance extended, divisible, and impenetrable" (1666, in SPP, p. 18). Matter being one, something else is required to account for the diversity we see in bodies. However, this additional principle need not consist of immaterial Aristotelian forms, but simply of the diverse motions that different parts of matter have with respect to each other. As Boyle puts it: "To discriminate the catholic matter into variety of natural bodies, it must have motion in some or all its designable parts; and that motion must have various tendencies, that which is in this part of the matter tending one way, and that which is in that part tending another" (Ibid.). Indeed, the actual division of matter into parts of different sizes and shapes is "the genuine effect of variously determined motion"; and "since experience shows us (especially that which is afforded us by chemical operations, in many of which matter is divided into parts too small to be singly sensible) that this division of matter is frequently made into insensible corpuscles or particles, we may conclude that the minutest fragments, as well as the biggest masses, of the universal matter are likewise endowed each with its peculiar bulk and shape" (SPP, p. 19). "And the indefinite divisibility of matter, the wonderful efficacy of motion, and the almost infinite variety of coalitions and structures that may be made of minute and insensible corpuscles, being duly weighed, I see not why a philosopher should think it impossible to make out, by their help, the mechanical possibility of any corporeal agent, how subtle or diffused or active soever it be, that can be solidly proved to be really existent in nature, by what name soever it be called or disguised" (Boyle 1674, in SPP, p. 145).

This view justifies Galileo's assertion that the universe is like a book open in front of our eyes in which anyone can read, provided that she or he understands the "mathematical language" in which it is written – for "its characters are triangles, circles, and other geometric figures without which it is humanly impossible to understand a single word of it" (1623, §6; EN VI, 232). It also implies the notorious distinction between the inherent "primary" qualities of bodies, viz., number, shape, motion, and their mind-dependent "secondary" qualities, viz., all the more salient features they display to our senses.[15] As Galileo explains further on in the same book:

---

[15] The distinction can be traced back to Democritus's dictum "By custom, sweet; by custom, bitter. By custom, hot; by custom, cold. By custom, color. In truth: atoms and void" (DK 68.B.9). But there are deep conceptual differences between ancient atoms and modern matter. Greek atomism is a clever and imaginative reply to Eleatic ontology: *Being* cannot change, but if allowance is made for Non-Being in the guise of the void, there is room for plurality and motion, and this is enough to account for the variety of appearances. Each atom is indivisible (*a-tomos*) precisely because it is a specimen of Parmenides's changeless Being. Modern matter is not subject to such ontological constraints. Descartes explicitly rejects atoms and denies the possibility of a void. Even Boyle, who invested much ingenuity and effort in pumping air out of bottles, was not committed to the existence of a true vacuum absolutely devoid of matter of any sort. And, although Boyle believed that matter is stably divided into very little bodies, he did not think that these *corpuscles* were indivisible in principle.

As soon as I conceive a matter or corporeal substance I feel compelled to think as well that it is bounded and shaped with this or that figure, that it is big or small in relation to others, that it is in this or that place, in this or that time, that it moves or rests, that it touches or does not touch another body, that it is one, or few, or many; and I cannot separate it from these conditions by any stretch of the imagination. But that it must be white or red, bitter or sweet, sonorous or silent, of pleasant or unpleasant smell, I do not feel my mind constrained to grasp it as necessarily attended by such conditions. Indeed, discourse and sheer imagination would perhaps never light on them, if not guided by the senses. Which is why I think that tastes, odors, colors, etc. are nothing but names with regard to the object in which they seem to reside, but have their sole residence in the sensitive body, so that if the animal is removed all such qualities are taken away and annihilated. But since we have bestowed on them special names, different from those of the other primary and real attributes, we wish to believe that they are also truly and really different.

(Galileo 1623, §48; EN VI, 347–48)

Less notorious but no less remarkable is the similarity between Galileian matter and the Aristotelian aether: Both are imperishable and unalterable, and capable only of change by motion. (Indeed, Galileo held at times that the natural motion of all matter is circular, though not indeed about the center of the universe.) Looking at a museum exhibit of lunar rocks we admire Galileo's hunch that they would turn out to be just sublunar stuff. But the very familiarity of that sight may cause us to overlook the main drift of his work. His aim was not to conquer heaven for terrestrial physics (a dismal prospect, given the state of the latter *c.* 1600), but rather to apply right here on earth exact mathematical concepts and methods such as those employed successfully in astronomy. The modern concept of matter made this viable.

The modern concept of matter also conferred legitimacy on experimental inquiry in the manner I shall now explain. If natural change involves the supervenience and operation of "forms" that scientists do not know, let alone control, and there is moreover an essential distinction between natural and forced changes, it is highly questionable that one can learn anything about natural processes by experiment. But if all bodies consist of a single uniform and changeless stuff, and all variety and variation results exclusively from the motion and reconfiguration of its parts and particles, the distinction between natural and forced changes cannot amount to much. If all that ever happens in the physical world is that this or that piece of matter changes its position and velocity, a scientist's intervention can only produce divisions and displacements such as might also occur without him. Experiments can be extremely helpful for studying – and mastering – the ways of nature because they are just a means of achieving faster, more often, and under human control, changes of the only kind that matter allows, viz., by "local motion, which, by variously dividing, sequestering, transposing, and so connecting, the parts of matter, produces in them those accidents and qualities upon whose account the portion of matter they diversify comes to belong to this or that determinate species of natural bodies" (Boyle 1666, in SPP, p. 69).

The most resolute and forceful spokesman for the modern idea of matter was Descartes (1641, 1644). He asked himself what constitutes the reality of any given body, for example, a piece of wax fresh from the honeycomb – not its color, nor its smell, nor its hardness, nor even its shape, for all these are soon gone if the piece of wax is heated, and yet the piece remains. But, says Descartes, when everything that does not belong to it is removed and we see what is left, we find "nothing but something extended, flexible, mutable" (AT VII, 31). Indeed, "extension in length, breadth and depth", its division into parts, and the number, sizes, figures, positions, and motions of these parts (AT VII, 50) are all that we can clearly and distinctly conceive in bodies and therefore provide the entire conceptual stock of physics. Obviously, *motion* is the sole idea that Cartesian physics adds to Cartesian geometry. Moreover, it is defined by Descartes in geometric terms:

Motion as ordinarily understood is nothing but *the action by which a body goes from one place to another.* [. . .] But if we consider what must be understood by motion in the light not of ordinary usage, but of the truth of the matter, we can say that *it is the transport of one part of matter or of one body, from the vicinity of those bodies which are immediately contiguous to it, and are regarded as being at rest, to the neighborhood of others.* By *one body* or *one part of matter* I understand all that is transported together, even if this, in turn, consists perhaps of many parts which have other motions. And I say that motion is *the transport*, not the force or action which transports, to indicate that motion is always in the mobile, not in the mover [. . .]; and that it is a property (*modus*) of it, and not a thing that subsists by itself; just as shape is a property of the thing shaped and rest of the thing at rest.

(Descartes 1644, II, arts. 24, 25; AT VIII, 53–54)

Matter as extension being naturally inert, the property of motion is bestowed on several parts of it by God; indeed the actual division of matter into distinct bodies is a consequence of the diverse motions of its different parts. While Descartes is emphatic that motion is just change of relative position, he was well aware that in collisions it behaves like an acting force. To account for this, Descartes developed the concept of a *quantity of motion*, which resides in the moving body and is transferred from it to the bodies it collides with according to fixed rules. According to Descartes, the immutability of God requires that the quantity of motion He conferred on material things at creation should remain the same forever. Descartes computes the quantity of motion of a given body by multiplying its *speed* by its *quantity of matter*. This notion led to the classical mechanical concept of momentum (*mass × velocity*), so I shall call it *Cartesian momentum*. Two important differences must be emphasized: (i) If extension is the sole attribute of matter, the quantity of matter can only be measured by its volume, so there is no room in Cartesian physics for a separate concept of mass. (ii) Cartesian momentum is the product of the quantity of matter by (undirected) *speed*, not by (directed) *velocity*, so, in contrast with classical momentum, it is a scalar, not a vector. This raises questions to which I now turn.

(i) A ball of solid gold can cause much more damage on impact that a ball of cork of the same size moving with the same speed. How does Cartesian physics cope with this fact? If matter coincides with extension, there are no empty interstices in the cork. However, the quantity of motion borne by either ball depends on their respective quantity of matter, that is, the volume of all the matter that moves together, and within the outer limits of each ball there are interstices filled with matter that does not move with the rest. This Cartesian solution is quaint, for one normally expects a moving sponge to drag the air in its pores, but it is not altogether absurd.

(ii) The only principle of Cartesian physics that still survives is the principle of inertia: A moving body, if not impeded, will go on moving with the same speed in the same direction. Here we have a universal tendency that – one would think – underlies the conservation of motion in a system of two or more bodies that impede each other by collision. Now, if direction is one of the main determinants of the persistent motion contributed by each colliding body, why did Descartes exclude it from his definition of the quantity of motion conserved in the system? This is not an easy question, as we shall now see.

The principle of inertia is embodied in two "natural laws" that Descartes derives "from God's immutability":

The *first* law is: Each thing, in so far as it is simple and undivided, remains by itself always in the same state, and never changes except through external causes. Thus if a piece of matter is square, we shall easily persuade ourselves that it will remain square for ever, unless something comes along from elsewhere which changes its shape. If it is at rest, we do not believe it will ever begin to move, unless impelled by some cause. Nor is there any more reason to think that if a body is moving it will ever interrupt its motion out of its own initiative and when nothing else impedes it.

· ·

The second natural law is: each part of matter considered by itself does not tend to proceed moving along slanted lines, but only in straight lines. [. . .] The reason for this rule, like that for the preceding one, is the immutability and simplicity of the operation by which God conserves motion in matter. For He conserves it precisely as it is at the moment when He conserves it, without regard to what it was a little earlier. Although no motion can take place in an instant, it is nevertheless evident that every thing that moves, at every instant which can be indicated while it moves, is determined to continue its motion in a definite direction, following a straight line, not any curved line.

(Descartes 1644, II, arts. 37, 39; AT VIII, 63–64)

With hindsight we scoff at Descartes for overlooking that an instant tendency to move in a particular direction may well be coupled with an instant tendency to change that direction in a given direction and still with a third tendency to change the direction of change, and so on, so that any spatial trajectory could result from a suitable combination of such directed quantities.[16] But this does not detract from the novelty and significance of his insight: Although motion cannot be carried out *in* an instant, it can exist *at* an instant, not as "the actuality of potential being" (whatever this might mean), but as a fully real directed quantity. Still, how could this insight be entirely forgotten in the definition of Cartesian momentum? It is true that vector algebra and analysis are creatures of the nineteenth century. But the addition

[16] For example, a particle moving at instant t with unit velocity $v(t)$ in a particular direction can also be endowed at that instant with, say, unit acceleration $v'(t)$. If $v'(t)$ is perpendicular to $v(t)$ and its rate of change $v''(t) = 0$, the particle moves with uniform speed on a circle of unit radius. In Newtonian dynamics, acceleration is always due to external forces and the Principle of Inertia is preserved, but this is not the outcome of a logical necessity, let alone a theological one, as Descartes claims (see §2.1).

of directed quantities by the parallellogram rule dates at least from the sixteenth century, and Descartes used it in his *Dioptrique* and subsequently discussed it at length with Fermat in 1637 in correspondence mediated by Mersenne (AT I, 357–59, 451–52, 464–74).[17]

Why then did he not resort to it for adding the motions of colliding bodies? In §1.5 we shall consider some devastating criticism of Cartesian physics by Leibniz and Huygens, which ultimately results from this omission. Some scholars think that Descartes could not combine motions by the parallellogram rule because he shared the Aristotelian belief that "each individual body has only one motion which is peculiar to it".[18] I cannot go further into this matter here, but there is one interesting consequence of the definition of Cartesian momentum as a scalar that I must mention. According to Descartes the human mind is able to modify – he does not say how – the direction of motion of small particles in the pineal gland although it cannot alter their quantity of motion. This ensures that a person's behavior can depend on her free will. This escape provision for human freedom is not available if the unalterable quantity of motion is a vector instead of a scalar.

1.4   Galileo on Motion

Galileo was 32 years older than Descartes and was already philosophizing about motion when the latter was born in 1596. Galileo's early writings criticize some Aristotelian tenets and show the influence of the impetus theory. According to this view, which was fathered by John Philoponus in late Antiquity and revived and further elaborated in the fourteenth century, the violent motion of missiles continues after they separate from the mover because the initial thrust impresses

---

[17] By this rule, if v and w are two directed quantities represented by arrows with a common origin p, their sum v + w is a directed quantity represented by an arrow from p to the opposite vertex of the parallellogram formed by the arrows representing v and w. Compare Newton's rule for the composition of forces, illustrated in Fig. 7 (§2.2).

[18] Descartes (1644, II art. 31), as cited in Damerow et al. (1992, p. 105). Taken in context, the passage does not, in my view, seem to support their opinion. Descartes wrote: "Etsi autem *unumquodque corpus habeat tantum unum motum sibi proprium*, quoniam ab unis tantum corporibus sibi contiguis et quiescentibus recedere intelligitur, participare tamen etiam potest ex aliis innumeris, si nempe sit pars aliorum corporum alios motus habentium" (AT VII, 57; I have italicized the sentence quoted by Damerow et al.).

a force or "impetus" on the missile that keeps it going until it gradually wears out. Although this conception is superficially similar to Descartes's idea of momentum transfer, it is not linked to a conservation principle and therefore is of little use for the quantitative study of motion. Galileo's fame as Newton's forerunner in the foundation of modern dynamics does not rest however on his early writings, but on the Latin treatise "On local motion" inserted in the Third Day of the *Discorsi* (1638). It begins with the analysis of uniform motion to which I referred in §1.1. It then sets up a mathematical model of free-fall, and finally tackles the motion of missiles, viewed as a combination of uniform motion and free-fall. Compared with the modern treatment of these matters, Galileo's suffers from some obvious limitations, for (i) in the proof of some key theorems he must make do without the infinitesimal calculus invented thirty years later by Newton and Leibniz, and (ii) he did not postulate, like Descartes and Newton, that unimpeded motion continues indefinitely at the same speed on a straight line in *every* case, but he argued for and used a "Law of Inertia" restricted to *horizontal* motion, that is, to motion near the surface of the earth that neither falls nor rises and therefore remains perpendicular to the local radius of the earth. But despite these shortcomings, "On local motion" provided a paradigm of mathematical physics that inspired the next generations and in a general way is still alive today.

Galileo defines *uniform motion* as one "in which the parts run through by the mobile in any equal times whatever are equal to one another" (EN VIII, 191). Since the impact of a freely falling body increases with the height from which it falls, it is clear that free-fall is not uniform motion. Galileo guesses that it is the simplest conceivable sort of accelerated motion (for free-fall is natural, and nature "habitually employs the first, simplest and easiest means" – EN VIII, 197), so he offers a precise definition of a type of motion meeting this requirement, produces geometrical proofs of several properties that this type of motion must have according to its definition, and leaves it to experiment to show that free-fall exhibits these properties. The definition is this: "We shall call that motion equably or uniformly accelerated which, departing from rest, adds on to itself equal momenta of swiftness (*momenta celeritatis*) in equal times" (EN VIII, 205). The expression 'momenta of swiftness' is later used interchangeably with 'degrees of speed', so one may assume that 'momentum' here simply means 'amount' or 'increment'; however, its Latin meaning, 'impulse,

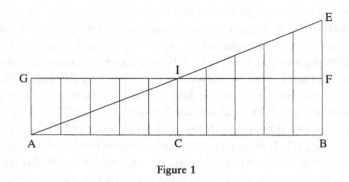

Figure 1

push', conveys the idea of a physical quantity that translates into impact on collision and whose actual presence in the mobile is what carries it forward at any given time.

This idea lurks in the proof of the fundamental Theorem I: "The time in which a space is traversed by a mobile in motion uniformly accelerated from rest is equal to the time in which the same space would be traversed by the mobile carried in uniform motion with a degree of speed (*velocitatis gradus*) equal to one-half the final and greatest degree of speed of the said uniformly accelerated motion". To prove it, we draw the lines AB, representing the duration of the uniformly accelerated motion, and BE, perpendicular to AB, representing the final speed v attained in that motion (Fig. 1).

Let F be the midpoint of BE, and draw the rectangle ABFG. Each parallel to GA drawn from a point on AB and contained in this rectangle represents the speed of the mobile moving uniformly with speed $\frac{1}{2}v$, at the time represented by that point. On the other hand, the speed attained at that time by the uniformly accelerated mobile is represented by the perpendicular to AB drawn from the said point to the line AE. Let I be the midpoint of FG and C the point where the perpendicular from I meets AB. The speed of the uniformly accelerated mobile equals $\frac{1}{2}v$ only at the instant represented by C. For each instant t before C in which the speed of the uniformly accelerated mobile falls short of $\frac{1}{2}v$ by a certain amount there is exactly one instant $f(t)$ after C in which the speed of the uniformly accelerated mobile exceeds $\frac{1}{2}v$ by precisely the same amount. Thus, "the deficit of the momenta in the first half of the accelerated motion – represented by the parallels in triangle AGI – is made up by the momenta represented by the parallels in triangle IEF"

(EN VIII, 209). Hence, the two mobiles will run through equal spaces in the time represented by AB.

It is worth noting that this argument does not rest only on the stated one-to-one correspondence between speed deficits at instants before C and speed excesses at instants after C. It holds good because, under the circumstances, the following metric condition is satisfied: given any time interval T before C, if $\mathcal{V}$ denotes the set of speeds less than $\frac{1}{2}v$ that the uniformly accelerated mobile successively sports during T, the matching set $f(\mathcal{V})$ of speeds greater than $\frac{1}{2}v$ is sported by the mobile after C during an interval $f(T)$ *of the same length* as T. Evidently, if $f(T) \neq T$, the contribution of $f(\mathcal{V})$ to the mobile's displacement cannot balance the deficit in the contribution of $\mathcal{V}$. The said contributions depend not only on the "momenta of swiftness" contained in $\mathcal{V}$ and $f(\mathcal{V})$, but also on the length of time during which their push is at work. A dolt would probably need the calculus to grasp this, but Galileo obviously did not.

By establishing a precise relation between uniformly accelerated motion and uniform motion, Theorem I enables us to use the mathematically manageable properties of the latter to calculate quantitative features of the former. Equation (1.2) entails that if mobiles M and M' run, respectively, through spaces s and s' in times t and t' with constant speeds v and v', $s/s' = vt/v't'$. From this elementary relation and Theorem I, Galileo infers that, "if a mobile descends from rest in uniformly accelerated motion, the spaces run through in any times whatever are to each other ... as the squares of those times" (Theorem II; EN VIII, 209). Let v_i denote the speed attained and s_i the space run through by the uniformly accelerated mobile in time t_i. According to Theorem I, s_i would also be run through in time t_i if the mobile were carried in uniform motion with speed $\frac{1}{2}v_i$. The concept of uniformly accelerated motion entails that v_1 has to v_2 the same ratio that t_1 has to t_2. (This can be read from Fig. 2, where the speeds attained in times AB and AC are represented, respectively, by the perpendiculars BE and CI, and BE:CI::AB:AC.) So

$$\frac{s_1}{s_2} = \left(\frac{v_1 t_1}{v_2 t_2} \right) = \frac{t_1^2}{t_2^2}$$

As a corollary we have that the spaces traversed by a uniformly accelerated mobile in the first, second, third, fourth, ... , nth units of time are to each other as the successive odd numbers $1:3:5:7:\ldots:2n-1$. (See Fig. 2, or apply the formula $n^2 - (n-1)^2 = 2n - 1$.)

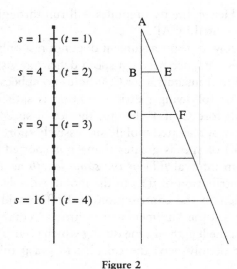

Figure 2

Galileo devised a clever way of testing this corollary.[19] He prepared a tilted plane with a groove down which a polished metal or marble ball could roll freely. He tied moveable gut frets around the plane as frets are tied on the neck of a lute. The rolling ball would make a sound as it passed each fret. The positions of the frets were adjusted until the sounds were on beat – this was something that Galileo's trained musical ear could ascertain to within 1/64 of a second. He could then verify that the distances between the frets – the spaces run through by the ball in equal times – satisfied the corollary. Galileo's plane had approximately a 3% tilt and was about 2 meters long. On a steeper plane the fall would be too fast for this kind of chronometry. However, Galileo's results are extended to any, even vertical, inclination by the following postulate introduced right after the definition of uniformly accelerated motion:

[19] See Drake (1978, pp. 86 ff.). Barbour (1989, p. 371) claims that in fact Galileo first discovered the corollary through the experiment that will be described in the main text, and then developed the theory from which he subsequently derived it. Although he does not cite any hard evidence for this claim, he may well be right. We ought however to bear in mind that, as Barbour himself emphasizes, "meaningful measurement is hardly possible without an underlying theoretical conception of what it is significant to measure" (1989, p. 376).

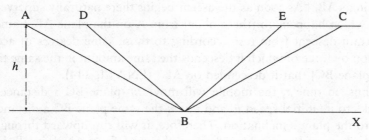

Figure 3

I assume that the degrees of speed acquired by a given mobile over differently inclined planes are equal whenever the heights of those planes are equal.

(Galileo EN VIII, 205)[20]

This postulate does not add any new qualifications to the concept of uniformly accelerated motion put forth by Galileo, but it does place a constraint on the physical realizations of this mathematical concept. Galileo claims that this concept is applicable to all cases of unimpeded free-fall, by which he means fall on inclined planes near the surface of the earth with all impediments removed.[21] The postulate prescribes a definite quantitative relation that must hold between such applications.

From Theorems I and II and the said postulate, Galileo derives numerous, in part testable, propositions concerning motion on diversely inclined planes. The postulate plays an essential role in Galileo's argument for his "Law of Inertia". Let us take a look at the curious reasoning leading to it. Consider a mobile freely falling from rest at A on an inclined plane AB of height h (Fig. 3). If it attains speed v in time t, the distance s it traverses in this time equals $\frac{1}{2}vt$. If the mobile then continues to move with *constant* speed v, it will traverse in time t the distance $vt = 2s$. Suppose however that during this second interval t the mobile is made to go up a plane BC with the same incli-

---

[20] The second, posthumous edition of the *Discorsi* (1655) contains a "proof" of this postulate that Galileo dictated before his death in 1642. It requires additional assumptions – which Galileo smuggled in from statics – for, as a moment's reflection shows, the behavior of the mobile on differently inclined planes cannot be extracted from the definition of uniformly accelerated motion.

[21] Note the difference between the classical notion of free fall and Galileo's. Classical physics treats the inclined plane as a partial impediment to fall.

nation as AB. "As soon as the ascent begins there naturally supervenes that which happens to [the mobile] from A on the plane AB, namely, a certain descent from rest according to those same degrees of acceleration by force of which it descends the same amount in the same time on [plane BC] that it descended on AB" (EN VIII, 244).

Thus, in time t, the mobile will move *up* plane BC a distance $2s$ thanks to its initial speed v, and *down* the same plane BC a distance s due to the plane's inclination. Therefore, it will run upward through a net distance s, thus reaching the same height h from which it fell. Had the mobile fallen from height h on a differently inclined plane DB, it would climb to height h on plane BE with the same tilt as DB. By the postulate the mobile attains at B the same speed v whether it falls on plane AB or DB. And, as we have just seen, once it has reached this speed it will climb back to height h, on plane BC or on plane BE, regardless of their particular inclinations. So, Galileo concludes, a mobile moving up a tilted plane with initial speed v will move forward, if all impediments are removed, until it reaches the same height h by falling from which it would have attained that speed. The smaller the tilt, the greater the space run through by the mobile and the longer the time it will employ to decelerate until it stops. In the limiting case in which there is no tilt at all, *the mobile will move forever with the same speed.*

This is the law of rectilinear uniform motion – the "Law of Inertia" – that Galileo will use in his theory of missiles. It is clear, from its derivation, that it is only meant to hold within the narrow range in which this theory is applicable, namely, within a few miles from a cannon's muzzle, under circumstances in which gravity is the only accelerating or retarding factor that counts. When the Aristotelian Simplicio points out that a surface which neither rises nor falls cannot be a plane, for its points must be equidistant from the center of the earth, Galileo's spokesman Salviati promptly concedes that he is right but also cites the example of Archimedes, who took it "as a true principle that the arm of a balance or steelyard lies in a straight line equidistant at all points from the common center of heavy things, and that the cords to which weights are attached hang parallel to one another" (EN VIII, 274). Such blatantly false assumptions lead only to negligible errors because "in actual practice our instruments and the distances we employ are so small in comparison with the great distance to the center of our terrestrial globe that we may well regard one minute of a degree of the meridian as if it were a straight line, and two verticals hanging from its extremities as if they were parallel" (Ibid., 275–76). The greatest distances occur in artillery shots, the longest of which "do not exceed four miles", *ca.* 1/1000 of the distance to the center. Indeed, the Galilean gunner is likely to err much more by considering air resistance to be negligible than by assuming the horizontal to be straight.

The keystone of Galileo's theory of missiles is his proof that a mobile endowed with uniform horizontal and uniformly accelerated vertical motions describes a semiparabola. The parabola had been studied by Apollonius in his treatise on conic sections (third century B.C.), but it had no place in a seventeenth-century gentleman's education. By appealing to such curves, the new physics wedded itself from the outset to what was then perceived as higher mathematics. For the benefit of Sagredo and Simplicio, his dialogue partners, Salviati defines a parabola as the curve at which the surface of a right cone is cut by a plane parallel to one of the cone's sides and orthogonal to the plane through that side and the cone's axis of symmetry. Such a curve has its own axis of symmetry, which it meets at the *apex*. The apex divides the parabola into two *semiparabolas*. Let P be a point on one of them. P's *amplitude* $x(P)$ is the length of the perpendicular from P to the axis. P's *altitude* $y(P)$ is the distance from the apex to the said perpendicular.[22] Galileo utilizes only two properties of the parabola, viz., (Π_1) if Q is another point on the semiparabola, $y(P)/y(Q) = x^2(P)/x^2(Q)$ – the altitudes are to one another like the amplitudes squared –, and (Π_2) if Q is a point on the axis, on the convex side of a parabola, at a distance from the apex equal to the altitude of P, the straight QP is tangent to the parabola.[23]

The proof in question follows at once from property (Π_1). Let a mobile move from a point O with constant horizontal velocity v, while falling vertically from rest with uniformly accelerated motion. Let x_i measure its horizontal advance and y_i its vertical descent after time t_i. Since $x_1/x_2 = t_1/t_2$ and $y_1/y_2 = t_1^2/t_2^2$ for any pair of times t_1 and t_2, the successive descents of the mobile are to one other like its advances

---

[22] Galileo speaks of the amplitude and altitude *of a given semiparabola*, by which he means, in my jargon, the amplitude and altitude of the point on which it meets the base of the cone. However, any parabola Γ forming the intersection of a finite cone K with a plane Φ is part of a longer parabola Γ' that forms the intersection of Φ with a cone K' that includes K as a proper part. Since the height of the cone employed for determining a given semiparabola is utterly irrelevant to Galileo's dynamics, it seems to me that my terminology, with mild anachronism, conveys Galileo's thinking more clearly than his.

[23] Salviati proves both. The modern reader will recognize in property (Π_1) the definition of a parabola in analytic geometry by the equation $y = ax^2$ (in terms of the coordinates of an arbitrary point in a Cartesian system originating at the apex).

squared. Descents and advances thus stand to one another, respectively, as the altitudes and amplitudes of the points on a semiparabola with its apex at O and a vertical axis. Therefore, the successive positions of the mobile must lie on such a semiparabola.

This beautiful piece of reasoning rests on the assumption – adopted by Galileo as a matter of course – that the mobile's constant forward thrust and its uniformly increasing downward thrust coexist and do their respective work without interfering.[24] In the above formulation it only applies to horizontal ("point-blank") shots, but it can be readily extended to missiles shot at any angle above the horizontal by analyzing the muzzle velocity into a horizontal and a vertical component and subtracting the constant magnitude of the latter from the growing downward velocity of fall (see Polya 1977, pp. 102–5). Galileo, however, does not handle such cases in this way but invokes instead considerations of symmetry. He tackles the gunner's problem of reaching a desired range by adjusting the gun's angle of elevation. After proving a few theorems he produces tables that display, for angles from 1° to 89°, the ranges and the maximum heights attainable by missiles shot with the same initial speed, and the different initial speeds required for attaining a given range. Such tables provide a large supply of testable predictions. The arguments leading to these results rest, as heretofore, on Galileo's ability to translate physical relations into ratios between geometrical magnitudes. To this effect he introduces a geometrical measure of a missile's initial speed, which he calls 'sublimity'. The sublimity $\sigma(v)$ is the *height* through which a mobile must fall from rest to attain speed v. Let x and y denote the amplitude and the altitude of an arbitrary point in the parabolic trajectory of a missile thrown horizontally with initial speed v. Galileo proves that $\frac{1}{2}x$ is the

---

[24] When Sagredo and Simplicio say that they are confused by the composition of motions, Salviati takes the said assumption for granted and merely discusses its algebra. He reaches the following precise statement of the rule for adding orthogonal vectors (cf. note 17): "When one must indicate the quantity of impetus resulting from two given impetuses, one horizontal and the other vertical, both being equable, one must take the squares of both, add them together, and extract the square root of their sum; this will give us the quantity of the impetus compounded from both." However, "when a motion enters the mixture, which starts from the greatest slowness and increases its speed as time goes by, it is necessary that the quantity of time shall manifest to us the quantity of the degree of speed at the given point. As for the rest, the impetus compounded from these two is (as in uniform motions) equal in the square to both components." (EN VIII, 289).

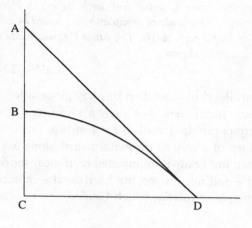

Figure 4

mean proportional between y and $\sigma(v)$. This enables him to calculate amplitudes from altitudes, for any given initial horizontal speed. Proposition VII and its remarkable "Corollary" pave the way for inferences concerning nonhorizontal shots.

> PROPOSITION VII. *In missiles which describe semiparabolas of the same amplitude, less impetus is required by the one which describes that whose amplitude is twice its altitude, than by any other.*

> (Galileo, EN VIII, 295)

Say that B and D are, respectively, the apex and an arbitrary point of a parabola, and let C denote the foot of the perpendicular from the axis to D (Fig. 4). Then, by Proposition VII, the requisite horizontal thrust for a missile to go from B to D is minimal if CD = 2BC. But 2BC is the distance from the horizontal through C (and D) to the point A where the axis meets the tangent to the parabola at D (by property Π_2 of parabolas). Thus the right triangle ACD is isosceles, and the tangent at D makes an angle of 45° with the horizontal. Coupled with a tacit but extremely bold appeal to symmetry, this simple reasoning yields the following

> COROLLARY. *From this it is clear that, conversely, the missile thrown from point D requires less impetus to describe the semiparabola DB, which has the tangent AD making one-half a right angle with the horizontal, than any other semiparabola having greater or smaller elevation than semiparabola DB. Hence it follows that if missiles are thrown from*

> *point* D *with the same impetus but with different elevations, the*
> *maximum throw, or amplitude of semiparabola . . . will be achieved with*
> *the elevation of half a right angle. The other throws, made at larger or*
> *smaller angles, will be shorter.*
>
> (Galileo, EN VIII, 296)

The title 'Corollary' bestowed on this key proposition is apparently meant to silence questioners. For such a symmetric correspondence between the semiparabola described by a missile thrown horizontally and the trajectory of a missile thrown upward along a tangent to the former's path can hold only if unimpeded rectilinear motion persists in *every* direction – and not just on the horizontal – in accordance with Descartes's principle of inertia, which Galileo never quite managed to embrace.[25]

1.5   Modeling and Measuring

In this section I shall propose three examples of seventeenth-century physics in the style of Galileo. In each one of them his method of mathematical representation of essentials while neglecting the negligible is brought to bear on a well-defined physical question under assumptions that were standard after Galileo and Descartes. In the first two examples, important theses of Cartesian physics are demolished and replaced. In the third one, the power of mathematical modeling yields the first estimate – to the right order of magnitude – of a speed so fast that its direct measurement by some commonsense method was utterly impracticable.

1.5.1   Huygens and the Laws of Collision

According to Descartes's definition of motion (cited in §1.3), a body B moves with speed v if (i) its place relative to its immediate surroundings S changes at that speed and (ii) S is regarded as being at rest. Let T denote the immediate outer surroundings of S. Suppose that the place of S relative to T is also changing with speed v, but in the opposite direction. One can then obviously regard S as moving and its total

---

[25] He comes close but stops short of it, for example, in the following statement: "Whatever degree of speed is found in the mobile, *this is by its nature indelibly impressed on it* when external causes of acceleration or retardation are removed, which happens *only on the horizontal plane*" (EN VIII, 243; my emphasis – Galileo's text is all in italics).

Figure 5

immediate surroundings $T \cup B$ as being at rest. Descartes's concept of motion is therefore completely relativistic (to a degree that Einstein's, despite his verbal allegations, is not). Indeed, his referral of motion to *immediate* surroundings is philosophically idle and only serves his desire of avoiding ecclesiastical odium (Descartes's earth is at rest *in the atmosphere*, although it circles the sun). Christiaan Huygens took this relativism seriously and used it for deriving true laws of collision to replace Descartes's rules.

Huygens assumes the validity of Descartes's principle of inertia. He postulates that two equal bodies colliding head-on at the same speed rebound after collision with the same speed. This is Descartes's first rule of collision (1644, II, art. 46) and should be evident on grounds of symmetry. By virtue of this postulate, Huygens's inquiry is restricted to *perfectly elastic collisions*. Descartes stated six more rules for collisions between bodies of different sizes or moving against each other with different speeds. By paying due attention to Descartes's definition of motion, Huygens shows that at least two of these rules must be false if the first one is true. He invites us to reflect on some imaginary experiments performed on a boat gliding along a quay – a daily occurrence in Huygens's Holland. The experiments are analyzed from the point of view of two persons who hold hands while standing, respectively, on the boat and on the quay, and assign different speeds to the colliding objects (Fig. 5). For the sake of brevity, I shall ignore this romantic setting.

Two equal balls, E and F, collide head on while moving in opposite directions with speed v relatively to boat B. Let $\mathbf{v}$ stand for the (directed) velocity of E, so that of F is $-\mathbf{v}$. According to Descartes's first rule, when they collide the balls in effect exchange velocities: E rebounds with velocity $-\mathbf{v}$ and F with velocity $\mathbf{v}$. Suppose that at the instant of collision B moves along Q with velocity $\mathbf{v}$. What are the velocities of E and F relative to Q before and after colliding? Huygens's commonsense answer is that E – which moves on B with the same velocity $\mathbf{v}$ that B has relative to Q – moves relatively to Q with velocity $2\mathbf{v}$, while F is at rest. Upon collision, the balls exchange velocities, as their motion relative to Q is computed by subtracting their new velocities on B from the latter's velocity $\mathbf{v}$ along Q (so E is at rest and F moves with velocity $2\mathbf{v}$). This is Huygens's first theorem on collisions: *If a body collides with an equal body at rest, after contact it will be at rest, while the latter will have acquired the velocity of the body that collided with it.* It openly contradicts Descartes's sixth rule.[26]

Suppose now that E and F collide while moving on Q with velocities $\mathbf{v}$ and $k\mathbf{v}$, respectively (where k is any positive or negative real number). Let B move along Q with velocity $\mathbf{v}' = \frac{1}{2}(\mathbf{v} + k\mathbf{v}) = \frac{1}{2}(1 + k)\mathbf{v}$. Then, right before collision, the velocity of E relative to B is $\mathbf{v} - \mathbf{v}' = \mathbf{v} - \frac{1}{2}(1 + k)\mathbf{v} = \frac{1}{2}(1 - k)\mathbf{v}$, while the velocity of F relative to B is $k\mathbf{v} - \mathbf{v}' = k\mathbf{v} - \frac{1}{2}(1 + k)\mathbf{v} = -\frac{1}{2}(1 - k)\mathbf{v}$. Thus, relative to B, balls E and F collide head-on with equal speeds. Descartes's first rule applies: E and F exchange velocities. So after collision, E moves relative to Q with velocity $-\frac{1}{2}(1 - k)\mathbf{v} + \mathbf{v}' = -\frac{1}{2}(1 - k)\mathbf{v} + \frac{1}{2}(1 + k)\mathbf{v} = k\mathbf{v}$, and F moves relative to Q with velocity $\frac{1}{2}(1 - k)\mathbf{v} + \mathbf{v}' = \frac{1}{2}(1 - k)\mathbf{v} + \frac{1}{2}(1 + k)\mathbf{v} = \mathbf{v}$. We have thus proved Huygens's second theorem on collisions: *If two equal bodies collide head-on with different velocities they shall move after collision with their velocities exchanged.* It contradicts Descartes's third rule.[27]

---

[26] Descartes (1644, II, art. 51). Here is a paraphrase: *Rule 6*: If a body collides while moving with speed v with an equal body at rest, it will rebound with speed $\frac{1}{4}v$ while the second body moves with speed $\frac{1}{4}v$ in the direction the first one had initially.

[27] Descartes (1644, II, art. 48). Here is a paraphrase. *Rule 3*: If two equal bodies collide head on while moving with speeds v and v' such that $v > v'$, after colliding they both move together with speed $v - v'$ in the original direction of the faster body. Huygens's second theorem also conflicts with Descartes's Rule 7 (Ibid., art. 53), insofar as it applies to equal bodies. This is a very complicated and seemingly arbitrary rule applicable to bodies moving in the same direction with different speeds that collide when one catches up with the other.

Huygens's second theorem entails both his first theorem and Descartes's first rule as special cases. By his masterful appeal to the relativity of motion Huygens succeeded in deriving the general rule from Descartes's first rule. One could also derive the latter and Huygens's second theorem from the first theorem, by using as a term of comparison a frame in which one of the balls is momentarily at rest at the time of collision. All three laws require the idealizing assumption of perfect elasticity, but will certainly be verified, within a plausible margin of error, on a billiard table. On the other hand, Descartes's rules 2–7 probably made him the laughing stock of billiard players who became acquainted with them. Note, by the way, that Huygens does not in any way assume the *constancy* of the velocities under discussion: The instantaneous velocity of each body right before and right after collision, with regard to each of two frames, is all that matters. It would seem however that the frames in question – a boat and a quay – do indeed move uniformly past each other. Not much later, Newton stated fundamental Laws of Motion under which this condition turns out to be indispensable.

1.5.2   Leibniz and the Conservation of "Force"

The German lawyer, philosopher, and mathematician, G. W. Leibniz, raised numerous objections against the physics and metaphysics of Descartes. In particular, he repeatedly argued that the 'quantity of motion' defined by Descartes as *mass × speed* cannot be conserved in nature. Strictly speaking, this quantity does not even exist, "for a whole never exists unless its parts coexist" (Leibniz, GM VI, 235), and motion, "being successive, perishes continually" (Leibniz, GP VII, 402). "Thus there is nothing real in motion save that which must consist of a force striving towards change", and "whatever there is in corporeal nature besides the object of geometry, or extension, must be reduced to this force" (GM VI, 235). "Force", in this sense, is the causal agent in motion. Its quantity must remain unchanged, for if it increased, "there would be an effect more powerful than its cause, that is, mechanical perpetual motion, capable of reproducing its cause plus something in addition, which is absurd"; and if it could decrease, "it would in the end perish utterly, because not being able to grow and yet being liable to diminish, it would always be set on a path of decay, which is no doubt contrary to the order of things" (GM VI, 220). The latter explanation reminds one of the bland optimism popularly associated with Leibniz since Voltaire's *Candide*. On the other hand, by his unhesitant dismissal of mechanical perpetual motion – the engineering equivalent of the economist's proverbial "free lunch" – Leibniz stands out clearly as the first one to adumbrate the modern concept of energy: "No machine, and thus not even the whole world can stretch its force without a new impulse from outside" (GM VI, 117).[28]

To prove his point Leibniz examines a particular example, but, he says, "one will find the same in any other example one might choose" (GM VI, 222). He assumes that the same "force" is required to lift one pound four feet as to lift four pounds one foot. (This is obvious, insofar as each operation can be analyzed into four operations of lifting one pound one foot.) Then, in the style of Galileo, he demands that all "external impediments shall be excluded or neglected, as if there weren't any".[29] He explains that, since we are here pursuing "les raisons des choses" and not any practical business, we may conceive of motion as taking place in a void, so there is no resistance of the medium; we may imagine that planes and spheres are perfectly smooth, so there is no friction, and so on. All this "in order to examine each thing separately with a view to combining them in practice". In virtue of this idealizing requirement Leibniz can sensibly assume that the entire force of a body can be transferred to another body without loss. The same, of course, must hold for Descartes's quantity of motion if this quantity is conserved. Finally, Leibniz accepts as "proven by others" that if any two bodies attain speeds v_1 and v_2 by falling freely from rest through heights h_1 and h_2, $h_1/h_2 = v_1^2/v_2^2$ (this is a straighforward consequence of Galileo's Theorem I on free-fall). And, of course, a pair of bodies moving with initial speeds v_1 and v_2 and meeting no

---

[28] Leibniz distinguishes between *physical* and *mechanical* perpetual motion. The former would be performed by a perfectly free pendulum "but this pendulum will never surpass its initial height, nor indeed will even return to it if on the way it operates or produces the slightest effect or overcomes the slightest obstacle; otherwise this would be a case of mechanical perpetual motion". The latter occurs when a system of bodies in motion "returns after some time to a state that is not only as violent as it was initially but even more so, because it is required that the device, besides restoring the first state, produces some mechanical effect or utility, without any external force contributing to it" (Leibniz in Costabel 1960, p. 98). Leibniz's use of 'mechanical' agrees with the original sense of the Greek verb μεχανάομαι, 'to contrive, procure for oneself'.

[29] This quotation and the next are from Leibniz's "Essay de dynamique" of 1692, published in Costabel (1960, p. 99).

air resistance will climb along perfectly smooth inclined planes to heights h_1 and h_2 such that $h_1/h_2 = v_1^2/v_2^2$.

From these premises we can reason as follows. Choose the unit of time so that a body falling from rest at a height of one foot acquires a speed of one foot per time unit. Suppose that we have the "force" to lift one pound four feet. If Descartes's quantity of motion is conserved, we can "stretch" this "force" to lift the same pound 16 feet! For we can certainly employ it to lift four pounds one foot. As they fall back to the starting point, our four pounds acquire a speed of one foot per time unit and thus a quantity of motion of $4 \times 1 = 4$. Under our idealizing assumptions, Descartes's conservation principle implies that this quantity of motion can be integrally transferred to a body of one pound. This body will then acquire a speed of four feet per time unit and will climb along a smooth inclined plane to a height of $4^2 = 16$ feet. Hence, Descartes's principle cannot hold. The quantity conserved in collisions and other such processes in which some bodies gain "force" at the expense of others is not proportional to *mass × speed* but rather to *mass × speed squared*. This quantity was usually referred to in eighteenth-century literature as "live force" (*vis viva*). If the factor of proportionality is set equal to $\frac{1}{2}$, the live force equals our kinetic energy, $T = \frac{1}{2}mv^2$.[30]

However, if the "quantity of motion" is equated, not with "the motion taken absolutely (without regard to its direction)", but with "the advance in a certain direction", then

> the total advance ... will be the sum of the particular quantities of motion, when the two bodies go in the same direction. But when they go in opposite directions, it will be the difference of their particular quantities of motion. And one shall find that *the same* quantity of advance is

---

[30] In 1807 Thomas Young introduced 'energy' as a term for "the product of the mass or weight of a body, into the square of the number expressing its velocity". However, it did not gain currency in physics until the 1850s, when it was adopted by Rankine and others as a general term for a body's capacity to do mechanical work. See the *Oxford English Dictionary* (*OED*), *s.v.* 'Energy', 6. Rankine distinguished between 'actual energy' or the power possessed by a body of doing work by virtue of its motion, and 'potential energy' or the power of doing work a body possesses by virtue of its position relative to other bodies. The latter name has stuck, but the former gave way to 'kinetic energy' (i.e., 'energy of motion'). Still, Einstein used 'lebendiger Kraft", – i.e., 'live force' – for 'kinetic energy' as late as 1902 (p. 433, line 7). As to the factor $\frac{1}{2}$, Kuhn (1977, p. 87, n. 47) says that it was introduced by Coriolis in 1829 to make the *vis viva* numerically equal to the work that it can produce.

conserved. But this should not be confused with the quantity of motion
in the ordinary [i.e., Cartesian – R. T.] sense. The reason for this rule
about advance can be shown in some way and it is reasonable that if
nothing supervenes from outside the whole (composed of bodies in
motion) will not impede its advancing as much as it did earlier.

(Leibniz in Costabel 1960, p. 105)

Thus, Leibniz recognized the conservation of *momentum* or *mass ×
velocity*, besides the conservation of the quantity he called "force",
which he deemed proportional to *mass × speed squared*, and also to
mass × height of fall.

1.5.3   Rømer and the Speed of Light

To the untutored eye light instantaneously fills any space exposed to
it. However, if light is a corporeal substance flowing from its source,
it is reasonable to believe, with Empedocles, that it takes time to spread
(DK 31.A.57). So Aristotle, ever anxious to reconcile natural philosophy with common sense, denied that light is an effluence of any sort
and conceived it as "the actuality of the transparent *qua* transparent"
(418^b10). The latter is a certain nature (*physis*) common to air, water,
and other things that possess it at least potentially even in darkness. It
becomes actual at a stroke due to the presence of fire and the like.
Therefore, Empedocles

> was wrong in saying that light travels and arrives at some time between
> the earth and that which surrounds it, without our noticing it. For this
> is contrary to the clear evidence of reason and also to the apparent facts
> (*ta phainomena*); for it might escape our notice over a short distance,
> but that it does so over the distance from east to west is too big an
> assumption.

(Aristotle, *De anima* II.7 418^b20–25, Hamlyn's transl.)

The instantaneous propagation of light remained the prevailing
opinion for two millenia, until almost the end of the seventeenth
century. Descartes explained it on the analogy of an impulse instantaneously transmitted from one end to another of a rigid body (AT VI,
83–85).[31] Galileo too expressed approval for it in *Il Saggiatore* (EN VI,
352); but in the *Discorsi*, Salviati questions the Aristotelian belief

---

[31] Surprisingly, this view did not prevent Descartes from trying to explain the dispersion of light on the analogy of a rain of pellets, whose angular velocities change in
different ways on penetrating a new medium.

voiced by Simplicio (EN VIII, 87). He proposes a simple experiment for measuring the speed of light:

> I would have two men each take one light, inside a lantern or other vessel, which each could conceal and reveal by interposing his hand, in sight of his companion. Facing each other at a distance of a few fathoms, they could practice revealing and concealing the light from each other's view, so that when either man saw a light from the other, he would at once uncover his own. After some mutual exchanges, this signaling would become so adjusted that without any sensible variation, either would immediately reply to the other's signal, so that when one man uncovered his light he would at the same time see the other man's light. This practice having been perfected at a short distance, the same two companions could place themselves with similar lights at a distance of two or three miles and resume the experiment at night, observing carefully whether the replies to their showings and hidings followed in the same manner as near at hand. If so, they could surely conclude that the expansion of light is instantaneous, for if light required any time at a distance of three miles, which amounts to six miles for the going of one light and the coming of the other, the delay ought to be quite noticeable. And if it were desired to make such observations at yet greater distances, of eight or ten miles, we could make use of the telescope.
>
> (Galileo, EN VIII, 88)

The idea behind this experiment underlies the very accurate methods for measuring the speed of light employed in the twentieth century (until it was fixed at 299,792,458 meters per second [m/s] by international convention in 1983), as well as Einstein's famous rule for the synchronization of distant clocks (§5.1). But Galileo's proposal for its actual performance grossly underestimates the speed of light compared with that of nervous signals from the experimenters' eyes to their hands. His spokesman acknowledges that *no delay* was observed when the experiment was tried over a distance of less than one mile.

In 1676, Ole Rømer reported to the French Academy a phenomenon which showed, in his analysis, that the speed of light is finite and provided a way of measuring it. At the time there was a desperate need for a reliable method of ascertaining longitude on a ship at sea. Someone figured out that Jupiter's satellite, Io, periodically hiding behind the planet once in every 40 hours or so, could be used for that purpose.[32] So Rømer, a junior astronomer in the Paris observatory,

---

[32] If I read in the nautical almanac that Io will emerge from hiding at midnight GMT and see it appear three hours before midnight, I can calculate that I am placed at lon-

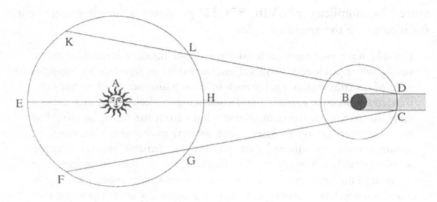

Figure 6

began painstakingly timing Io's eclipses. Prompted by observed irregularities or by his understanding of kinematics (or, more probably, by a combination of both), he lighted on the very bright idea that I shall now explain with the aid of Fig. 6. Let the circle through EFGHLK represent the orbit of the earth about the sun at A, while the circle through CD represents the orbit of Io about Jupiter at B. At C, Io enters Jupiter's shadow, from which it emerges at D.

Rømer estimated Io's period to be 42.5 hours. He realized, however, at that if Io is seen to emerge when the earth is at L and its next emergence occurs when the earth is at K, the interval between these two events must be greater than 42.5 hours for one must add the time that the light from Io takes in traversing the chord LK. Likewise, if two successive emergences are observed from F and G, the interval will be less than 42.5 hours. Rømer reckoned that if light took one second to traverse a distance equal to the diameter of the earth, it would need 3.5 minutes to go from L to K or from G to F, so the satellite's observed period would be seven minutes shorter when the earth approaches Jupiter than when it recedes from that planet. Since no such difference was recorded he concluded that light takes less than one second to travel through one terrestrial diameter (so its speed is greater than 12,000 km/s).[33] This does not mean, however, that it takes no time at

---

gitude 45°W. Indeed in the seventeenth century clocks were not accurate enough for such methods to work properly.

[33] This is the only numerical estimate of the speed of light ever published by Rømer. But our teaching profession has rushed in to fill the gap. Andrzej Wróblewski (1985)

all.[34] Although Rømer could not detect a time difference when comparing two single satellite periods, he found that 40 consecutive periods observed from side F are sensibly shorter than 40 consecutive periods observed from side L – no matter what the position of Jupiter – and that the difference amounts to 22 minutes for the earth's orbital diameter HE.

On the strength of his idea and of eight years of observations, Rømer announced in September 1676 that in November of that year Io would be emerging 10 minutes later than one would have expected from observations performed in August. When Io's emergence was observed in Paris on 9 November 1676 at 17h 34m 45s, confirming his prediction, he submitted to the French Academy a paper containing the above analysis of the delay. He stressed that the inequality could not be explained by Io's excentricity or any of the other causes usually adduced to account for the irregularities of the moon and the planets. For although he had observed that Io is excentric and that its periods are shorter or longer depending on whether Jupiter approaches the sun or recedes from it, the inequalities arising from these causes were not sufficient to filter out the inequality due to the finite speed of light.

Rømer's idea can be concisely expressed as follows: A signal issued at fixed intervals and propagating with the same finite speed will be

has compiled an amusing table of values of the speed of light, as diverse as 193,120 km/s and 351,000 km/s, for which Rømer is held responsible in standard physics textbooks. The dates given for Rømer's discovery in these textbooks range from 1656 – when he was a mere child – to 1876 – when, if still alive, he would have been long past the age of discovery – but they tend to cluster in the interval 1673–78.

[34] Robert Hooke – a distinguished English experimentalist remembered mainly for Hooke's law of elastic force, for his admirable drawings of tiny creatures seen under the microscope, and for his priority dispute with Newton over the discovery of universal gravitation – thought otherwise. Commenting on Rømer's work he wrote in 1680: "So far he thinks indubitable that [light] moves a Space equal to the Diameter of the Earth, or near 8000 Miles, in less than one single Second of the time, which is as short time as one can well pronounce 1, 2, 3, 4: And if so, why it may not be as well instantaneous I know no reason" ("Lectures on Light" in Hooke, PW, p. 78; cited in Wróblewski 1985, p. 625). Other contemporary scientists, including Rømer's boss, Cassini, rejected his discovery. But both Huygens and Newton embraced it wholeheartedly. The former reproduced Rømer's reasoning in his Traité de la lumière (submitted to the French Academy in 1678 and published in 1690); based on it, he estimated that light traveled more than 600,000 times faster than sound. Using for the speed of sound Huygens's value of 180 toises/s ≈ 330 m/s, this gives a speed of light > 198,000 km/s.

perceived at shorter intervals by an observer moving toward its source and at longer intervals by one moving away from it. This is known as the *Doppler effect*, for the German physicist who rediscovered it in the nineteenth century. It can be heard as a sudden tone change when a police siren speeds by you. The variation depends on the velocities of the signal and the observer (relative to the source), and if one of them is known, we can calculate the other. Let v denote the fixed frequency with which the signal is issued and Δv the difference, positive or negative, between the observed frequency and v; let u and c be, respectively, the observer's speed and the signal's speed relative to the latter's source. If we straightforwardly add velocities in the manner of Huygens (§1.5.1) and classical kinematics, $\Delta v/v = \pm(u/c)$, with the plus sign applying if the observer approaches the source and the minus sign if it recedes from it. If any three of the four quantities in this equation are given, it is an easy matter to compute the fourth one.

CHAPTER TWO

✦

Newton

"Natural Philosophy consists in discovering the frame and operations of Nature, and reducing them, as far as may be, to general Rules or Laws, – establishing these rules by observations and experiments, and thence deducing the causes and effects of things . . .".[1] Sir Isaac Newton wrote this in the program he proposed to the Royal Society after he became its president in 1703. It is likely that by "the frame of nature" he and his readers meant its ultimate ingredients, the original components of bodies.[2] One is tempted, however, to take the phrase as referring to the conceptual frame required for the mathematical description and explanation of natural phenomena. Newton himself put forward one such frame in the introductory sections of his masterpiece, the *Principia* of 1687 (Newton 1726, pp. 1–27). Without it one cannot make sense of the single, simple mathematical law by which he accounts in one breath for heavenly motions and free-fall. Newton's resolve to make explicit the structure underlying his physics sets him apart from the other founding fathers of modern physics; we must go back to Aristotle to find something of comparable breadth and depth. But Newton's conceptual frame, in stark contrast with Aristotle's, involves quantifiable, measurable attributes of things and was designed to fit

---

[1] "Scheme for establishing the Royal Society" (University Library Cambridge, *Add. MS* 4005.2); quoted in Westfall (1980, p. 632).

[2] That is apparently the meaning of the phrase in the following passage: "Perhaps the whole frame of Nature may be nothing but various Contextures of some certaine aethereall Spirits or vapours condens'd as it were by praecipitation, much after the manner that vapours are condensed into water or exhalations into grosser Substances, though not so easily condensible; and after condensation wrought into various formes, at first by the immediate hand of the Creator, and ever since by the power of Nature" (Newton, *Correspondence*, I, 364; quoted by Westfall 1971, p. 364).

the needs of mathematics and experiment. It remained the universal frame of physical inquiry until the advent of Einstein's Relativity, and, as we shall see in Chapter Three, it supplied a substantial part of the subject matter of Kant's critique of reason. It is still a paradigm of what such frames must be, which all who seek to outdo Newton bear in mind. It is a moot question whether a "frame of nature" in this sense is not always an *invented* frame *for* nature, but Newton and his immediate successors tended to think that he had *discovered* it.[3]

The elements of Newton's conceptual frame are *mass*, *force*, *time*, and *space*. Let us take them up by pairs.

2.1   Mass and Force

'Mass' is short for the *quantity of matter*, which, according to Definition I in the *Principia*, "arises jointly from its density and its magnitude".[4] The mass of a body is proportional – at a given location – to its weight. If m is the mass of a body moving with velocity v, its *quantity of motion* is measured by the product mv (Def. II). This definition recalls Descartes's yet differs from it in two essential respects: the body's mass is not equated with its bulk, and velocity is consistently treated as a *directed* quantity.

The word 'force' (Lat. 'vis', It. 'forza', Fr. 'force') was loosely used by seventeenth-century natural philosophers to mean an effective source of activity. Thus we hear about the force of oars and the force of a magnet, the force of a spring and the force of a blow, "the force of the wheels pulled by the weight" of a pendulum clock and "the law . . . that bodies conserve the force which causes their center of gravity to rise to the height from which it descended". Descartes protested that by "force" he did not mean "the power called the *force* of a man when

---

[3] In his reply to Cotes's letter of 18 February 1713, Newton wrote that "the first Principles or Axioms which I call the laws of motion . . . are deduced from Phaenomena and made general by Induction" (quoted in Koyré 1965, p. 275). But, of course, phenomena must be *described* and therefore *conceived* if they are to be used as premises in a deduction; and, as we shall see in §§2.1 and 2.2, Newton's description of the phenomena of motion essentially depends on the analysis expressed in the said "Principles or Axioms".

[4] Some have objected that this definition is circular, for 'density', they say, must be understood in terms of 'mass'. However, I do not see why Newton, who invented the calculus, could not have conceived density as a scalar field (a real-valued function of location in space) and mass as the integral of density over volume. I dare say that Definition I furnishes eloquent evidence that he did so, even if there is no other.

we say that such a man has more *force* than another" (AT II, 432); but quite obviously he often meant just that.[5] We saw in §1.5.1 that Leibniz sought to fix the meaning of the word in a physically significant way and designated by 'force' what we now call 'kinetic energy' (or perhaps 'energy' in general). Newton too struggled to attach a precisely quantifiable meaning to 'force' or 'vis'. He succeeded so well that the concept of force that is taught to fledgling students of physics even to this day can be traced directly to his.[6]

In an unfinished manuscript, "On the gravity and equilibrium of fluids" (c. 1668), Newton proposes the following definition:

> Def. 5. Force is the causal principle of motion and rest. And it is either an external principle that generates or destroys or in some way changes the motion impressed in a body; or an internal principle by which the motion or rest vested (*inditam*) on a body is conserved, and by which any being endeavors to persevere in its state and fights back if hindered (*et impeditum reluctatur*).
>
> (Hall and Hall 1962, p. 114)

This is echoed in *Principia*, Definitions III and IV, which characterize the *vis insita* (*inherent force*) of matter as that "by which every body, insofar as it is up to it, perseveres in its state of rest or of uniform straightline motion", and the *vis impressa* (*impressed force*) as "an action, exercised on a body, to change its state of rest or of uniform straightline motion" (1726, p. 2). Newton's remarks on these two kinds of force deserve close attention. The *vis insita* in a body is always proportional to its quantity of matter or *mass* (which Newton, in opposition to Descartes, distinguishes from its bulk). It differs from the *inertia* of mass only in the way in which it is conceived. "Due to the inertia of matter it is difficult to perturb the state of rest or motion of any body." Hence, the inherent force may also be called *vis inertiae* ('force of sluggishness'). A body exercises *this* force only when its state

---

[5] All the examples are culled from the appendices on the use of the term 'force' by Galileo, Descartes, and Huygens in Westfall (1971). The first two quotations are from Huygens (OC, XVIII 95) and (IX 456), respectively. Westfall says that "'force' in Borelli's usage was more an intuitive term, which referred to any apparent source of activity, than a precise technical one" and that it is often impossible to distinguish 'force' from 'strength' (1971, pp. 542–43); I should say that this is also true of the other authors mentioned. Of course, the core meaning of Latin *vis* and its equivalents in romance languages comprises both 'force' and 'strength'.

[6] For a concise and most illuminating study of seventeenth-century concepts of force, see de Gandt (1995, Ch. 2).

is changed by *another* force impressed on it. The exercise of *vis iner-tiae* is, under different respects, both resistance and impulse: "resis-tance, insofar as the body, to preserve its state, fights (*reluctatur*) the impressed force; impulse, insofar as the same body, yielding with difficulty to the force of a resisting obstacle, tends to change the state of this obstacle." As to the *vis impressa*, it "consists in the action only, and does not persist in the body after the action." Hence, the new state induced by the impressed force is maintained in the body solely by the force of inertia. *Vis impressa* has diverse origins, such as blows, pres-sure, or "centripetal force", that is, that force "by which bodies are drawn or impelled or in any way tend from any place towards a certain point, as to a center" (1726, p. 3).

Newton's remarks may not seem altogether consistent, for first he says that a body's *vis inertiae* is exercised only when it resists other things and reacts on them, and later he adds that the *vis inertiae* is what continually keeps the body in its current state of rest or motion. But perhaps such uneventful state preservation is not what Newton meant by 'exercising a force'. If, for a moment, one ignores the refer-ence to centripetal force, one may feel inclined to understand Newton's basic scheme as follows: Each body is endowed with an inherent force that is a principle of motion and rest that keeps it moving in a fixed direction with constant speed ≥ 0; when two bodies collide, the inher-ent force of each translates into an impulse that is impressed on the other. This is reminiscent of the Aristotelian world in which some bodies are compelled to move in a forced way as a consequence of the inherent motion of others; although, of course, for Newton as for Descartes, any inherent motion is uniform and rectilinear. But if Newton ever toyed with such an idea, he must soon have dropped it, for it is beset with difficulties. The resistance that a Newtonian body opposes any attempt to change its state of motion or rest depends on the body's mass, *not* on its velocity. On the other hand, the effect that such a body has on another body that stands in its way depends both on its mass and on the velocity with which it moves relatively to the other – indeed, as Leibniz showed, some effects are proportional to the *squared* velocity. In the light of these facts it is very difficult to view even the impulsive forces that show up in collisions as an expression of the *vis inertiae* of the colliding bodies. Moreover, the definition of force in "On the gravity . . ." suggests that already at its fairly early date Newton thought of the *external* principle of motion as primary, while the *internal* principle, the *vis inertiae*, merely preserves what a body has received from the former. This is clearly stated in one of the queries added by Newton to the 1717 edition of his *Opticks*:

> The *Vis inertiae* is a passive Principle by which Bodies persist in their Motion or Rest, receive Motion in proportion to the Force impressing it, and resist as much as they are resisted. By this Principle alone there never could have been any Motion in the World. Some other Principle was necessary for putting Bodies into Motion.
>
> (Newton, *Opticks*, p. 397)

Indeed, only by regarding impressed forces as primary – and not necessarily rooted in some inherent property of matter – was Newton free to conceive his revolutionary idea of a centripetal force that pulls a distant body to a mere point in space.

If inertia is "a passive principle", it is no wonder that physicists stopped referring to it as a *force*. 'Force' – in current parlance, 'Newtonian force' – was promptly equated with what he called 'impressed force'. And even this was not understood by his followers exactly as in his text. To see the difference we must turn to the "Axioms, or Laws of Motion" in which Newton specifies the relation between force and motion. In *Principia* these Axioms are very sensibly placed after the Scholium in which he explains the notions of time and space, for without these notions the Axioms are unintelligible. However, after 200 years of relentless public education, Newtonian space and time have become part and parcel of common sense, at least in the countries where the present book will circulate, so I may well wind up my discussion of force before tackling the Scholium. Indeed, as I hope to show, the Scholium itself only makes the right sense in the light of the Axioms, so there is some advantage in stating them first. Here is Newton's text, in Clifford Truesdell's translation:

> Law I. *Every body continues in its state of rest, or of uniform motion straight ahead, unless it be compelled to change that state by forces impressed upon it.*
>
> Law II. *The change of motion is proportional to the motive force impressed; and it takes place along the right line in which that force is impressed.*
>
> Law III. *To an action there is always a contrary and equal reaction; or, the mutual actions of two bodies upon each other are always equal and directed to contrary parts.*
>
> (Truesdell 1968, pp. 88–89; cf. Newton 1726, pp. 13–14)

Except for the "unless" clause, Law I is none other than the principle of inertia that we met in §1.3. The "unless" clause is a timely reminder of a remarkable fact: In a Newtonian world *no body* can ever continue, even for a very short while, in a state of rest or of uniform motion in a straight line,[7] unless indeed it is infinitely distant from all other bodies. For every body is attracted toward every other body by a force inversely proportional to their distance squared. Law I does not therefore reflect observations of the uniform rectilinear motion of bodies that are not acted on by forces – according to Newton, there are no such bodies to be seen – but rather a *decision* to analyze every *actual* motion into two contributing factors: the *present* velocity along the tangent to the observed trajectory, and the *change* that it is undergoing. The former is conceived as a state that the mobile possesses and will naturally keep unless disturbed, while the latter is to be accounted for by an external force. But why should the present velocity be rectilinear? Descartes argued that at any given instant a body can be determined to move in a particular direction but not along a particular curve (1644, II §39; AT VIII, 64; quoted in §1.3). Commenting on Descartes's text, I noted that a body endowed with a suitable set of instantaneous tendencies to move in a given direction, to change that tendency at a certain rate, to change that rate of change at a certain rate, and so on, can thereby be determined to move along a fixed trajectory of any conceivable shape. Newton's incisive analysis disposes of this objection: All those tendencies may indeed be alive at a given instant, but only the first is to count as the body's actual state of motion; the second displays the present action of external forces, while the third, fourth, and so on, must be accounted for by the changing configuration of such forces. The decision to see things in this way committed physics to find those forces or, at the very least, the equations linking their presence and evolution to observable physical quantities. To endorse Newton's analysis was therefore tantamount to signing an enormous promissory note. But Newton's first installment payment was so

---

[7] Descartes asserted that the same was true in his world; but the reason he gave does not hold water in his physics. He noted that, since there is no void, a body B_1 cannot move unless another body B_2 yields its place to it. B_2 in turn will take the place yielded by B_3, etc. Such replacements ultimately require that a body B_n (for some finite index n) takes up the place that B_1 is vacating. Therefore, concludes Descartes, in the real world all motion is by circulation. But this conclusion is quite unnecessary if matter, as postulated by Descartes, extends indefinitely in every direction.

magnificent and subsequent contributions by his followers so successful that his scheme of analysis was generally accepted as the naked
truth of the matter.

Except in history books, Newton's Law II is never stated today in
his own terms. Consider a body of mass m that is now moving with
velocity $\mathbf{v}$. Let $\mathbf{p}$ stand for the (quantity of) motion $m\mathbf{v}$. According to
Newton's text, if the motion presently changes to $\mathbf{p} + \Delta\mathbf{p}$, the change
is proportional to and collinear with the force impressed on the body:
$\Delta\mathbf{p} \propto \mathbf{F}$. This approach to motion change is well suited to the case
of collisions in which a body's motion is altered by the seemingly
instantaneous action of another body. However, it works rather poorly
if the change is brought about by a centripetal force, acting continually. Evidently, the magnitude of the change in motion effected by
a force of this kind depends on the length of time Δt during which
it acts. It is more reasonable, therefore, to put $\mathbf{F} = \mathbf{f}\Delta t$ and to conceive the force as the (directed) quantity represented by the vector $\mathbf{f}$.
In these terms, $\mathbf{f} \propto \Delta\mathbf{p}/\Delta t$. Indeed, since the centripetal force on a body
B changes its direction from instant to instant as B moves, the stated
relation is ambiguous or perhaps meaningless unless one puts on the
right-hand side, not the quotient $\Delta\mathbf{p}/\Delta t$, but the *limit* $\dot{\mathbf{p}}$ to which that
quotient converges as Δt decreases indefinitely: thus alone will the left-
hand side stand for the actual force acting on B at any given moment
(cf. Newton 1726, p. 38; $\dot{\mathbf{p}}$ is written dp/dt in the now current notation invented by Leibniz). If, by a good choice of units, we make the
constant of proportionality equal to 1, we obtain the now current formulation of Law II:

$$\mathbf{f} = \dot{\mathbf{p}} = m\dot{\mathbf{v}} \qquad (2.1)$$

The last term displays matter's resistance to change in motion: The
acceleration $\dot{\mathbf{v}}$ due to a given force $\mathbf{f}$ is inversely proportional to the
mass m. Equation (2.1) is appropriate also in the case of collisions, as
a moment's reflection will show. Since Newton and his contemporaries
did not assume that a colliding body's motion $\mathbf{p}$ can suffer a finite variation in no time at all, $\Delta\mathbf{p}$ must be regarded as the cumulative effect
of a force acting during a finite interval Δt, be it ever so short. So one
should write $\Delta\mathbf{p} = \mathbf{f}\Delta t$ or, making allowance for a time-dependent force,
$\Delta\mathbf{p} = \int_0^{\Delta t}\mathbf{f}(t)dt$. The latter quantity – which I previously designated by $\mathbf{F}$
– is commonly known as *impulse*. But there is no question that Newton
himself called it *force* (*vis*). Take, for instance, this passage from the
Scholium to the Laws of Motion:

When a body is falling, its uniform gravity (*gravitas uniformis*), acting
equally in equal time intervals, impresses equal forces (*vires*) on that
body, and generates equal velocities; and in the whole time impresses
a total force (*vim totam*) and generates a total velocity proportional to
the time.

(Newton 1726, p. 21)[8]

Newton's choice of words suggests that the force generating the body's
velocity is gradually impressed on the body and stored inside it. But
this suggestion is clearly incompatible with Newton's remark that the
impressed force "consists in the action only, and does not persist in the
body after the action" (1726, p. 2; quoted above). This remark suits
well the quantity I called **f**, made manifest by the *instantaneous rate of
change of the motion* as in eqn. (2.1). There are passages in *Principia*
where 'force' (*vis*) can only mean **f**. Take the first sentence of the proof
of Book II, Prop. XXIV: "For the velocity which a given force (*vis*) can
generate in a given matter in a given time is directly proportional to
the force and the time and inversely proportional to the matter" (1726,
p. 294). This entails that the force is directly proportional to the matter
and the velocity generated and inversely proportional to the time.
Obviously here 'force' denotes **f**, not **F**. With the symbols we used
above, the last statement can be written as $f \propto m\Delta v/\Delta t$, which yields
eqn. (2.1) at the limit $\Delta t \rightarrow 0$.[9]

The argument leading to eqn. (2.1) also motivates the standard
understanding of Law III as a statement about force (**f**) and counterforce (−**f**). Note that Law III speaks of *actions* exerted on bodies *by*
bodies.[10] Thus it is devoid of application if, as some learned commen-

---

[8] In Motte's translation, 'gravitas uniformis' is rendered as 'the uniform force of its
gravity'. Since 'vires' and 'vim totam' are rendered by 'forces' and 'whole force',
respectively, Newton is made to appear using the one word 'force' to name the two
quantities **f** and **F** in a single sentence.

[9] Here is another example. In the corollaries to Lemma X of Book I, Sect. I (1726, p.
34) we read that the distances traversed by bodies urged by different forces are among
themselves, "at the very beginning of the motion, as the products of the forces and
the squares of the times" (Corol. 3); so "the forces (*vires*) are directly as the spaces
described at the very beginning of the motion and inversely as the squares of the
times" (Corol. 4). Proportionality to s/t^2 holds of course for |**f**| (the magnitude of **f**)
but not for |**F**|.

[10] Cf. the following draft of Law III, first published by Herivel (1965, p. 307): "Corpus
omne tantum pati reactione quantum agit in alterum" ("As much as a body acts on
another so much does it experience reaction").

tators argue on the basis of other texts, Newtonian matter is incapable of acting. Anyway, gravity, the one force of nature that Newton was able to conceive at work in the phenomena of motion, shows up according to him as an action of each piece of matter on all other pieces of matter, which in turn react on the former in accordance with Law III.[11] If, as Newton's text openly suggests, Laws II and III hold in the same way for all forces, every observed acceleration **a** must be matched by an observable acceleration **b** collinear with **a** but in the opposite direction. However, **b** might not be easy to see if **a** is due to the action of many scattered bodies, so that **b** is in turn the acceleration of the *system* composed of the latter, resulting from the reactions suffered by each of them. Let m_k be the mass of a body labeled k and $\mathbf{a}_{hk}$ the acceleration experienced by body k due to the action of body h. Then, by Laws I and II, in the case of two mutually interacting bodies labeled 1 and 2, we have that

$$m_1 \mathbf{a}_{21} = -m_2 \mathbf{a}_{12} \tag{2.2}$$

In particular, if $\mathbf{a}_{21} = -\mathbf{a}_{12}$, evidently $m_1 = m_2$. Thus, in Newton's scheme of things, two bodies are said to have the same mass if and only if, when they interact, they impart on each another quantitatively equal accelerations. If a given body b is agreed to have unit mass, another body B has mass m if and only if, when interacting with b, it imparts on it an acceleration that is equal to the acceleration imparted by b on B multiplied by the scalar factor $-m$. Ernst Mach (1868) regarded this as the only viable *definition* of the Newtonian concept of *mass* (see §4.4.3).

After the Laws of Motion, Newton states six "Corollaries". The name suggests that they can be easily inferred from the Laws, but this is not wholly obvious. Corollary I says that "a body, acted on by two forces simultaneously, will describe the diagonal of a parallelogram in the same time as it would describe the sides by those forces separately" (1726, p. 14). Of course this can hold only if the lines along which the forces act lie on the same plane, but Newton does not mention this condition, perhaps because he tacitly assumed that the forces are both applied at the same point, as indeed they must be if they act on a dimensionless body (a "particle" in the sense defined in

---

[11] I shall therefore occasionally say that a gravitational force is exerted on a body *by* another body. Readers who feel uncomfortable with this manner of speech should substitute 'from' for 'by'.

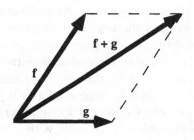

Figure 7

§2.2).[12] It follows that two such forces, represented by vectors **f** and **g**, have the same effect on the body as the force represented by their vector sum **f** + **g**, constructed as in Fig. 7.

Corollaries V and VI are quoted and discussed in §§2.2 and 2.3, respectively.

2.2 Space and Time

In a much quoted passage of the Preface to *Principia* (1687), Newton says that the whole task of philosophy consists in this: "From the phenomena of motion to investigate the forces of nature, and then from these forces to demonstrate the other phenomena". The Laws of Motion set up the links required for performing the first part of this task: Any body that deviates from rest or uniform motion in a straight line bears witness to the action of external forces, as well as to their magnitude and direction. However, without further specifications, it is impossible to ascertain, let alone to measure, such deviations.

For the sake of clarity and conciseness I shall introduce a few terms that became familiar in subsequent discussions of Newtonian mechanics. By a *particle* I mean a body of negligible length, width, and depth. By a *rigid body* I mean a collection of at least four particles, not all on the same plane, whose mutual distances do not vary. Newton's Laws of Motion must be understood as referring primarily to particles; indeed, they cannot be applied without more ado to nonrigid bodies. A particle is said to be *free* when it is not subject to impressed forces.

[12] Lagrange (1788, p. 6) states the principle of composition of forces without saying that for it to make sense the forces must be coplanar. But then on p. 50 he explains that "up to now we have considered bodies as points".

So Law I says in effect that *a free particle which is not at rest moves uniformly in a straight line.*

Talk of uniform motion makes no sense without a standard of time by which successive intervals can be pronounced equal. The Greeks found such a standard in the rotation of the firmament about the poles;[13] but for Newton and his contemporaries that motion merely reflected the actual rotation of the earth, which – if governed by Newtonian mechanics – achieves *very nearly* but not *perfectly* uniform speed.[14] So Newton made it clear that time, as understood in his Laws of Motion, is "absolute, true, and mathematical time," which "in itself, and from its own nature, flows equably without relation to anything external," and should be carefully distinguished from "any sensible and external (whether accurate or inequitable) measure of duration by means of motion" (Newton 1726, p. 6).[15] Still, "true time" cannot play a role in our physics unless it is displayed in a definite way by the phenomena of motion. Now, Law I provides just that: A free particle in motion traverses equal distances in equal times. This does not mean, however, that the uniform motion clause of Law I is merely a convention bestowing physical meaning on Newton's mathematical time. The Law is supposed to hold for *all* free particles. Hence, if *one* free particle is conventionally chosen to display real time, the other free particles will bear witness to the Law's validity. This view of Law I was put forward by Carl Neumann (1870). It raises an important question, which was noted by James Thomson. The arrival of a free particle at successive equidistant points on its path marks the completion of successive equal time intervals. This information, however, should be available everywhere. How can we collect it, say, at the origin of the

---

[13] Greek scientists were so sure about it that when they realized that the circumpolar motions of the firmament and of the Sun do not keep a constant but a periodically fluctuating ratio, they unhesitatingly concluded that the Sun's apparent motion was not uniform.

[14] The angular momentum of the earth is being very slowly but steadily eroded as a consequence of tidal friction. In the short run, of course, the earth's angular momentum is practically constant, so its angular velocity must fluctuate if there are small changes in the distribution of the earth's mass about its axis; this happens all the time, as a significant part of the earth's water gets reapportioned among the polar caskets, the oceans, and the atmosphere.

[15] Note, however, that Newton's absolute time, although free from any link to particular motions, is nevertheless structurally indistinguishable from the time of Greek mathematical astronomy and so, from the Euclidian straight line, the covering space of "the circle above" (Sophocles's term for the firmament at *Philoctetes* 815).

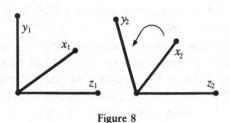

Figure 8

motion? In other words, how can one determine which events at the
origin are respectively simultaneous with the particle's arrival at the
said points? Evidently, this can only be done by signals transmitted
between distant points. Thomson comments: "The time required in the
transmission of the signal involves an imperfection in human powers
of ascertaining simultaneity of occurrence in distant places. It seems,
however, probably not to involve any difficulty of idealising or imag-
ining the existence of simultaneity" (1884, p. 569). Newton, at any
rate, took it for granted. "Every moment of time – he wrote – is diffused indivisibly throughout all spaces" (Hall and Hall 1962, p. 104).[16]
    Talk of straight line motion must also be referred to some standard
of rest or spatial frame of reference, apart from which it is meaningless.
Consider a rigid body B_1, consisting of three mutually perpendicular
straight edges, x_1, y_1, z_1, that meet at one point. Let B_2 be a copy of B_1,
with edges x_2, y_2, z_2, such that z_2 is always aligned with z_1. Suppose that,
when viewed from B_1, B_2 appears to rotate about the z_1–z_2-axis (Fig. 8).
Obviously, if Law I holds when referred to one of our bodies it does not
hold if referred to the other. For if a free particle moves with constant
speed, say, along the straight line containing x_1, it cannot describe a
straight line at rest relative to B_2. Newton resolved such uncertainties
by assuming that his Laws of Motion describe displacements in
"absolute space", which, "without relation to anything external,
remains always similar and immovable". Absolute space must not be
confused with any relative space "defined by our senses by its position
with respect to bodies" (1726, p. 6). Newton claims that, when thus
understood, his Laws provide of themselves the means for distinguishing between "true" motions in absolute space and "relative" motions in

[16] Curiously enough, when Thomson raised the question of distant simultaneity, Michelson had just invented a laboratory device that, by increasing beyond all contemporary expectation the precision of optical measurements, would wreak havoc with
Thomson's facile solution. See §5.1.

relative spaces. Specifically, absolute and relative motion are distinguished from one another "by the forces of receding from the axis of circular motion", which do not exist if the circular motion is merely relative, but are proportional to the quantity of the motion if it is true and absolute (1726, p. 10; cf. the quotation in §4.4.3). Thus, if our body B_1 is at rest in absolute space, B_2 will suffer stress along x_2 and y_2 while x_1 and y_1 remain unstressed, although they are rotating about the z_1–z_2–axis in the relative space defined by B_2.

Newton's clever appeal to his own Laws to fix the space for which they are meant to hold does not work quite as he says. Suppose that B_2 rotates about the z_1–z_2–axis as above while this axis is parallelly translated with constant speed in absolute space. Then, according to Newton's Laws, x_2 and y_2 will suffer exactly the same stresses as before, and x_1 and y_1 will be unstressed even though they are moving together with z_1. This follows at once from Corollary V to Newton's Laws of Motion:

> *The motions of bodies included in a given space are the same among themselves, whether that space is at rest, or moves uniformly forwards in a straight line without any circular motion.*

> (Newton 1726, p. 20)

Corollary V embodies Newton's principle of relativity,[17] by virtue of which the Laws of Motion hold equally well when referred to any member of an infinite family of frames – known as 'inertial frames' – moving uniformly in straight lines past one another. Ludwig Lange (1885; see also J. Thomson 1884), defined them on the analogy of Neumann's time standard: Three free particles, A, B and C, traveling from a point in three noncollinear directions, determine an inertial frame $\mathcal{F}$ on which one may define polar coordinates with origin at A, axis through A and B, and meridian $\varphi = 0$ on the plane ABC.[18] The First Law holds then *by definition* for particles A, B, *and* C but as *a matter of testable fact* for every other free particle. Moreover, it can be proved

---

[17] Better known as the Principle of *Galileian* Relativity, presumably because Galileo eloquently argued that by observing the phenomena of motion in a closed room one cannot tell whether one is standing on land or on a ship smoothly sailing on a tranquil sea (EN VII, 212–14). However, an essential feature of Galileo's comparison – as *he* saw it – is that the ship moves uniformly in a constant direction on the *spherical* surface of the earth. Cf. the various references to Galileo's "Law of Inertia" in §1.4.

[18] Lange does not specify that the trajectories of the three particles must not be collinear, but his scheme does not work without this requirement (made explicit by Robertson and Noonan 1969, p. 13). On the other hand, the condition that the three trajectories be *noncoplanar* – stipulated by von Laue (1955, p. 3) – is needlessly strong.

by sheer mathematical argument that any frame moving uniformly and rectilinearly with respect to $\mathcal{F}$ is also inertial by this definition.

It appears therefore that Newton's notion of absolute space, which has raised so much dust in philosophical debates, is actually otiose. However, in Newton's time mathematical discourse was not nimble enough to articulate the idea of inertial frames, that is, of relative spaces definable by their positions with respect to rigid bodies but extending to infinity and moving uniformly in one another with every conceivable velocity. On the way to this idea, the quaint conception of absolute space, which is traceable to late medieval thinkers[19] and was well entrenched in Cambridge c. 1660, served in effect as an excellent crutch. Indeed, what Newton himself had to say about the nature of space is, as we shall now see, very helpful for understanding frames (and much else that is essential to mathematical physics).

In the medieval ontology taught in seventeenth-century schools, everything was either a *substance* or a so-called 'accident', that is, an *attribute* or *relation* of substances. A substance in the world could be created or annihilated by God without regard to the existence of other worldly substances. On the other hand, attributes rested on the substances to which they belonged. Since a real interdependence between created substances would restrict God's power to act on one without touching the others, there was a strong tendency to think that relations among such substances existed only in the eye of the beholder. Contrary to all expectations, Newton maintains that space "has its own manner of existence, which fits neither substances nor accidents" (Hall and Hall 1962, p. 99). For space is incapable of acting like a substance – that is, of thinking like a mind or moving like a body – and yet it is not an attribute of a particular substance but rather something that is shared by all, "a disposition of being *qua* being"; for "no being exists or can exist which is not related to space in some way: God is everywhere, created minds are somewhere, and body is in the space that it

---

[19] Bradwardine (†1349), Oresme (†1382), Crescas (c.1340–c.1411). F. M. Cornford (1936) equated modern absolute space with the Greek atomists' void, which, as he noted, made its appearance simultaneously with the science of geometry. However, Greek geometry is not the science of infinite space, but of finite figures. And there are some deep differences between the atomists' void, existing from eternity *outside* bodies, and the space of Newton (and medieval theologians) *in which* God placed all bodies at creation. Anyway, it is likely that *cinquecento* natural philosophers like Bruno – who wholeheartedly embraced absolute space – drew inspiration from the atomist Lucretius, who referred to the void as 'spatium inane'.

fills" (p. 103). A deeper intimation of the ontological peculiarity of space is contained in Newton's proof of its immobility:

> Just as the parts of duration are individuated by their order, so that (for example) if yesterday could change places with today and become the later of the two, it would lose its individuality and would no longer be yesterday, but today; so the parts of space are individuated by their positions, so that if any two could exchange their positions, they would also exchange their identities, and would be converted into each other *qua* individuals. It is only through their reciprocal order and positions (*propter solum ordinem et positiones inter se*) that the parts of duration and space are understood to be the very ones that they truly are; and they do not have any other principle of individuation besides this order and position.
>
>   (Hall and Hall 1962, p. 103)

Newton's conception of a multitude of entities that are precisely the individual entities they are only by virtue of their mutual relations flies in the face of medieval ontology.[20] On the other hand, it is essential to modern mathematical physics, for any realization of a mathematical structure will in a way satisfy this conception. Now, a structure's relational system does indeed individuate its elements, but only up to isomorphism. This provides all of the individuation that we are likely to need if there are no internal isomorphisms besides the identity.[21]

[20] Of mundane things, that is, for the Christian Trinity, if at all conceivable, surely must come under some such conception.

[21] Let S and T be two realizations of the same type of structure. An *isomorphism* ϕ: $S \to T$ is a structure-preserving one-to-one mapping of S onto T, that is, roughly speaking, an assignment of an element $\phi(x)$ of T to every element x of S such that (i) each element of T corresponds to one and only one element of S, and (ii) if x, y, z, ... $\in S$ have a certain structural property or hold a certain structural relation to one another, $\phi(x)$, $\phi(y)$, $\phi(z)$, ... $\in T$ have the homologous property or hold the homologous relation. The isomorphism is *internal* if $T = S$. The standard term for internal isomorphism is 'automorphism'. Obviously, the *identity* mapping $x \mapsto x$, which assigns each $x \in S$ to itself, is an automorphism of S. As a simple example of a structure that admits no automorphisms besides the identity consider the ring of integers modulo 3, i.e., the set $\{0, 1, 2\}$, with addition defined by $0 + a = a + 0 = a$, $1 + 1 = 2$, $1 + 2 = 2 + 1 = 0$, $2 + 2 = 1$, and multiplication defined by $0 \times a = a \times 0 = 0$, $1 \times 1 = 1$, $1 \times 2 = 2 \times 1 = 2$, $2 \times 2 = 0$. To see this, note that if ϕ is an isomorphism of this ring onto itself, $\phi(0) = 0$ (if $\phi(0) = 1$, $\phi(1) \neq 1$, and $\phi(0) \times \phi(1) = \phi(1) \neq \phi(0)$, so ϕ is not an isomorphism; and the same would hold, *mutatis mutandis*, if $\phi(0) = 2$). Thus, unless ϕ is the identity, $\phi(1) = 2$ and $\phi(2) = 1$, in which case $\phi(2) \times \phi(2) = 1 \neq \phi(1)$, so ϕ is not an isomorphism. On the other hand, the *group* of integers modulo 3, i.e., the set $\{0, 1, 2\}$, with addition defined as above but without multiplication, admits the automorphism $\phi(0) = 0$, $\phi(1) = 2$, $\phi(2) = 1$, as the reader can easily verify.

However, Newtonian – that is, Euclidian – space admits an infinity of distinct internal isomorphisms, namely, by *translation* (of each point by the same distance along parallel directions), *rotation* (of each point by the same angle about the same center), *reflection* (mirror-imaging respect to a given plane), or any combination of mappings of one or more of these three kinds. Therefore, a single realization of Newtonian space contains infinitely many copies of itself and its points are individuated by their mutual relations only in the context of one or the other of those copies. In particular, if we designate one of these copies by $\mathscr{E}$ and we represent by the vector **v** a translation of each point of $\mathscr{E}$ in the direction of **v** by a distance equal to **v**'s length, then, if the parameter t ranges over the real numbers, the translations t**v** yield the successive positions of a frame $\mathscr{E}_v$ moving through $\mathscr{E}$ with constant velocity **v**. If $\mathscr{E}$ is inertial in Lange's sense, so is $\mathscr{E}_v$. Since $\mathscr{E}$ and **v** are arbitrary, in this – admittedly anachronistic – approach all inertial frames are equivalent.[22]

Newton's remarks on the ontology of space are contained in the manuscript "On the gravity . . .", first published in 1962. That is why his understanding of space as a self-contained relational system has been traditionally ascribed to Leibniz.[23] Newton did however communicate to the public his thoughts on the link of space to God. In the General Scholium appended to the second edition of *Principia* he says that God "endures forever, and is everywhere present; and, by existing always and everywhere, he constitutes time and space" (1726, p. 528). In Queries 28 and 31, at the end of the *Opticks*, he is more specific. He resorts to the medieval notion of the *sensorium* of animals, which he describes as "that place to which the sensitive Substance [the perceiving soul – R. T.] is present, and into which the sensible Species [aspects – R. T.] of Things are carried through the Nerves and Brain,

---

[22] We reach a more natural understanding of inertial frames and their equivalence in Newtonian physics if, with even bolder anachronism, we start not from Newtonian space *and* time, but from neo-Newtonian spacetime. See, for example, Ellis and Williams (1988, pp. 6–13), Friedman (1983, pp. 71–86), or Torretti (1983, pp. 20–31).

[23] But Leibniz would not dare to think that a relational system has "its own manner of existence which fits neither substances nor accidents", and maintained that space "is a mere ideal thing", which men conceive by forgetting bodies and paying attention only to their abstract order of coexistence. See *Mr. Leibnitz's Fifth Paper*, §47, in *The Leibniz-Clarke Correspondence* (Alexander 1956, pp. 69–72).

that there they may be perceived by their immediate presence to that Substance"; and he goes on to speak of "a Being incorporeal, living, intelligent, omnipresent, who in infinite Space, as it were in his Sensory, sees things themselves intimately, and throughly (sic) perceives them, and comprehends them wholly by their immediate presence to himself" (*Opticks*, p. 370). This "powerful, ever-living Agent [. . .] being in all Places, is more able by his Will to move the Bodies within his boundless uniform Sensorium, and thereby to form and reform the Parts of the Universe, than we are by our Will to move the Parts of our own Bodies" (*Opticks*, p. 403). I cannot say that I understand these theological pronouncements. I mention them only because we shall find them echoed, with an anthropic twist, in the philosophy of Kant.

2.3   Universal Gravitation

Newton's most celebrated achievement is his discovery of universal gravitation, or, in plain English, the heaviness of everything. But Newton does not understand heaviness (Lat. *gravitas*) in the Aristotelian sense, as the natural tendency of a body to move with accelerated motion toward the center of the universe. Newtonian gravity is doubly universal in that, by virtue of it, every body is accelerated toward every other body at once. Thus bluntly expressed the idea sounds incredibly confusing, but placed in the Newtonian frame described in §§2.1 and 2.2 it is not so. By Law II, the acceleration a_{21} of a given particle p_1 toward another particle p_2 discloses the action of a force f_{21} impressed on p_1 and pulling it toward p_2, in accordance with eqn. (2.1):

$$f_{21} = m_1 a_{21} \qquad\qquad (2.1^*)$$

Newton unhesitatingly treats f_{21} as a force *issuing* from p_2 that is therefore matched, according to Law III, by a counterforce f_{12}, issuing from p_1 and impressed on p_2, such that

$$f_{12} = m_2 a_{12} = -m_1 a_{21} = -f_{21} \qquad\qquad (2.2^*)$$

Still, this (anachronistic) display of vector algebra would be a false pretence had Newton not produced a "law of force" stating precisely how f_{12} depends on the positions and the masses of p_1 and p_2. Equation (2.3), commonly employed for expressing this law, is not found in Newton's writings in any even remotely similar form, but there is no doubt that it faithfully conveys the sense of his words, as illustrated by his practice, if Laws II and III are correctly represented by eqns. (2.1) and (2.2). Let m_i denote the mass of particle p_i and $r_i(t)$ its position at time t in the chosen inertial frame of reference ($i = 1,2$). For simplicity's sake I formulate eqn. (2.3) for a time t at which p_2 is momentarily located at the origin of our frame, so that $r_2(t) = 0$. $r_1(t)$ is therefore a vector pointing from p_2 to p_1 and its length $|r_1(t)|$ equals the distance between the two particles at t. Then, the gravitational force exerted by p_2 on p_1 at time t is given by:

$$f_{21}(t) = -Gm_1 m_2 \frac{r_1(t)}{|r_1(t)|^3} \qquad (2.3)^{24}$$

G is a constant of proportionality that can be made equal to 1 by a clever choice of units (but, of course, if natural phenomena are subject to several distinct "laws of force", one cannot expect all such constants to be simultaneously eliminable in this way). Substituting the right-hand side of eqn. (2.3) for the left-hand side of (2.2*) we see at once that m_1 occurs on both sides and may be canceled. In this way, Newton's law of gravity does justice both to the ordinary feeling that heaviness goes hand in hand with massiveness and to the experimental fact that all bodies fall with the same acceleration at any given place on earth.

Using Corollary I to the Laws of Motion (§2.1 *ad finem*), we can infer from eqns. (2.3) and (2.2) the acceleration $a_k(t)$ experienced by each member p_k of a system of n particles due to the gravitational forces exerted on it at time t by the other $n - 1$ particles:

$$a_k(t) = -G \sum_{i \neq k} m_i \frac{r_k(t) - r_i(t)}{|r_k(t) - r_i(t)|^3} \qquad (2.4)^{25}$$

If the only forces impressed on each particle are the gravitational forces issuing from the others, $a_k(t)$ is the total acceleration of p_k, and is there-

---

[24] We write r_1 in the numerator to convey the direction of the force f_{21}; but by so doing we make this force proportional to the distance $|r_1|$ between the particles. So, to express that f_{21} is inversely proportional to $|r_1|^2$ we must write $|r_1|^3$ in the denominator.

[25] The Greek letter Σ (sigma) indicates summation; the summands are obtained by replacing the index i in the term on the right of Σ by each number in the range indicated under Σ, which in this case is every number from 1 to n except k (so there are $n - 1$ summands).

fore equal to $\ddot{\mathbf{r}}_k(t)$, the second time derivative of $\mathbf{r}_k(t)$ (that is, the instantaneous rate of change of the instantaneous rate of change of p_k's posi-
tion at t). Equation (2.4) holds, of course, for any time t in which the
said conditions hold. Letting k range over $\{1, \ldots, n\}$, we obtain a
system of n second order ordinary differential equations in n unknowns
$\mathbf{r}_1, \ldots, \mathbf{r}_n$:

$$\ddot{\mathbf{r}}_k = -G\sum_{i \neq k} m_i \frac{\mathbf{r}_k - \mathbf{r}_i}{|\mathbf{r}_k - \mathbf{r}_i|^3} \qquad (2.5)^{26}$$

It is a property of such systems that they have at most one solution for
each set of initial conditions. In other words, if solutions exist on a
time interval I for certain given values of the masses m_i and of the posi-
tions and velocities $\mathbf{r}_i(t)$ and $\dot{\mathbf{r}}_i(t)$ at a particular time $t \in$ I, eqns. (2.5)
uniquely determine the values of the $\mathbf{r}_i$'s and the $\dot{\mathbf{r}}_i$'s at all times in that
interval. Of course, such determinations can be translated into exact
predictions and retrodictions only if the initial data are exact and
eqns. (2.5) are solved exactly.

For all their power, eqns. (2.5) would be of little use – and it would
have been virtually impossible for anyone to light on them – were it
not for the following features of our natural environment as it is
captured in the Newtonian frame:

1. The nongravitational forces acknowledged by Newton and his successors (to this day) do not make a contribution to the observable
   accelerations of the major bodies – planets, planetoids, satellites,
   comets – that circulate around the sun.
2. The action of nongravitational forces on bodies falling near the
   earth can be attributed to air resistance or else brought under the
   concept of constraints (e.g., to roll along an inclined plane or to
   swing at the end of a string; more is said on constraints in §2.5.3);
   Galileo taught how to filter out these two kinds of action in the
   study of terrestrial gravitation.
3. In a remarkable theorem (*Principia* Bk. I, Prop. LXXVI), Newton

---

[26] Roughly speaking, a second–order ordinary differential equation in one unknown is
an equation relating known quantities to the second derivative $d^2\varphi/dt^2$ of an unknown
function φ of a single independent variable t and possibly also to its first derivative
$d\varphi/dt$ and to φ itself. The equation is supposed to hold for the entire domain of the
function, that is, for every value of t for which φ is defined. In the notation that
Newton began to use in his manuscripts *c.* 1690, if φ is a function of time, the first
and second derivatives of φ are denoted by $\dot{\varphi}$ and $\ddot{\varphi}$, respectively.

established that, under his Law of Gravitation, the total attractive
force between any two disjoint, spherically symmetric masses M_1
and M_2 is equal to the force with which the masses M_1 and M_2
would attract each other if each was concentrated at the center of
the respective sphere. Thus, if the density of a spherical body
depends only on the distance from the center and not on the direction, one can study its gravitational action on an external particle
near its surface without paying any attention to the fact that some
parts of the body are closer to the particle than others. Fortunately,
the earth, the sun, and other major bodies of interest for the study
of gravitation can, to a good approximation, be treated as spherically symmetric in this sense.

4. Although there are much more than two bodies in the world, the
   overwhelming preponderance of the earth over the bodies that fall
   on it, and of the sun over the bodies that move around it makes the
   study of gravitation in two-body systems a realistic option. Thus,
   the acceleration of gravity at a particular latitude can be studied on
   a pendulum while ignoring the gravitational attraction of small
   bodies nearby. Thus, also, Kepler's results on planetary motion can
   be accounted for by Newton's Law of Gravity if – and only if – each
   planet is viewed as forming with the sun an isolated two-body
   system (see notes 28 and 29). This was essential for the discovery
   and initial confirmation of the Law, for even today eqns. (2.5) can
   only be solved exactly if $n = 2$.

5. Corollary VI to the Laws of Motion entitles one to deal with a two-
   body system as if it were isolated even when it is subjected to the
   strong gravitational action of other bodies, so long as this action is
   constant in intensity and direction over the region occupied by the
   two bodies. For Corollary VI states that

> *If bodies, moved in any manner among themselves, are urged in the*
> *direction of parallel lines by equal accelerative forces, they will all*
> *continue to move among themselves, after the same manner as if they*
> *had not been urged by those forces.*

> (Newton 1726, p. 21)

Thanks to Corollary VI, the simple Keplerian laws derivable for isolated two-body systems from the Law of Gravity apply to the system
formed by the earth and the moon, even though they are jointly
accelerated toward the sun, and to the system formed by the sun
and any single planet or comet, even though they jointly suffer the
gravitational pull of distant stars.

Newton claimed that the Law of Gravity was found by "deduction from phenomena" and generalized by "induction" (1726, p. 530), in compliance with the Rules of Philosophy set forth in *Principia* at the beginning of Book III. Before discussing these Rules (in §2.4; see the text quoted there), it will be good to watch him follow them. In the first place we must note that the six "phenomena" listed right after the Rules and on which Newton's induction rests are not such as one may observe by opening one's eyes and paying attention to what one sees.[27] They describe the behavior of diverse components of the Solar System as reconstructed by Kepler and his successors by laborious analysis – supplemented with geometrical interpolation – of numerous astronomical observations performed by Tycho Brahe with the naked eye and by seventeenth-century astronomers with the telescope. Kepler elicited from Tycho's data his so-called laws:[28]

I Every planet travels on an ellipse with the sun at one focus (*law of elliptical orbits*).[29]

---

[27] The title "Phænomena" was first used in the second edition (1713), which also introduced Phenomenon II. In the first edition, the other five phenomena were listed under the label "Hypotheses", together with the first two Rules of Philosophy and two other statements that later were either moved or removed. 'Hypothesis' is used here, as in Euclid, for the assumptions invoked as premises in the proof of the subsequent theorems. This usage differs from that of the General Scholium of the third edition, where – after proudly declaring: "I do not contrive hypotheses" – Newton defines the term as follows: "Whatever is not deduced from phenomena is to be called an hypothesis" (1726, p. 530; quoted in context in §2.5.1).

[28] Although Kepler's laws merely describe planetary motion, they are a far cry from sense appearances, as can be seen by the fact that astronomers from Eudoxus to Tycho did not even dream of them. Indeed, in Newton's own time not all astronomers acknowledged them. Huygens only accepted Law III and Newton himself was probably unaware of Law II – or rejected it – as late as 1679 (de Gandt 1995, pp. 84, 283). "Hence it was an unusual and very daring step to erect an astronomical system encompassing Kepler's three laws, as Newton did"; and they "gained a real status in exact science" only through him (Cohen 1980, p. 229). For an illuminating account of Kepler's achievement, see Barbour (1989, pp. 264–351).

[29] Newton was well aware that, because the Solar System comprises many interacting bodies, the law of elliptical orbits cannot hold exactly. In the tract *De Motu*, containing draft versions of definitions and propositions of *Principia*, he notes that, even if one treats the Solar System as isolated, so that its common center of gravity is either at rest or moves inertially in real space, this center does not coincide with the Sun, although it lies within it or is always very close to it. "Due to this deviation of the Sun from the center of gravity the centripetal force does not always tend to that immobile center, and hence the planets neither move exactly in ellipse nor revolve twice in the same orbit. There are as many orbits to a planet as it has revolutions, as in the

II The straight line joining the planet to the sun sweeps out equal areas
in equal times (*area law*).

III For each planet, the square of the time of revolution is proportional
to the cube of its mean distance to the sun (*harmonic law*).[30]

Newton's Phenomena I–IV are reformulations of the area law and
the harmonic law as applying (*a*) to Jupiter's moons in relation to
Jupiter (I); (*b*) to Saturn's moons in relation to Saturn (II); and (*c*) to
Mercury, Venus, Mars, Jupiter, and Saturn in relation to the sun (III
and IV). Phenomenon VI is that the area law is fullfilled by the moon
with respect to the earth, and Phenomenon V is that neither the area
law nor the harmonic law are obeyed by any of the named planets with
respect to the earth. Newton does not mention the law of elliptic orbits
among the phenomena, but it was evidently in Newton's mind as he
developed the mathematical theory of motion under the action of a
centripetal force in *Principia*, Book I.

Modern textbooks often show how to derive Kepler's three Laws
from eqn. (2.3).[31] Newton proceeded in the opposite direction. We
cannot go into details here, but I shall sketch the general drift of his
argument.[32]

Consider a body *B* moving around a fixed point O, to which it is

motion of the Moon, and each orbit depends on the combined motion of all the
planets, not to mention the action of all on one other. But to consider all these causes
of motion at once and to define these motions by exact laws allowing of convenient
calculation exceeds – if I am not mistaken – the force of every human intellect. Ignoring those minutiae, the simple orbit which is the mean among all errors will be the
ellipse previously discussed." (Herivel 1965, p. 297; transl. on p. 301).

[30] Kepler's harmonic law fails even for two-body systems, but it too can be recovered
by ignoring minutiae. The situation is as follows: If two bodies P_1 and P_2 move under
Newton's Law of Gravity around a third body S, then, if we neglect the interaction
between the first two bodies and denote by μ_1 and μ_2 the ratio of their respective
masses to the mass of S, their periods of revolution T_1 and T_2 stand in this relation
to the average distances r_1 and r_2 between each body and S:

$$\frac{(r_1/T_1)^{3/2}}{1+\mu_1} = \frac{(r_2/T_2)^{3/2}}{1+\mu_2}$$

The harmonic law holds exactly only if $\mu_1 = \mu_2$.

[31] See, for example, Courant (1936, pp. 422 ff.) or Pollard (1976, Ch. 1).

[32] Chandrasekhar (1995) provides a detailed commentary of Newton's argument for
universal gravitation, designed for today's reader. Unfortunately, this splendid book
was not available when I planned and wrote the present chapter, so I have used it
only to improve my exposition here and there.

drawn by a force. Newton proves that the areas swept by the moving radius OB all lie in the same plane and are proportional to the times it takes to sweep them; in other words, the body B obeys Kepler's law of areas (Bk. I, Prop. I; see the appendix at the end of this section). He proves next that if O is fixed or moves inertially and B moves around it in accordance with Kepler's law of areas, B is urged by a centripetal force directed to O (Prop. II). Moreover, if O is the center of another body C, moving in any way, and B moves around O according to the law of areas, B is urged by a force compounded of a centripetal force directed to O plus the sum of all the accelerative forces acting on C (Prop. III; a consequence of Cor. VI). If B moves with constant angular velocity in a circle, the centripetal force is directed to the center of the circle; and if several bodies move in this way and their periods of revolution are as the $\frac{3}{2}$th power of the radii of the respective circles (cf. Kepler's harmonic law), the centripetal forces will be inversely as the squares of the radii, and conversely (Prop. IV and its Cor. 6). Suppose now that B moves in an ellipse, urged by a centripetal force directed to a point O inside it. Can we say something about the magnitude of that force if O is (i) the center of the ellipse or (ii) one of its foci? Newton shows that in case (i) the force is proportional to the distance OB (Prop. X), and in case (ii) the force is inversely proportional to OB squared (Prop. XI). These results are extended to parabolic and hyperbolic trajectories. In particular, if the center of the ellipse of case (i) is removed to an infinite distance, "the ellipse degenerates into a parabola, the body will move in this parabola, and the force, now tending to a center infinitely remote, will become constant: this is *Galileo's* theorem," that is, his law of free-fall (Newton 1726, p. 54). Take several bodies moving in ellipses around a point O and urged by centripetal forces directed to O and inversely proportional to the square of their respective distances from O; these bodies obey Kepler's harmonic law, or, as Newton puts it, "the periodic times in the ellipses are as the $\frac{3}{2}$th power of their greater axes" (Prop. XV). "Therefore, the periodic times in ellipses are the same as in circles whose diameters are equal to the greater axes of the ellipses" (1726, p. 61; cf. Prop. IV and its Cor. 6).

In the middle sections of Book I, Newton proves a series of propositions on the motion of bodies attracted toward a motionless or inertially moving center of force. Many of them tackle from diverse angles the important astronomical problem of finding a trajectory from a few given points. Others draw conclusions from hypothetical assumptions to eventually show contrapositively that such assumptions do not hold in nature (e.g., Prop. XLV Cor. 1, on the motion of the apsides of the elliptical trajectory of a body urged by a centripetal force varying inversely as the rth power of its distance to the center, where $r \neq 2$). Then, at the beginning of Section XI, he notes that "very probably there is no such thing in nature" as a motionless center of attraction, "for attractions are made towards bodies, and the actions of the bodies attracted and attracting are always reciprocal and equal, by Law III; so that if there are two bodies, neither [one] is truly at rest, but both, being as it were mutually attracted, revolve about a common center of gravity;" and if there are several bodies, "which either are attracted by one body, which is attracted by them again, or which all attract each other mutually, these bodies will be so moved among themselves, that their common center of gravity will either be at rest, or move uniformly forwards in a straight line" (1726, p. 160). However, as Newton now goes on to prove, the abstract theory of centripetally urged motion is applicable in more realistic settings. Among other things, he proves the following: (*a*) If two bodies move in any way under mutual attraction, "their motions will be the same as if they did not at all attract each other, but were both attracted with the same forces by a third body placed in their common center of gravity; and the law of the attracting forces will be the same in respect of the distance of the bodies from the common center, as in respect of the distance between the two bodies" (Prop. LXI). (*b*) Bodies urged by centripetal forces that are inversely proportional to the square of their mutual distances "can move among themselves in ellipses, and by radii drawn to the foci describe areas very nearly proportional to the times" (Prop. LXV). (*c*) If three bodies attract each other with forces that are inversely proportional to the square of their mutual distances and the two lesser ones revolve about the largest one, the lesser body closest to the largest one describes areas around it that are more proportional to the times and a figure more approaching to that of an ellipse having one focus at the largest body if the mutual attractions between any two bodies are equal and opposite, in accordance with Law III, than if the largest body is not attracted at all by the lesser ones, or is attracted very much more or very much less than it attracts them (Prop. LXVI). (*d*) If any two bodies A and B attract each other and a system of other bodies C, D, and so on, with an acceleration that is inversely proportional to the square of the distance from the attracted to the attracting body, the forces issuing from A and B are proportional to their respective masses (Prop. LXIX).

Section XII proves several theorems on the attraction of spherical bodies, including these: If to the several points of a given sphere there tend equal centripetal forces decreasing as the square of the distances from the points, a particle placed within the sphere is attracted to the center of the latter by a force that is proportional to its distance from it (Prop. LXXIII), and a particle situated outside the sphere is attracted to the center of the latter by a force that is inversely proportional to the square of its distance from it (Prop. LXXIV). On the other hand, if a particle placed outside a homogeneous sphere is attracted to the center of the latter by a force that is inversely proportional to the square of the distance, and the sphere consists of attractive particles, the force of each particle will vary inversely as the square of the distance from it (Cor. 3 to Prop. LXXIV). I referred previously to Prop. LXXVI.

The astronomical phenomena listed at the beginning of Book III can be readily viewed as instances of the mathematical theory of Book I.[33] The phenomena are described and the theory itself was developed with a view to just this application. As noted earlier, the phenomena concern what we may call the Jovial system (Jupiter and its moons), the Saturnal system, the SP system (the sun and the five major planets known from Antiquity), and the EM system (earth and moon). By applying Props. II or III and Cor. 6 of Prop. IV (in Bk. I) to Phenomena I, II and V, Newton concludes that the Jovial, Saturnal, and SP systems are held together by centripetal forces tending, respectively, to Jupiter, Saturn, and the sun, which are inversely proportional to the squared distances from the circumambulating bodies to the respective centers. In the case of the SP system, the proportion between the centripetal forces and the squared distances can also be inferred from the stability of the orbital apsides, for "the slightest deviation" from the said proportion "should produce – by Cor. 1 of Prop. XLV in Bk. I – a motion of the apsides detectable in a single revolution, enormous in many" (1726, p. 395). From Phen. VI and Props. II or III in Bk. I it follows that the force by

---

[33] Book II of *Principia* develops a mathematical theory of the motion of bodies in resisting mediums, partly to show that Descartes's vortex theory of gravity is physically impossible (at any rate within the Newtonian frame). It marks the beginning of fluid mechanics.

which the moon is retained in its orbit tends to the earth. That this force is inversely as the square of the distance from the position of the moon to the earth's center is inferred through a detailed calculation from the very slow motion of the moon's apogee (i.e., the orbital apside furthest from the earth). Newton takes next the bold step that joined heaven and earth: he claims that the force that continually draws the moon off from a rectilinear motion and retains it in its orbit is none other than *gravity*, the force that causes ordinary heavy bodies to fall. The claim rests on a calculation: Given the moon's distance from the center of the earth R and its centripetal acceleration G, the earth's radius r, and the gravitational acceleration of a heavy body on the earth's surface g, Newton shows that, to a satisfactory approximation, $g/G \approx R^2/r^2$. To better appreciate the import of this calculation, Newton bids us imagine that the earth has several moons, and that the lowest of them is very small and so near the earth as almost to touch the tops of the highest mountains. The calculation implies that the centripetal force that retains this little moon in its orbit would be very nearly equal to the weight of any terrestrial bodies found on those mountain tops. "Therefore if the little moon should be deprived of the motion by which it advances in its orbit, then, lacking the centrifugal force by which it persists therein, it would descend to the earth; and that with the same velocity, with which heavy bodies actually fall on the tops of those mountains, due to the equality of the forces that make them descend. [. . .] Therefore since the force on heavy bodies and the one on the moons are directed to the center of the earth, and are similar and equal between themselves, they both have (by Rules I and II [see §2.4]) the same cause" (1726, p. 398).

From this point on, Newton's reasoning flies unhampered. Since the revolutions of Jupiter's and Saturn's moons about Jupiter and Saturn, and of the planets about the sun, "are phenomena of the same kind as the revolution of the moon about the earth", and it has been shown that the forces that keep the said bodies in their orbits tend, respectively, to the centers of Jupiter, of Saturn, and of the sun, and decrease with distance in the same way as the terrestrial gravity decreases in receding from the earth, he concludes that there is a force of gravity tending to all the planets, that all planets gravitate toward one another, and that every body gravitates toward each planet (1726, pp. 399, 400). By filling wooden boxes with equal weights of gold, silver, lead, glass, sand, common salt, wood, water, and wheat, and letting them oscillate as pendulums from equal threads, 11 feet long, Newton had satisfied himself that terrestrial gravity acts equally on all materials.[34] Having established that it is no different from the gravity that tends to the other heavenly bodies, he took these experimental results as sufficient evidence that "the weights of bodies towards any one planet, at equal distances from the center of the planet, are proportional to the quantities of matter which they severally contain" (Bk. III, Prop. VI). The next proposition proclaims universal gravitation:

PROP. VII. *Gravity is exerted towards all bodies; it is proportional to the quantity of matter in each.* [...] Cor. 2. *Gravitation towards the several equal parts of a body is inversely proportional to the square of the distance from the particles.*

For this twofold conclusion Newton argues as follows: He has proved that all planets gravitate toward each other and that the gravitational attraction toward each, considered separately, is inversely proportional to the square of the distance from the center of the planet to the place where the pull is exerted. Hence, by Bk. I, Prop. LXIX, the gravity tending toward the planets is proportional to their respective masses. Now, all the parts of any planet A gravitate toward any other planet B, and the gravity of each part is to the gravity of the whole as the mass of the part is to the mass of the whole. Since every action is matched by an equal reaction (by Law III), planet B will gravitate toward all the parts of planet A, and its gravity toward each part will be to its gravity toward the whole as the mass of the part is to the mass of the whole. Corollary 2 flows immediately from Bk. I, Prop. LXXIV, Cor. 3 (quoted on p. 65).

Appendix

To give a foretaste of Newton's style I shall paraphrase here his proof of Proposition I of *Principia*, Book I.[35] It says that *the areas which a revolving body describes by radii drawn to an immovable center of*

---

[34] According to Newton, his results could be wrong at most by 1 part in 1,000. Braginsky and Panov (1972) confirmed them for solar gravity to 1 part in 1,000,000,000,000. For a recent survey of this matter, see Ciufolini and Wheeler (1995, pp. 91–97).

[35] I wish I could also give the proof of Prop. XI but I lack the room. I heartily encourage all readers to try to work their way through it. This has now been made easy by de Gandt's lucid explanation (1995, pp. 38–41; Prop. XI of *Principia* Book I = Problem 3 of Newton's *De Motu*).

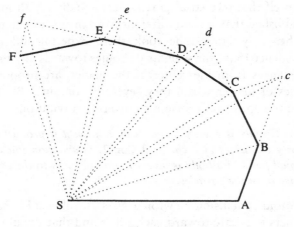

Figure 9

force lie in the same immovable plane and are proportional to the times in which they are described. To prove it, Newton bids us suppose the time to be divided into equal parts, in the first of which the body by its innate force describes the straight segment AB (Fig. 9). In the second part of the time it would, if unhindered, proceed directly to *c* along the segment B*c* = AB. Choose an arbitrary point S, outside the straight line through A, B, and *c*. ΔABS = ΔB*c*S (they have equal bases and equal heights). Thus, if the body moves inertially, the radii joining it to S sweep equal areas in equal times. Assume now that when the body reaches B a centripetal force pulls it suddenly toward S, so that the body instantly deviates from its original trajectory and continues to move along the straight line toward C. Draw *c*C parallel to BS. Then, by Cor. I to the Laws of Motion, at the end of the second part of the time the body will be at C, in the same plane with ΔABS. ΔSBC = ΔSB*c* = ΔABS, so the radii joining the body to S sweep equal areas in equal times. By a similar argument, if the centripetal force drawing the body toward S acts suddenly when it reaches C, D, E, and so on, causing it to describe the straight segments CD, DE, EF, and so on, its trajectory will lie on the same plane and ΔSEF = ΔSDE = ΔSCD = ΔSBC. Therefore, in equal times, equal areas are described in one immovable plane; and, by composition, any sums SABCDS, SABCDEFS, of those areas are to each other as the times in which they are described. These conclusions do not depend on the length of the equal parts into which we divided time in the preceding argument and therefore hold good if we make such parts arbitrarily short. In the limit, the trajectory of the body becomes a curved line and the centripetal force by which it is drawn back from the tangent to this curve acts continually. Still, the areas described by the radii joining the body to the center of force S lie on one plane and are proportional to the times of description.

2.4   Rules of Philosophy

Newton could only jump to such stupendous conclusions by following inference rules of his own devising. He called them 'Rules of Philosophy'.[36] Only the first two appeared in the first edition of *Principia*; Rule III was added in the second and Rule IV in the third. In Newton's parlance, 'philosophy' designates what we would now call 'physics'; on the other hand, to excogitate and to phrase such rules as his would be described, in current usage, as a philosophical activity. In all likelihood Newton found them not by musing on truth and reason in the abstract but by reflecting on his own intellectual practice. So the statement of Newton's Rules of Philosophy, which was in a way the founding act of the modern philosophy of physics, followed on the foundation of modern physics itself. The Latin text of the Rules can be rendered into English as follows:

RULE I. *One shall allow no more causes of natural things than are both true and sufficient to explain their phenomena.*
RULE II. *Therefore, to natural effects of the same kind* (ejusdem generis) *one shall – as far as possible – assign the same causes.*
RULE III. *The qualities of bodies which cannot intensify or weaken and belong to every body on which it is possible to perform experiments, shall be held to be qualities of all bodies* (corporum universorum).
RULE IV. *In experimental philosophy, any propositions gathered by induction from phenomena shall be held to be true – either accurately or to the best available approximation* (quamproxime) *– notwith-*

---

[36] In Latin, *Regulae philosophandi*, that is, properly, 'rules for *doing* philosophy'. This title was introduced in the second edition; in the first, as I pointed out in note 27, Rules I and II are labeled 'hypotheses' and lumped together with five statements later labeled 'phenomena' and two more statements, one of which was subsequently deleted, while the other – "That the center of the system of the world is immovable" – was renumbered as 'Hypothesis I' and placed after Prop. X of Bk. III.

*standing any contrary hypotheses, until other phenomena occur by
which they may either be made more accurate or liable to exceptions.*

(Newton 1726, p. 387)[37]

These Rules are admirably tailored to vindicate the boldest steps in
Newton's argument. After proving from his mathematical theory and
detailed computation from astronomical data that the moon M is accelerated continually by a centripetal force directed toward the center of
the earth E and proportional to $(r_{EM})^{-2}$ (where r_{EM} stands for the distance from E to M), Newton calculated that if r_{EM} were the earth's
radius, the moon's acceleration would be equal to that of heavy bodies
close to the earth's surface (remember his thought experiment with
"little moons" – which is now performed every time an orbiting spaceship is braked in preparation for landing). So the moon's acceleration
is "of the same kind" as the acceleration of heavy bodies and must
therefore, by Rules I and II, have the same cause, that is, gravity (heaviness). But Newton had also shown – again by combining astronomical data with his mathematical theory of motion under centripetal
forces – that the Jovial, Saturnal, and SP systems are held together by
forces "of the same kind" as the one acting on the moon, that is, by
gravity. By ably wielding the equality of action and reaction (Law III),
he concluded that the force of gravity must issue from every part of
these bodies, since every part experiences it. Since this is shared by all
the bodies on which we can perform experiments,[38] we must say, by
Rule III, "that all bodies whatsoever gravitate towards each other"
(1726, p. 388). And, by Rule IV, this result of induction must be held
true in the face of any speculative hypotheses contrived to account differently for the same phenomena, until it is improved or overturned by
further inductions.

Newton's Rules of Philosophy usually make a poor impression
on philosophically trained readers. Yet modern physics could only

---

[37] A draft for "Rule V" has been found in Newton's manuscripts. It is reproduced in
Koyré (1965, p. 272). Newton probably desisted from publishing it because it is not
really a rule of inference, but a characterization of "hypotheses" such as – by Rule
IV – one ought not to allow to prevail over propositions gathered from phenomena.
An equivalent although much shorter characterization was printed in the General
Scholium (see note 27).

[38] In Book III, Newton studies two additional grounds for his induction, viz., ocean
tides and the behavior of comets (but his gravitational account of tides is not altogether satisfactory).

get going on some such principles. Philosophers' misgivings about Newton's Rules arise, I dare say, from mistaking their purpose. If, following the lead of the Councils of the Christian Church, we consider truth to be a matter of dogma, that is, of neat definitive pronouncements not liable to correction or qualification, we cannot sensibly expect that Newton's Rules will direct us to find the truth about nature. But, although Newton very probably shared this dogmatic conception of truth with almost everybody else in his century, the text of the Rules, especially of the fourth one, clearly indicates that they are not meant to yield statements which in that sense *are true*, but statements that one should for the time being *hold to be true*. Of course Descartes, under the spell of the said dogmatic conception, had decided not to admit anything as true for which the evidence was not as overwhelming as the evidence he had of his own existence. But Descartes himself was not completely faithful to this stringent standard, under which physics would have been stillborn. So Newton, inspired, as I have already suggested, by his own practice, developed rules for *holding* true – and therefore for building on – statements of whose truth "in God's eye" he could not be sure.

Before looking more closely into this, I must try to clarify some key ideas involved in the Rules. As we have seen, the term 'phenomena' is not used here by Newton in the sense made familiar by "phenomenalist" philosophers. A Newtonian phenomenon is not an eddy in someone's stream of thought, let alone a congeries of so-called sensations, but rather a class of processes or states of affairs that, through the intelligent reading of many observations, we have come to recognize as a real feature of our environment. The six phenomena proposed by Newton himself under this designation were discovered after long, painstaking, intellectual work. But some conceptual ordering of perceptions and some judgments are also required for establishing more obvious phenomena, for example, that the point in the horizon where the sun rises every morning at a given place moves year after year back and forth between two fixed limits. Clearly, if the constitution of phenomena proceeds according to some general principles – concerning, for example, the structure of space and time – these principles are not reached by inference *from* phenomena pursuant to Newton's Rules.

Rules I and II concern the assignment of *causes* to natural effects. Here again we have a term that students of philosophy, standing in the shadow of David Hume, normally take in a sense that is very different from Newton's. In Hume's sense, an object *c* is the cause of an object *e* only if objects of the same kind as *e* are always contiguous with and immediately preceded by objects of the same kind as *c*. But the universal force of gravity – the one cause that emerges in *Principia* from the application of Rules I and II – neither *precedes* the acceleration of falling bodies nor can be properly described as *contiguous* to it. Naturally, Newton would not use the term 'cause' in 1687 in a sense developed fifty years later by Hume to fit his own narrow outlook. Newton took it either (i) in the common anthropomorphic sense in which the cause of *e* is the agent that is to blame for *e*'s existence, or (ii) in the traditional Aristotelian sense in which a cause of *e* is anything that contributes to explain *e*. (Aristotle counted among the *causes* of a statue not just the sculptor who made it, but also the purpose for which, the material from which, and the design after which it was made.) Both (i) and (ii) fit Newton's force of gravity in its relation to the phenomena of motion, but the unqualified reference to explanation in Rule I suggests that Newton understood 'cause' in sense (ii) (but not that he expected every explanation to fit into one of the four pigeonholes – agent, goal, matter, form – of Aristotle's scheme).

Rule I states a ban: No more causes should be allowed than are true and sufficient for explaining the phenomena. Now, it is clear that if a cause is known to be untrue, that is, not really operative in the case one desires to explain, it cannot be allowed. On the other hand, if a cause is known to be true, one should certainly allow it, even if a sufficient set of causes is already available without it. But the rules are supposed to guide us in ascertaining the true causes, or at any rate those that may be held to be true, so the inclusion of truth among the *criteria* for this rule's *application* seems somehow incongruent. What we have here is really an injunction to stop our inquiry when the explanation achieved is *sufficient*, coupled with a warning that it must restart if the explanation turns out to be *false*. To justify this rule, Newton invokes the old saying that Nature does nothing in vain, which he takes to mean that she will not do with more what can be done with less. One wonders why the omnipotent God whom Newton believed to be the author of Nature should submit to this principle of economy. Indeed, even if He chooses to follow it – say, because overexpenditure is so vulgar – we have no inkling of what is actually *less* for Him; it might well be that He finds it costlier to restrain His exuberance than to go on effortlessly multiplying things beyond necessity. Although useless for determining which are the true causes in God's eye, Rule I is perfectly apt as a guide for physical research: If we have found an explanation that, as far as we can see, *leaves nothing to be desired*, it would be wrong and confusing for us to seek any further.

Rule II is presented as an obvious consequence of Rule I, but it brings an additional idea into play: if we are not to accept any more causes than are sufficient to account for a given effect, we must ascribe a single (kind of) cause to all effects *of the same kind*. To use this rule one must certainly be able to classify effects into kinds. Now this appears to be purely a question of tact, or – if you wish – of genius, for there are plainly no clear, firm rules by which to do it, nor can we simply abide by the classifications inherited from the Stone Age. At any rate Newton does not abide by them, but analyzes in a novel way a body's state of motion into inertial velocity and acceleration, and proceeds to argue, in a fantastic feat of scientific imagination, that the actual acceleration of any moon or planet does not differ in quantity or direction from the acceleration that would be experienced by any heavy body placed in its stead. From this he concludes that the accelerations of planets, moons, and falling bodies are effects of the same kind, so that, by Rule II, all should be assigned the same cause. Rule II is flawless from a human standpoint, but it cannot lead to incorrigible truth if the classification of phenomena is open to revision.[39]

Rule III is the mainstay of physical induction: *Certain* properties must be attributed to *all* bodies if they are found in every body *within our reach*. The properties in question must not be liable to intensify or weaken. This condition is not so restrictive as it seems. Of course, most observable physical properties can be graded or quantified. But any such property P displayed by a body B may be conceived as the particular value taken by a function f in the circumstances of B, and the property of being a seat – or, as mathematicians say, an argument – of f is not liable to increase or decrease. Thus, while the property of being attracted with a given force to a given body at a given distance cannot be generalized to all bodies by Rule III for it admits variation, the property of being susceptible to attractive forces governed by a given functional relation – for example, eqn. (2.3) – is not a matter of degree and therefore should be held to be a universal physical property if it is seen to "belong to every body on which it is possible to perform experi-

---

[39] As we shall see in §5.4, when Einstein saw that it was impossible to accommodate Newton's Theory of Gravity within the new kinematics he had developed to cope with electromagnetic phenomena, he reclassified free-fall as a form of inertial motion, imaginatively anticipating the now familiar spectacle of weightless astronauts.

ments". This last condition confirms, by the way, that Rule III cannot be relied on in a quest for certainty. The bodies on which we experiment and the properties our experiments disclose in them are not a fixed, or a large, or even a random sample of the whole, and no induction can be secure which rests on such a ground. Rule III, like the preceding two, yields, not glimpses of God's worldview, but stepping stones for human inquiry.

If the reader still has any doubt about this, the text of Rule IV should dispel it. By virtue of it, the conclusions of our inductive inferences must be held to be true *aut accurate aut quamproxime* ("either exactly or as close to it as possible") until newly discovered phenomena compel us to revise them. Thus, Newton himself makes it clear that his rules are not meant to lead us to timeless truth, but to a provisional fixation of belief, such as we need to keep research going. Central to Rule IV is the adverb *quamproxime* – literally 'as near as possible' – which is also found in other passages of *Principia*. Evidently, the generalizations obtained by induction from experiment can only be as good as our measurements, which are accurate only within a margin of error. Let me illustrate with a few examples some of the implications of this fact. Take eqns. (2.5). You will not clash with any of the observations that support Newton's Law of Gravity if in the denominator of the right-hand side you substitute $3 + \varepsilon$ for the exponent 3, provided that you take the arbitrary real number ε sufficiently close to 0. You can also replace the constant G with a function of time $G(t)$, provided that the derivative dG/dt is close to 0, so that $G(t)$ practically behaves like a constant. Moreover, you may add to the left-hand side a polynomial in the time derivatives of r_k up to the nth, for some arbitrarily large integer n, provided that you pick sufficiently small coefficients to multiply each term. That no such thing is ever done can of course be explained by the physicists' faith in the so-called simplicity of nature (or by their belief that, as Einstein put it, "the Lord is sophisticated, but not nasty"). But a more likely explanation is that the actual practice of physics will gain nothing by such changes if their overall result still agrees within the admissible margin of error with the simpler original formula.[40]

---

[40] On the other hand, changes in the Law of Gravity like the first two described above have at times been proposed to deal with otherwise unexplained phenomena. A slight change in the exponent of the denominator in the right-hand side of eqn. (2.5) was tried as one way of coping with Mercury's anomalous perihelion advance. A varying

2.5   Newtonian Science

2.5.1   The Cause of Gravity

From the phenomena of planetary motion and free-fall, set in the Newtonian frame, it is possible to infer by the Rules of Philosophy that
every bit of matter attracts and is attracted by every other bit of matter
with a force obeying Newton's Law of Gravity. It would seem, therefore, that the property of exerting such a force and of responding to
it is a universal property of matter. Such, at any rate, was the reading
of Newton's contemporaries, both friend and foe. It was reasonable
to expect that hitherto unsuspected properties should accrue to the
concept of matter with the progress of inquiry, if matter was designed
and created by God and can therefore comprise everything that God
judged useful for the fulfillment of His plans (cf. §1.3). Thus John
Locke insinuated in his *Essay* that matter could even have the property of thinking, which Cartesians so studiously ruled out ([Locke]
1690, IV.iii.6). And yet in the same book, published three years after
Principia, Locke asserted that it is "impossible to conceive that body
should operate *on what it does not touch* (which is all one as to imagine
it can operate where it is not), or when it does touch, operate any other
way than by motion" (II.viii.11). But later he deleted this passage
because, as he explained to Stillingfleet,

> I have been convinced by the judicious Mr. Newton's incomparable book
> that there is too much presumption in wishing to limit the power of God
> by our limited conceptions. The gravitation of matter toward matter in
> ways inconceivable to me is not only a demonstration that God, when

gravitational "constant" was suggested in the 1930s to bring the age of rocks calculated from the statistics of radioactive minerals into harmony with the time available
for the Earth to form in our expanding, formerly very hot universe. My last example
apparently has no counterpart in history, perhaps because raising the order of their
differential equations is the last thing that physicists would wish to do. When Einstein was searching for a new law of gravity he explicitly decided not to look for
equations of a higher order than Newton's because "it would be premature to discuss
such possibilities in the present state of our knowledge of the physical properties
of the gravitational field" (Einstein and Grossmann 1913, p. 234). However, a term
multiplied by a devilishly small coefficient λ was added by Einstein (1917b) to the
field equations of gravity (1915i) to secure a model of the universe that was both
finite and static. (When he became convinced that the universe was not static but
expanding he described this move as his greatest mistake ever.)

it seems to Him good, can put into bodies powers and modes of acting which are beyond what can be derived from our idea of body or explained by what we know of matter; but it is furthermore an incontestable instance that He has really done so.

(Locke 1699, p. 468)

This was generally the position in England. Indeed, Roger Cotes, who prepared the second (1713) edition of *Principia* under Newton's supervision, placed gravity among the "primary" properties of bodies, on a par with extension, mobility, and impenetrability.[41] But this was precisely what the older generation of continental *savants*, educated in Cartesian austerity, would not countenance. Thus Huygens, while expressing admiration for Newton's achievement in brushing aside difficulties previously associated with Kepler's laws and in destroying Descartes's planetary vortices, said that he did not agree

> with a Principle according to which all the small parts that we can imagine in two or several different bodies mutually attract each other or tend to approach each other. That is something I would not be able to admit because I believe that I see clearly that the cause of such an attraction is not explainable by any of the principles of Mechanics, or of the rules of motion.

(Huygens, *Discours sur la cause de la pesanteur* [1690]; OC, XXI 471)

And on 18 November 1690 he wrote to Leibniz that Newton's "Principle of Attraction" seemed "absurd" to him (Huygens OC, XXI 538). Leibniz's own opposition to Newtonian attraction was just as strong. In his view, Newton had revived, with a vengeance, the discredited occult qualities of medieval and renaissance science. For, as he wrote to Hartsoeker on 6 February 1711,

> the ancients and moderns, who admit that gravity is an *occult quality*, are right, if they mean by it that there is a certain mechanism unknown to them, whereby all bodies are pushed towards the center of the earth. But if their opinion is that the thing is performed without any mechanism, by a simple primitive quality, or by a law of God, who produces that effect without employing any intelligible means, it is an unreasonable occult quality, and so very occult, that it is impossible that it should

---

[41] Preface to Newton 1713 (in Newton 1934, p. 27). In a draft Cotes had written that gravity is an "essential property" of bodies, but he weakened the epithet in response to criticism by Clarke (see Koyré 1965, p. 159).

ever become clear, though an Angel – not to say God himself – should wish to explain it.

(Leibniz GP, III, 519)

And again, in a letter to Conti (November or December 1715):

> It is not sufficient to say: God has made such a law of Nature, therefore the thing is natural. It is necessary that the law should be capable of being fulfilled by the nature of created things. If, for example, God were to give to a free body the law of revolving around a certain center, he would have either to join to it other bodies which by their impulsion would make it always stay in a circular orbit, or to put an Angel at its heels; or else he would have to concur extraordinarily in its motion.

(Quoted in Koyré 1965, p. 144)

Newton was infuriated by the suggestion that he traded in occult qualities. Among his several pronouncements on this point, the following (from *Opticks*, Query 31) is particularly clear. In his view, the particles of matter do not just have "a *vis inertiae*, accompanied with such passive Laws of Motion as naturally result from that Force, but [are also] moved by certain active Principles, such as is that of Gravity, and that which causes Fermentation, and the Cohesion of bodies".

> These Principles I consider, *not as occult Qualities*, supposed to result from the specifick Forms of Things, *but as general Laws of Nature*, by which the Things themselves are form'd; their Truth appearing to us by Phænomena, though their Causes be not yet discover'd. For these are manifest Qualities, and their Causes only are occult. [. . .] To tell us that every Species of Things is endow'd with an occult specifick Quality by which it acts and produces manifest Effects, is to tell us nothing: But to derive two or three general Principles of Motion from Phænomena, and afterwards to tell us how the Properties and Actions of all corporeal Things follow from those manifest Principles, would be a very great step in Philosophy, though the Causes of those Principles were not yet discover'd: And therefore I scruple not to propose the Principles of Motion above-mention'd, they being of very general Extent, and leave their Causes to be found out.

(Newton *Opticks*, pp. 401f.; my italics)

But Newton emphatically denies that the power of attraction is an "inherent" property of matter. On this he appears to be much closer to Huygens than to Locke or Cotes. As early as 1693, he wrote to the great Hellenist Richard Bentley:

It is inconceivable that inanimate brute matter should, without the mediation of something else which is not material, operate upon and affect
other matter without mutual contact [. . .]. That gravity should be innate,
inherent, and essential to matter, so that one body may act upon another
at a distance through a vacuum, without the mediation of anything else,
by and through which their action and force may be conveyed from one
to another, is to me so great an absurdity that I believe no man who has
in philosophical matters a competent faculty of thinking can ever fall
into it. Gravity must be caused by an agent acting constantly according
to certain laws, but whether this agent be material or immaterial I have
left to the consideration of my readers.

(Newton 1974, p. 54)

This stance probably inspired Newton's dignified words in the General
Scholium he added to *Principia* in 1713:

Thus far I have explained the phenomena of the heavens and of our sea
by the force of gravity, but have not yet assigned the cause of gravity.
This force must arise in any case from some cause that penetrates to the
very centres of the sun and planets, without suffering the least diminution of its power; which acts not according to the area of the *surfaces*
of the particles acted upon (as is usual with mechanical causes), but
according to the quantity of *solid* matter; and whose action extends on
all sides to immense distances, decreasing always as the inverse square
of the distances. [. . .] But hitherto I have not been able to deduce from
phenomena the reason for these properties of gravity, and I do not contrive hypotheses. For whatever is not deduced from the phenomena is to
be called an *hypothesis*; and hypotheses, whether metaphysical, or physical, or of occult qualities, or mechanical, have no place in *experimental philosophy*. In this philosophy propositions are inferred from the
phenomena, and afterwards rendered general by induction. Thus it was
that the impenetrability, the mobility, and the impulsive force of bodies,
and the laws of motion and of gravitation, became known. And it is
enough that gravity really exists, and acts according to the laws which
we have explained, and sufficiently accounts for all the motions of the
celestial bodies and of our sea.

(Newton 1726, p. 530)

The *cause* of gravity that Newton has in mind is certainly not a type
of event that is contiguous with and immediately precedes every exercise of gravitational attraction (and thus not something that a Humean
philosopher would recognize as a cause). As I noted in connection with

Rule I (§2.4), on the meaning of 'cause' Newton oscillates between old common sense and Aristotle: The *cause of gravity* is either the *agent*, material or immaterial, that brings about the mutual attraction of all bits of matter (thus at the end of the above quotation from the letter to Bentley), or it is the *reason*, broadly conceived, why matter behaves as it does, that is, in agreement with Newton's mathematical law of attraction (thus in the famous abjuration of hypotheses, halfway through the last quotation from *Principia*).

With his talk about the cause of gravity, did Newton mean to say that something was wanting in his theory, which another more fortunate scientist might find? I do not think so. His remark that the force of gravity does not operate like the usual mechanical causes seems designed to warn us that the phenomena effectively *preclude* the kind of explanation that his adversaries foolishly *demanded*. And the curt *satis est* ("it is enough") in the last sentence quoted does not encourage any further search for the missing cause. In this matter – as in his endorsement of tenability *quamproxime*, in contrast with Descartes's commitment to incontestable certainty – Newton quite resolutely sets the path of future science while still paying lip service to the notions of his time. A century later, Auguste Comte classified the search for causes – in the sense here at stake – as typical of the prescientific "metaphysical" age of intellectual history, as opposed to the "positive" scientific age, busy with the search for laws.[42] Another century had not wholly passed when Bertrand Russell cheekily asserted that "the reason why physics has ceased to look for causes is that, in fact, there are no such things".[43]

[42] Comte was strongly influenced by Joseph Fourier, whose masterpiece on heat begins with the sentence: "The primary causes are unknown to us; but they are subject to simple and constant laws that can be discovered by observation, the study of which is the purpose of natural philosophy" (1822, p. i). A similar sentiment is found in Ampère (1827, p. 177): "To establish the laws of [electrodynamic] phenomena I have only consulted experience, from which I have deduced the formula that alone can represent the forces to which they are due. I have not inquired into the cause one might assign to these forces, being convinced that every inquiry of this sort should be preceded by purely experimental knowledge of the laws, and by the determination, based solely on the laws, of the value of the elementary forces...".

[43] Russell (1917, p. 180). The quoted sentence immediately precedes Russell's famous *bon mot*: "The law of causality, I believe, like much that passes muster among philosophers, is a relic of a bygone age, surviving, like the monarchy, only because it is erroneously supposed to do no harm."

2.5.2 Central Forces

Newton's Theory of Gravity was criticized not only because it postulated forces that acted at a distance without intermediaries, but also
because it vested those forces on centers of mass, that is, dimensionless points, which could be located in a void. A Newtonian might
indeed reply that the gravitational pull toward the center of mass of a
body or of a system of bodies was always the *resultant* of forces issuing
from pieces of matter. But a critic could still counter that the real forces
that actually manifest themselves through measurable effects are
always the resultants, not the components into which our thinking analyzes them.

A stronger line of defense, and one with which the Newtonians
finally won the day, is that the little we know about the workings of
nature does not entitle us to dismiss a theory whose predictive accuracy and breadth of coverage so much exceed that of every earlier
product of natural philosophy. As a matter of fact, we do not understand how bodies act on each other by contact any better than we
understand their interaction at a distance. We are not surprised when
a billiard ball communicates its motion to another because we have so
often seen it happen, but we do not have a clearer notion of impulsive
force and the mutual exclusion of bodies than of Newtonian attraction.[44] Why, it could even be that the impenetrability of solid bodies
and the transmission of motion in collisions are themselves the manifestation of a *repulsive* force. Indeed, Boskovic (1758, 1763) argued
that it could not be otherwise, if, as everyone assumed, Nature never
jumps and all transitions are continuous. For let a body A moving with
speed u catch up and collide with a body B moving in the same direction as A with speed $v < u$. After colliding, both bodies continue to
move together with speed $(u + v)/2$. Immediately upon collision the
speed of A has decreased and the speed of B has increased by $u - v$.
This change can be gradual only if it begins to occur *before* the two
bodies come into contact, that is, if the acceleration of B and the decel-
eration of A are caused by repulsive forces acting at a distance.
Boskovic's matter consists of dimensionless particles that act on one
another with a force that, as the distance between the particles varies,
alternatively becomes repulsive and attractive.

[44] This point was eloquently made by Maupertuis (1732); see Arana (1990, p. 141).

A theory of matter along similar lines had been proposed by Immanuel Kant in his *Monadologia physica* (1756) as an example of the "joint use of metaphysics and geometry in natural philosophy". His declared aim was to reconcile the philosophical conception of bodies as ultimately composed of indivisible elements with the geometric truth of the infinite divisibility of space. But there are indications that he was also motivated by a desire to tackle the intractable problem of the relations between mind and body that modern philosophy had inherited from Descartes. Kant describes the physical world as an aggregate of simple substances – designated by the Leibnizian term *monads* – located each at a point in space, some of which are human souls. Each monad exerts on all the others a repulsive and an attractive force. Both forces depend on the distance between the locations of the interacting monads. Over short distances, the repulsive force prevails over the attractive force, but it decreases with distance at a faster rate than the latter. In Kant's view, the interplay of both kinds of forces ensures that each monad takes up – or, as he says, "occupies" – a definite volume in space, which cannot be penetrated by the volumes "occupied" by other monads. However, Kant's claim is not backed by precise mathematical arguments, nor does he articulate any known phenomenon into a testable model of his theory. Kant lost all hope that his monads could be used for solving the mind–body problem when he realized that, by the said interplay of forces, they would be liable to be amassed into balls; his philosophical good sense would not let him countenance a "clod of souls" (Kant 1766, Ak. II 321). Nevertheless, the conception of the human mind as open to modification by the direct physical action of other created entities persisted as a subtext in his critical writings, although in these writings he forbade all inquiry into the ultimate constitution of reality.

The speculations of Kant and Boskovic had little impact on the development of physics. But the concept of central forces acting at a distance found effective application outside the field of planetary motion and free-fall. It was known since the thirteenth century that the north pole of a magnetized body (i.e., the extreme of it that tends to point to the north) repels the north pole and attracts the south pole of any other such body. In 1733, Charles du Fay published his discovery that electrified bodies – that is, bodies that, after being rubbed or being brought into contact with a body already electrified, behave like briskly rubbed *electron* (Greek for 'amber') – fall into two classes, so that those belonging to each class repel one another and attract those of the other class. In the 1780s Coulomb showed, by means of his torsion balance, that these phenomena of electric and magnetic attraction and repulsion can be accounted for – within the Newtonian frame – by forces directly proportional to certain quantities characteristic of the interacting electrified or magnetic bodies and *inversely proportional to their distance squared*.[45]

Volta's invention of the pile *c*. 1800 made it possible to experiment with steady electric currents and led to Oersted's discovery, in 1820, of the following action of electricity on magnets: If a straight wire is suspended above a magnetic needle at rest and parallel to it, the flow of electric current through the wire causes the needle to turn so that the pole under that part of the wire which receives electricity most immediately from the negative end of the pile declines toward the west, the declination angle being smaller if the distance between the needle and the wire is greater, and also if the battery is less efficient.[46] The phenomena suggest the presence of a magnetic force that is *perpendicular both* to the direction of the wire *and* to the shortest line from the wire to the magnet. This is hardly the orientation one would expect a central force to have. Nevertheless, André-Marie Ampère, after an extraordinary bout of experimentation and mathematical theorizing, succeeded in explaining – or so it was thought – electromagnetic interaction by central forces acting at a distance. The explanation involves Ampère's experimental discovery – communicated one week after he

[45] The torsion balance measures the angle through which a fine metal wire is twisted, and enables one to calculate very exactly the torque on the wire. One such device had been built by John Michell *c*. 1750 and used by him to establish the inverse square law of magnetostatic force before Coulomb. An improved version of Michell's torsion balance was employed by Cavendish to measure the Newtonian attraction of large leaden balls on small bodies (and thereby to calculate the values of the gravitational constant and the mass of the Earth). Cavendish published this result in 1798. He had also used the torsion balance to establish Coulomb's law of electrostatic force before Coulomb, but this and many other important findings of Cavendish were not published until the late nineteenth century. Coulomb's electrostatic law had also been anticipated by Priestley, who inferred it from the fact that a hollow electrified sphere exerts no action on bodies placed inside it; this shows that electric force satisfies the proposition proved in Book I, Section xii of Newton's *Principia* (quoted in §2.3) and must therefore obey an inverse square law.

[46] Oersted published his findings in a pamphlet in Latin, privately distributed to scientists and scientific societies on 21 July 1820. An English translation is reproduced in Shamos (1959, pp. 123–27).

learned of Oersted's – that two parallel wires attract each other if they carry electric currents in the same direction and repel each other if they carry currents in opposite directions. It also resorts to the hypothesis that magnets consist of well-aligned, minuscule, current-carrying circuits (such circuits are also postulated in nonmagnetic bodies, where they neutralize each other because they are not aligned). Ampère accounts for all electrodynamic and electromagnetic phenomena known to him by a force acting at a distance between infinitesimal elements of electric current, along the line joining them. According to Ampère's Law, any two such elements interact with a force directed along the line joining them and proportional to the product of their current intensities multiplied by a function of their directions and divided by their distance squared. Thus, Ampère believes, he has satisfied in this field the Newtonian requirement that "all motions in nature must be reducible by calculation to forces acting always between two material particles along the straight line that joins them, so that the action exercised by one upon the other is equal and opposite to that which the latter exercises at the same time upon the former" (Ampère 1827, pp. 175f.). The compliance must be taken with a pinch of salt, however, for Ampère's force depends on the *direction* of the currents and thus, in effect, on the relative motion of electrified particles. One naturally expects this of a force generated by such motion. But *real* forces dependent on *relative* motion do not fit well in the Newtonian frame, and their admission will eventually wreak havoc with it (§5.1). Still, Ampère's "Mathematical theory of electrodynamic phenomena deduced from experience alone" (1827) was perceived as a major triumph of the Newtonian paradigm of central forces acting at a distance. Acceptance of the paradigm peaked two decades later, in a paper by Hermann von Helmholtz, who describes it as "the condition of the complete intelligibility of nature" (1847, p. 6; this passage is quoted in context in §4.3.1). That was 160 years after the publication of *Principia*, and only 10 before the emergence of a completely different style of physical explanation in Maxwell's paper "On Faraday's lines of force" (1855/56).[47]

---

[47] In the 1870s Helmholtz led the last big fight for an action-at-a-distance theory of electrodynamics in the style of Ampère, against Maxwell's field theory. Ironically, it was Helmholtz's assistant, Heinrich Hertz, who secured Maxwell's triumph by building the first rudimentary radio transmitter.

2.5.3   Analytical Mechanics

During the century following the publication of *Principia* several
remarkable mathematicians – most notably Leonhard Euler – contributed to the progress of mathematical analysis based on the differential and integral calculus invented by Newton and Leibniz, and used
it for the formulation and solution of mechanical problems within
Newton's conceptual framework. Their work led to the analytical
mechanics of Joseph-Louis de Lagrange, which proposes general
methods for the solution of all mechanical problems. This is not
the place for a detailed, historically accurate, abstract of Lagrange's
Mécanique analitique (1788), but I shall try to explain, with mild
anachronism, the gist of his approach.[48] This will enable me to introduce some notions that we shall need later.

In a Newtonian setting one naturally conceives a material system as
a collection of finitely or infinitely many particles acted upon by forces
issuing from the remaining particles and perhaps also from sources
external to the system. Taking *dynamis* as Greek for 'force', a system
thus conceived is called *dynamical*. Suppose that we have n particles
labeled with the positive integers from 1 to n. As before, I denote by
$\mathbf{r}_i$ the radius vector from the origin of an inertial frame of reference to
the location of the ith particle and by m_i the mass of this particle. Let
$\mathbf{f}_{ki}$ be the force with which the kth particle acts on the ith particle and
$\mathbf{f}_i^e$ the resultant of all external forces acting on the ith particle. The
workings of the system are then completely described by n second-
order ordinary differential equations, the system's *equations of motion*:

$$m\ddot{\mathbf{r}}_i = \mathbf{f}_i^e + \sum_{k=1}^{n} \mathbf{f}_{ki} \qquad (i = 1, \ldots, n) \qquad (2.6)$$

The conceptual simplicity of these equations is very impressive, but as
n grows they soon become unmanageable. Also, while the external
forces on a dynamical system under study are often well known, especially when they are introduced by the experimenter, one is usually
ignorant of the detailed interaction between the system's parts. Moreover, the subsequent evolution of physics has shown that the interac-

---

[48] For a judicious sketch of the development of analytical mechanics in the eighteenth
and early nineteenth centuries see Dugas (1955, pp. 232–85, 323–408). Before
attempting my own presentation I consulted a few recent treatises. In the end, I think
I profited most from Pars (1965).

tion between small parts of matter at close distances cannot even be described properly within the Newtonian frame.

However, Lagrange and his predecessors realized that it was not always necessary to know the internal forces f_{ki} to predict the evolution of a dynamical system under the Laws of Motion. In most cases of interest the system's freedom of motion is drastically curtailed by constraints, which we believe are due to the interaction of the system's parts, but which are described and taken into consideration in the study of the system's motion without our having the slightest inkling of how that interaction works. Consider, for instance, a billiard ball of unit radius confined to the top of a table 30 units long and 20 units wide. Choosing one of the corners of the table as the origin of our Cartesian coordinate system (§1.1) and identifying the z-axis with the vertical through the origin, the constraints are fully expressed by the inequalities $1 \leq x_c \leq 29$, $1 \leq y_c \leq 19$, $1 \leq z_c$ (where x_c, y_c, z_c denote the position coordinates of the ball's center). If a slab that is parallel to the surface of the table prevents the ball from jumping ever so little, the third inequality must be replaced by the equation $z_c = 1$. As a second example, consider a bead that is free to slide along a thread hanging between two walls and swinging backward and forward. In this case the exact description of the constraints is more troublesome but still can be given in purely kinematic terms, without making any assumptions about the forces that maintain them.

Among the forces acting on a dynamical system Lagrange distinguished the *forces of constraint*, which hold the constraints in place, from all the rest, which we shall call the *impressed forces*. Then, by deftly wielding an idea of d'Alembert's, he eliminated the forces of constraint from the equations of motion. To explain how this is done, it is convenient to drop the vector notation used in eqns. (2.6) and to refer our n particles to a Cartesian coordinate system. To simplify notation, I write the coordinates of the rth particle as $\langle x_{3r-2}, x_{3r-1}, x_{3r} \rangle$. Its mass will, depending on context, be represented by one of the three symbols m_{3r-2}, m_{3r-1}, or m_{3r} (all of which stand of course for the same quantity). Each force acting on a particle will be viewed as the resultant of three component forces, parallel to the coordinate axes. If x_k is the ith coordinate of a given particle ($i = 1, 2, 3$), I denote, respectively, by X_k and Ξ_k the impressed force and the force of constraint which act on it in the direction of the ith axis.[49] With this notation we may replace

---

[49] For greater precision, let me note that, if $[q]$ denotes the integral part of the positive rational number q – i.e., if $[q]$ stands for the greatest integer less than or equal to q

the n equations of motion (2.6) with the following system of $N = 3n$ equations:

$$m_k\ddot{x}_k = X_k + \Xi_k \qquad (k = 1, \ldots, N) \qquad (2.7)$$

Let $\mathbf{x} = \{x_1, \ldots, x_N\}$ be the coordinates of the system at a given moment and consider the variations they may simultaneously experience given the constraints. As we gathered from the previous examples, such variations are restricted in definite ways, no matter what the impressed forces. If (as in our first example) the constraints do not depend on time, any set of such variations $\delta\mathbf{x} = \{\delta x_1, \ldots, \delta x_N\}$ corresponds to a possible displacement of the system. We call it a *virtual displacement*.[50] However, if the constraints change with time (as in our second example), we reserve this name for such displacements as *would be possible* if the constraints were frozen at the moment in question. More precisely: In a constrained system with n particles, the possible infinitesimal displacements $d\mathbf{x} = \{dx_1, \ldots, dx_N\}$ are subject to K ($< N = 3n$) conditions of constraint:

$$\sum_{k=1}^{N} A_{hk}(\mathbf{x}, t)dx_k + A_h(\mathbf{x}, t)dt = 0 \qquad (h = 1, \ldots, K) \qquad (2.8)$$

where the coefficients $A_{hk}(\mathbf{x},t)$ and $A_h(\mathbf{x},t)$ are real-valued differentiable functions defined on a suitable region of $\mathbb{R}^{n+1}$. A definite value of $\langle \mathbf{x},t \rangle$ specifies a particular system configuration $\mathbf{x}$ at a time t. Any solution $\{\delta x_1, \ldots, \delta x_N\}$ of the equations

$$\sum_{k=1}^{N} A_{hk}\delta x_k = 0 \qquad (h = 1, \ldots, K) \qquad (2.9)$$

for the corresponding values of the A_{hk}'s is a virtual displacement available from that configuration at that time. If conditions (2.8) do not involve time, they agree with (2.9) for each configuration $\mathbf{x}$, and the class of virtual displacements coincides with the class of possible displacements.

Lagrange assumed that *in virtual displacements the forces of con-*

–, then, by our conventions, X_k acts on the $[(k + 2)/3]$-th particle in the direction of the $(k - 3[(k - 1)/3])$-th axis.

[50] The term 'virtual displacement' originated in statics. If a system is in equilibrium, it experiences no *real* displacements. However, we may inquire what simultaneous displacements will be permitted by the constraints on the system in case the equilibrium is broken. Thus, in a balance with equal sides one scale will climb from height h to $h + \delta h$ if and only if the other simultaneously descends from h to $h - \delta h$. This combination of movements constitutes a *virtual* displacement.

straint do no work; in other words, he assumed that for any virtual displacement $x_k \mapsto x_k + \delta x_k$ $(k = 1, \ldots, N)$,

$$\sum_{k=1}^{N} \Xi_k \delta x_k = 0 \qquad (2.10)$$

If we now multiply by δx_k both sides of eqns. (2.7), add up all the resulting equations, substitute from (2.10), and reshuffle terms, we obtain:

$$\sum_{k=1}^{N} (m_k \ddot{x}_k - X_k)\delta x_k = 0 \qquad (2.11)$$

Equation (2.11) is often called, somewhat improperly, *d'Alembert's Principle*.[51] It provides a general framework for the formulation and solution of all mechanical problems concerning systems for which eqn. (2.10) holds. It must be emphasized that not every conceivable system that is subject to Newton's Laws of Motion falls under this category.[52] Lagrange's theory is strictly stronger than Newton's, so its scope is strictly narrower. It is a sign of Lagrange's genius that he was ready to give up the universality of eqns. (2.6) or (2.7) in exchange for the real problem-solving power of eqn. (2.11).[53] Indeed, the best known and most impressive corollaries of Lagrange's theory are even more restricted in scope, as we shall now see.

To derive them, we recall first that when the constraints do not depend on time, every possible displacement of the system counts as a virtual displacement, so its actual velocity is a virtual velocity. One may therefore substitute $\dot{x}_k$ for δx_k in eqn. (2.11), and obtain

$$\sum_{k=1}^{N} m_k \dot{x}_k \ddot{x}_k = \sum_{k=1}^{N} X_k \dot{x}_k \qquad (2.12)$$

Now, the left-hand side of this equation is the time derivative dT/dt of the kinetic energy $T = \frac{1}{2}\sum_{k=1}^{N} m_k \dot{x}_k^2$ (cf. §1.5.2), so

$$\frac{dT}{dt} = \sum_{k=1}^{N} X_k \dot{x}_k \qquad (2.13)$$

[51] Compare d'Alembert (1758, pp. 72ff). Lindsay and Margenau (1957, pp. 103ff.) discuss d'Alembert's original approach and explain its connection with Lagrangian mechanics.

[52] Of course, all exceptions vanish if we take eqn. (2.10) as the *definition* of 'forces of constraint'. But then some of the unperspicuous internal forces that hold our system together may still turn up in eqn. (2.11).

[53] This power is beautifully displayed in Pars (1965), and also in more elementary textbooks of analytical mechanics (e.g., Goldstein 1950).

We now return to eqn. (2.11) and focus our attention on cases in which the force components $X_1, \ldots, X_N$ depend only on the position coordinates $x_1, \ldots, x_N$, not on their time derivatives $\dot{x}_1, \ldots, \dot{x}_N$ or on the time t (remember the quotation from Helmholtz at the end of §2.5.2). In many situations of this type there exists a scalar function V (i.e., a real-valued function depending on position alone), such that

$$\sum_{k=1}^{N} X_k dx_k = -dV \qquad (2.14)$$

V is called the potential energy function or simply the *potential*. For reasons that will now become apparent, such systems – and the impressed forces acting on them – are said to be *conservative*. Equation (2.14) implies that

$$\frac{dV}{dt} = \sum_{k=1}^{N} \frac{\partial V}{\partial \dot{x}_k} \dot{x}_k = -\sum_{k=1}^{N} X_k \dot{x}_k \qquad (2.15)$$

If the constraints on a conservative system do not depend on time, eqn. (2.13) holds. By adding eqns. (2.13) and (2.15) we obtain the *energy equation*,

$$\frac{d}{dt}(T + V) = 0 \qquad (2.16)$$

Thus, in a conservative system subject to time-independent constraints the sum of the kinetic and the potential energy does not change: *Total energy is conserved.*

In a system subject to constraints the position coordinates $\{x_1, \ldots, x_N\}$ are linked to each other and cannot vary independently. In many cases the K conditions of constraint (2.8) can be integrated to yield K equations:

$$f_h(x_1, \ldots, x_N, t) = 0 \qquad (h = 1, \ldots, K) \qquad (2.17)$$

If the constraints meet this condition, the system is said to be *holonomic*.[54] Equations (2.17) can then be used to eliminate K coordinates by expressing them in terms of the remaining $N - K$. More generally, the N interlinked coordinates $\{x_1, \ldots, x_N\}$ can be expressed as differentiable functions of $N - K = m$ independent variables $\{q_1, \ldots, q_m\}$:

---

[54] Heinrich Hertz, who apparently invented the term 'holonomic', explains it as follows: "The name indicates that such a system obeys integral (ὅλος) laws (νόμος), while material systems in general are subject only to differential laws" (1894, §123).

$$x_k = x_k(q_1, \ldots, q_m, t) \qquad (k = 1, \ldots, N) \tag{2.18}$$

The q_j's are called *generalized coordinates*; in contrast with the x_k's, they do not usually measure physical distances. m is known as the number of *degrees of freedom* of the system.

It follows that, for each $k \in \{1, \ldots, N\}$,

$$\dot{x}_k = \sum_{j=1}^{m} \frac{\partial x_k}{\partial q_j} \dot{q}_j + \frac{\partial x_k}{\partial t} \tag{2.19}$$

and

$$\delta x_k = \sum_{j=1}^{m} \frac{\partial x_k}{\partial q_j} \delta q_j \tag{2.20}$$

The second term of eqn. (2.11) becomes

$$\sum_{k=1}^{N} X_k \delta x_k = \sum_{k=1}^{N} \sum_{j=1}^{m} X_k \frac{\partial X_k}{\partial q_j} \delta q_j = \sum_{j=1}^{m} Q_j \delta q_j \tag{2.21}$$

The quantities $Q_j = \Sigma_{k=1}^{N} X_k (\partial X_k / \partial q_j)$ are called *components of the generalized force*.[55]

Substituting eqn. (2.20) into the first term of (2.11), it is clear that:

$$\sum_{k=1}^{N} m_k \ddot{x}_k \delta x_k = \sum_{j=1}^{m} \sum_{k=1}^{N} m_k \ddot{x}_k \frac{\partial x_k}{\partial q_j} \delta q_j \tag{2.22}$$

Now

$$\sum_{k=1}^{N} m_k \ddot{x}_k \frac{\partial x_k}{\partial q_j} = \sum_{k=1}^{N} \left[\frac{d}{dt} \left(m_k \dot{x}_k \frac{\partial x_k}{\partial q_j} \right) - m_k \dot{x}_k \frac{d}{dt} \left(\frac{\partial x_k}{\partial q_j} \right) \right] \tag{2.23}$$

In the light of eqn. (2.19) it is clear that

$$\frac{d}{dt} \left(\frac{\partial x_k}{\partial q_j} \right) = \sum_{i=1}^{m} \frac{\partial^2 x_k}{\partial q_j \partial q_i} \dot{q}_i + \frac{\partial^2 x_k}{\partial q_j \partial t} = \frac{\partial \dot{x}_k}{\partial q_j} \tag{2.24}$$

Moreover,

$$\frac{\partial x_k}{\partial q_j} = \frac{\partial \dot{x}_k}{\partial \dot{q}_j} \tag{2.25}$$

inasmuch as x_k depends only on $q_1, \ldots, q_m$ and t, and $\partial \dot{q}_i / \partial \dot{q}_k = 0$ if $i \neq k$. Substituting from eqns. (2.24) and (2.25) into (2.23) we obtain:

---

[55] Note that the Q_j need not have the dimensions of force. However, each product $Q_j \delta q_j$ has the dimension of work, i.e., mass × (space/time)^2.

$$\sum_{k=1}^{N} m_k \ddot{x}_k \frac{\partial x_k}{\partial q_j} = \sum_{k=1}^{N} \left[\frac{d}{dt}\left(m_k \dot{x}_k \frac{\partial \dot{x}_k}{\partial \dot{q}_j} \right) - m_k \dot{x}_k \frac{\partial \dot{x}_k}{\partial q_j} \right] \quad (2.26)$$

Evidently

$$\sum_{k=1}^{N} \frac{d}{dt}\left(m_k \dot{x}_k \frac{\partial \dot{x}_k}{\partial \dot{q}_j} \right) = \frac{d}{dt}\left[\frac{\partial}{\partial \dot{q}_j}\left(\sum_{k=1}^{N} \frac{1}{2} m_k \dot{x}_k^2 \right) \right] \quad (2.27)$$

and

$$\sum_{k=1}^{N} m_k \dot{x}_k \frac{\partial \dot{x}_k}{\partial q_j} = \frac{\partial}{\partial q_j}\left(\sum_{k=1}^{N} \frac{1}{2} m_k \dot{x}_k^2 \right) \quad (2.28)$$

As we know, the polynomial $\sum_{k=1}^{N}\frac{1}{2}m_k\dot{x}_k^2$ that occurs in both eqns. (2.27) and (2.28) represents the system's kinetic energy T. Substituting from (2.27) and (2.28) into (2.26), and from (2.21) and (2.26) into (2.11), we obtain "d'Alembert's Principle" for holonomic systems:

$$\sum_{j=1}^{m} \left\{ \left[\frac{d}{dt}\left(\frac{\partial T}{\partial \dot{q}_j} \right) - \frac{\partial T}{\partial q_j} \right] - Q_j \right\} \delta q_j = 0 \quad (2.29)$$

Since the δq_j's are arbitrary and not tied by constraints, eqn. (2.29) can only hold if all the factors of the δq_j's vanish, that is, if, for $j = 1, \ldots, m$,

$$\left[\frac{d}{dt}\left(\frac{\partial T}{\partial \dot{q}_j} \right) - \frac{\partial T}{\partial q_j} \right] - Q_j = 0 \quad (2.30)$$

If the system is conservative, there is a potential V such that, by eqns. (2.21) and (2.14),

$$\sum_{i=1}^{m} Q_i dq_i = \sum_{k=1}^{N} X_k dx_k = -dV \quad (2.31)$$

Hence,

$$\frac{\partial V}{\partial q_j} = -Q_j \quad \text{and} \quad \frac{\partial V}{\partial \dot{q}_j} = 0 \quad (2.32)$$

Substituting from eqns. (2.32) in (2.30) and putting $T - V = L$ we derive the *Lagrange equations of motion:*[56]

---

[56] As a brief calculation shows, eqns. (2.33) also hold for a holonomic system that is conservative in the following *extended sense*: There is a function U, depending on the q_j *and* their time derivatives $\dot{q}_j$, such that

$$Q_j = \frac{\partial U}{\partial q_j} + \frac{d}{dt}\left(\frac{\partial V}{\partial \dot{q}_j} \right)$$

$$\frac{\mathrm{d}}{\mathrm{d}t}\left(\frac{\partial L}{\partial \dot{q}_j}\right) - \frac{\partial L}{\partial q_j} = 0 \qquad (j = 1, \ldots, m) \tag{2.33}$$

The function L is the *Lagrangian* of the system.

A few remarks are in order before closing this subject.

(A) A smooth one–one mapping of the domain of coordinates $\mathbf{q} = \{q_1, \ldots, q_m\}$ onto a suitable subset of $\mathbb{R}^m$ yields new generalized coordinates $\mathbf{q}' = \{q'_1, \ldots, q'_m\}$. If L' stands for the Lagrangian expressed as a function of the new coordinates, it is fairly easy to show that eqns. (2.33) entail the following:

$$\frac{\mathrm{d}}{\mathrm{d}t}\left(\frac{\partial L'}{\partial \dot{q}'_j}\right) - \frac{\partial L'}{\partial q'_j} = 0 \qquad (j = 1, \ldots, m) \tag{2.33'}$$

This property is described by saying that the Lagrange equations are preserved under arbitrary coordinate transformations, or that they are *generally covariant*.

(B) The generalized coordinate set $\mathbf{q}(t) = \{q_1(t), \ldots, q_m(t)\}$ and its time derivative $\dot{q}(t) = \{\dot{q}_1(t), \ldots, \dot{q}_m(t)\}$ provide a full description of the system's state of motion at time t. As t varies, say, from a to b, $\mathbf{q}(t)$ describes a path in $\mathbb{R}^m$. The mapping $t \mapsto \mathbf{q}(t)$ is a curve in the *configuration space* $\mathbb{R}^m$, which provides an accurate representation of the evolution of the system in the time interval (a,b). This method of representing the successive states of a dynamical system with m degrees of freedom ($m > 3$) as tracing out a trajectory in m-dimensional space prepared the way for even bolder modes of representation, like the $2m$-dimensional *phase space* (see D) and the infinite-dimensional *Hilbert space* of Quantum Mechanics (§6.2.4).

(C) Readers acquainted with the calculus of variations will recall that eqns. (2.33) state necessary and sufficient conditions for the action $S = \int L\mathrm{d}t$ to be stationary. In other words, the Lagrange equations of motion are logically equivalent to the variational principle:

$$\delta S = \delta \int_{t_0}^{t_1} L\mathrm{d}t = 0 \tag{2.34}$$

What this means is that the curve $\gamma:[t_0,t_1] \to \mathbb{R}^m$, which represents the evolution of our system in configuration space from time t_0 to time t_1,

---

and $L = T - U$. In Maxwellian electrodynamics forces depend on positions *and* velocities but are derivable from such a generalized potential function U, so the Lagrange equations – in the extended sense – can also be used in the context of this theory.

is a solution of eqns. (2.33) if and only if the integral $\int_\gamma L dt$ along γ is either less than or greater than the integral $\int_\eta L dt$ along each neighboring curve $\eta : [t_0, t_1] \to \mathbb{R}^m$ that coincides with γ at the endpoints but differs slightly from it in between. Equation (2.34) is *Hamilton's Principle*, one among several variational principles of mechanics proposed in the eighteenth and early nineteenth centuries. Like the others, it suggests that mechanical processes are governed by final causes, inasmuch as the value reached by the integrals $\int L dt$ along different trajectories at the *end* of a time interval determines *right from the beginning* the choice of one of these trajectories. But the matter can, of course, also be viewed in this way: The actual trajectory is determined gradually, at each instant, by the differential equations (2.33), so that in the end – by virtue of a mathematical theorem – the integral $\int L dt$ takes an extremal value along the trajectory thus generated.

(D) The Lagrangian L can of course be expressed as a function of our original Cartesian coordinates $x_1, \ldots, x_N$. Clearly,

$$\frac{\partial L}{\partial \dot{x}_k} = \frac{\partial T}{\partial \dot{x}_k} - \frac{\partial V}{\partial \dot{x}_k} = \frac{\partial}{\partial \dot{x}_k}\left(\sum_{k=1}^{N} \frac{1}{2} m_k \dot{x}_k^2\right) - 0 = m_k \dot{x}_k \qquad (2.35)$$

So, if x_k is the ith coordinate of a given particle ($i = 1,2,3$), $\partial L/\partial \dot{x}_k$ is that particle's momentum component along the ith Cartesian axis. On this analogy, the quantity

$$p_j = \frac{\partial L}{\partial \dot{q}_j} \qquad (2.36)$$

is called the *generalized momentum canonically conjugate with* the generalized coordinate q_j. From eqn. (2.33) we see at once that

$$\dot{p}_j = \frac{d}{dt}\frac{\partial L}{\partial \dot{q}_j} = \frac{\partial L}{\partial q_j} \qquad (2.37)$$

Let $(\mathbf{q}, \mathbf{p})$ stand for the list of numbers $\langle q_1, \ldots, q_m, p_1, \ldots, p_m \rangle$. The $2m$-dimensional $(\mathbf{q}, \mathbf{p})$-space is the system's *phase space* to which I alluded above. The m second-order Lagrange equations (2.33) translate into an extremely elegant set of $2m$ first-order equations in the q's and p's, which determine the system's trajectory in phase space. We introduce first the *Hamiltonian* function H:

$$H = \sum_{j=1}^{m} \dot{q}_j p_j - L \qquad (2.38)$$

For an arbitrary variation $\{\delta q_1, \ldots, \delta q_m, \delta \dot{q}_1, \ldots, \delta \dot{q}_m\}$, or, equivalently, $\{\delta q_1, \ldots, \delta q_m, \delta p_1, \ldots, \delta p_m\}$, with $\delta t = 0$, the corresponding variation of H is given by:

$$
\begin{aligned}
\delta H &= \sum_{j=1}^{m} \left(p_j \delta \dot{q}_j + \dot{q}_j \delta p_j - \frac{\partial L}{\partial q_j} \delta q_j - \frac{\partial L}{\partial \dot{q}_j} \delta \dot{q}_j \right) \\
&= \sum_{j=1}^{m} \left(\dot{q}_j \delta p_j - \frac{\partial L}{\partial q_j} \delta q_j \right)
\end{aligned}
\tag{2.39}
$$

For each $i \in \{1, \ldots, m\}$, the partial derivative of H with respect to q_i – or p_i – is calculated by allowing only q_i – or, respectively, p_i – to vary while all the other $(\mathbf{q},\mathbf{p})$-coordinates remain fixed. Therefore, eqn. (2.39) implies that

$$
\frac{\partial H}{\partial p_i} = \dot{q}_i \qquad \frac{\partial H}{\partial q_i} = -\frac{\partial L}{\partial q_i}
\tag{2.40}
$$

So, by eqns. (2.37),

$$
\dot{q}_i = \frac{\partial H}{\partial p_i} \qquad \dot{p}_i = -\frac{\partial H}{\partial q_i} \qquad (i = 1, \ldots, m)
\tag{2.41}
$$

Equations (2.41) are the *Hamilton equations of motion*, which were introduced in 1835 by W. R. Hamilton. They can also be derived directly from Hamilton's principle (2.34), which, by using eqn. (2.38), can be rewritten as:

$$
\delta S = \delta \int \left(\sum_{j=1}^{m} \dot{q}_j p_j - H \right) \mathrm{d}t = 0
\tag{2.42}
$$

Equation (2.42) also entails the *Hamilton-Jacobi equation*, which I give here for future reference:

$$
\frac{\partial S}{\partial t} + H(\mathbf{q},\mathbf{p},t) = 0
\tag{2.43}
$$

Dynamical systems governed by the Hamilton equations (2.41) – or, equivalently, by the Lagrange equations (2.33) – generally meet the mathematical conditions under which these equations have unique solutions. When this is so, each point in the system's phase space lies on one and only one solution of eqns. (2.41). If the system's state at any given moment t_0 is represented, say, by point $(\mathbf{q}_0, \mathbf{p}_0)$, the unique solution through this point is a curve representing the complete history of the system before and after t_0. This is ground enough for Laplace's renowned vision of universal determinism (if the entire universe is in effect a conservative dynamical system and the equations that govern it meet the conditions for the existence and uniqueness of solutions):

> An intelligence that knew, for a given instant, all the forces acting in nature, as well as the positions of all the things that constitute it, and who was capable of subjecting these data to analysis, would embrace in a single formula the motions of the largest bodies and those of the lightest atom. For her nothing would be uncertain, and the future, like the past, would be present to her eyes.
>
> (Laplace 1795, in OC, vol. VIII, pp. vi–vii)

Appendix

The following simple example illustrates some of the foregoing ideas and the use of vector notation to convey them. Consider a particle of mass m moving in a potential V. We denote its position by the Cartesian coordinates $\langle x_1, x_2, x_3 \rangle$, or, in vector notation, by $\mathbf{x}$. The Lagrangian is

$$L(\mathbf{x}, \dot{\mathbf{x}}, t) = \frac{1}{2} m \dot{\mathbf{x}}^2 - V(\mathbf{x}, t) \tag{2.44}$$

so the momentum components are, by eqns. (2.36),

$$p_1 = \frac{\partial L}{\partial \dot{x}_1} = m\dot{x}_1 \qquad p_2 = m\dot{x}_2 \qquad p_3 = m\dot{x}_3 \tag{2.45}$$

or, in vector notation, $\mathbf{p} = m\dot{\mathbf{x}}$

By eqn. (2.38), the Hamiltonian $H = \dot{\mathbf{x}} \cdot \mathbf{p} - L$, or, substituting from eqns. (2.44) and (2.45),

$$H(\mathbf{x}, \mathbf{p}, t) = \frac{\mathbf{p}^2}{2m} + V(\mathbf{x}, t) \tag{2.46}$$

Clearly H is the total energy of the particle.

The Lagrange equations are

$$\frac{d}{dt}\left(\frac{\partial L}{\partial \dot{x}_i}\right) - \frac{\partial L}{\partial x_i} = 0 \qquad (i = 1, 2, 3) \tag{2.33*}$$

or, substituting from eqns. (2.44) and (2.45),

$$\frac{dp_1}{dt} = -\frac{\partial V}{\partial x_1} \qquad \frac{dp_2}{dt} = -\frac{\partial V}{\partial x_2} \qquad \frac{dp_3}{dt} = -\frac{\partial V}{\partial x_3} \qquad (2.47)$$

In vector notation, and by using the "nabla" operator

$$\nabla = \left\langle \frac{\partial}{\partial x_1}, \frac{\partial}{\partial x_2}, \frac{\partial}{\partial x_3} \right\rangle$$

eqns. (2.47) can be compressed into one:

$$\dot{\mathbf{p}} = -\nabla V \qquad (2.48)$$

which is, of course, none other than Newton's Second Law (2.1) with the force expressed, on the right-hand side, as the gradient of the potential.

Substituting from eqn. (2.46) into the Hamilton–Jacobi equation (2.43) we have that

$$\frac{\partial S}{\partial t} + \frac{\mathbf{p}^2}{2m} + V = 0 \qquad (2.49)$$

which – as I will show next – is tantamount to

$$\frac{\partial S}{\partial t} + \frac{(\nabla S)^2}{2m} + V = 0 \qquad (2.50)$$

This is the Hamilton–Jacobi equation for a single particle to which I refer in §6.4.2.

To prove that eqn. (2.49) is equivalent to (2.50) it is enough to show that $\mathbf{p} = \nabla S$, that is, that $p_i = \partial S/\partial x_i$ for $i = 1, 2,$ and 3. To evaluate the derivatives $\partial S/\partial x_i$ we compare the action S along neighboring trajectories that begin at the same point x_0 at time t_0 but pass through different points at time t_1. The change in S is given by

$$\delta S = \int_{t_0}^{t_1} \left[\sum_{i=1}^{3} \left(\frac{\partial L}{\partial x_1} \delta x_i + \frac{\partial L}{\partial \dot{x}_1} \delta \dot{x}_i \right) \right] dt \qquad (2.51)$$

We note that $\delta \dot{x}_i = \delta(dx/dt) = d(\delta x)/dt$. Integrating

$$\int \frac{\partial L}{\partial \dot{x}_1} \frac{d(\delta x_i)}{dt} dt$$

by parts, we obtain:

$$\delta S = \left[\sum_{i=1}^{3} \frac{\partial L}{\partial \dot{x}_i} \delta x_i \right]_{t_0}^{t_1} + \int_{t_0}^{t_1} \left[\sum_{i=1}^{3} \left(\frac{\partial L}{\partial x_i} - \frac{d}{dt} \frac{\partial L}{\partial \dot{x}_i} \right) \delta x_i \right] dt \qquad (2.52)$$

Since the paths of actual motion satisfy the Lagrange equations (2.33*), the integrand in the last term equals 0. In the other term, of course, $\delta x_i(t_0) = 0$ for all indices i (the paths compared begin all at the same point), so, writing simply δx_i for $\delta x_i(t_1)$, we have that

$$\delta S = \sum_{i=1}^{3} \frac{\partial L}{\partial \dot{x}_i} \delta x_i = \sum_{i=1}^{3} p_i \delta x_i \qquad (2.53)$$

Therefore, by the same reasoning that took us from eqn. (2.39) to (2.40), we obtain the desired result:

$$\frac{\partial S}{\partial x_1} = p_i \qquad (2.54)$$

Since our derivation is based on general principles, and does not in any way depend on the special conditions, number of degrees of freedom, or choice of coordinates in our particular problem, it is clear that eqn. (2.54) holds, with full generality, for any pair $\langle p_i, q_i \rangle$ of canonically conjugate momentum and position coordinates.

✦

CHAPTER THREE

Kant

The specter of determinism and its implications for moral responsibility acted as a powerful motive on Kant's "critical" investigation of the structure of human reason and the limits of human knowledge. He was convinced that mathematical physics was on the right path and constituted an example that all natural sciences ought to follow. "I assert" – he wrote in 1786 – "that each special discipline concerning nature (*besondere Naturlehre*) can contain only so much *genuine* science as it contains mathematics."[1] But he stoutly opposed the facile opinion that modern physics can yield metaphysical conclusions concerning the subjects of greatest interest for mankind: God, freedom, and immortality. On such matters he "found it necessary to deny knowledge in order to make room for faith" (1787, p. xxx). Kant's faith was a distillation of Christianity. He understood it, however, not as a supernatural gift, but as the natural response of our "theoretical reason" to the living fact of "practical reason", the cognitive echo of the voice of duty, so to speak.

Pious Christians had voiced qualms about modern natural philosophy since its inception. So Blaise Pascal, after making splendid con-

---

[1] Preface to *Metaphysical Principles of Natural Science* (Ak. IV, 470). Kant clarifies the meaning of this statement by means of the following example: "As long, therefore, as no constructible concept has been found for the chemical actions of matters on one another, that is, no law of *the approximation and removal of its parts by which, say, in proportion to their densities and the like, their motion and its consequences can be made a priori intuitive and represented in space* [. . .], chemistry will be no more than a systematic craft [*Kunst*] or experimental study, but never a genuine science [*Wissenschaft*], for its principles are merely empirical [. . .], and therefore do not make the *possibility* of chemical phenomena in the least understandable, because they do not admit the application of mathematics." (Ak. IV, 470; my italics).

tributions to geometry and physics, wrote *c*. 1660 about Cartesian mechanicism: "One ought to say in general: 'It happens by figure and motion'; for that is true. But to say which, and to compose the machine, is ridiculous, for it is useless and uncertain and wearisome. And if it were true, we do not think that all philosophy is worth an hour of trouble" (*Pensées*, no. 192; ed. Chevalier). Christian writers heaped abuse on sweet Spinoza, who, long before mathematical physics had anything substantial to show in favor of determinism, had, in his mock-geometric *Ethics* (1677), proclaimed its universal rule. Religious worries may also have prompted some ideas of Leibniz, which set imprecise but unsurpassable limits to physics, and they certainly inspired Berkeley's invention of positivism. I shall deal briefly with Leibniz and Berkeley in §3.1. After that, leaving aside the question of motives, I shall dwell at some length on Kant's conception of the sources and scope of Newton's conceptual frame, for it was the first full-blown philosophy of physics and remains to this day the most significant.

3.1 Leibniz and Berkeley on the Scope of Mathematical Physics

3.1.1 The Identity of Indiscernibles

Leibniz was persuaded that every creature of God is an irreplaceable individual, not merely a particular realization of a common blueprint. The omniscient Creator has, of course, full knowledge of each even before creating it, for He certainly knows what He is doing when he decides to create it. Moreover, He has the power to annihilate any one of them – or many, or all but one – irrespective of the rest. These few and – for a Christian – fairly obvious theological assumptions account for the main traits of Leibniz's philosophy. There is a complete individual concept of each individual creature, containing all its properties and its entire – eternal – history. All true statements are *either* (i) analytic statements in Kant's sense, that is, statements of the form '*S* is *P*', where *P* is a predicate contained in the complete individual concept of the subject *S*, *or* (ii) analytic statements in a sense closer to Frege's, viz., logical consequences of statements of type (i) and the laws of logic. Two creatures must differ in some respect and cannot share the same complete concept. Thus God can tell between any two of His creatures merely by surveying their respective concepts. This is Leibniz's Principle of the Identity of Indiscernibles.[2] In reply to an objection by Clarke, Leibniz stressed that this principle is not a logical truth but a consequence of "divine wisdom"[3] inasmuch as there is no reason why there should exist several creatures with exactly the same properties. And yet one wonders how God could even *conceive* the plan of creating two or more indiscernible creatures to reject it as unwise. Of course, our human concepts are usually common to many entities that we nevertheless succeed in distinguishing; but that is because we supplement the clarity of the shared concept with confused and obscure ideas of sense, which a perfect intellect is able to analyze into clear and distinct sets of notions, a different one for each entity.

Be that as it may, what matters here are not so much Leibniz's grounds for upholding the Identity of Indiscernibles as the implications that the principle has for physics. Leibniz talked uninhibitedly about them in a letter to the Dutch physicist de Volder on 20 June 1703:

Things which are different must differ in something, that is, they must have in themselves some specifiable diversity. It is surprising that this most obvious axiom has not been applied by men, like so many others. But the generality of men are content to satisfy their imaginations and do not care about reasons [. . .]. Thus they commonly use only incomplete and abstract (or mathematical) concepts, which thought supports but which nature does not know in their bare form; such notions as that of time, also of space or of what is only mathematically extended, of merely passive mass, of motion considered mathematically, etc. Such concepts men can easily fancy to be diverse without diversity – for example, two equal parts of a straight line, since the straight line is something incomplete and abstract, which is worth considering only for the sake of theory. But in nature any straight line is distinguished from any other by its contents. Hence it cannot happen in nature that two bodies are at once perfectly similar and equal. Also things which differ in position must express their position, that is, their surroundings, and so must

---

[2] Some twentieth-century philosophers have tried to render the Identity of Indiscernibles innocuous by treating '*is identical with* S' as one of the predicates contained in the complete concept of any given individual S. They argue that, if P stands for the conjunction of all the other predicates of S, there could well be another subject S* such that S* is P and yet differs from S, inasmuch as S is P and identical with S, while S* is P and identical with S*. Leibniz was too clever and too busy to indulge in such ploys.

[3] *Mr. Leibnitz's Fifth Paper*, §25, in Alexander (1956, p. 62).

not be distinguished only by their location or by an extrinsic denomi-
nation, as these are commonly understood. So in nature there can be no
bodies as they are commonly conceived, like the atoms of the Dem-
ocriteans and the perfect globules of the Cartesians, and these are
nothing but the incomplete thoughts of philosophers who do not sufficiently look into the natures of things.

(Leibniz GP, II, 249–50)[4]

It follows at once that modern matter, as conceived in the seventeenth
century (§1.3), is just a fiction, serviceable for the intellectual endeavors of finite minds who cannot have *adequate* thoughts about anything
real. And the same holds, of course, for more recent views, in which
nature is made to consist ultimately of a few irreducible homogeneous
"elements". Since mathematical structures can only specify their realizations up to isomorphism and experiments can teach nothing unless
they are repeatable, it is clear that in a Leibnizian world the
mathematico-experimental physics of Galileo and Newton and of
Leibniz himself cannot claim to properly know things as they actually
are. The Identity of Indiscernibles accounts in part for Leibniz's
repeated statement that physics busies itself with "well-founded
appearances" or "true phenomena". He wrote the following to
Arnauld on 9 October 1687:

Matter, considered as the mass in itself, is just a sheer phenomenon or
well-founded appearance, as are space and time also. It does not even
have the precise and definite qualities which could make it pass for a
determined being [...] because in nature even the figure which is essential to a limited extended mass is never, strictly speaking, exact or determined, due to the actual division of the parts of matter to the infinite.
There is never a globe without irregularities, or a straight line without
intermingled curvings, or a curve of a particular finite nature which is

---

[4] Compare the following passage from *First Truths* (c. 1680–84):

Perfect similarity occurs only in incomplete and abstract concepts, where
matters are conceived, not in their totality, but according to a certain single
viewpoint, as when we consider only figures and neglect the figured matter. So
geometry is right in studying similar triangles, even though two perfectly
similar material triangles are never found. And although gold or some other
metal, or salt, and many liquids, may be taken for homogeneous bodies, this
can be admitted only as concerns the senses and not as if it were true in an
exact sense.

(Leibniz OFI, p. 519)

not mixed with some other, and this in its small parts as in the large
ones; so that far from being constitutive of the body, figure is not even
an entirely real and determinate quality outside of thought. One can
never assign a definite and precise surface to any body, as could be
done if there were atoms. I can say the same thing about magnitude and
motion, namely, that these qualities or predicates are phenomenic, like
colors and sounds, and though they contain more distinct knowledge,
they can no more sustain a final analysis. As a consequence, extended
mass [. . .] which consists only in these qualities is [. . .] a mere phenomenon like the rainbow.

(Leibniz GP, II, 118–19)[5]

The Identity of Indiscernibles does not impose any precise boundaries
on physical inquiry, but it certainly precludes it from reaching metaphysical conclusions.

3.1.2 Mentalism and Positivism

George Berkeley is best known for his contention that bodies do not
exist without the minds that perceive them. He argued that a body's
primary qualities of shape, extension, and impenetrability are inconceivable apart from its so-called secondary qualities of color, hardness
or softness, warmth or coldness, and so on. Since the latter were admittedly mind-dependent, the same was true of the former. Modern matter

---

[5] Cf. the following passage from *First Truths* (cf. note 4). The paragraph in brackets
was crossed out by Leibniz.

> *There is no actual determinate figure in things*, for none can satisfy infinitely
> many impressions. So neither a circle nor an ellipse nor any other line defin-
> able by us exists except in our understanding, or if you will, before the lines
> are drawn or their parts separated.
>
> {Space, time, extension and motion are not things but modes of consider-
> ation which have some ground [*modi considerandi fundamentum habentes*]}.
>
> Extension, motion, and bodies themselves, insofar as they consist in exten-
> sion and motion alone, are not substances but true phenomena, like rainbows
> and parhelia.

The two passages I have quoted to illustrate Leibniz's phenomenalism are comparatively early, but he continued to express this sentiment until the last years of his life;
thus, on 11 February 1715 he wrote to Remond: "Matter itself is nothing but a phenomenon, though a well-grounded one, resulting from monads" (GP, II, 636; *monad*
– i.e., 'unit' – is Leibniz's term for an individual being).

– colorless, odorless, insipid – is sheer fiction, like Locke's abstract
idea of a triangle that is neither equilateral, nor isosceles, nor scalene.
Philosophers are prey to such fancies because they fail to understand
the power of signs, by means of which one is able to refer separately
to an aspect of a concrete object while forgetting all the other aspects
that go with it. (Thus the word 'triangle' refers to the triangularity of
any triangular figure, and not to the comparative size of its respective
sides, which, however, is well defined for each.) Berkeley was emphatic
that he was arguing only against certain aberrations of philosophers,
and had no quarrel at all with the commonsense beliefs of ordinary
people, "not yet debauched by learning" (1710, §123).

> I do not argue againt the existence of any one thing that we can appre-
> hend either by sense or reflection. That the things I see with my eyes and
> touch with my hands do exist, really exist, I make not the least ques-
> tion. The only thing whose existence we deny is that which philosophers
> call matter or corporeal substance. And in doing of this there is no
> damage done to the rest of mankind, who, I dare say, will never miss it.
>
> (Berkeley 1710, §35)

Berkeley's denial of matter motivates his views on the scope and the
purpose of physics. This is not to discover the nature of things, but
only to ascertain the regularities of phenomena:

> There are certain general laws that run through the whole chain of
> natural effects; these are learned by the observation and study of nature
> and are by men applied as well to the framing artificial things for the
> use and ornament of life as to the explaining the various phenomena –
> which explication consists only in showing the conformity any particu-
> lar phenomenon has to the general laws of nature or, which is the same
> thing, in discovering the *uniformity* there is in the production of natural
> effects.
>
> (Berkeley 1710, §62)

Thus, for Berkeley a particular phenomenon is scientifically explained
by conceiving it as an instance of a type that in turn regularly plays a
specific role in a typical series of phenomena. This is not to be confused with *causal* explanation, which consists, of course, in finding the
cause of the phenomenon, that is, the agent that produced it. According to Berkeley, only minds can act as causes, and the vast majority of
natural phenomena should be ascribed directly to the activity of God.
Physics does not look for causal explanations.

Though it be supposed the chief business of a natural philosopher to trace out causes from the effects, yet this is to be understood not of agents but of principles, that is, of component parts, in one sense, or of law or rules, in another. In strict truth, all agents are incorporeal, and as such are not properly of physical consideration.

. . . .

There is a certain analogy, constancy, and uniformity in the phenomena or appearances of nature, which are a foundation for general rules: and these are a grammar for the understanding of nature, or that series of effects in the visible world whereby we are enabled to foresee what will come to pass in the natural course of things.

(Berkeley 1744, §§247, 252)

As I noted at the end of §2.5.1, a full century after Berkeley the French philosopher Auguste Comte proclaimed that explanation by laws, *not* causes, was the distinctive feature of mature, "positive" science.

This is not the only way in which Berkeley anticipated the conception of science of latter-day positivism. In the dialogues against free-thinkers he published anonymously in 1732, Berkeley stresses that words and other signs need not evoke ideas to be useful in scientific discourse. For example, "in casting up a sum, where the figures stand for pounds, shillings, and pence," it is obviously unnecessary to form in each step, "throughout the whole progress of the operation," ideas of pounds, shillings, and pence; "*it will suffice if in the conclusion those figures direct our action with respect to things*" (1752, VII, §5; my italics). Likewise, future and past events can be calculated from present data through long chains of reasoning consisting for the most part of words to which no ideas are attached. After recalling the disagreement and confusion surrounding the term 'force' in the scientific literature of that time, Berkeley's spokesman, Euphranor, continues:

And yet, I presume, you allow there are very evident propositions or theorems relating to force, which contain useful truths: for instance, that a body with conjunct forces describes the diagonal of a parallelogram in the same time that it would the sides with separate. Is not this a principle of very extensive use? Doth not the doctrine of the composition and resolution of forces depend upon it, and, in consequence thereof, numberless rules and theorems directing men how to act, and explaining phenomena throughout the Mechanics and mathematical philosophy? And if, by considering this doctrine of force, men arrive at the knowledge of many inventions in Mechanics, and are taught to frame engines, by

means of which things difficult and otherwise impossible may be performed; and if the same doctrine which is so beneficial here below serveth
also as a key to discover the nature of the celestial motions; shall we
deny that it is of use, either in practice or speculation, because we have
no distinct idea of force?

(Berkeley 1752, VII, §7)

As the reader may recall, twentieth-century positivists – after vainly
struggling to secure the thoroughgoing "empirical meaning" of scientific language – finally reached a position akin to Berkeley's on the scientific utility of terms without referent, which they euphemistically
dubbed "theoretical terms" (cf. Carnap 1956; see also §§7.1, 7.2).

3.2 Kant's Road to Critical Philosophy

Kant's interest in the physical sciences is clear from his first two publications, the M.A. dissertation on the problem of live forces (1746)
and *The Natural History and Universal Theory of the Sky* (1754). In
the former he deals with the dispute between Cartesians and Leib-
nizians on whether the physical quantity conserved in elastic collisions
is proportional to mv or to mv^2 (see §1.5.2). In the latter he put
forward a Newtonian hypothesis concerning the formation and
evolution of the solar system, and introduced the now current view
that most of the nebulae apparently scattered among the stars are in
fact gigantic star systems, separated from the Milky Way by enormous
distances.

In §2.5.2 I referred to Kant's *Monadologia physica* (1756), where
he presents the world as a collection of simple substances (*monads*),
which are interactive centers of force and also centers of perception.
The actual physical interaction of monads was of course denied by
Leibniz and his more faithful followers, but it was countenanced by
Martin Knutzen, with whom Kant studied philosophy in Königsberg.
Knutzen presumably taught him also to view sense perception in the
orthodox Leibnizian way, that is, as an unclear and indistinct form of
intellection. The rejection of this view and the recognition of sensibility as a peculiar and independent source of knowledge was probably
the "great light" that Kant said had dawned on him in 1769, precipi-
tating the development of his mature philosophy (Ak. XVIII, 69). The
autarchy and even the primacy of feeling and sensation were asserted
in the eighteenth century, with increasing self-assurance, from Shaftes-

bury to Sade, and Kant himself, a fervent admirer of Rousseau,[6] was certainly touched by contemporary thinking on morality and aesthethics. But his explicit acknowledgement of sensibility as an epistemic faculty with its own principles occurred in connection with a problem that is native to modern physics and its philosophy, namely, the debate on the nature and ontological status of space.

On this issue, Kant initially held that the geometric system of spatial relations is grounded on and abstracted from the actual interaction of things. In particular, the fact that space has just three dimensions is a consequence of the inverse square law of gravitation (1746, §10).[7] This relationist view should not be equated with Leibniz's, for whom things did not interact at all. On the other hand, it somehow anticipates Riemann's view, which in turn strongly influenced Einstein's (see §§4.1, 5.3). But Kant himself discarded it. He was probably impressed by Euler's defense of Newton's absolute space as a prerequisite for a satisfactory description of the phenomena of motion,[8] but he went much further. In a short paper published in 1768 in a local magazine he argued that bodies depend for their very essence on their relation to space.

Kant's argument is best explained through an example. Consider two screws, one of them of the regular kind that will enter into a wall if driven with a screwdriver rotating clockwise, the other an exact mirror image of the former, which therefore advances only if the screwdriver is rotated counterclockwise. Suppose that you sell such screws and receive a written order in which one of them is described in terms of the mutual distances and relative positions of its parts. No matter how detailed the description, you cannot know what sort of screw is being requested, unless you are told how it is set – or, as Kant said, *oriented* – in the surrounding space. The ambiguity is not removed by referral to a room, or

---

[6] In a private annotation, reproduced in Ak. XX, 58f., Kant draws a remarkable parallel between Newton and Rousseau. The rational order in nature and in human society depends on the laws discovered, respectively, by each of them. "After Newton and Rousseau, God is vindicated (*gerechtfertigt*)".

[7] Obviously the inverse square law can only be stated as such with respect to a previously given geometry. So we ought to understand Kant as meaning something like this: The "order of coexistence" of things that interact according to Newton's Law of Gravity must possess the structure of a *three*-dimensional Euclidean space in which the gravitational force between any two mass points is inversely proportional to the *square* of their distance.

[8] See Euler (1748; 1765, ch. II, §§78–94; 1768–74).

a city, or even a galaxy, for such relative spaces can in turn be oriented one way or the other. So the character of each screw rests in the end on its peculiar relation to infinite space. The same holds of course for the many species of plants and animals – notably molluscs – which display spirals oriented in a definite, hereditary sense. Kant concludes that space is required for the full determination and hence for the existence of bodies and so cannot be derived from them by abstraction.

In Kant's judgment, this conclusion forced him either to regard space as a self-subsisting infinite entity indistinguishable from God himself (as in Spinoza's *Ethics*) or to produce a wholly new way of conceiving it. He chose the latter alternative. He claimed that differences in spatial orientation – that is, between right and left, or between clockwise and counterclockwise – cannot be intellectually grasped through general concepts but are directly sensed through the positioning of our own body. He saw this as an indication that sensibility is an independent source of knowledge, irreducible to intellect, and that space is intimately related to it. The same holds for time, which Kant – like Newton, Leibniz, and Euler – unhesitatingly assimilated to space. For a while he thought that by keeping the understanding (*intellectus, Verstand*) separate from sensibility – in effect, by purging it from any notions involving space or time – we would finally succeed in establishing metaphysics as a solid science of God and the soul. But further research dispelled this illusion, and Kant's doctrine of sensibility, space, and time became the foundation of his outright denial of the very possibility of such a science.

The first work published by Kant after the "great light" of 1769 was the inaugural dissertation he submitted to the Faculty of Philosophy in Königsberg when he took possession of the Chair of Logic and Metaphysics (1770). Time and space are described there as the "forms" of the world of sense (*mundus sensibilis*), in a sense of 'form' that I shall now explain. Kant characterizes a 'world' as a *whole* that is not in turn a *part* (1770, §1). The *matter* of a world consists of its parts, "which here we assume to be *substances*" (§2 I), while its *form* consists in "the coordination of substances". This he conceives as something "real and objective", indeed as "the principle of possible [mutual] influences of the substances constituting the world".

> For the identity of a whole is not secured by the identity of its parts but requires the identity of its *characteristic* composition. Above all, this rests on *a real ground*. For the nature of the world, which is the internal first

principle of any variable determinations pertaining to its state, cannot be opposed to itself and is therefore naturally, i.e. of itself, unchangeable. Hence in any world there is a constant, invariable form attributable to its nature, which is the perennial principle of any contingent and transitory form that belongs to the world's state. Those who disdain this reflection are defeated by the concepts of *space* and *time*, as self-given original conditions, by virtue of which many actuals relate to one another as parts belonging together [*uti compartes*] and constitute a whole.

(Kant 1770, §2 II)

Besides *matter* and *form*, Kant's elucidation of 'world' includes one more item, which he calls *universitas* and explains as "the absolute totality of parts belonging together". Though seemingly easy, this idea is "a crux for the philosopher", for "it is hard to conceive how the *never ending series* of states of the universe *eternally* succeeding each other can be reduced to a *whole* comprising absolutely all changes [. . .]. For nothing succeeds the whole series, and yet, given a series of successive items, only the last is succeeded by none: so there must be a last item in eternity, which is absurd." The difficulty arises also in the case of simultaneous infinity. "For simultaneous infinity offers eternity an unexhaustible material for successively progressing to infinity through its countless parts, which series would yet be actually given, complete in all numbers, in simultaneous infinity, so that a series which can never be completed by successive addition could nevertheless be given in its entirety." According to Kant there is, however, a clear way out of this "thorny question" since neither the successive nor the simultaneous coordination of a multitude "belongs to the *intellectual* concept of a whole, but only to the conditions *of sense intuition* [*intuitus sensitivi*]," for both modes of coordination "rest upon concepts of time" (§2 III).

The meaning and weight of this remark become clear in the light of Kant's revival of the ancient distinction between the intelligible and the sensible world. The former he conceived as the totality of things as they are "in themselves", which – he believed in 1770 – is accessible to our intellect. The latter is the gathering of all appearances (*apparentia, phaenomena*) displayed by things through our sensibility. He describes sensibility as the receptivity of the mind, through which its state of awareness can be affected in a definite way by the presence of an object (§3). Things do not strike our senses with their form; so, to gather the manifold of sense affections into a whole "an internal principle of the mind is required by which that manifold acquires an aspect according to stable and innate laws" (§4). This principle is the form of the sensible world. One expects that such a principle of unity should be unitary, and indeed Kant often describes it in the singular, for example, as "a certain law inherent in the mind for coordinating among themselves the sensa that issue from the object's presence" (§4), or as "a definite law of the mind, by which it is necessary that everything which (through its qualities) can be an object of the senses be seen as *necessarily* belonging to the same Whole". It turns out, however, that there are, according to Kant, *two* "absolutely primary universal formal principles of the phenomenal world, which are like the schemata and conditions of everything else that is sensual [*sensitivi*] in human knowledge", namely, *time* and *space*. In the *Critique of Pure Reason* (1781, 1787), the duality of these "primary principles" acquires a semblance of justification from Kant's doctrine that time is the form of the "inner sense" through which I appear to myself, while space is the form of "outer sense" through which I get to know the presence of other things. But in the inaugural dissertation I find no trace of this – in my view disastrous – teaching. On the other hand, at least two passages of the work make a bold suggestion that, if carried through, would secure the oneness of Kant's "formal principle of the phenomenal world". The first of them was quoted at the end of the foregoing paragraph; in it Kant takes for granted that "simultaneous coordination" – which is displayed as spatial relations – is based on a time concept (1770, §2 III). The second passage is the footnote to §14.5, which again takes up the subject of simultaneity and which, among other interesting things, says the following: "If time is represented by a straight line extended to infinity and simultaneous things at each point of time by lines serially applied to the former, the surface which is thus generated will represent *the phenomenal world* [. . .]." Obviously, this representation would be much more adequate if we attach a full copy of space to each point of the line representing time. The suggestion clearly is that we treat three-dimensional space as an aspect or, more precisely, a substructure of the four-dimensional world (or "spacetime"). Kant, however, did not pursue this suggestion. Instead, he kept space separate from time, eventually associating each with a different "side" of sense.[9]

---

[9] Kant had no use for the fusion of space and time in a single four-dimensional continuum. He had long been familiar with the idea of n-dimensional space for $n > 3$ (cf. Kant 1746, §10); but he presumably conceived it as a metric space governed by the n-dimensional analogue of Pythagoras's Theorem. Now, within Newtonian

Kant's arguments for his conception of space and time stress three main points: (i) our ideas of space and time are not intellectual – for they do not relate to particular spaces and times like a concept to its instances, but like a whole to its parts –; nevertheless, (ii) they are not acquired through the senses – for our knowledge of space geometry and the linear order of time is not liable to correction and improvement through experience –; and (iii) they involve requirements – expressed in the said incorrigible knowledge – that every object of the senses must comply with.[10] Kant concludes that time and space originate in our own sensibility. Therefore, although things as known through our senses are thoroughly impregnated with spatial and temporal properties and relations, these need not belong to things as they are in themselves. In Kant's view, this solves the paradoxes of completed infinity (see refs. to Kant 1770, §2 III). The paradoxes depend on the assumption that existing things are fully determined in every respect.[11] But appearances are not thus determined, so their potential infinity need not be actually completed. Of course, the solution works only if things in themselves are neither spatial nor temporal. Now, a moment's reflection shows that if we eliminate *all* explicitly or implicitly spatial and temporal properties and relations from our description of bodies and events, there will be *nothing* left to say about them. So Kant's new conception of space and time and his solution of the paradoxes of infinity imply that our senses do not teach us anything at all about things as they are in themselves.

In 1770, Kant thought that this result was very promising for metaphysics. By keeping our understanding clean of any ideas stemming from our sensibility – and this includes, of course, every reference to space or time – we would soon be able to know the core of things. It is baffling that Kant should ever have entertained such hope, for he had *equated* sensibility with our capacity to receive information from objects. Hence the purified understanding could only know what it

mechanics there is no room for such a spacetime metric. Indeed, the application of Pythagoras's Theorem to Minkowski's relativistic spacetime involves the rather untidy trick of using time coordinates that are imaginary numbers.

[10]   The arguments are found in their original form in Kant (1770, §§13–14). Kant (1787, §§3–7) contains the final version.

[11]   This assumption probably has to do with the Christian belief that, apart from God himself, who lacks nothing, all existing things are God's creatures, and therefore their totality has everything it was meant to have.

drew out of itself. The first clear evidence that he realized this difficulty and changed his program for metaphysics accordingly is his letter to Marcus Hertz of 21 February 1772. Kant asks, *what is the ground of the relation of an idea to its object?* Let me call this 'the question of 1772'. Kant observes – naïvely yet plausibly – that if the idea contains the way in which the subject is affected by the object, the idea is related to its object as an effect to its cause. "But our understanding is not the cause of the object through its ideas [. . .], nor is the object the cause (in a real sense) of the ideas of the understanding. The pure concepts of the understanding cannot therefore be abstracted from sensations or express the receptivity of ideas through the senses, but they must have their sources in the nature of the soul [. . .]" (Ak. X, 130). Kant recalls that in the inaugural dissertation of 1770 he had said that "the ideas of sense present things as they appear, and intellectual ideas present them as they are". But, "if such intellectual ideas rest on our own internal activity, wherefrom comes the agreement that they are supposed to have with objects which, however, are not produced by them?" (Ak. X, 131). The question of 1772 is thus closely connected with the epistemological problem that lies at the center of the *Critique of Pure Reason*, viz., how can we possess information about objects that is not supplied by the action of those objects on our senses?[12] Kant did not doubt that such information was contained in arithmetic, geometry, and the more general statements of physics (e.g., the principle that the quantity of matter remains constant in all natural processes). Moreover, the metaphysical textbooks used in German universities by him and his colleagues purported to give information about things in general (general metaphysics or ontology, the science of being *qua* being) and about God, the human soul, and the totality of creatures (special metaphysics, comprising natural theology, rational psychology and philosophical cosmology), which, by its very nature, cannot be based on sense experience. However, the erratic performance of these supposedly scientific disciplines made Kant wary of them.

[12] In Kant's parlance, an informative statement is said to be *synthetic*, and a statement not based on sense data is said to be *a priori*. Hence, Kant's epistemological problem can be concisely formulated as follows: *How are synthetic a priori statements possible?* The possibility of *analytic* statements – i.e., statements that do not convey any information about the objects that they mention, but merely speak about what is implicit in the meaning of their terms – does not constitute a problem for Kant. (Analytic statements, of course, are not – and need not be – supported by sense experience and are consequently all a priori.)

The *Critique of Pure Reason* delivers Kant's solution to the question of 1772 in the so-called transcendental deduction of the pure concepts of the understanding (1781, pp. 95–130; 1787, pp. 129–69).[13] I cannot reproduce here Kant's convoluted argument, but I shall try to summarize its main steps. A manifold of successive sense impressions can be apprehended only if it is retained, reproduced in (short-term) memory, and recognized as such. All this involves a mental activity of unifying and binding together (Kant's term is "synthesis"). To recognize parts of a manifold as remaining the same and belonging together the active mind must of course be aware of its own identity. "Thus all manifold of intuition has a necessary relation to the 'I think' in the same subject in which this manifold is found." This relation requires, however, that "all my ideas (even if I am not conscious of them as such) meet the condition under which alone they *can* stand together in one universal self-consciousness" (1787, p. 132). This condition amounts to the *possibility* of being combined into a single interconnected system, governed by certain principles. The human understanding is characterized by Kant as the power to effect such combinations. So the said principles must reflect the rules under which the understanding operates on the manifold of sense. Indeed, according to Kant, the primary pure concepts of the understanding (or "categories") are just abstract representations of those rules and "contain nothing more than the unity of reflection about phenomena, insofar as they must necessarily belong to a possible empirical consciousness" (1787, p. 367).[14]

---

[13] The text of 1787 is quite different from that of 1781, which, however, according to Kant, should still be read as a complement of the new version (cf. 1787, p. xlii). Note that 'deduction' in this context simply means 'justification' or 'proof of legitimacy' (1781, p. 84). The adjective 'transcendental' applies in Kant's parlance to knowledge that "busies itself not with objects, but with our way of knowing objects, insofar as this ought to be possible a priori" (1787, p. 25). I discussed both versions of the transcendental deduction in Torretti (1967, pp. 262–385) (in Spanish). For a detailed commentary in English, see Paton (1936, I, 313–585).

[14] Kant took special pride in his alleged discovery of the complete list of the categories. Although his views on this matter have almost never been shared by others (see however Reich 1932), I shall explain them briefly in this note. Kant contended that traditional formal logic knew enough about the operations of the human understanding to provide a full classification of the acts of judgment by which different ideas are brought under a single concept. According to him, "the same function which gives unity to the different ideas *in a judgment* also gives unity to the bare synthesis of different ideas *in an intuition*" (1781, p. 78f.; 1787, p. 105). Since the categories express in the abstract the several modes in which this unifying function is exercised

Thus the original and necessary consciousness of one's own identity is at the same time consciousness of an equally necessary unity of the synthesis of all phenomena according to concepts, that is, according to rules, which not only make phenomena necessarily reproducible but also in so doing determine *an object* for the intuition of them, *i.e. the concept of something wherein they are necessarily interconnected.*

                                                  (Kant 1781, p. 108; my italics)

Therefore, in Kant's mature thinking, even sense impressions owe to the understanding their reference to objects. The agreement between such objects and the pure concepts of the understanding is no longer puzzling, for these concepts express the rules by which the understanding organizes the manifold of sense into a system of objects. Such agreement, however, is confined to the objects of sense. As Kant puts it: "The categories [. . .] do not afford us any knowledge of things except through their possible application to *empirical intuition*; that is, they serve only for the possibility *of empirical knowledge*," or "experience" (1787, p. 147).[15] Consequently, there can be no metaphysical

on the manifold of sense, their full list should match exactly the logical classification of the ways of exercising the same function in judgment. To secure this match, Kant distinguished two kinds of judgment of the form '*S* is *P*' besides those admitted in traditional logic, viz., the *singular* judgments, in which the predicate *P* is arbitrary but the subject *S* is the proper name of an individual, and the *indefinite* judgments, in which the subject *S* is arbitrary but the predicate *P* expresses a privation. (They allegedly correspond to the categories of *unity* and of *limitation*, respectively.) But even with Kant's doctoring the match remains doubtful. Thus, in spite of Kant's lengthy explanations, few have been privileged to see how the category of *community* or *reciprocity between agent and patient* matches the class of *disjunctive* judgments (of the form '*S* is either P_1 or P_2 or . . . P_n'). More plausible and yet, in my opinion, quite perverse is Kant's view that the intellectual operation yielding a causal statement of the form '*A* causes *B*' exercises precisely the same function as hypothetical judgments of the form 'If *p*, then *q*'.

[15] This important conclusion is further elucidated in the following text:

Space and time, as conditions of the possibility that objects be given to us, are valid no further than for objects of the senses, and therefore only for objects of experience. Beyond these limits they represent nothing, for they are only in the senses and have no existence outside them. The pure concepts of the understanding are free from this limitation, and extend to objects of intuition in general, be it like ours or not, provided that it be sensible and not intellectual. But this further extension of concepts beyond *our* sensible intuition is of no help to us in anything. For as such they are empty concepts of objects, which do not even enable us to judge whether those objects are possible or not. They are mere forms of thought without objective reality, because we have no intu-

science of theology, psychology, or cosmology, and our interest in God, immortality and freedom, should seek satisfaction from sources other than our cognitive powers. On the other hand, a science of being in general or ontology is indeed possible, but only insofar as it deals with the conditions that all *phenomenal* objects must meet to become integrated in our experience. Such a study ultimately provides a metaphysical foundation for mathematical physics. In the next three sections I shall discuss some aspects of Kant's work in this area.

3.3   Kant on Geometry, Space, and Quantity

Since its Greek beginnings, geometry was a paradigm of secure knowledge. Plato stressed its independence from sense experience; indeed, he even held that sense appearances were inherently unable to satisfy geometric relations with perfect accuracy. According to him, the certainty and precision of geometry show that our human souls are rooted in another world; they give us intimations of immortality and warrant our capacity to achieve no less reliable knowledge of right and wrong. Modern thinkers conceived geometry as the science of space, which they either identified with matter (Descartes) or regarded as an eternal precondition for matter's existence (Newton), so they had no doubts as to the exact realization of geometric truths in the physical world. But they continued to see geometry as a purely intellectual achievement, and even the arch-empiricist Locke dreamt of a deductive science of morality on the analogy of mathematics.

Kant's new conception of space radically changed this state of affairs. Ethics now has to go it alone and can draw no comfort from the success of geometry. Geometry, on the other hand, is now bound to physics more than ever, inasmuch as its truth rests wholly on the fact that it spells out conditions of the possibility of our knowledge of physical objects. (Of course, since such objects are just phenomena, i.e., objects *for us*, the conditions of the possibility of our knowing them are also conditions of the possibility of the physical objects themselves.) Our notion of space is *pure* – for it is presupposed by our external percep-

ition at hand to which the synthetic unity of apperception – which constitutes the whole content of these forms – could be applied so that they might determine an object. Only *our* sensible and empirical intuition can procure them meaning and reference.

(Kant 1787, pp. 148–49)

tions and therefore cannot be extracted from them (1770, §15A) – and *intuitive* – for it is the notion of a unique object, of which particular spaces are *parts*, not a common concept of which particular spaces are *instances*. According to Kant, the axioms of geometry bear witness to this pure intuition. "That space has no more than three dimensions, that between two points there is single straight line, that about a given point on a plane a circle can be described with a given radius, are not conclusions inferred from some universal notion of space, but can only be discerned, so to speak, concretely in space itself" (1770, §15C). Indeed Kant maintained that even geometric demonstrations do not proceed by sheer logical deduction but must resort to intuition at every significant step.[16] Moreover, according to him, those who try to develop a nonstandard geometry labor in vain, for they are forced to use the standard – intuitive – notion of space in support of their fictions.[17]

Summing up, for Kant "space *is not something objective and real*, not a substance, not an attribute, not a relation, but as it were a scheme for coordinating together absolutely everything that is externally sensed, which is *subjective* and ideal and issues from the nature of the

---

[16] Kant (1770, §15C, last sentence); cf. (1781, pp. 716f., 734f., 782f.). Modern axiomatics belies Kant's claim, but in his time elementary geometric demonstrations did in fact resort to intuitive facts that were not expressed in the axioms and postulates. According to Jaakko Hintikka (1967), when Kant spoke of the incessant appeal to intuition in geometric demonstrations he referred to the use of existential instantiation in proper geometric reasoning ("consider this particular – yet arbitrarily chosen – triangle . . ."), not to the use of figures to fill in the gaps of reasoning. Hintikka's interpretation clears Kant from the – to my mind, venial – sin of not foreseeing the achievements of Pasch (1882) or Hilbert (1899), but I am not persuaded by it: Proof by existential instantiation works for predicates and relations displayed by the particular individual brought to mind only if they are also made explicit in the general premise that is being instantiated, and are thus seen to lie within one's conceptual grasp.

[17] Kant (1770, §15 E). Kant's friend, Johann Heinrich Lambert, was one of the forerunners of non-Euclidian geometry. In his posthumously published *Theory of Parallels* (1786) he suggests that trigonometric relations on a sphere with radius $\sqrt{-1}$ provide a model for a two-dimensional geometry in which Euclid's Postulate V is false (see §4.1.1). Lambert's tract was written *c.* 1766, but there is no evidence that Kant ever saw the manuscript. Nor does the extant correspondence between both men contain any reference to nonstandard geometries, but Kant may have learned about them in conversation. Strictly speaking, Lambert's sphere with an imaginary radius is not a figure of standard geometry. Still, conservative philosophers of a later age were wont to see the mathematicians' appeal to such para-Euclidean models as a sign that non-Euclidean geometries cannot stand on their own feet.

mind according to a stable law" (1770, §15D). Although with respect to things in themselves space is something "imaginary, . . . *with respect to anything sensible* not only is it *most true*, but it is the foundation of all truth in external sensibility. For things cannot appear to our senses under any aspect except through the mind's power of coordinating all sensations according to a stable law inherent in its nature" (1770, §15E). Therefore, nature is exactly subject to the precepts of geometry "not according to some fabricated hypothesis, but by virtue of an intuitively given subjective condition of all phenomena which nature can ever manifest to our senses" (Ibid.). Because of it, geometry is the paradigm and the vehicle of all scientific evidence, for, "as geometry studies *the relations of space*, whose notion contains the very form of all sense intuition, nothing can be clear and perspicuous in what is perceived externally except through the same intuition whose contemplation is the business of geometry" (1770, §15C).

In Kant's later writings this doctrine of space – and the parallel doctrine of time – is on the whole preserved. Indeed, in 1791 he described it as one of the two hinges on which true philosophy turns, the other one being "the reality of the concept of freedom" (Ak. XX, 311). However, his innovative views on the role of the understanding in the constitution of the subject matter of physics inevitably produced a shift in his thinking about intuition. The shift shows up in a small change introduced by Kant in the second edition of the *Critique of Pure Reason*. In both the first and the second editions space is said to be "nothing but the form of all phenomena of outer sense" (1781, p. 26; 1787, p. 42), but the notion of the *form* of a phenomenon is explained first as "that which makes that the manifold of the phenomenon is intuited [as] ordered in certain relations" (1781, p. 20),[18] and later as "that which makes that the manifold of the phenomenon can be ordered in certain relations" (1787, p. 34; cf. Ak. XVII, 639). This change is necessary because all coordination and hence all relational order in a manifold are now considered to be the work of the understanding.[19]

[18] I insert the word 'as' to meet the requirements of English syntax; but its German equivalent 'als' does not occur in Kant's text. The following rendering is perhaps more accurate: "that which makes that the manifold of the phenomenon is intuited in a certain relational order". However, it makes the changes in the wording of the second edition look greater than they really are.

[19] "The binding together [*Verbindung*] of a manifold in general can never come to us through the senses and cannot therefore be already contained in the pure form of sensible intuition. For it is an act [*Actus*] of the spontaneity of the cognitive faculty,

Enlightened by this view, Kant distinguishes between (i) the *form of intuition*, that is, the condition inherent in our sensibility that constrains us to perceive everything as embedded in the structures of space and time, and (ii) the cognitive grasp of those structures by what he now terms *formal intuition*.

> Space, represented as an *object* (as we actually need to do in geometry) contains more than the mere form of intuition, for it contains a *grasping together* [*Zusammenfassung*] in one *intuitive* representation of the manifold given according to the form of sensibility. Thus the *form of intuition* gives just a manifold, but *formal intuition* gives unity of representation. This unity [. . .] presupposes a synthesis, which does not belong to the senses but through which all concepts of space and time first become possible. [. . .] Through it – as the understanding determines sensibility – space and time are first *given* as intuitions.
>
> (Kant 1787, pp. 160–61n.)

The active contribution of the understanding to the geometric ordering of things is not properly highlighted in the *Critique* and is usually ignored by commentators, but §38 of *Prolegomena* leaves no doubt about it.[20] This §38 is meant to illustrate Kant's contention that "*the understanding does not draw forth from nature its* (a priori) *laws, but prescribes them to her*" (1783, §36, last sentence). As an example of such a law he proposes the following: If two chords AB and CD meet inside a circle at P, then, no matter how the chords are chosen, the rectangles formed from the segments of either chord are always equal (AP × PB = CP × PD). Kant asks himself:

> Does this law lie in the circle or in the understanding? That is, does this figure, independently of the understanding, contain in itself the ground of the law; or does the understanding, having constructed the figure according to concepts of its own (namely, of the equality of the radii), introduce thereby into the figure this law of chords intersecting each other in geometrical proportion? When we follow the proofs of this law we soon see that it can only be derived from the condition on which the understanding based the construction of the figure, namely, the equality of the radii.
>
> (Kant, Ak. IV, 321)

and since this [spontaneity], to distinguish it from sensibility, must be called understanding, so is all binding together [. . .] an action [*Handlung*] of the understanding" (Kant 1787, pp. 129–30).

[20] On *Prolegomena*, §38, see Friedman (1992, Ch. 4).

Kant proceeds to examine a generalization of his example and then asks again:

> Do these laws of nature lie in space, and does the understanding learn them merely by exploring the rich store of meaning which resides in space? Or do they lie in the understanding, and in the way in which it determines space according to the conditions of the synthetic unity to which all its concepts point? Space is something so uniform and so indeterminate with regard to all particular properties, that we should certainly not seek in it a treasury of laws of nature. Instead, what determines space to assume circular shape, or the figures of a cone and a sphere, is the understanding, insofar as it contains the ground of the unity of their constructions. The mere universal form of intuition, called space, is thus indeed the substrate of all intuitions which can be determined [as referring] to particular objects, and the condition of the possibility and the variety of such intuitions certainly lies in it. But the unity of objects is entirely determined by the understanding, and indeed according to conditions which lie in its own nature.
>
> (Kant, Ak. IV, 321–22)

Kant's strong statement about the indeterminacy of space "with regard to all particular properties" must, however, be taken with a pinch of salt, for he continued to think that the understanding is required by "the mere universal form of intuition, called space" to bestow on it a Euclidian structure. This can be seen in the section on "Axioms of intuition" in the *Critique of Pure Reason* (1787, pp. 202–207). It concerns the principle governing the application of the categories of quantity to the manifolds displayed in sense awareness. The principle is: "All *intuitions* are *extensive magnitudes*". Since "all phenomena contain, as to their form, an intuition in space and time," they can only be grasped "by the synthesis of the manifold through which the representations of a determinate space or time are generated, i.e. by putting together the homogeneous manifold and becoming conscious of its synthetic unity" (1787, pp. 202–203). Thus, "I cannot represent to myself a line, however small, without drawing it in thought, that is, generating from a point all its parts one after the other. [. . .] Similarly with all times, however small: I think only of the successive advance from one instant to another, whereby through all parts of time and their addition a determinate time-magnitude is finally generated" (1787, p. 203). An extensive magnitude is one "in which the representation of the parts makes possible the representation of the whole" (Ibid.), so that a line and a time interval can only be conceived as extensive magnitudes, and the same evidently holds for any more complex spatio-temporal configuration. This conclusion is fairly weak and seemingly uncontentious. But Kant jumps from it to a much stronger one, viz., that "this successive synthesis of productive imagination in the generation of figures" is the foundation of geometry and its axioms, for the latter "express the conditions of sense intuition a priori under which alone" the pure concepts of quantity can be applied to an external phenomenon (1787, p. 204). To illustrate this assertion, Kant judiciously picks two axioms that are not exclusively Euclidian, viz. (i) Two points are joined by a single straight line, and (ii) Two straight lines never enclose an area. But there can be little doubt that, under interrogation, he would have placed on a par with them all the axioms required for proving Pythagoras's Theorem, and hence for characterizing Euclidian distance. (After all, without a distance function, a geometry can hardly be seen as dealing with extensive magnitudes.) How, despite the uniformity and indeterminacy of space, the understanding is constrained by "the conditions of sense intuition a priori" to performing only Euclidian constructions is, I am afraid, anything but perspicuous.

Appendix

Kant does not view geometry as the only application of the categories of quantity. The following remarks deal briefly with two further aspects of his thinking on this matter.

Some followers of Kant – among them the great Irish mathematician W. R. Hamilton – fell into the temptation of conceiving arithmetic, on the analogy of geometry, as based on the a priori intuition of *time*. But Kant astutely resisted it.[21] On 25 November 1788, he wrote to Pastor Schultz that "time, as you correctly note, has no influence on the properties of numbers [. . .] and the science of number is – despite the succession required by every construction of magnitude – a purely intellectual synthesis, which we represent to ourselves in thought" (Ak. X, 557). Of course, the existence of such a "purely intellectual synthesis" is hard to reconcile with Kant's contention that no cognitive

---

[21] Despite his baffling characterization of number as "nothing else but the unity of the synthesis of the manifold of an homogeneous intuition in general by means of my generating time itself in the apprehension of the intuition" (1781, pp. 142–43).

synthesis is purely intellectual, and that in any case it must involve either time or space. Thus, it turns out that – contrary to the conventional wisdom – Frege's attempt to establish arithmetic as a chapter of quantificational logic (i.e., the theory of singular, existential, and universal propositions – the formal logical counterpart of Kant's categories of quantity) came in fact to Kant's rescue in this matter. For if the truths of arithmetic are actually truths of logic, they are, in Kant's terms, a priori but not synthetic, and therefore need not rest on one of the forms of intuition.

According to Kant only extensive magnitudes, such as volumes or durations, in which the whole is grasped as compounded from multiple parts, are conceived under the categories of quantity, viz., One, Many, and All; while intensive magnitudes or degrees belong under the categories of quality, viz., Reality, Negation, and Limitation. He explains this as follows:

> What corresponds in empirical intuition to sensation is reality (realitas phaenomenon); what corresponds to the lack of it is negation = 0. Now, every sensation is capable of diminution, so it can decrease and gradually vanish. Thus, between phenomenal reality [Realität in der Erscheinung] and negation there is a continuous connection through many possible intermediate sensations, the difference between which is always less than the difference between the given one and zero or complete negation.
>
> (Kant 1787, p. 209)

Consider two actual sensations, a and b, such that a is equal to one of the possible sensations intermediate between b and its complete negation. We say then that the realities corresponding to a and b sport the same quality to a different degree, and that the degree disclosed by a is less than that disclosed by b.

As a description of ordinary usage and its underlying rationale the above is fairly accurate. But Kant goes further. The statement that "in all phenomena, the real which is an object of sensation, has an intensive magnitude, i.e. a degree" (1787, p. 207) expresses, according to him, a condition of the possibility of experience, without which it is not possible to grasp an object in space and time. He rests this claim on the allegation that it is always possible for an empirical consciousness to change gradually "so that the real in it vanishes completely and there remains a merely formal consciousness (a priori) of the manifold in space and time". Whence, conversely, it is possible to synthesize "the generation of the magnitude of a sensation",[22] from its beginning, the pure intuition = 0, to an arbitrarily chosen intensity of it (1787, p. 208). This is Kant's principle of "anticipations of perception". It accounts for the fact – first noted by Hume (THN, I.I.i, p. 6) – that we sometimes can anticipate qualities we have never sensed, for example, a nuance of blue intermediate between two given samples of paint.

3.4   The Web of Nature

In Kant's thinking, the constitution of our experience of objects in space and time depends chiefly on the use of the categories of relation, viz., substance-and-attribute, cause-and-effect, and community of interaction. Kant states three principles – called by him "Analogies of Experience" –, which govern, respectively, the application of each category, plus a general principle of the Analogies, which sets the stage for the others. He offers proofs of all four. These principles raise different problems that still engross the philosophy of physics. I shall discuss them separately in the next four subsections.

3.4.1   Necessary Connections

In 1781, pp. 176–77, the general principle of the Analogies read thus: "All phenomena are subject a priori, with regard to their existence, to rules concerning the determination of their mutual relations in *one* time." In 1787, p. 218, it had been changed to: "Experience is possible only through the representation of a necessary connection of perceptions." The earlier version expresses in a general way the function of the three special principles. On the other hand, Kant's argument for the principle – added in 1787 – resulted in the second version.

The argument runs as follows. In experience, perceptions come together casually, and do not by themselves disclose a necessary connection between them. "Perceptual grasp [*Apprehension*] is just a putting together of the manifold of empirical intuition", and it does not include "a representation of the necessity of the connected existence in space and time of the phenomena it puts together". However, because experience is knowledge of *objects* through perceptions, rela-

---

[22] Kant says that "eine Synthesis der Grössenerzeugung einer Empfindung" – literally, "a synthesis of the magnitude-generation of a sensation" – is possible. My paraphrase involves a deliberate – and probably failed – attempt to make Kant's meaning clearer.

tions touching the existence of the manifold must be represented in experience not as the manifold is merely put together in time, but as it is in time *objectively*. Since time itself cannot be perceived, "the determination of the existence of objects in time can only occur through their being combined in time in general, and therefore only through a priori connecting concepts". Such concepts "always carry with them necessity, so that experience is only possible through a representation of the necessary connection of phenomena" (1787, p. 219).

The purport of this argument will, I hope, become clearer in the light of Kant's treatment of causality (§3.4.3). But we need not wait for it to inquire what sort of necessity must, by Kant's argument, be present in experience, or, more precisely, what are the source and the scope of such necessity. A careful look at Kant's reasoning prompts us to distinguish three sorts of necessity. (i) By the transcendental deduction, it is *necessary₁* that the manifold of sense be unifiable by the understanding in a single, coherent, self-conscious spatio-temporal experience. (ii) If Kant's table of categories is final and complete, then it is *necessary₂* that the synthesis of the manifold of sense which is thus necessarily possible should yield instances of his categories or of concepts derived from them. (iii) If any entity – be it an object, an event, or a system of objects or events – is an instance of a specific concept, it is *necessary₃* that the said entity shall possess all the attributes entailed by this concept. In particular, if the entity e is manifested through a variety of phenomena $f_1, \ldots, f_n$, any connection λ_{hk} between, say, f_h and f_k that is required for e to be an instance of a concept C is of course a *necessary* connection if e *actually is* such an instance.

Note that *all* concepts carry with them necessity of this third kind, whether or not they derive from Kant's table of categories. Thus, if something is correctly diagnosed as a healthy mammal, it *must* contain a healthy heart; if an event is correctly described within current particle physics as a collision between a negative and a positive electron, it *must* be a source of radiation. *Necessity₃* is often scorned as merely "verbal" necessity, in part no doubt because it is so obvious and pervasive, but also because, being conditional on the right application of concepts, it does not protect us against uncertainty. Commonsense concepts are often fuzzy and have a feeble grip on their instances. A mammal's health can fail suddenly, for example, if its heart stops. Science tries to work with concepts that are well-defined and stable. If two objects are in fact what is known in current physics as two differently charged electrons, it is *inevitable* that, upon meeting, they fuse into a spurt of radiation. Yet even in this case uncertainty persists, for the conceptual systems of physics are subject to seemingly endless revision.

I expect that after finishing this book the reader will be persuaded that natural necessity, as understood in modern physics, is simply necessity$_3$, and that the appearance of its being something stronger and harder to conceive is due to the richness of physical concepts, the complexity of their instances, and the firm grip that the former get on the latter through experiment and measurement. Kant, however, was not of this persuasion. As we shall see below, he argues in detail that the constitution of experience in time must be presided by the three categories of relation. In this way he claims necessity$_1$ not for concepts in general, but for these concepts specifically. As members of Kant's list, these categories possess moreover necessity$_2$, and admit no substitutes. The necessity that they carry with them, and which is displayed in experience by their instances, is of course necessity$_3$, but Kant apparently thought that it would be not be securely grounded unless the concepts from which it stems are specifically required for the possibility of experience (are necessary$_1$) and belong to the small permanent inventory of primary concepts of the human understanding (are necessary$_2$).

3.4.2 Conservation of Matter

The First Analogy or "Principle of the Permanence of Substance" reads thus: "In all change of *phenomena*, *substance* persists, and its quantum does not increase or decrease in nature" (1787, p. 224).[23] Kant's proof is quite remarkable: "All phenomena are in time", in which alone we can represent their coexistence and succession. Time therefore remains unchanged, while succession and coexistence "can only be represented as determinations of it". Time, however, "cannot be perceived by itself". Hence, "in the objects of perception, i.e. in phenomena, there must be found the substrate which represents time in general, and in which all change or coexistence can be perceived through the relation of phenomena to it" (p. 225). Kant explains that "the substrate of all that is real, i.e. belonging to the existence of things, is the *substance*".

---

[23] The statement of the First Analogy in the first edition made no mention of a fixed quantum: "All *phenomena* contain the permanent (*Substance*) as the object itself, and the changeable as its mere determination, i.e., as a way in which the object exists" (1781, p. 182).

"Consequently, the permanent, in relation to which alone all time relations of phenomena can be determined, is the substance in the phenomenon, i.e. what is real in it, which as substrate of all change remains
always the same. As this cannot change in existence, its quantum in
nature cannot increase or decrease" (p. 225).

This argument raises a problem for Kant interpreters. This is not the
place to discuss it, but allow me to state it. Throughout the *Critique*,
Kant maintains the existence of things-in-themselves, independent of
human experience, whose true nature cannot be known from their
appearances in space and time. "In fact, when we consider the objects
of sense, rightly, as mere phenomena, *we grant thereby that a thing in
itself lies at their foundation*, although we know not how it is constituted in itself, but only [know] its appearance, that is, the way in which
our senses are affected by this unknown something" (1783, §32; my
italics). But Kant's proof of the First Analogy neatly disposes of this
claim: To be "read" as experience, the changing sense appearances
must be "spelled" as manifestations of an underlying substrate;
however, such a substrate is just the conceptual representative of time
– the universal form of our sensibility –, and is nothing apart from its
role as a steady referent "in relation to which alone all time relations
of phenomena can be determined".[24]

I turn now to other questions that are more directly relevant to
physics. Suppose that the constitution of experience in time does indeed
require that we tie the kaleidoscopic flow of phenomena to a permanent substance. Still, one might ask, why should its permanence show
up as a conserved quantity?[25] The only answer I can find lies in Kant's
identification of "the substance in the phenomenon" with "what is real
in it". That "the real" should have an intensive magnitude in all phenomena was for Kant a condition of the possibility of experience

[24] For the "spell/read" metaphor, see Kant (1781, p. 314; 1783, §30). Kant's proof of
the First Analogy of Experience agrees well with – and further clarifies – Locke's well-
known remark on our "notion of pure substance in general": Whoever examines it
"will find he has no other idea of it at all, but only a supposition of he knows not
what *support* of such qualities which are capable of producing simple ideas in us"
(1690, II.xxiii.1). Obviously, such an idea cannot stem from sense impressions.

[25] As I said in note 23, a quantum of substance was not mentioned in the 1781 formulation of the First Analogy. However, already in 1781 (p. 185) Kant illustrated the
First Analogy with the story of the chemist who figured out the weight of smoke by
subtracting the weight of the ashes left over from the weight of the wood burnt. (This
story was already documented in late Antiquity; see Lucian, *Demonax*, 39.)

(1787, p. 207). This answer, however, raises other questions. *First*, the First Analogy asserts the conservation of a single physical quantity, which is cautiously described as "the quantum of substance", but which, as I intend to show in the next paragraph, can be none other than the *quantity of matter*, as measured by Newtonian *mass*; yet in Kant's discussion of intensive magnitude, sense qualities of different sorts are mentioned as having degrees, which take their places along different scales, so there must be more than just one physical quantity in this sense (viz. temperature, sound pitch, color saturation, etc.). One can reply that these are degrees of *sensation*, not of *reality*. The latter is "that which corresponds to sensation" (1787, p. 182), and Kant repeatedly calls it "matter" (1787, pp. 34, 182, 609f.; 1786, in Ak. IV, 481). Indeed, the aim of mechanistic natural philosophy was to pin all variety and variation in sensation on the distribution and redistribution of matter alone; and Kant, in his critical writings, certainly appears to have shared this aim.[26] Still, one may wonder, *secondly*, whether the argument for intensive magnitudes is not undermined by the First Analogy's conservation principle. For that argument – summarized in the Appendix to §3.3 – rested on the gradual variability of empirical consciousness *and hence* of the real that it grasps. So if the real is necessarily fixed, the reason given for assigning it a quantity does not seem to hold. This difficulty is solved, I think, by distinguishing between (i) the degree to which matter is present on a particular occasion at a point in space, and (ii) the quantity of matter in (*a*) a finite region V, or (*b*) in nature as a whole. We met this distinction already at the beginning of §2.1, in the definition of Newtonian mass: (i) is *density*, conceived as a time-dependent scalar field (Chapter Two, note 4), whose value can vary from point to point and may grow with time at a given point from 0 to any finite value; (ii) is *mass*, conceived (*a*) as the integral $\int_V \rho \, dV$ of the density ρ over the finite region V, or even (*b*) as the inte-

[26] However – in stark opposition to seventeenth-century mechanicism – he conceived matter as being solely a source of force, and he thought that there were natural forces of several kinds. He says in 1786: "All that is real (*alles Reale*) in the objects of the external senses, which is not a mere determination of space (place, extension and shape), must be regarded as motive force; so that the so-called solidity or absolute impenetrability is expelled from physics as a vacuous concept, and is replaced with repulsive force, while on the other hand the true immediate attraction is defended against the sophistries of a metaphysics which misunderstands itself, and is declared necessary, as fundamental force, for the very possibility of the concept of matter." (Ak. IV, 523).

gral $\int_{\mathscr{E}} \rho dV$ of the density ρ over the whole of Newtonian space $\mathscr{E}$. There is no question that both (ii*a*) and (ii*b*) can retain a constant value while the density ρ varies in time and space. The First Analogy requires the constancy of (ii*b*). It is doubtful, however, that it makes any sense to speak of the constancy of $\int_{\mathscr{E}} \rho dV$ unless the integral converges, which it can only do under stringent additional conditions. In physics, of course, 'conservation of mass' normally means something else, namely, that the value of $\int_V \rho dV$ over any finite region V at the end of a time interval τ is always equal to its value at the beginning of τ *plus the net inflow of mass across the boundaries of the region* during τ. I am not sure that the principle thus understood would meet Kant's requirements.

I still have to justify my assertion that the quantum of substance that, according to Kant, can neither increase nor decrease in nature, is precisely Newton's *quantitas materiae*. Although this is not made explicit in the proof of the First Analogy or in the comments that follow it, Kant does refer to chemical calculations based on the conservation of mass as a telling example of the First Analogy's use (1781, p. 185; see note 25). Then, in 1786, after defining "the quantity of matter" or "mass" as "the aggregate of what is moveable in a definite space (*die Menge des Beweglichen in einem bestimmten Raum*)" (Ak. IV, 538), he derives the following "law", which he links directly to the First Analogy:

> *First Law of Mechanics.* In all changes of corporeal nature, the quantity of matter in the whole remains the same, unincreased and undiminished.
>
> (Kant, Ak. IV, 541)

Why, then, does Kant studiously avoid the term 'matter' in his discussion of the First Analogy? Kant neatly distinguishes (i) "transcendental philosophy", which receives no information from the senses, and deals, independently of any particular object of experience, with "the laws which make possible the concept of nature in general", from (ii) the "metaphysics of corporeal nature", which takes "the empirical concept of matter" for granted and investigates the extent to which its object can be known a priori (1786, in Ak. IV, 469f.). The First Analogy belongs to transcendental philosophy; the "First Law of Mechanics" is its application to the metaphysics of bodies. So Kant would be right in leaving 'matter' out of the former. But is the *abstract* concept of matter involved in these considerations a truly *empirical* concept? By Kant's definition, matter can only be known through sensation, that is, empirically; but this does not imply that the *concept* 'matter' is empirical. Indeed, there is an almost perfect match between Kant's definition of the *matter* of phenomena as "that which corresponds to sensation" (1787, p. 34) and his characterization of *reality* in phenomena as "that which, in empirical intuition, corresponds to sensation" (1787, p. 209). And, as we saw above, the proof of the First Analogy turns on the identification of the permanent with "what is real in [the phenomenon] (*das Reale derselben*), which as substrate of all change remains always the same" (1787, p. 225). So his avoidance of the term 'matter' may be somewhat disingenuous after all. Still, one may wish to distinguish between the fully abstract concept of matter in the definition just quoted and the narrower concept that the "metaphysics of corporeal nature" takes as its starting point. Matter is characterized there as "that which in external intuition is an object of sensation" and is specified further as "the moveable" that "fills space" and "*qua* moveable, possesses motive force" (1786, in Ak. IV, 481, 480, 496, 536). Perhaps this concept of matter as the object of *outer* sense is in effect empirical. Still, according to Kant, only matter *thus understood* can afford the permanent reality demanded by the First Analogy. For, as he argues in the "refutation of idealism",

> All determination of time presupposes something *permanent* in perception. But this permanent cannot be an intuition in me. For all grounds of determination of my existence which can be found in me are representations and require, as such, a permanent [something] distinct from them, in relation to which their change, and so my existence in the time wherein they change, may be determined.
>
> (Kant 1787, p. 276, as corrected on p. xxxix)

So "we have nothing permanent which we can place as intuition under the concept of a substance, except only *matter*" (1787, p. 278). This is to be understood, of course, as the object of outer sense.

Besides the principle of the conservation of mass, mathematical physics has countenanced, since its inception, several other conservation principles. Due to their global scope, empirical evidence for them has often seemed inadequate, so there has been a tendency to vindicate them by reason alone. We have seen Descartes derive the conservation of motion from the immutability of God (§1.3), and Leibniz argue for the conservation of "force" (i.e., energy) on the ground that one cannot get something for nothing (§1.5.2). By the 1780s three conservation principles were well entrenched in mechanics, regarding (α) linear and (β) angular momentum, as well as (γ) mechanical energy. By a remarkable mathematical theorem due to Emmy Noether (1918), these principles can be shown to follow respectively from the symmetries of Euclidian space (α,β) and Newtonian time (γ).[27] From an orthodox Kantian standpoint Noether's theorem yields a perfect a priori justification for these three principles. Nowadays, however, one would rather argue, by means of it, *for* the adoption of this or that particular symmetry group, *from* whatever empirical evidence can be mustered for the corresponding conservation principles.

Although Kant's thoughts on the constitution of human experience were chiefly directed toward the rational foundation of physics, they were meant to apply also to prescientific experience as described in ordinary discourse. Now, in everyday talk – at any rate in English and other European languages – we do refer the quickly varying aspects of sense experience to more or less stable things. But these are many, not just one, as Kant's argument for the First Analogy would seem to suggest.[28] And, although they usually last a good deal more than their fugitive states, they are certainly not everlasting. Thus, ordinary usage

---

[27] The conservation of linear momentum is a consequence of the homogeneity of space (invariance of spatial relations under translation in any direction), the conservation of angular momentum is a consequence of the isotropy of space (invariance of spatial relations under rotation about any point), and the conservation of mechanical energy – in a conservative mechanical system – is a consequence of the homogeneity of time (invariance of temporal relations under translation). See Landau and Lifschitz (1960, Ch. II). For a very readable proof of Noether's theorem, see Lovelock and Rund (1975, pp. 201–206).

[28] Throughout most of the section on the First Analogy, Kant uses the word 'substance' (*Substanz*) in the singular. It occurs in the plural (*Substanzen*) only toward the end, in the following two sentences: "Alteration can therefore be perceived only in substances", and "Substances (in the phenomenon) are the substrate of all determinations of time" (both in 1787, p. 231). The next occurrence of 'substance', at the beginning of the proof of the Second Analogy, is strongly monistic: "All phenomena in the succession of time are one and all only *alterations*, i.e. a successive being and not-being of the attributes of the substance, which abides" (1787, p. 232). But in the section on the Third Analogy, Kant persistently speaks of interacting *substances* (in the plural). In fact, his proof of the First Analogy is quite compatible with a plurality of substances if *the permanent* ultimately consists of ingenerate and indestructible indivisible corpuscles or *atoms*. But in his maturity Kant rejected all forms of atomism, including the dynamic monadology of his youth. He conceived matter as a continuum whose presence at each point of space is represented by a variable density scalar. Such a view of matter can no doubt be expressed in pluralistic terms, but a monistic formulation sounds much more natural.

supports the philosophical category of substance-and-attribute, but it does not agree with Kant's handling of it. He went so far as to write that if some substances are born and others perish, this "would remove the sole condition of the empirical unity of time, and phenomena would then relate to two different times, in which existence would flow in parallel streams, which is absurd" (1787, pp. 231f.). I find it hard to think of a more preposterous philosophical claim. Durable things can stand proxy, as in Kant's argument, for the one permanent time we cannot perceive, even if they are not eternal. It is enough that they mutually interact within partially overlapping sets that, so to speak, take turns at guarding the identity and continuity of time. I do not expect my watch to last forever; but the time it keeps need not break down with it; it can be kept further by my next watch, provided that there is a third one that coexists for a while with each, was seen to agree with the former, and is used for setting the latter.

3.4.3   Causality

Even if we do not subscribe to Kant's theory of the human under-standing, we may readily agree that the category *substance-and-attribute* is deeply entrenched in our everyday language – where it shows up in the subject–predicate structure of simple sentences –; that it is also found, with little or no modification, in the classical physi-cist's conception of matter whose quantity is conserved through its manifold changes of state; and that Kant's contention that *substance* functions in the constitution of experience as a tangible representative of time is, if not true, at least reasonable. When we come to the cate-gory *cause-and-effect*, things are less straightforward. It is clear that we continually use some such category in the analysis and interpreta-tion of ordinary events, for describing which we have a rich stock of causal verbs.[29] On the other hand, we often hear of a 'principle of causality' that ruled over mathematical physics until Quantum Mechanics allegedly dethroned it, but which might be restored to its former glory if the de Broglie–Bohm "causal interpretation" of Quantum Mechanics is successful. However, as we shall see, between the ordinary concept of a causal relation and the applications of the

---

[29] The following "small selection" is due to Anscombe: *scrape, push, wet, carry, eat, burn, knock over, keep off, squash, make* (e.g., noises, paper boats), *hurt* (1971, p. 9).

said principle of physics there are incongruities that prompt one to think that 'causality' in the name of the latter is a misnomer (and a source of confusion, given that physics, in its experimental practice, cannot do without the ordinary concept). Due in part to these incongruities, the Second Analogy of Experience, conceived by Kant as a common foundation of the use of both the physical principle and the ordinary concept of causality, in the end accounts for neither. To justify these assertions, I shall first go over Kant's argument for the Second Analogy, and then discuss its relations to the ordinary notion of cause-and-effect and to the so-called principle of causality of classical physics.

In the second edition of the *Critique of Pure Reason*, the Second Analogy bears the title of "Principle of Succession in Time in Accordance with the Law of Causality" and is stated as follows: "All alterations happen in accordance with the law of the connection of cause and effect" (1787, p. 232). In the first edition it was called "Principle of Generation (*Erzeugung*)" and read thus: "Everything that happens (begins to be) presupposes something which it follows *in accordance with a rule*" (1781, p. 189). Despite the enormous change in wording, Kant apparently believed that both texts were semantically equivalent, for he did not rewrite the first edition's discussion of the Analogy, but merely placed in front of it two new paragraphs, headed by the word "Proof". In these paragraphs he recalls that, as a consequence of the First Analogy, all phenomena that succeed each other in time are merely *alterations*, "i.e., a successive being and not-being of the attributes" of a permanent substance. He then argues for the Second Analogy (version of 1787) as follows: When I perceive that two phenomena A and B succeed each other I connect together two perceptions in time. This connection is not a gift of sense intuition, but "the product of a synthetic faculty of imagination" (1787, p. 233). This can connect the phenomena so that A precedes B or so that B precedes A. Since time itself is not perceived, it is not possible to ascertain which of the two phenomena came first "in the object" by comparing them both with time. Since the objective order of succession does not depend on the way the phenomena are given to the senses, and it cannot be derived from their relation to time, it must lie in our conceptual grasp of it: "the relation between the two states must be so *thought* that it is *thereby* determined as *necessary* which of them must be placed before and which after, instead of the other way around" (p. 234; my italics). The rest of the proof merely applies the argument for the general prin-

ciple of the Analogies (§3.4.1) to the special case of succession in time:
"The concept which carries with it a necessity of synthetic unity can
only be a pure concept of the understanding, which does not lie in perception, and here it is the concept of the *relation of cause and effect*,
the former of which determines the latter in time, as its consequence"
(p. 234).

This understanding of causality as the relation between two events
(i.e., two transient phenomena or "alterations", to use Kant's term),
one of which *necessarily* follows the other in accordance with a *rule*,
comes straight out of Hume. Hume had argued that, since the idea of
necessary connection is not obtained from sense impressions, it can
only reflect our compulsive tendency to think of the cause in the presence of the effect and to think of the effect in the presence of the cause
due to the habit of perceiving them together. To counter this degrada-
tion of physical to psychological necessity Kant resorted to his doctrine
concerning the constitution, by the human understanding, of a web of
objective time relations in nature, but he accepted Hume's analysis of
the causal relation almost unmodified.[30] Yet that analysis has little to
do with our ordinary concept of cause-and-effect. Humean causation
is a relation between *two events*, but our causal verbs normally take
persons or *things* as subjects.[31] In ordinary causation, the cause usually
exists before the effect, but it can very well come into being together
with it (like the newborn child and his first act of breathing, according to some definitions of birth). Everyday causal inquiries usually ask
(*a*) *who* – or *what* – is to blame for some, usually unwelcome, change
in our environment, or (*b*) which effects a *person* must cause to achieve
some desired state of affairs. Inquiries of type (*a*) appeal to the regu-

[30] We should recall, however, that according to Hume the proximate cause must be contiguous to its effect. Kant wholly disregards this requirement, presumably because it
rules out instantaneous action-at-a-distance.

[31] See note 29. Our word 'cause' is derived from the Latin *causa*, which primarily meant
'legal case' – whence also 'plea', 'excuse', 'pretext', 'motive', 'purpose', and 'reason'
– but which occurs in philosophical literature as the standard Latin equivalent of the
Greek word αἰτία. Now, αἰτία was Aristotle's word for 'cause' in a sense so broad
that some modern translators render αἰτία (in Aristotle) as 'explanation'; but in ordinary Greek αἰτία primarily meant 'responsibility', mostly in a bad sense, that is,
'blame' (but also 'merit' – Aeschylus, *Septem*, 4). The noun αἰτία was closely related
to the adjective αἴτιος, 'responsible', 'culpable', and the verb αἰτιάομαι, 'accuse',
'censure', or 'lay to one's charge', 'impute'. αἰτία is first attested in Pindar and
Herodotus, but αἴτιος and αἰτιάομαι are old, common words, that were already
found in Homer.

larities of nature (e.g., paternity inquiries resort to the laws of genetics), but only for guidance. A person X can be held responsible for a change Y even if effects such as Y do not follow from actions such as X's *necessarily* or even *regularly*. U is guilty of V's death if V died from one bullet shot at her by U, even if four other bullets missed the target or caused only minor wounds and a sixth bullet, shot at V's guest, failed to kill him because he reacted better to medical care. Inquiries of type (b) do, of course, depend decisively on the regularity of phenomena, but for which they would rarely be of any use. And yet even in this case necessity is not an essential feature of causation. Consider an agent U who, in the light of such an inquiry, causes $V_1, \ldots, V_n$ to achieve W. Surely the connection between U and the V's is not a necessary one, or else he would not have had to inquire about them. As for the connection between the V's and W, U surely would like it to be necessary, but ordinarily this will not be the case. Typically, there might be several, partially overlapping, sets of means available, $\mathcal{V}_1 = \{V_1, \ldots, V_b\}, \mathcal{V}_2, \ldots, \mathcal{V}_r$, and U will choose to execute one of them after estimating their respective *cost* and *probability of success*. If we never settled for less than a necessary connection between means and ends, there is precious little we could get done.

So much for the similarity between Kant's Humean category of cause-and-effect and the homonymous concept one meets in life. But perhaps what Kant – and Hume – had in mind was not the anthropomorphic concept of causation we inherited from our prehistoric ancestors, but the scientific concept involved in the "principle of causality" that is said to govern classical physics (and which allegedly broke down with the advent of Quantum Mechanics). As we shall now see, this is admirably suited for the role assigned by Kant to cause-and-effect in the construction of an objective time order. Still, due perhaps to Kant's desire to keep his categories in touch with folk concepts,[32] his "law of the connection of cause and effect" does not quite agree with the physicists' "principle of causality". The latter is in fact only a short – but high-sounding – way of referring to the following characteristic of classical physical systems: Their evolution in time is governed by a system of differential equations that, by virtue of its mathematical properties, normally has unique solutions, in which case, if the value of certain

---

[32] Folk concepts form the "massive central core of human thinking which has no history – or none recorded in histories of thought" (Strawson 1959, p. 10). They give firmer support to Kant's view of changeless reason than the fickle notions of science.

physical quantities at a particular instant is given exactly, the state of the system is fixed for all times (see §2.5.3, after eqns. (2.43)). Thus, the physicists' "principle of causality" is in effect a principle of determinism,[33] and therefore, according to our discussion in the foregoing paragraph, it is quite foreign to ordinary causal thinking. Of course, scientists are wont to use words in different, often quite disparate, senses (think of 'vector' as used in physics and in epidemiology). However, since commonsense causal thinking is a central feature of laboratory life, it is convenient to avoid any confusion between 'causation' and 'determinism' – at any rate, a physicist's interventions in his experimental equipment must surely be up to him, and not fixed since time immemorial by the evolution of a physical system that embraces them both.

There are other discrepancies between the concept of cause-and-effect, in both the ordinary and the Kantian acceptance, and the idea of a physical system's deterministic evolution under differential equations with unique solutions. Suppose we try to describe the evolution of such a system S in causal terms. To do so we must consider causality, with Kant and Hume, as a relation between *events*, the relata being in this case the states of S at different times. The state s_1 of S at any given time t_1 certainly "follows in accordance with a rule" on its state s_0 at some earlier time t_0. But could one say, without sounding artificial, that s_0 is the cause of s_1? Should I say, for example, that the present angular momentum of the earth and its position relative to the fixed stars (from which I see the sun right over my head, moving westward at 15° per hour) *cause* the angular momentum and the position that the earth will have 18 hours from now (from which I shall see the sun rise in the east)? One may feel tempted to see this as a case of indirect causation, today's state causing tomorrow's *through* all the states that the system will have in between. However, in ordinary conversation we would never assert that A indirectly causes B unless we believe that there is some effect C that A causes directly and which in turn brings about B. Thus, if we say that the Luddite terrorist Mr. U in Sacramento indirectly caused the death of Ms. V, a communications engineer in New York, with a mailbomb, we imply that he intervened personally at a definite point in the process leading to the explosion of the bomb

---

[33] For example, Heisenberg (1927, p. 197), gives a "sharp formulation of the Law of causality (*Kausalgesetz*)" as follows: "When we know the present exactly we can calculate the future". See also Frank (1932, pp. 30ff.) and Hopf (1948, pp. 1–2).

that killed her, either by packaging it and mailing it, or by ordering his assistant to do so, or in some other way. But in the evolution under differential equations of our system S there is no state s_i that is caused directly by an earlier state. Between a given state s_1 and any earlier state s_0 there must be an uncountable infinity of states, or else their succession could not be governed by differential equations. Since the same is true again of s_0 and *each* state between it and s_1, none of these states can be singled out as a direct effect of s_0. The contrast between the discreteness of causal chains, as ordinarily understood, and the continuity of evolution under differential equations moved J. R. Lucas to present the latter not as a mere application, but rather as a "generalization" – in fact, a creative extension – of commonsense causal thinking (1984, chapter X). Still, this approach does not take care of the biggest discrepancy. If a closed system S evolves under differential equations with unique solutions, the state s of S at any particular time t determines *every* other state s' of S, no matter whether s' follows or precedes s. And, of course, s' also determines s. In other words, the binary relation 'x determines y', where x and y are different states in the evolution of a physical system subject to differential equations, is a *symmetric* relation; whereas 'x causes y' is *antisymmetric*: If x causes y, it is certainly false that y causes x.[34]

On the other hand, the physicist's understanding of successive phenomena as states of a system that evolves under a set of differential equations with unique solutions bestows necessity on the temporal relations between those phenomena in a manner that is unrivaled by any other form of thought. This is necessity₃ in the sense of §3.4.1: State s_1 necessarily follows state s_0 after time $t_1 - t_0$ (if $t_0 < t_1$) or is followed by it after time $t_0 - t_1$ (if $t_0 > t_1$), because both states lie on the same solution of the said set of equations and correspond respectively to times t_1 and t_0. If we represent the solutions of our set of equations – as indeed we may – by curves in a space of sufficiently large dimension number, we see that s_1 follows or precedes s_0, at the stated time intervals, with the same kind of necessity that constrains two straight lines on a Euclidian plane Π to meet at some point of Π, unless there

---

[34] Antisymmetry holds for both common sense and Humean causality. The former must be antisymmetric because the relata are heterogeneous: If x is to blame for y, y cannot be to blame for anything, and x is not the sort of thing that something else could be blamed for. Humean causation is a binary relation between entities of the same (most general) kind, viz., events, but the effect must succeed the cause in time, and temporal succession is of course antisymmetric.

is a third straight line that meets them both at right angles, or that makes the distances from the foci of an ellipse to any point on it add up to the length of the diameter through the foci. If s_1 did not stand in precisely that temporal relation to s_0, s_0 and s_1 could not be what the physicist takes them to be, viz., just those states of just that system. The strength and the scope of this kind of necessity are more readily acknowledged in the case of simple geometric figures in three-dimensional space because the mathematical concept of a physical system is so much more complex and it is so much harder to ascertain whether and in what terms a given set of phenomena should be brought under it. Anyway, the necessity$_3$ that the classical concepts of physical systems carry with them serves the demand for an objective ordering of phenomena in time (put forward by Kant as necessary$_1$) much better than the presumptive necessity$_2$ of the category of cause-and-effect. For the latter is not only doubtful or liable to exceptions, as we have seen; but even if this category were applicable in Kant's sense, it would constitute only an arbitrary and unintelligible connection between its relata. Kant concedes as much when – commenting on his Analogies of Experience – he compares the meaning of 'analogy' in mathematics and in philosophy. A mathematical analogy states the equality of two quantitative relations (ratios), so that when three of the four quantities involved are given the fourth can be "constructed" (if $x{:}a{::}b{:}c$, then $x = ab/c$). But a philosophical analogy asserts the equality of two *qualitative* relations, so that "from three given terms I can only know the *relation* to a fourth one, but not *this* fourth *term* itself; yet I have a rule for seeking it in experience, and a mark by which to find it there" (1787, p. 222). This is a good deal less than what mathematical physics can do for us. Just think of astronomers who, after a few sightings of a newly discovered comet, merely by conceiving it as part of the Solar System (regarded as a practically closed Newtonian gravitational system), are able to construct its trajectory for the next six months and to understand why it cannot be otherwise.

3.4.4 Interaction

The Third Analogy governs the application of the category of interaction. Since this category is Kant's own creature, we do not have to consider whether his treatment of it agrees with ordinary usage. Our attention will go to the category's prototype in Newtonian physics and to the role that Kant assigned to it in the constitution of experience.

The concept of interaction is meant to fit the relation between two bodies that attract each other in accordance with Newton's Law of Universal Gravitation. This relation is perfectly symmetric, although, if the bodies have unequal masses, it has different effects on each.[35] Kant's words suggest at times that interaction is just two-way causation.[36] Can this be right? Causation is antisymmetric: If x causes y, then y does not cause x. Thus, 'x interacts with y' cannot, under pain of self-contradiction, mean the same as 'x causes y and y causes x'. Of course, if x and y stand for bodies exerting a Newtonian gravitational pull on one another, 'x interacts with y' is not intended to mean that 'x causes y and y causes x', but rather that 'x causes a state (viz. of acceleration) or a change (viz. of velocity) in y while y causes a (similar) state or change in x'. This explication, however, will not work if 'x causes y' is given the Humean sense of a relation between *events*, for an event is not the sort of thing that suffers changes or abides in states. So, if Kant's concept of causation agrees on this point with Hume's, it cannot be used in this way for explicating his concept of interaction. It is preferable to take seriously his description of interaction as a *category*, that is, as a basic, irreducible concept, that cannot be understood solely in terms of other concepts.[37]

Throughout the section devoted to the Third Analogy, Kant speaks of interaction as a relation between *substances*. This agrees well with what I have just said, but it might seem to raise a problem with Kant's proof of the Analogy. The Third Analogy, like the first two, was rewritten for the *Critique* of 1787, but in this case the proof added in 1787 suits the text of 1781 quite well. The earlier version was: "All substances, insofar as they exist simultaneously (*zugleich*), stand in thoroughgoing community (i.e. mutual interaction)" (1781, p. 211). This was replaced by: "All substances, insofar as they can be perceived in

---

[35] A crumb of bread and the planet earth attract each other with forces of exactly the same magnitude, but only the former experiences a significant acceleration as a result of this, because $F = ma$, and the earth's m is so very much larger than the crumb's. Interaction in Kant's sense obviously covers Coulomb attraction or repulsion between electric charges and also momentum exchange in elastic collisions. Indeed, the Third Analogy translates, in the *Metaphysical Principles of Natural Science*, into the following version of Newton's Third Law of Motion: "*Third Law of Mechanics.* In all communication of motion, action and reaction are always equal" (1786 in Ak. IV, 544).

[36] For instance, in the paragraph beginning on p. 261 of Kant (1787).

[37] Cf. Kant's letter to Johann Schultz of 17 February 1784 (Ak. X, 366–68).

space as simultaneous, are in thoroughgoing interaction" (1787, p. 256). The proof recalls that two things are said to be simultaneous when the perception of one of them can follow that of the other, *and* vice versa. "Thus I can direct my perception first to the Moon and then to the Earth, or, conversely, first to the Earth and then to the Moon; and because the perceptions of these objects can alternatively follow each other, I say that they exist simultaneously" (1787, p. 257). However, we cannot perceive time and learn that two things exist simultaneously by observing their placement in it. All we can gather from our temporal grasp of things is that each perception is present in the subject when the other one is not, and vice versa, but not that the objects are simultaneous, that is, that when one exists the other one also exists at the same time, and that this is necessary in order that the perceptions can alternate as they do.

> Consequently, in order to say that the alternating sequence of perceptions is grounded in the object and thereby to represent simultaneous existence as objective, we require a pure concept of the alternating sequence of the properties of these things existing simultaneously outside each other. Now, the relation between substances one of which has properties whose ground is contained in the other is the relation of influence, and when each, reciprocally, contains the ground of properties in the other, this is the relation of community or interaction. Therefore, the simultaneous existence of substances in space cannot become known in experience except on the assumption of their mutual interaction. Consequently, this is also the condition of the possibility of the things themselves as objects of experience.
>
> (Kant 1787, pp. 257f.)

The problem with this argument is that, in the light of Kant's discussion of the First Analogy, if there is more than one substance, they must all be coeval, for the birth of some while others perish would destroy the unity of time (§3.4.2, last paragraph). Hence, all substances exist simultaneously all the time, whether they interact or not. The First Analogy thus undermines Kant's argument for the Third and makes the latter, as formulated in 1781, completely idle. However, the text of 1787 – "All substances, *insofar as they can be perceived* in space *as simultaneous*, are in thoroughgoing interaction" – might avoid this reproach if, with some hermeneutic good will, one seeks the key to its meaning in the words I have italicized. We perceive substances only through their transient states, and so – presumably – we can perceive them as existing at the same time only if the states through which we perceive them occur simultaneously. From this perspective, Kant's argument for the Third Analogy may be reconstructed as follows: The simultaneous occurrence of state a of substance A and state b of substance B will be established as a matter of objective fact only by grounding a on B's being b and b on A's being a; therefore, unless all the substances are in thoroughgoing interaction, the simultaneous occurrence of their states cannot be *known* and, consequently, the simultaneous existence of the substances themselves cannot be *perceived*.

If this argument is valid, the simultaneity of distant events presupposes instantaneous distant interaction. In this way, by dint of Kant's philosophical ingenuity, the most objectionable feature of Newtonian gravity is turned into a precondition of the empirical knowledge of spatial objects in time. Indeed, Kant has nothing practical to say about the synchronization of distant events, which certainly could not be carried out in his time – or even now – by observing gravitational interactions. However, by mentioning the need for a physical foundation of objective simultaneity he probably contributed to motivate Einstein's more fruitful handling of this question in 1905 (§5.1).[38]

The Third Analogy rounds off the Kantian construction of nature as a field of human experience. In Kant's own eloquent words:

> By nature (in the empirical sense) we understand the connection of phenomena, as regards their existence, according to necessary rules, that is, according to laws. There are certain laws, indeed a priori laws, which first make nature possible. Empirical laws can operate and be discovered only through experience, and indeed in consequence of those original laws through which experience itself first becomes possible. Our analogies therefore properly represent the unity of nature in the connection of all phenomena under certain characters which only express the relation of time (insofar as time comprises all existence) to the unity of apperception, which can only be achieved in synthesis according to rules. So, taken together, the analogies say that all phenomena lie, and must lie, in

---

[38] Kant expressly mentions light as a vehicle for the propagation of simultaneity, although he must have known that it travels with finite speed: "From our experiences one may easily gather that only the continuous influences in all points of space can lead our senses from one object to another; that light, playing between our eye and the heavenly bodies, effects a mediate community between us and them, and thereby demonstrates the simultaneous existence of the latter" (Kant 1787, p. 260).

one nature, because without this a priori unity no unity of experience, and therefore no determination of objects in it, would be possible.

(Kant 1787, p. 263)

3.5   The Ideas of Reason and the Advancement of Science

As the final item in our selection of Kantian themes, I propose to deal briefly with the Ideas of reason and their significance for the philosophy of physics.

Having shown that all purported knowledge of God, freedom, or immortality is illusory, Kant did not dismiss the pseudoscience of metaphysics as a manifestation of man's silliness and conceit, but ascribed its origin to a natural illusion of reason, a "transcendental mirage" that will not vanish merely because the critique of reason has exposed its vacuity ("e.g. the mirage in the statement: 'the world must have a beginning in time'" – Kant 1787, p. 353). This explanation of metaphysics as a necessary evil is closely related to Kant's partition of the intellectual powers of man among two "faculties", viz., *understanding* (*Verstand*), which *constitutes* experience by articulating sense appearances as objective phenomena, and *reason* proper (*Vernunft*), which guides the understanding in this process and therefore may be said to *regulate* experience. Reason performs this function by setting certain unattainable *goals* that will keep the human understanding busy forever. Every such goal is represented in thought by what Kant calls an *Idee*, that is, a "necessary concept of reason, such that no object congruent with it can be given to the senses" (1787, p. 384; cf. 1783, §40; 1790, §57, Anm. I). I render this Kantian term as 'Idea' with a capital 'I' to distinguish it from the ordinary English word 'idea' (which is closer to Kant's *Vorstellung*). Although Ideas cannot determine any object, "they can serve the understanding as a canon for its extended and consistent employment; the understanding does not hereby get to know any more objects than it would by its own concepts, but is better and further conducted in this knowledge" (1787, p. 385).

The conception of reason as a guide of life can be traced back to the ancient Stoics, who called it *to hegemonikon*, 'the guiding [principle]'. Kant's originality lies in considering it as a guide *of science*, and indeed in the thought that science can use a guide. Such a thought was out of the question while 'science' designated the repository of all truth, shining forever in God's mind, of which our human science was a tiny – yet otherwise unadulterated – portion. What was needed then was a

guide of *ignorance*, that is, a method for progressively getting rid of it
and increasing one's share in divine science. But Kant, as we have seen,
radically separated human from divine science, the former being confined to phenomena while the latter presumably embraces things-in-
themselves. In fact, "intellectual intuition" – the sort of knowledge that
God, if He exists, has of everything – is introduced by Kant only as a
negative Idea: a paradigm of what human knowledge cannot be. By contrast, human science is essentially an *enterprise*, forever unfinished and
thoroughly drenched with ignorance, so it is no wonder that it should
need guidance. Specifically, the understanding, in all areas of experience, faces conditioned aspects of phenomena, whose conditions it must
determine, for instance, by locating them in spatial and temporal surroundings, or by analyzing them into parts, or by finding their causes.
Typically, such conditions are conditioned in turn: Locations have their
own surroundings, parts are analyzable wholes, causes are events
effected by other causes. In each line of inquiry, reason prescribes the
search for the *totality of conditions* of every conditioned feature of
things. Such totality, of course, will never be given in experience and
therefore can only be represented by an Idea. Thus, the Ideas

> of totality in the synthesis of conditions are necessary – and grounded
> in the nature of human reason – at any rate as *tasks* for carrying through
> the unity of the understanding, where possible, up to the unconditioned;
> even though these transcendental concepts otherwise lack any suitable
> application in concreto, so that their sole utility lies in setting the under-
> standing on such a course that its employment is both extended to the
> uttermost and made thoroughly consistent with itself.
>
> (Kant 1787, p. 380)

Kant is emphatic that reason is never directly concerned with an
object, but only with the understanding. "It does not, therefore, *create*
any concepts (of objects), but only *orders* them, and gives them that
unity which they can have in their widest possible extension, that is,
with respect to the totality of series" (1787, p. 671). Reason

> directs the understanding towards a certain goal, with a view to which
> the guiding lines of all the latter's rules converge to a point. This is indeed
> only an Idea (focus imaginarius), that is, a point from which the con-
> cepts of the understanding do not actually proceed, for it lies wholly
> outside the limits of possible experience; but it serves however to procure
> them maximal unity together with maximal scope. From this we get the
> illusion that the lines issue from an object outside the field of empirically

possible knowledge (just as we see objects behind the surface of a mirror). Though we can hinder this illusion from deceiving us, it is nonetheless inevitably necessary if besides the objects in front of our eyes we also want to see those which lie far behind our backs, i.e., in our case, if we want to direct the understanding beyond every given experience (a part of the whole possible experience), and so also towards the furthest and greatest possible enlargement.

(Kant 1787, pp. 672–73)

Different forms of this illusion generate the three branches of *metaphysica specialis* as delineated by Christian Wolff, viz., natural theology, rational psychology, and philosophical cosmology. The first two rest, according to Kant, on fallacious inferences, which he explains and refutes in some of the more readable portions of the *Critique of Pure Reason* (1781, pp. 341–406, 571–642; 1787, pp. 399–432, 600–670). In the remainder of this chapter I shall deal only with the contradictions of philosophical cosmology, better known as Kant's antinomies.[39]

The four cosmological questions leading to the antinomies concern, (I) the temporal origin and spatial boundary of the world, (II) the divisibility of bodies, (III) the existence of an uncaused initial cause in every causal series, and (IV) the thoroughgoing necessity or utter contingency of physical events. They are age-old problems. In the *Monadologia physica* (1756) Kant valiantly attempted to solve the second one (as I noted in §2.5.2). And in his inaugural dissertation he presented the first as an apparently insuperable obstacle to the very notion of *world* as an "absolute totality of parts belonging together" (see §3.2, after the indented quotation from Kant 1770, §2 II). In 1772 or 1773, while working on the *Critique of Pure Reason*, he lighted on the idea that we had here a conflict of reason with itself – neatly articulated in four theses and antitheses to match the fourfold table of categories – which can be overcome only by admitting that bodies and processes in space and time are mere phenomena and not things-in-themselves.

The thesis of the First Antinomy is that the world (*a*) has a begin-

---

[39] The title of the relevant chapter is "The Antinomy of Pure Reason", in the singular, meaning the state of opposition (*anti*) to its own law (*nomos*) in which reason finds itself by virtue of the cosmological contradictions. However, Kant subsequently refers to the contradictions themselves – and/or the purportedly valid arguments that lead to them – as "the Antinomies" (first through fourth). This impropriety has been universally adopted in the literature and is probably the source of the misnomer 'antinomy' applied to all sorts of contradictions and paradoxes.

ning in time and (b) is limited in space. To prove it, Kant argues that, if it were not so, (a) the present would be at the end of an eternal series of events, yet the notion of such a completed infinity is absurd; and (b) an infinite time would be required for synthesizing a spatially infinite world, which, by the same token, is absurd as well. The antithesis denies both parts of the thesis. If such denial were false, Kant argues, then (a) there would be a first instant in the history of the world, preceded by empty time, but nothing can be born in an empty time;[40] and (b) since the whole world can be spatially limited only by empty space, there would be a relation between the world, that is, the absolute totality of spatial objects, and the absolute absence of them, "but such a relation, and consequently the limitation of the world by empty space, is nothing" (1787, p. 457).

In the Second Antinomy the existence of ultimate indivisible parts of matter faces the infinite divisibility of bodies. The latter follows, of course, from the infinite divisibility of space: The smallest body can be divided, at least in thought, into two smaller bodies (each, say, with one-half its volume). But it runs against the following difficulty: If every body is composed of other smaller bodies, without end, then *nothing* remains when the relation of composition is removed; and yet a composite can only subsist on the strength of the reality of its parts. So some parts must be indivisible, even in thought. But this clashes with the divisibility of space.

According to Kant, the First and the Second Antinomies arise from the assumption that physical objects are fully determinate, so that, in either case, the argument proving the falsehood of the thesis necessitates the truth of the antithesis, and vice versa. He believes that this assumption holds for things-in-themselves.[41] However, if physical objects are *phenomena*, which are being gradually determined in the

---

[40] "Because no part of such a time possesses, as compared with any other, a distinguishing condition of existence rather than of non-existence, and this applies whether the thing is supposed to arise of itself or through some other cause" (Kant 1787, p. 456; Kemp Smith translation).

[41] This belief was common among modern Christian metaphysicians, who apparently thought that nature would not measure up to the Creator if any determinables remained undetermined. But it was certainly not shared by Aristotle, who maintained that bodies, although indefinitely *divisible*, are not therefore infinitely *divided* (*Phys.* 263a29). Why Kant, notwithstanding his devastating critique of Leibniz–Wolffian metaphysics, continued to uphold the thoroughgoing determinacy of things-in-themselves is a mystery that I am unable to unravel.

course of the progressive construction of experience by the human understanding, the two theses and the two antitheses can all be false at once, and therefore the contradictions vanish.

The Third and Fourth Antinomies can be seen as variants of the problem of the groundless ground, which is aptly illustrated by an ancient myth. To the question "Why doesn't the earth fall?" the myth replies "Because it stands on top of an elephant". The elephant, in turn, stands on the back of a tortoise, which stands on top of another elephant, and so on. Evidently, the piled-up creatures will not provide the required support unless there is one at the bottom that is able to float freely. But if something can possess this property, why not the earth itself? In Kant's book the problem unfolds in two antinomies, corresponding, respectively, to the relational category of cause-and-effect and to the modal category of necessity. This ensures the correspondence between the system of antinomies and the table of categories, but it also allows Kant to deal separately with two different principles adduced by metaphysicians as groundless grounds – uncaused causes – of events in the world, namely, God and human freedom. The proof of the thesis of the Third Antinomy is a cosmological argument for the existence of freedom; the proof of the thesis of the fourth foreshadows the cosmological proof of the existence of God. Note that, despite the allegedly *modal* character of the Fourth Antinomy, the concept of *cause* occurs prominently in the formulation and the proofs of its thesis and antithesis.

We need not go further into the Fourth Antinomy, which does not mean much for the philosophy of physics (and does not show Kant at his best). On the other hand, there are two points in the discussion of the Third that deserve our attention.

(i) The thesis is: "Causality in accordance with the laws of nature is not the only one from which the phenomena of the world can all be derived; to explain them it is necessary to assume also causality through freedom" (1787, p. 472). In the Observation following its proof, Kant countenances a conception of nature as woven from causal chains, some of which begin with causeless acts of freedom.

> If, for instance, I at this moment arise from my chair, in complete freedom, without being necessarily determined thereto by the influence of natural causes, a new series, with all its natural consequences *in infinitum*, has its absolute beginning in this event, although as regards time this event is only the continuation of a preceding series. For this resolution and act of mine do not form part of the succession of purely natural

effects, and are not a mere continuation of them. In respect of its happening, natural causes exercise over it no determining influence whatsoever. It does indeed follow upon them, but without arising out of them; and accordingly, in respect of causality though not of time, must be entitled an absolutely first beginning of a series of appearances.

                                                        (Kant 1787, p. 478)

This is not a conception that many scientifically minded philosophers are prepared to endorse, and Kant himself emphatically rejects it – in the Observation on the antithesis – because, if allowed, "the connection of phenomena determining one another with necessity according to universal laws, which we call nature, and with it the criterion of empirical truth, whereby experience is distinguished from dreaming, would for the most part disappear" (1787, p. 479). And yet the existence of deterministic developments with free beginnings – surely the most shocking feature of the said conception – is taken for granted in the daily practice of laboratory physics. The experimenter sets up and sets going, one assumes, of her own free will an almost closed physical system whose evolution is governed to a good approximation by this or that system of differential equations, and lets it run until she intervenes, again freely, to practice some measurement or to willfully alter its course. The scientist's freedom is no less essential than the system's determinism – or quasideterminism – to the epistemic purpose of this exercise. As Hawking and Ellis pointedly note, "the whole of our philosophy of science is based on the assumption that one is free to perform any experiment" (1973, p. 189; quoted in context in §7.1).

(ii) The antithesis is: "There is no freedom, and everything in the world happens solely in accordance with the laws of nature" (1787, p. 473). In the Observation following its proof, Kant assumes that a thoroughgoing causal concatenation of events is essential for the objective articulation of experience. This claim, already adumbrated in his discussion of the Second and Third Analogies, is put forward again in subsequent sections concerning the solution of this antinomy. In them Kant describes the "principle of the thoroughgoing connection of all events in the sensible world according to unchangeable natural laws" (1787, p. 564) as "a law of the understanding from which no deviation is allowed and no phenomenon can be exempted, on any pretext whatsoever", for it is only by virtue of it that "phenomena can constitute a *nature* and yield objects of experience" (1787, p. 570). We must bear in mind that these strong statements refer to phenomena in space and time. As we saw above, Kant's solution of the first two antinomies rests on the fact that phenomena are *not* fully determined: Before, after, beyond, and within *any* known collection of phenomena one must expect to find others, not yet known, preceding, succeeding, surrounding, or articulating the former. Consequently, any definite set of events must sport vacant slots for connection "according to unchangeable natural laws" with still other events.[42] No system of lawful connections among given events can presently be thoroughgoing. Thus, Kant's claim is purely programmatic: Faced with any phenomenon, the scientist should try to ascertain all its causal slots and to fill them by making out both the phenomena on which the given phenomenon depends and those that depend on it. This is a task without end, for every time a slot is filled new vacant ones are exposed (in the phenomena adduced to fill the former). So what shall we make of Kant's admonition that no phenomenon can be exempted "on any pretext whatsoever" from the principle of thoroughgoing connection? Evidently this is an exhortation not to yield in the quest for causal links, no matter how difficult it is to find them. But what if a particular phenomenon resists all our efforts to connect it through and through to the universal web of nature? Surely this will not, *pace* Kant, bring about the breakdown of experience. For experience is not right now – and never has been – a thoroughly connected system of phenomena, but very much rather a thoroughly fragmented one, despite our success in capturing vast segments of it in a few – different, not always mutually compatible – conceptual nets. Thoroughgoing connection may be – perhaps – the goal of experience, but, contrary to Kant's suggestion, it certainly is not its prerequisite.

The Third Antinomy has brought us back to the theme with which this chapter began. The thesis is made to stand for moral freedom, the antithesis for universal determinism. Kant embraces the latter without exception for phenomena in space and time, while insisting that free initiatives can conceivably be attributed to things-in-themselves. I do not know whether this solution is sufficient to rescue morality, but it is surely worthless for the physicist who expects to be free to intervene

---

[42] Just as the development of a tumor, observable with the naked eye, turns out to depend on cellular changes, visible under the microscope, the latter, too, probably depend in turn on processes still unknown occurring at the atomic and nuclear levels. Our understanding of the cellular processes must leave an opening, so to speak, for determination by such other processes, or it is bound to be wrong.

in her experiments here and now, in the light of incoming results, and not just in the timeless realm where her moral character is supposedly chosen once and for all.

✦

I cannot go into Kant's *Metaphysical Principles of Natural Science* (1786), and his posthumous manuscript on *The Transition from the Metaphysical Principles of Natural Science to Physics* (Ak. XXI – XXII), lest this chapter should grow out of proportion. Friedman (1992, Chapters 3 and 5) provides excellent guidance on both. I shall only mention one significant addition that Kant made in 1786 to his teachings on space, and which is an apt and persuasive application of his notion of Idea. Kant emphasizes that absolute space "cannot be an object of experience, for space without matter is not an object of perception, and yet it is a necessary concept of reason, and so nothing more than a mere *Idea*" (Ak. IV, p. 558). He elaborates further:

In order that motion be given, even if only as phenomenon, an empirical representation of space is required with respect to which the moveable is to change its relation. However, the space, which must be perceived, has to be material and so, pursuant to the concept of matter in general, itself moveable. To think of it as moving one need only to conceive it as contained in a broader space and to assume that the latter is at rest. But the same consideration can be applied to such a broader space, and so on and on, without ever attaining through experience a motionless (immaterial) space with respect to which one could absolutely attribute motion or rest to any matter [. . .]. Whence it is clear: *First*, that all motion and rest is merely relative and that none can be absolute, i.e. that matter can be conceived as being in motion or at rest merely in relation to matter, never with respect to sheer space, so that absolute motion, i.e. motion conceived without any reference of one matter to another, is absolutely impossible. *Secondly*, that for this very reason a concept of motion or rest in relative space which is valid *for every phenomenon* is not possible, but one must conceive a space in which this [relative space] can be thought to be in motion and which does not again depend on another empirical space and so is not conditioned in turn; i.e., an absolute space to which all relative motions can be referred, so that everything empirical is moveable in it [. . .]. Absolute space is therefore necessary not as the concept of an actual object, but as an idea which ought to serve as a rule for regarding every motion in it as purely relative; and all motion and rest must be referred (*reduziert*) to absolute space, if their phenomena are to be transformed into a definite concept of experience (which unifies all phenomena).

(Kant 1786, in Ak. IV, pp. 558–60)

At first sight, this seems to contradict Kant's doctrine about the "formal intuition" (1787, p. 160n.; quoted in §3.3), by which space "is represented as an infinite *given* magnitude" (1787, p. 39; Kant's italics). One should, however, bear in mind that we are dealing here with kinematics, not geometry. According to Kant (1768; 1770, §15.C; 1783, §13), in geometry the intuitive presence of absolute space shows up in the distinction between incongruous counterparts (such as the two differently oriented screws mentioned in §3.2). But to make space available as a frame of reference for motion there must be some way of effectively identifying four of its points, not all on one plane, throughout the duration of the motion. As a matter of fact, this can be done only by marking those points on a rigid body that is assumed to be at rest. Whence Kant's conclusion "that matter can be conceived as being in motion or at rest merely in relation to matter, never with respect to sheer space", and that absolute space, understood as the ultimate kinematic frame of reference, is just a regulative idea.

CHAPTER FOUR

✦

The Rich Nineteenth Century

In the North Atlantic countries that have been the main stage of our story, the nineteenth century was a time of enormous production, not only of manufactured goods but also of art and literature, science, and philosophy. From the great wealth of new ideas in nineteenth-century mathematical physics we can consider only a very small part, chosen mainly for their impact on twentieth-century physics and philosophy. I begin with the new geometries (§4.1), whose emergence is often treated as a chapter in the history of mathematics and its philosophy but which in fact attracted some of the mathematicians who developed them – notably Riemann – chiefly for their potential significance for physics (which Einstein subsequently made good in a wholly unexpected way). The next two sections deal with the concept of field, especially in electrodynamics (§4.2), and with the introduction of chance into thermal physics (§4.3). Finally, we shall take a glance at some nineteenth-century philosophies, which set the tone of twentieth-century debates (§4.4)

4.1 Geometries

4.1.1 Euclid's Fifth Postulate and Lobachevskian Geometry

I can still recall my frustration when, early in my first term of high school geometry, the teacher "proved" – such was his word – Euclid's Theorem I.29 "by parallel transport", in effect by sliding a wooden square along a steel ruler pressed on the blackboard. I had been fascinated by his hortative talk about geometry's breadthless lines and the power of deductive proof and saw the tottering advance of the square from one precarious chalk band to another as one more example of

147

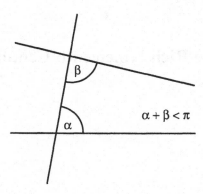

$$\alpha + \beta < \pi$$

Figure 10

the seemingly unlimited capacity of adults to fail their promises. I learned later that Euclid inferred that theorem from his Postulate V, an unproven assumption that can be paraphrased as follows (see Fig. 10):

> Any two coplanar straight lines which are intersected by a transversal straight line meet on that side of the transversal on which the internal angles they form with the transversal add up to less than two right angles.[1]

'Postulate' is the standard translation of *aitema*, which literally means a 'demand' or 'request' that the audience must grant before geometry can get going. Of course, a request is anything but self-evident,[2] so many mathematicians sought to *prove* Postulate V. Some succeeded in

---

[1] On each side of a straight line lies a region of the plane, which has no points in common with the region on the other side. The *internal* angles formed by the transversal with one of the lines are the two angles that lie on the same side of this line as the intersection of the transversal with the remaining line. On each side of the transversal there are *two* internal angles. Obviously, the pair on one side adds up to less than two right angles if and only if the pair on the other side adds up to more than two right angles. If each pair adds up to two right angles, then, by Euclid's Postulate V, the straight lines do not meet, i.e., they are parallel lines. Thus, Postulate V entails (and is also entailed by) the following proposition, known as Playfair's Axiom: Given a straight line λ and a point P not on λ, there is on the plane (λ,P) one and only one straight line through P that is parallel to λ.

[2] This is nicely illustrated by Euclid's Postulate III, "To draw a circle with any center and any radius", which is not only not self-evident but also downright impossible if, as most contemporaries of Euclid believed, the world is wholly contained within a finite surface.

deriving it from other, equally strong propositions, which were also in need of proof. Saccheri (1733), who tried to prove it by *reductio*, that is, by inferring a contradiction from its negation, obtained from the latter a surfeit of surprising but not inconsistent theorems, but finally stopped their flow and forced a contradiction by sleight of hand. Saccheri's theorems were later independently rediscovered by Gauss, Lobachevsky, and Bolyai, who, unbeknownst to each other, treated them as propositions of a new geometry. Given that Lobachevsky was the first to publish, the system of geometry that negates Postulate V but agrees with Euclid on every question which does not depend on that postulate is properly called *Lobachevskian geometry*.[3]

The key difference between Euclidian and Lobachevskian geometry is that in the latter two figures can have the same shape only if they are equal in size. To be specific, consider two convex polygons $A_1 A_2 \ldots A_n$ and $B_1 B_2 \ldots B_n$, and designate the angle at A_i by α_i and the angle at B_i by β_i ($1 \leq i \leq n$). If $\alpha_i = \beta_i$ for each index i, then both polygons are similar and there is a constant factor k such that $A_n A_1 = k B_n B_1$ and, for every $i < n$, $A_i A_{i+1} = k B_i B_{i+1}$. In Euclidian geometry k can be any real number, but in Lobachevskian geometry similarity can hold only if $k = 1$, that is, if the polygons are *congruent*. This implies that on a Lobachevskian plane there are no rectangles, and of course no squares; nor are there any cubes or rectangular parallellepipeds in Lobachevskian 3-space. The sum σ of the internal angles of a Lobachevskian triangle is always less than two right angles: $\sigma = \pi - \delta$, the *defect* δ being proportional to the area of the triangle. Since $0 \leq \sigma$, $\delta \leq \pi$ and the area of a Lobachevskian triangle has an upper bound equal to π times the constant ratio between the area and the defect.

Consider a straight line m and a point P outside it. P and m determine a plane (m,P). If (m,P) is Lobachevskian, there is more than one straight line on it that goes through P and never meets m. Let h be the distance from P to m. Let the perpendicular from P meet m at M. By

---

[3] Gauss, born in 1777, was already convinced by 1799 that Postulate V could never be proved from more perspicuous premises; *c*. 1813 he began work on what he initially called "anti-Euclidian geometry", but published nothing, for fear of "the outcry of Boeotians", as he later said. Lobachevsky, born in 1793, announced his findings in French in 1826 in a public lecture at the University of Kazan and first published them in 1829–30, in Russian, in four installments carried by a local journal. Bolyai, born in 1802, cryptically alluded to his wonderful discoveries in a letter to his father of 23 November 1823, but he published them only in 1832, in a 26-page appendix to his father's *Tentamen in elementa matheseos*.

definition, PM = h. Let n be a straight line through P on the plane (m,P). Let α_n be the smallest angle that n forms with PM. There is an angle $\Pi(h)$, – the *angle of parallellism* for h – such that m meets n if and only if $\alpha_n < \Pi(h)$. If $\alpha_n = \Pi(h)$, we say that n is *parallel* to m.[4] In Euclidian geometry, $\Pi(h)$ is a right angle, no matter what the value of h. In Lobachevskian geometry, $\Pi(h)$ is an acute angle, which decreases as h increases.[5] In each instance of Lobachevskian space there is a definite distance κ, such that $\Pi(\kappa) = \pi/4$ (half a right angle). If PM = κ, the two straight lines through P that make with PM an angle equal to $\pi/4$ are parallel to m and perpendicular to one another. The constant κ can be taken as an absolute unit of length, characteristic of the instance in question.

There is a legend that Gauss tried to test the physical truth of Lobachevskian geometry by measuring the defect of a large triangle formed by the tops of three mountains in Germany.[6] However, as early as 1819 Gauss had told a correspondent that "in the light of our astronomical experience, the constant [κ] must be enormously larger than the radius of the earth" (WW VIII, 182). To someone acquainted with this fact, the legendary attempt should have seemed preposterous. On the other hand, Lobachevsky did try to evaluate the constant κ by measuring the defect of the triangle formed by three well-known stars. According to his calculations, the defect amounted to 3.7-millionths of a second of arc, well within the range of observational error. So Lobachevsky concluded that "all lines subject to our measurements, even the distances between heavenly bodies, are too small in comparison with the line which plays the role of a unit in our theory, so that the usual equations of plane trigonometry must still be viewed as correct, having no noticeable error" (ZGA, I, 22).

More interestingly perhaps, Lobachevsky contemplated the use of

---

[4] Under this definition of parallellism, which was independently adopted by Gauss, Lobachevsky, and Bolyai, there are at most two parallels to a given straight line through a given point outside it. It differs from the standard definition: The straight lines m and n are parallel to each other if they lie on the same plane and do not meet. (See the next note.)

[5] Continuing with the preceding note, we can now see that on a Lobachevskian plane there are, under the new definition of parallellism, exactly two parallels to a given straight line through a given point outside it but infinitely many straight lines through that same point that never meet that line.

[6] Cf. Torretti (1978, p. 381, n. 40), for a brief comment on the origin of this legend and references to A. I. Miller, who first exposed it.

two or more incompatible geometries in physics. He maintains that in nature we only know movement, "without which sense impressions are impossible"; and that "all other concepts, e.g. geometrical concepts, are generated artificially by our understanding, which derives them from the properties of movement" (ZGA, I, 76). "In nature there are neither straight nor curved lines, neither plane nor curved surfaces, [but] only bodies; so that all the rest is created by our imagination and exists solely in the realm of theory" (p. 82). Since our geometry constructs its concept of space from an experience of bodily motion due to physical forces, we might well make allowance for more than one geometry, corresponding to different kinds of natural forces.

> To explain this idea, we assume that [. . .] attractive forces decrease because their effect is diffused upon a spherical surface. In ordinary geometry the area of a spherical surface of radius r is equal to $4\pi r^2$, so that the force must be inversely proportional to the square of the distance. I have found that in imaginary [i.e., Lobachevskian – R. T.] geometry the surface of a sphere is equal to $\pi(e^r - e^{-r})$; such a geometry could possibly govern molecular forces, whose variations would then entirely depend on the very large number e.
>
> (Lobachevsky, ZGA, I, 76)

Lobachevsky also brought up the question of the consistency of the new geometry, which Gauss and Bolyai apparently took for granted. He derived the fundamental equations of trigonometry, which, as in the Euclidian case, suffice to determine all metric relations in his system. He noted that any contradiction that might eventually emerge in a theorem of the new geometry must therefore be implicit in the said equations. But "these equations become [the familiar] equations of spherical trigonometry as soon as we substitute $a\sqrt{-1}$, $b\sqrt{-1}$, $c\sqrt{-1}$ for sides a, b, c" (ZGA, I, 65; recall Lambert's suggestion cited in Chapter Three, n. 17). Consequently, if any contradiction can be derived in Lobachevskian geometry, a matching contradiction must occur in standard geometry. This is a proof of relative consistency: Lobachevskian geometry is consistent unless ordinary spherical trigonometry is inconsistent. The proof rests on the purely formal agreement between two sets of equations, so it does not require, like the better known proofs of Beltrami and Poincaré, a contrived interpretation of curved surfaces and lines in one system as planes and straights of the other. Moreover, it proves the consistency of Lobachevskian geometry, including the negation of Postulate V, relative to a part of Euclidian geometry that does not depend on Postulate V. Thus, any contradiction that might emerge in the new geometry will not stem from the negation of Postulate V but must be contained already in the assumptions shared by both systems.

4.1.2   The Proliferation of Geometries and Klein's Erlangen Program

Lobachevskian geometry did not become a subject of public debate until the 1860s, when Gauss's favorable view of it became known through the publication of his correspondence with Schumacher. On the other hand, a much more radical revision and extension of standard geometry was being actively pursued by mathematicians since the publication of Poncelet's *Traité des propriétés projectives des figures* (1822). The new geometry, known as *projective* geometry, enriches each Euclidian straight line *m* with an ideal point at infinity where *m* meets its parallels (i.e., every straight coplanar with *m* that does not meet *m* at an ordinary point). By this seemingly innocuous trick many proofs are made easier and many theorems simpler. But, while the relations between neighboring ordinary points remain unaltered, the addition of the points at infinity completely disrupts the global neighborhood structure of Euclidian space. *Every* neighborhood of the point at infinity on a straight *m* contains ordinary points from *both* extremes of *m*; thus, the projective straight has the neighborhood structure of a Euclidian circle! Readers who are acquainted with the rudiments of topology will appreciate the deep difference between Euclidian and projective space if I say that the latter is compact while the former is not.[7]

The success of projective geometry led to a multiplication of geometric systems. Felix Klein sought to make sense of their diversity from a single unifying point of view in the booklet, generally known as the Erlangen Program, that he published as he joined the faculty at the University of Erlangen (1872). Klein's proposal turns on the concept of a *group of transformations*. 'Group' is taken here in the sense familiar to students of algebra (and to all those who still remember Rubik's cube). As this fairly simple notion is central to much of recent physics, the reader who is not acquainted with it should make a resolute effort to understand the following explanation. Consider an arbitrary set *G*

---

[7] Some elementary information about topological spaces is given in Supplement III.1.

(its elements can be numbers, or rotations of a cube, or anything else). Assume that there is a "multiplication table" that assigns a definite element of G to each pair of elements of G. (The element assigned to the pair $\langle a,b \rangle$ is called the *product* of a and b and is denoted by ab; note that ab need not be the same as ba.) Assume that "multiplication" is associative, in other words, that for any three elements a, b, and c of G, $a(bc) = (ab)c$. Assume that there is an element e of G such that, for every element a of G, $ae = ea = a$. We call e the *zero* or *neutral* element of G. Assume, finally, that for every element a of G there is an element a^* of G such that $aa^* = a^*a = e$. a^* is called the *inverse* of a. If these four assumptions are fulfilled, the set G possesses the structural properties that characterize it as a *group*. Let me give two simple examples:

(i) A *permutation* is a one–one mapping of a set onto itself. There are six possible permutations of the set $\{1,2,3\}$, formed by the first three positive integers. For reference, I shall designate them with the first six letters of the alphabet. I write beneath each integer the integer assigned to it by each permutation:

$$a = \begin{pmatrix} 123 \\ 132 \end{pmatrix} \qquad b = \begin{pmatrix} 123 \\ 321 \end{pmatrix} \qquad c = \begin{pmatrix} 123 \\ 213 \end{pmatrix}$$

$$d = \begin{pmatrix} 123 \\ 312 \end{pmatrix} \qquad e = \begin{pmatrix} 123 \\ 123 \end{pmatrix} \qquad f = \begin{pmatrix} 123 \\ 231 \end{pmatrix} \tag{4.1}$$

The product xy of permutations x and y is the permutation resulting from carrying out first permutation y, and then permutation x. Thus, $ab(1) = a(b(1)) = a(3) = 2$; $ab(2) = 3$ and $ab(3) = 1$, so that $ab = f$. The reader should compute the rest of the multiplication table and verify that it defines a group, that the neutral element is e, and that the inverses of the other five elements are, respectively, $a^* = a$, $b^* = b$, $c^* = c$, $d^* = f$, and $f^* = d$. Note that the three permutations e, d, and f form a group by themselves, a *subgroup* of the larger group, formed by a, b, c, d, e, and f.

(ii) Take the set $\mathbb{Z}$ of all integers and view the *sum* of any two of them as their product (in the sense introduced above). Addition of integers is, of course, associative. For any integer a, $a + 0 = 0 + a = a$. Finally, for every integer a there is an integer $-a$, such that $a + (-a) = (-a) + a = 0$. So $\langle \mathbb{Z},+,0 \rangle$ is a group.

A group's structure is fully specified by its multiplication table, regardless of the particular nature of its elements. For example, the group of permutations of $\{1,2,3\}$ is structurally identical with the group formed by the following operations on an equilateral triangle with vertices A, B, and C: flippings about the triangle's three heights and clockwise rotations about its center of gravity by 120°, 240° and 360°. To see this, denote the three rotations by d, f, and e, respectively; by a the flipping about the height through A, and so on. Evidently, the three rotations constitute a subgroup, with the multiplication table

$$
\begin{array}{c|ccc}
 & d & e & f \\
\hline
d & f & d & e \\
e & d & e & f \\
f & e & f & d \\
\end{array}
$$

To explain what Klein means by a *group of transformations*, I note that 'transformation' is just another word for permutation, which mathematicians prefer to use when the set that is being mapped onto itself is what they call a *space* – that is, a structured set at least remotely resembling Euclidian space. If φ and ψ are two transformations of a space Σ, their product $\varphi\psi$ is the transformation of Σ that results from carrying out ψ first, followed by φ. This operation is clearly associative. The identity mapping I_Σ: $x \mapsto x$, which maps each element x of Σ to itself, is of course a transformation. Moreover, for every transformation φ of Σ there is a transformation φ^{-1}, such that $\varphi^{-1}\varphi = \varphi\varphi^{-1} = I_\Sigma$ (φ^{-1} – the *inverse* of φ – assigns to each element x of Σ the element to which x was assigned by φ).[8] Consider now the set of all transformations of Σ that satisfy some specific condition. This set constitutes a *group of transformations* of Σ if and only if it contains the identity mapping I_Σ, the inverse of every member of the set and the product of any two members of the set. Take for example the isometries of Euclidian space, that is, the transformations that map every segment onto a segment of the same length. This means that, if φ is an isometry and p and q are any two points in space, the distance $\delta(p,q)$ between these points equals $\delta(\varphi(p),\varphi(q))$. Obviously, the identity mapping is an isometry, and so is the inverse of any isometry. And, of course, if φ and ψ are isometries, $\delta(\varphi\psi(p),\varphi\psi(q)) = \delta(\psi(p),\psi(q)) = \delta(p,q)$, so that $\varphi\psi$ is an isometry too. Thus the isometries form a group.

The last example also illustrates the crucial concept of *invariance*. Let R be an n-ary relation on Σ (in other words, 'R' is a predicate that

---

[8] We have that, for each x in Σ, there is one and only one y in Σ such that $x = \varphi(y)$. Thus, $y = \varphi^{-1}(x)$.

can be meaningfully applied to lists of n objects in Σ).[9] We say that a transformation φ of Σ *preserves* R and that R is *invariant* under φ if R is true of the list $\langle\varphi(x_1), \ldots, \varphi(x_n)\rangle$ whenever it is true of $\langle x_1, \ldots, x_n\rangle$. If R is invariant under every transformation of a group G of transformations of Σ, we say that G preserves R and that R is G-invariant. Thus the distance between point pairs is preserved by the group of isometries, and so are the size of angles, the area of surfaces, and the volume of spatial figures.

Klein perceived that the relations studied by different geometries were in effect the invariants of different transformation groups and proposed the following formulation for the most general problem of geometry:

> Let there be given a manifold and a group of transformations in it. To investigate the configurations belonging to the manifold with respect to such properties as remain invariant under the transformations of the group.

(Klein 1893, p. 67)[10]

Klein adds that "this is the universal problem which spans not only ordinary geometry but also, in particular, the new geometric methods to be mentioned later and the different treatments of manifolds with arbitrarily many dimensions". Since the transformation group adjoined to the manifold can be arbitrarily chosen, every way of dealing with the general problem – that is, each of the new forms of geometry mentioned in Klein's booklet, and many others still awaiting development – is equally justified (Ibid.).

---

[9] To avoid verbal complications I speak of 'objects in Σ', meaning not just the points of Σ but also the sets, sets of sets, and so on, that can be constructed from those points. A mapping $\varphi\colon \Sigma \to \Sigma$, that assigns points to points, induces mappings on the domains formed by such sets, sets of sets, etc. These mappings are customarily designated by the same name φ. Thus, if $\triangle ABC$ is the triangle with vertices A, B, and C, $\varphi(\triangle ABC)$ is the triangle with vertices $\varphi(A)$, $\varphi(B)$, and $\varphi(C)$.

[10] The word 'manifold' translates the German 'Mannigfaltigkeit', which was used by Klein's contemporary Georg Cantor to designate what he later called 'Menge', i.e., a wholly arbitrary, unstructured set. However, the word was also used in a narrower sense, for structured sets of real or complex numbers or of lists (ordered n-tuples) of such numbers, with the standard topologies inherited from the real number field $\mathbb{R}$ or the complex number field $\mathbb{C}$ (for more on algebraic fields, see Supplement I.2). Klein's sense is probably somewhat wider than this narrower sense but essentially akin to it. On this point, cf. Torretti (1978, pp. 137f.). For the current use of 'manifold' in mathematics, see Supplement III.

This last remark was probably motivated by an important discovery concerning Lobachevskian and Euclidian geometry that Klein published at about the same time as the Erlangen Program and that, he thought, resolved the disputed question as to "the true geometry" (Klein 1871, 1873, 1874; cf. Klein 1890). I cannot give here an accurate description of it, but the following rough indications should be sufficient for our purposes.[11] To make things easier I consider only the two-dimensional case. By a *collineation* I mean a transformation of the projective plane that maps each triple of collinear points onto a triple of collinear points. Collineations form a group, and plane projective geometry studies precisely the invariants of this group. One such invariant is the function known as the *cross-ratio*, which assigns a real number to every list of four collinear points. An interesting family of figures in projective – as in ordinary – geometry are the curves called *conics*. A conic on the projective plane can be defined analytically – as in ordinary geometry – as the locus of points whose coordinates satisfy some definite quadratic equation or geometrically by its behavior under certain transformations. Following an idea of Cayley, Klein considered the cross-ratio of point quadruples $\langle P_1, P_2, P_3, P_4 \rangle$ such that P_3 and P_4 lie on a given conic ζ. Since P_3 and P_4 must be the points where the straight through P_1 and P_2 meets ζ, the said cross-ratio may be regarded as depending only on P_1 and P_2, that is, as a function of point pairs. The collineations that map a given conic onto itself form a group, and the said function is clearly an invariant of this group. Let d_ζ denote the principal value of the natural logarithm of this function. Klein showed that the restriction of d_ζ to a well-chosen region $\mathfrak{R}_\zeta$ of the projective plane – with $\mathfrak{R}_\zeta$ bounded by or in some other way dependent on ζ – behaved like an ordinary distance function. Depending on the nature of the conic ζ, the structure $\langle \mathfrak{R}_\zeta, d_\zeta \rangle$ – that is, the region $\mathfrak{R}_\zeta$, endowed with the distance function d_ζ – satisfies all theorems in the plane geometries of Euclid or Lobachevsky or in a third kind of geometry discovered by Klein, which he termed 'elliptic' (his names for the other two systems were 'parabolic' and 'hyperbolic', respectively). Thus, depending on whether ζ belongs to one or the other of three types of conic, the group of collineations that map ζ onto itself is structurally identical with one of the three groups of Lobachevskian, Euclidian, or elliptic isometries. Similar results hold for the three-dimensional

---

[11] For a passable sketch, see Torretti (1978, pp. 125–32) (pp. 110–25 contain preliminary explanations). For a detailed modern exposition, see Rédei (1968).

case, with ζ a quadric surface instead of a conic. In the light of them one sees better why Klein thought that the different geometries had "equal rights" or "equal justification" (*gleiche Berechtigung* – 1893, p. 67).

4.1.3 Riemann on the Foundations of Geometry

In a lecture delivered to the Faculty of Philosophy at Göttingen in 1854, Bernhard Riemann put forward a general view of geometry that is broader and deeper than Klein's.[12] He noted that geometry has hitherto taken for granted the notion of space and the basic concepts for constructions in space. These assumptions of geometry are spelled out in nominal definitions and axioms that throw no light on their mutual relations, so we cannot see whether and to what extent their combination is necessary or even possible. This is due to the fact that "the general concept of multiply extended magnitudes", of which space is an instance, has not been worked out. He proposes therefore to construct such a concept "from general concepts of magnitude":

> It will ensue that a multiply extended quantity admits diverse metric relations, and that space is therefore only a special case of a triply extended quantity. A necessary consequence of this is that the theorems of geometry cannot be inferred from general concepts of quantity, but that those properties which distinguish space from other conceivable triply extended quantities can only be obtained from experience. Thus arises the task of inquiring after the simplest facts from which the metric relations of space can be specified; a task which by its very nature is not completely determined, for several systems of simple facts can be proposed which suffice to determine the metric relations of space – the most important one for our present purpose being that laid down by Euclid. Like all facts, these facts are not necessary but have only empirical certainty: they are hypotheses. One can therefore study their probability – which within the bounds of observation is anyway very large – and thereupon judge the admissibility of extending them beyond the bounds of observation, both in the direction of the immeasurably large as in that of the immeasurably small.
>
> (Riemann 1867, pp. 133–34)

---

[12] Although Riemann's lecture, "On the Hypotheses that lie at the Foundation of Geometry" preceded Klein's program by 18 years, it was not printed until 1867 (after Riemann's death). This may help to explain why Klein paid so little attention to Riemann's conception and came up with a program that fails to cover it.

An n-fold extended magnitude ($n = 1, 2, 3, \ldots$) as conceived by Riemann is substantially the same as what we now call a real n-dimensional smooth manifold.[13] For brevity, I shall say 'n-manifold'. Since contemporary cosmology and celestial mechanics conceive the world as a 4-manifold of uncertain shape, it is important to get the gist of Riemann's idea. I shall try to explain it as intuitively as I can. Then I shall consider his proposals concerning metric relations on an n-manifold.

Any smooth surface, such as a sphere, a Möbius strip,[14] or the surface of a sculpture by Henry Moore, illustrates the concept of a 2-manifold. Note that in these three examples a neighborhood of each point can be mapped continuously one-to-one onto a flat piece of paper – say, a page of an atlas – but no such mapping is possible for the surface as a whole. Analogously, a neighborhood of each point can be mapped continuously one-to-one on an open set of $\mathbb{R}^2$ (the set of all ordered pairs of real numbers, with the standard topology generated from open rectangles).[15] Such a mapping is called a *coordinate system* or *chart*. A collection of charts of parts of a given surface $\mathcal{S}$ constitutes an *atlas* for $\mathcal{S}$ if (i) each point of $\mathcal{S}$ lies in the domain of at least one chart, and (ii) given any two charts g and h in the collection, the composite mappings $g \circ h^{-1}$ and $h \circ g^{-1}$ are differentiable wherever they are defined. The mappings $g \circ h^{-1}$ and $h \circ g^{-1}$ are *coordinate transforma-*

---

[13] Riemann's "n-fold extended magnitude" is the direct source of our notion of a smooth, or differentiable, manifold. We make a distinction between real and complex manifolds, depending on the nature of the coordinates employed for identifying their points, and between manifolds of finite or infinite dimension, depending on the number of coordinates assigned to each point. Riemann contemplated manifolds of infinite dimension (1867, last two sentences of I.3) but went on to discuss only n-manifolds with real-valued coordinates. Still, the subsequent introduction of complex manifolds was true to his spirit.

[14] To form a Möbius strip, take a rectangular strip of paper, twist it, and paste together the two short sides so that the upper left-hand corner coincides with the lower right-hand corner and the lower left-hand corner coincides with the upper right-hand corner. Note that the resulting surface can be said to have only *one* side, in the following intuitive sense: Any point on the surface that can be reached from another by *piercing* the surface can also be reached by sliding a pencil *over* the surface, without ever touching its edges.

[15] 'Open set' and 'topology generated from . . .' are defined in Supplement III.1. An open rectangle in $\mathbb{R}^2$ is a set of ordered pairs $\langle x,y \rangle$, such that x and y are real numbers satisfying the inequalities $a < x < b$, $c < y < d$, for any given a, b, c, and d with $-\infty \leq a$, $c < \infty$ and $-\infty < b$, $d \leq \infty$.

tions; they map number pairs to number pairs, so it makes good sense to say that they are differentiable.[16] The surface $\mathscr{S}$ endowed with an atlas $\mathscr{A}$ is a 2-manifold. An *n*-manifold is defined similarly, substituting $\mathbb{R}^n$ for $\mathbb{R}^2$ in the definition of 'chart'. If $\langle \mathfrak{M}_1, \mathscr{A}_1 \rangle$ is an *n*-manifold, $\langle \mathfrak{M}_2, \mathscr{A}_2 \rangle$ is an *m*-manifold, and *f* is a mapping of $\mathfrak{M}_1$ into $\mathfrak{M}_2$, we say that *f* is *differentiable* at a point *p* of $\mathfrak{M}_1$ if there is a chart *h* defined at *p* and a chart *g* defined at *f(p)* such that $g \circ f \circ h^{-1}$ is differentiable at *h(p)*. *f* is differentiable *tout court* if it is differentiable at every point of $\mathfrak{M}_1$. As one readily sees, every *n*-manifold is like $\mathbb{R}^n$ *locally*, on a neighborhood of each point, but it can differ widely from $\mathbb{R}^n$ *globally*. Atlases provide the means of subjecting manifolds to the power of mathematical analysis while also keeping track of their variegated overall shapes.[17]

I shall now define the *tangent space* at a point of an *n*-manifold. This concept does not occur in Riemann's lecture. Yet, combined with two more ideas that I shall mention in the next paragraph (which were even further from his thought), it has turned out to be indispensable for understanding his proposal on metric relations as well as the mathematical treatment of physical fields (§4.2). Let us concentrate once more on 2-manifolds. It is intuitively obvious that, if *p* is any point of our smooth surface $\mathscr{S}$, there is a plane tangent to $\mathscr{S}$ at *p*. We seek a way of extending this structure to less tangible manifolds. Consider a smooth curve on $\mathscr{S}$. We think of it as being drawn by a point that moves over $\mathscr{S}$ during a period of time represented by an open – finite or infinite – real interval $\mathscr{I}$. So we identify our curve with a mapping $\gamma: \mathscr{I} \to \mathscr{S}$, which assigns to each instant *t* in $\mathscr{I}$ the position $\gamma(t)$ of the moving point at that instant. The collection of these positions – the range of γ – we call a *path*; the variable *t* that, so to speak, regulates the deployment of γ along its path is known as the curve's *parameter*. To simplify my exposition I shall hereafter assume that all curves are injective, that is, that each point of the path corresponds to one and only one value of the parameter. A given path can be the range of many

---

[16] By 'differentiable' I mean differentiable to every order.

[17] I ought perhaps to mention that *any* atlas $\mathscr{A}$ for a manifold $\mathfrak{M}$ determines a unique maximal atlas $\mathscr{A}_{max}$. It is the collection of every conceivable chart *x* such that, for every *y* in $\mathscr{A}$, the coordinate transformations $x \circ y^{-1}$ and $y \circ x^{-1}$ are differentiable wherever they are defined. $\mathscr{A}_{max}$ obviously contains $\mathscr{A}$. The topology generated from the domains of the charts in $\mathscr{A}_{max}$ is the *manifold topology* of $\mathfrak{M}$ (see Supplement III.2). Henceforth every manifold mentioned is endowed with its manifold topology, unless otherwise stated.

curves, defined on different intervals and drawn at different paces; such curves are said to be *reparametrizations* of each other. If we regard $\mathcal{S}$ as a part of ordinary space, with its familiar metric relations, and we assume that the point's motion is smooth, with no brusque changes of speed or direction, we can assign to the curve γ, at a given instant t_0, a definite velocity, which can be represented by a vector tangent to $\mathcal{S}$ at the corresponding point $p = \gamma(t_0)$. The vectors representing the velocity at p of each curve through p generate a two-dimensional vector space[18] that is tangent to $\mathcal{S}$ at p. This is the prototype for the concept of tangent space that I shall define. However, since n-manifolds are not endowed from the outset with metric relations, I must proceed more deviously. The remainder of this paragraph will probably be too tough for some readers and yet, I hope, quite useful for others. Consider the collection $\mathcal{F}(p)$ of all differentiable real-valued functions defined on some neighborhood of point p. With the ordinary operations of function addition and multiplication by a constant, $\mathcal{F}(p)$ has the structure of a vector space.[19] Each function φ in $\mathcal{F}(p)$ varies with t, along the path of γ, in some neighborhood of p. Its rate of variation at $p = \gamma(t_0)$ is properly expressed by the derivative $d(\varphi \circ \gamma)/dt$ at $t = t_0$. As φ ranges over $\mathcal{F}(p)$, the value of $d(\varphi \circ \gamma)/dt$ at $t = t_0$ is apt to vary in $\mathbb{R}$. So we have here a mapping of $\mathcal{F}(p)$ into $\mathbb{R}$, which I denote by $\dot{\gamma}_p$ (also by $\dot{\gamma}(u)$, if $p = \gamma(u)$). It is in fact a linear function[20] and therefore a vector in the dual space $\mathcal{F}^*(p)$ of real-valued linear functions on $\mathcal{F}(p)$.[21] The vectors

---

[18] Supplement I provides elementary information about vector spaces.

[19] If φ and ψ are any two functions in $\mathcal{F}(p)$, their sum $(\varphi + \psi)$ assigns the value $\varphi(q) + \psi(q)$ to each point q in the intersection of their respective domains. If φ is in $\mathcal{F}(p)$ and a is any real number, their product $a\varphi$ assigns the value $a\varphi(q)$ to any q in the domain of φ.

[20] Linear functions on vector spaces are defined in Supplement I.5. It is clear that, for any real numbers a and b, and any two functions φ and ψ in $\mathcal{F}(p)$, $\dot{\gamma}_p(a\varphi + b\psi) = a\dot{\gamma}_p(\varphi) + b\dot{\gamma}_p(\psi)$.

[21] The foregoing explanation throws light on a piece of notation that the reader may have seen and which I shall use in the sequel. Consider a chart x defined on a neighborhood U_x of our point p that assigns to each point q in U_x the coordinates $x^1(q)$ and $x^2(q)$. There is a path through p along which x^1 changes while the other coordinate remains fixed. The curve that assigns each point on that path to the successive values of coordinate x^1 is the *parametric curve* of that coordinate through p. Consider its tangent vector at p. According to our definition, it is a linear mapping that assigns to each function φ in $\mathcal{F}(p)$ its rate of variation at p as x^1 changes while x^2 remains fixed. In other words, it assigns to φ the partial derivative $\partial\varphi/\partial x^1|_p$. It is therefore customary to denote the tangent vector to the parametric curve of coordinate x^1

$\dot{\gamma}_p$ corresponding to all the curves γ whose paths contain the point p span a two-dimensional subspace of $\mathcal{F}^*(p)$. This subspace is, by definition, the tangent space of $\mathcal{S}$ at p. The *tangent space* $T_q\mathfrak{M}$ at a point q of an n-manifold $\mathfrak{M}$ is defined in exactly the same way, substituting n for 2 in all occurrences of the dimension number.

Two further complications that Riemann presumably never had in mind have contributed in the twentieth century to the final clarification of his philosophy of geometry. *First,* the tangent spaces at all points of an n-manifold $\mathfrak{M}$ can be bundled together into a single $2n$-manifold $T\mathfrak{M}$ in a wholly natural way. The *projection* mapping $\pi\colon T\mathfrak{M} \to \mathfrak{M}$ assigns to each tangent vector v in $T\mathfrak{M}$ the point $\pi(v)$ at which it is tangent to $\mathfrak{M}$. The structure $\langle T\mathfrak{M}, \mathfrak{M}, \pi \rangle$ is the *tangent bundle* over $\mathfrak{M}$.[22] *Second,* any vector space $\mathcal{V}$ is automatically associated with other vector spaces, such as the dual space $\mathcal{V}^*$ of linear functions on $\mathcal{V}$, the spaces of so-called covariant multilinear functions on $\mathcal{V}$, "contravariant" multilinear functions on $\mathcal{V}^*$, and "mixed" multilinear functions on both.[23] This holds, of course, for each tangent space of a manifold $\mathfrak{M}$. Moreover, there is a natural way of bundling together into a k-manifold (for some suitable integer k) all the vector spaces of a definite type associated with the tangent spaces of $\mathfrak{M}$ (e.g., all the spaces of bilinear functions on $T_p\mathfrak{M}$ for every p in $\mathfrak{M}$).

Armed with such post-Riemannian notions we can now turn to Riemann's revolutionary views on metric relations in physical space. He observes that concepts of magnitude or quantity (*Grössenbegriffe*) can only be applied to a general concept that admits different modes. Depending on whether the transition from one mode to another is continuous or not, we have a continuous or a discrete variety. In the latter

---

at p by $\partial/\partial x^1|_p$. Note, by the way, that $\partial/\partial x^1|_p$ and $\partial/\partial x^2|_p$ span the tangent space at p. The mapping $p \mapsto \partial/\partial x^k|_p$ is a differentiable mapping of U_x into the tangent bundle TU_x (see the next paragraph and note 22). The vector it assigns to each point p belongs precisely to the tangent space at p. A mapping of this sort, defined on a manifold $\mathfrak{M}$, constitutes what is known as a *vector field* on $\mathfrak{M}$. These notions and notations are readily extended, *mutatis mutandis,* to charts in n-manifolds.

[22] The natural manifold structure of $T\mathfrak{M}$ can be readily seen if we recall that each tangent space is an n-dimensional real vector space and therefore isomorphic to the vector space $\mathbb{R}^n$. So one can pick, for each p in $\mathfrak{M}$, an isomorphism $h_p\colon T_p\mathfrak{M} \to \mathbb{R}^n$. Given a chart g of $\mathfrak{M}$ defined on a suitable set U of $\mathfrak{M}$ we can define a chart h on $\pi^{-1}(U) \subseteq T\mathfrak{M}$ as follows: For each v in $\pi^{-1}(U)$, $h(v) = \langle g(\pi v), h_{\pi v}(v) \rangle$. Clearly, $h(v)$ is a list of $2n$ real numbers.

[23] Multilinear functions are defined in Supplement I.5.

case, the size of different parts can be compared by counting the modes that they contain. But in the continuous case this can only be done by measurement. As described by Riemann, measurement is effected by superposition of the quantities to be compared and therefore "requires a means of transporting one quantity to be used as standard for the other" (1867, p. 135). In the case of physical space, the behavior of standards of measurement under transport depends of course on the forces of nature and can only be learned by experience. At the human scale, metric relations in space agree well with the theorems of Euclidian geometry. However, since all measurements are approximate, such agreement does not warrant the validity of Euclidian geometry in the very large and in the very small. Indeed, "the empirical concepts on which the metric determinations of space are grounded, viz., the concept of a solid body and that of a light ray, would seem to fail in the infinitely small; so it is quite conceivable that the metric relations of space in the infinitely small do not satisfy the hypotheses of geometry, and in fact one should assume this if one can thereby explain the phenomena in a simpler way" (p. 149). To decide such matters one ought to start from the well-corroborated conception of phenomena laid down by Newton and gradually rework it under the pressure of facts that cannot be explained by it. However, to ensure "that this work is not hindered by the narrowness of concepts and that the progress of knowledge of the connection of things is not obstructed by traditional prejudices", a purely mathematical investigation of the conceivable alternatives will be helpful (p. 150). With this purpose in mind, Riemann undertakes such an investigation in Part II of his lecture.

The main obstacle obstructing the progress of geometry in Riemann's intellectual environment was the prejudice that spatial measurements must be carried out with rigid bodies (cf. Ueberweg 1851; Helmholtz 1866, 1868).[24] This is possible only in a space in which every figure can be moved about undeformed. As we shall see, Riemann spelled out the condition under which this requirement is met and described a spectrum of geometries that meet it, neatly defining their place in his own scheme of things. But he took a much broader view of geometry. He agreed that "metric determinations require that mag-

---

[24] For a short report in English about these works, see Torretti (1978, pp. 260–64, 155–71). The said prejudice motivated the rejection of Riemannian geometry by Poincaré (1891, p. 773) and Russell (1897, §§23, 143–57; see his *mea culpa* in Russell 1959, pp. 39f.).

nitudes be independent of place," but pointed out that "this can happen in more than one way" (1867, p. 138). The one he chooses for further elaboration in his lecture rests on the assumption that *the length of lines is independent of the way they lie in space, so that every line is measurable by every other.* Under this assumption, spatial measurements require inextensible cords rather than rigid rods. Riemann understood that the tailor's tape is more versatile than the clothier's yardstick. If the said assumption holds for an *n*-manifold $\mathfrak{M}$, the length of a path in $\mathfrak{M}$ must be an intrinsic property of the path, belonging to it as a 1-manifold embedded in $\mathfrak{M}$, regardless of its relation to the points outside of it.

Traditionally, the length of an arbitrary line is figured out by inscribing in it a series of polygonal lines of ever shorter and more numerous sides and calculating the limit to which the series of their total lengths converges as the number of sides grows beyond all bounds. To be specific, if γ is a smooth curve in Euclidian space $\mathscr{E}$ – that is, if γ is a differentiable mapping of a real interval (a,b) into $\mathscr{E}$ – and $x = \langle x^1, x^2, x^3 \rangle$ is a Cartesian coordinate system defined on $\mathscr{E}$, the length of γ's path is given by the integral

$$\int_a^b \left| \frac{\mathrm{d}}{\mathrm{d}u} (x \circ \gamma(u)) \right| \mathrm{d}u \qquad (4.2)$$

Here the integrand is, of course, the limit to which the length of a chord drawn from $\gamma(u)$ to a neighboring point $\gamma(u + h)$ converges as h goes to 0. Thus conceived, the length of a contorted line does not stand on its own feet, but depends on the availability of straight-sided polygons. However, the concept of a tangent space explained above furnishes the means for overcoming this limitation. Let the curve γ be a differentiable mapping of the real interval (a,b) into an *n*-manifold $\mathfrak{M}$. Suppose that, for each p in $\mathfrak{M}$, there is a real-valued mapping on the tangent space $T_p\mathfrak{M}$ that assigns a length $\mu_p(v)$ to each vector v in $T_p\mathfrak{M}$. Suppose, moreover, that the correspondence $p \mapsto \mu_p$ is a differentiable mapping of $\mathfrak{M}$ into some suitable bundle over $\mathfrak{M}$. One can then take the length $\mu_{\gamma(u)}(\dot{\gamma}(u))$ of the vector $\dot{\gamma}(u)$, tangent to the path of γ at point $\gamma(u)$ in $\mathfrak{M}$, as an index of the curve's advance as it goes through that point. The full length of γ's path can then be defined as the value of the integral

$$\int_a^b \mu_{\gamma(u)}(\dot{\gamma}(u)) \mathrm{d}u \qquad (4.3)$$

As required above, this definition is intrinsic to the path of γ because (i) it can be shown that the length thus defined is the same for all reparametrizations of γ, and (ii) for each $u \in (a,b)$, the vector $\dot{\gamma}(u)$ spans the one-dimensional tangent space at $\gamma(u)$ of the one-manifold constituted by the path itself, so the integral does not in any sense depend on the way this 1-manifold lies in $\mathfrak{M}$.

A mapping that assigns lengths to the vectors of a vector space $\mathcal{V}$ is called a *norm* in $\mathcal{V}$. One naturally expects that

N1. the length of the 0-vector is 0;
N2. the length of any other vector is a positive real number;
N3. if two vectors add up to the 0-vector, their lengths are equal; and
N4. the length of the sum of any two vectors is equal to or less than the sum of their lengths.

These conditions, however, can be met by a wide variety of norms. Riemann pursues one method of defining them, which he calls the "simplest case" (1867, p. 140). It is especially worth looking at – given the practical success of Euclidian geometry at the human scale – because metric relations determined on a 3-manifold by this method agree well[25] with the familiar Euclidian ones on a small neighborhood of each point. Riemann's simplest case is based on Gauss's formula for computing the length of arcs on any curved surface in Euclidian space in terms of the surface's own intrinsic geometry (i.e., without paying attention to the surrounding space). Riemann applies Gauss's approach to n-manifolds. This is how he states the gist of it: In the integral $\int ds$, which is equal to the length of a given curve, the integrand "ds = the square root of an everywhere positive homogeneous function of the second degree in the [coordinate differentials], in which the coefficients are continuous functions of the [coordinates]" (p. 140). Therefore, in a region U of an n-manifold $\mathfrak{M}$ charted by the coordinate functions x^1, ..., x^n,

$$ds = \sqrt{\sum_{h,k} g_{hk} dx^h dx^k} \qquad (1 \le h, k \le n) \qquad (4.4)$$

where the coefficients g_{hk} vary continuously with the coordinates. Let the curve in question be given – as in the discussion preceding eqn. (4.3) – by a mapping $u \mapsto \gamma(u)$ of a real interval (a,b) into U. Then, according to eqn. (4.4), its length is given by this integral:

$$\int_a^b \sqrt{\sum_{h,k} g_{hk} \frac{dx^h}{du} \frac{dx^k}{du}} du \qquad (4.5)$$

By using the post-Riemannian ideas introduced above, this can be elucidated as follows. Let $\mathfrak{M}$ be an n-manifold and $\mathbf{g}$ a differentiable

---

[25] To first order in the coordinate differentials.

mapping of $\mathfrak{M}$ into the bundle of bilinear functions on its tangent spaces, such that g assigns to each point p in $\mathfrak{M}$ a bilinear function $\mathbf{g}_p$ on $T_p\mathfrak{M}$. A mapping of this sort is called a tensor field. g is a *Riemannian metric* on $\mathfrak{M}$ if, for each $p \in \mathfrak{M}$, the function $\mathbf{g}_p$ is (i) *symmetric*, that is, $\mathbf{g}_p(v,w) = \mathbf{g}_p(w,v)$ for all vectors v and w in $T_p\mathfrak{M}$; (ii) *positive definite*, that is, $g_p(v,v) > 0$ unless v is the 0-vector; and (iii) *nondegenerate*, that is, $\mathbf{g}_p(v,w) = 0$ for all vectors w in $T_p\mathfrak{M}$ if and only if v is the 0-vector. The mapping $v \mapsto \sqrt{g_p(v,v)}$ is clearly a norm in $T_p\mathfrak{M}$. This is the norm we employ for defining the length of curves in agreement with Riemann's "simplest" method. A manifold in which lengths – and with them all metric relations – are defined in this way is called *Riemannian*. Thus, if the curve γ in eqn. (4.3) lies in a Riemannian n-manifold $\langle\mathfrak{M},\mathbf{g}\rangle$, the integral giving its length can be rewritten as:

$$\int_a^b \sqrt{g_{\gamma(u)}(\dot{\gamma}(u),\dot{\gamma}(u))}\,\mathrm{d}u \tag{4.6}$$

To recover eqn. (4.5) from (4.6) we must confine the range of γ to the domain U of coordinate functions $x^1, \ldots, x^n$. As is indicated in note 21, for each coordinate function x^k there is a vector field $\partial/\partial x^k$ on U. In the light of the preceding explanation it is clear that, for each pair $\langle h,k\rangle$ satisfying the inequality in (4.4), the mapping

$$p \mapsto g_p\left(\left.\frac{\partial}{\partial x^h}\right|_p, \left.\frac{\partial}{\partial x^k}\right|_p\right)$$

is a differentiable real-valued function on U. Let us denote it by g_{hk}. The tangent space $T_{\gamma(u)}\mathfrak{M}$ is spanned by the vectors

$$\left.\frac{\partial}{\partial x^1}\right|_{\gamma(u)}, \ldots, \left.\frac{\partial}{\partial x^n}\right|_{\gamma(u)}$$

so the vector $\dot{\gamma}(u)$ is equal to a linear combination of them, namely,

$$\dot{\gamma}(u) = \left.\frac{\mathrm{d}x^1}{\mathrm{d}u}\right|_{\gamma(u)} \left.\frac{\partial}{\partial x^1}\right|_{\gamma(u)} + \cdots + \left.\frac{\mathrm{d}x^n}{\mathrm{d}u}\right|_{\gamma(u)} \left.\frac{\partial}{\partial x^n}\right|_{\gamma(u)} \tag{4.7}[26]$$

By substituting from eqn. (4.7) into (4.6) and writing g_{hk} for the value of $\mathbf{g}_{hk}$ at $\gamma(u)$, we obtain the integral (4.5). The definition of the length

---

[26] To see that this equality holds, try the following exercise. Imagine that γ is the parametric curve through $\gamma(u)$ of a coordinate function belonging to some suitable chart. (This fantasy is permissible because we have assumed that γ is injective.) Obviously $\dot{\gamma} = \partial/\partial u$. Consider a differentiable function φ defined on a neighborhood of $\gamma(u)$ contained in the domain of the said chart and in that of chart x. Clearly

$$\left.\frac{\partial\varphi}{\partial u}\right|_{\gamma(u)} = \left.\frac{\partial\varphi}{\partial x^1}\right|_{\gamma(u)} \left.\frac{\mathrm{d}x^1}{\mathrm{d}u}\right|_{\gamma(u)} + \cdots + \left.\frac{\partial\varphi}{\partial x^n}\right|_{\gamma(u)} \left.\frac{\mathrm{d}x^n}{\mathrm{d}u}\right|_{\gamma(u)} \tag{4.8}$$

of a curve leads at once to the notion of *geodesic* (or straightest) curves, which are characterized by the fact that their length is *extremal*, that is, either the greatest or the shortest among all the curves that follow neighboring paths between the same two points.

In his study of curved surfaces in Euclidian space, Gauss introduced a real-valued function, the *Gaussian curvature*, which measures a surface's local deviation from flatness in terms of the surface's intrinsic geometry, without regard to the way it is embedded in space. From this point of view surfaces such as a cone or a cylinder, which can be flattened without stretching or tearing, are no less flat than the plane, and their Gaussian curvature is everywhere equal to 0. On the other hand, the surface of an egg is nowhere flat, and it evidently deviates most from flatness at one of its tips; its Gaussian curvature is everywhere positive, and greatest at this point. Consider, finally, the surface of a saddle; some points on it are the intersection of some curves curled upward and other curves curled downward; the Gaussian curvature is so defined that it takes a negative value at such points. Riemann extended this concept of curvature to Riemannian n-manifolds. He observed that each geodesic through a point in such a manifold is fully determined by its tangent vector at that point. Consider a point p in a Riemannian n-manifold $\langle \mathfrak{M}, \mathbf{g} \rangle$ and two linearly independent vectors v and w in $T_p\mathfrak{M}$. The geodesics determined by all linear combinations of v and w form a 2-manifold about p, with a definite Gaussian curvature $k_p(v,w)$ at p. The real number $k_p(v,w)$ measures the curvature of $\mathfrak{M}$ at p "in the surface direction" given by v and w (Riemann 1867, p. 144). Riemann (1861) conceived a global mapping on $\mathfrak{M}$, depending on the metric $\mathbf{g}$, that yields the said values $k_p(v,w)$ on suitable arguments $\langle p,v,w \rangle$. This object – or rather its twentieth-century rational reconstruction – is now known as the *Riemann tensor*. In post-Riemannian jargon, we describe it as a differentiable mapping that assigns to each p in a Riemannian n-manifold $\mathfrak{M}$ a 4-linear function on the tangent space $T_p\mathfrak{M}$.[27] It is a tensor field like the metric $\mathbf{g}$

---

[27] The requirement of differentiability makes sense because the Riemann tensor maps the manifold $\mathfrak{M}$ into the bundle of 4-linear functions on its tangent spaces. Let the Riemann tensor be $p \mapsto R_p$; then

$$k_p(v,w) = R_p(v,w,v,w) / \left(\mathbf{g}_p(v,v)\mathbf{g}_p(w,w) - (\mathbf{g}_p(v,w))^2 \right).$$

For readers who have some inkling of the subject I will note further that in a Riemannian manifold $\langle \mathfrak{M}, \mathbf{g} \rangle$ there is a canonic isomorphism between each tangent space

(although one of rank 4, while the metric is of rank 2). Given the above definition of $k_p(v,w)$, it is clear that, if $n = 2$, the Riemann tensor reduces to the Gaussian curvature function.

By using his extended concept of curvature, Riemann was able to characterize with great elegance the metric manifolds in which all figures can be freely moved around without changing their size and shape. These are the manifolds in which measurements can be performed with rigid rods and they are, of course, a very peculiar subspecies of Riemann's "simplest" case. They are the Riemannian manifolds of *constant curvature*, in which $k_p(v,w)$ has exactly the same numerical value at each point p for every local pair of directions v and w. "For obviously the figures in [such manifolds] could not be arbitrarily displaced and rotated if the curvature was not the same in every direction at every point" (1867, p. 145). This idea can be nicely combined with Klein's classification of metric geometries. Regarded as Riemannian 3-manifolds, Euclidian or parabolic space has constant zero curvature, Lobachevskian or hyperbolic space has constant negative curvature, and elliptic space has constant positive curvature. A different kind of space of constant positive curvature occurs in Einstein's earliest cosmological solution of his gravitational field equations. In this "Einstein world" each maximal set of simultaneous events constitutes a 3-manifold – with the neighborhood structure of the 3-sphere S^3 – in which the global spacetime metric induces a Riemannian metric of constant curvature > 0.[28] The spatial geometry thus defined has

---

$T_p\mathfrak{M}$ and its dual, the cotangent space $(T_p\mathfrak{M})^*$, that assigns to each v in $T_p\mathfrak{M}$ the linear function v^* in $(T_p\mathfrak{M})^*$ defined by $v^*(w) = g_p(v,w)$ for any w in $T_p\mathfrak{M}$. v and v^* are usually regarded as the "contravariant" and the "covariant" form of one and the same geometric object. By using the canonic isomorphism we can readily match the "covariant" or (0,4) form of the Riemann tensor, as characterized above, with other forms of it found in the literature. For example, one defines its (1,3)-form, contravariant in the first argument and covariant in the other three, by stipulating that if R_p' stands for its value at p, for each p in $\mathfrak{M}$, while R_p denotes the value at p of the said covariant form, then, for every quadruple $\langle a,b,c,d \rangle$ of vectors in $T_p\mathfrak{M}$

$$R_p(a,b,c,d) = R_p'(a^*,b,c,d) \qquad (4.9)$$

[28] S^3 is the topological space formed by the set of all real number quadruples $\langle x,y,z,w \rangle$ that satisfy the equation $x^2 + y^2 + z^2 + w^2 = 1$, with the subset topology induced by the standard topology of $\mathbb{R}^4$. The spatial metric of constant positive curvature in the said Einstein world agrees with the metric induced in S^3 by the usual flat metric of the 4-manifold $\mathbb{R}^4$ in which it is embedded.

sometimes been called "Riemannian" in a narrow sense, but it is preferable to call it *spheric*. Pursuant to the Erlangen Program, each of these geometries of constant curvature is characterized by its own group of isometries. But Klein's conception is too narrow to embrace all Riemannian geometries. Indeed, in the general case, the group of isometries of a Riemannian n-manifold is the trivial group consisting of the identity alone, and therefore it utterly fails to capture the peculiarity of the respective geometry.

4.2 Fields

The physico-mathematical concept of a field can be traced back to eighteenth-century work in fluid dynamics by Euler and others. A fluid is represented as a continuum characterized by physical quantities that may vary smoothly from point to point and from moment to moment. Take, for instance, the distribution of matter in the fluid. This is represented by the density function ρ, which, at any given moment, assigns to each point in the continuum a definite real number of mass units per unit of volume. The total mass m of the fluid is given by the integral $\int_V \rho \, dV$ taken over the fluid's total volume V. In the case of water and other such liquids, ρ is usually regarded – with mild idealization – as constant in time and uniform in space; therefore $m = \rho\int_V dV = \rho V$, and V does not increase or decrease. But this idea of an incompressible fluid of uniform density is obviously inadequate in the case of gases, in which ρ must be thought to change gradually over space and time. Still, in either case the density function is a constant or smoothly varying assignment of real numbers to the points of a continuum, that is, a so-called *scalar field*.[29] Suppose now that the fluid is in motion, relatively to an appropriate rigid reference frame F. Let x^1, x^2, x^3 be Cartesian coordinates affixed to F. If p is a particular point of the fluid,

[29] As is explained in Supplement I.3, a *scalar* is an object that, when "multiplied" by a vector, changes its size but not its direction, and so re*scales* it. The most common vector spaces take real numbers as scalars, whence the use of '*scalar field*' for a smooth assignment of real numbers to points and '*vector field*' for a smooth assignment of vectors. In these expressions, the English word 'field' translates the French 'champ' and the German 'Feld'. Unfortunately, the algebraic structure from which the scalars of a vector space are drawn is also called a 'field' in English ('corps' in French and 'Körper' in German). So a scalar field on a space S is a smooth mapping of S into a field of scalars – the word 'field' being used in completely disparate meanings in its two occurrences in this sentence.

its coordinate $x^i(p)$ ($i = 1,2,3$) is changing in any given instant t at the rate expressed by the derivative $dx^i(p)/dt|_t$. The assignment to each p in the fluid of velocity components

$$\left\langle \frac{dx^1(p)}{dt}\bigg|_t, \frac{dx^2(p)}{dt}\bigg|_t, \frac{dx^3(p)}{dt}\bigg|_t \right\rangle$$

suitably describes the motion of the fluid at time t, relative to the Cartesian system $\langle x^1, x^2, x^3 \rangle$. Of course, the motion is a relation between the fluid and the frame F, and it is not affected by the particular coordinate system affixed to F that we choose for describing it. In view of this, the motion of our fluid at time t is represented – since the late nineteenth century – by the assignment of a coordinate-free – or *geometric* – object to each p, viz., the velocity vector $\mathbf{v}_p(t)$. Both methods of description are directly related, as follows: Let $\mathbf{e}_i$ be a unit vector at p, pointing in the direction in which the coordinate x^i increases while the other two remain fixed ($i = 1,2,3$); then

$$\mathbf{v}_p(t) = \frac{dx^1(p)}{dt}\bigg|_t \mathbf{e}_1 + \frac{dx^2(p)}{dt}\bigg|_t \mathbf{e}_2 + \frac{dx^3(p)}{dt}\bigg|_t \mathbf{e}_3 \qquad (4.10)$$

The assignment $p \mapsto \mathbf{v}_p(t)$ – where p ranges over the points of a fluid and $\mathbf{v}_p(t)$ is the momentary velocity of p in the chosen frame of reference – is, presumably, the earliest example of a *vector field*.[30] This method of representation can be naturally extended to the forces exerted on each point and the resulting accelerations.

Early in the nineteenth century Laplace and Poisson described the action-at-a-distance of central gravitational and electrostatic forces by means of what we now would say are vector fields. Given a static distribution of electric charge in empty space, we consider its action under Coulomb's Law[31] on a test charge, that is, on a point charge so small that it does not significantly contribute to the effect of the charge distribution. If the test charge stands motionless at point p, we denote by $\mathbf{E}_p$ the force per unit charge exerted on it by the charge distribution. Relative to a Cartesian coordinate system $\langle x^1, x^2, x^3 \rangle$ affixed to the dis-

---

[30] One ought to ask oneself, to what vector space does $\mathbf{v}_p$ belong? The only answer that makes sense to me is that it is a vector in the tangent space at the point momentarily occupied by p in the space defined by the reference frame. But this answer became possible only much later. While the underlying space was Euclidian, nobody thought of distinguishing it from its tangent spaces.

[31] See page 82.

tribution's rest frame, $\mathbf{E}_p$ can – like $\mathbf{v}_p$ above – be analyzed into components. We write:

$$\mathbf{E}_p = E_1(p)\mathbf{e}_1 + E_2(p)\mathbf{e}_2 + E_3(p)\mathbf{e}_3 \qquad (4.11)$$

with $\mathbf{e}_1$, $\mathbf{e}_2$, $\mathbf{e}_3$ as in eqn. (4.10). The mapping $p \mapsto \mathbf{E}_p$ is the electric field generated by the given charge distribution. I deliberately say 'is', not 'represents', for, at this stage, the electric field itself was just a convenient mathematical representation of a physical reality consisting of the charges. Indeed the vector notation, which so eloquently wraps up the three coordinate-dependent components E_1, E_2, E_3 into the one geometric object $\mathbf{E}$, was not invented until much later, after Maxwell had conceived the electric and magnetic fields as the actual physical support of electric, magnetic, and optical phenomena. But before turning to him, I shall refer to an important mathematical idea that came up together with the introduction of central force fields.

To introduce a slightly different perspective, I shall link it to the gravitational field generated by an arbitrary distribution of point masses. To get a proper characterization of it, one need only substitute in the preceding paragraph 'mass' for 'electric charge' and 'charge', $\mathbf{G}$, G for $\mathbf{E}$, E, and 'Newton' for 'Coulomb'. Recall Leibniz's discussion of the conservation of "live force" (§1.5.2). In the light of it, it is clear that – in an ideal situation in which only gravitational forces count and friction may safely be ignored – the mechanical work done to move a test particle against the gravitational field forces is fully recovered by letting the particle fall back to the starting point. This perfect balance does not depend on the shape of the upward and downward paths nor on the time spent on them. In vector notation,

$$\oint \mathbf{G} \cdot \mathrm{d}\mathbf{s} = 0 \qquad (4.12a)$$

no matter what the closed path along which this integral is evaluated. In component notation, relative to a Cartesian system $\langle x^1, x^2, x^3 \rangle$ affixed to an inertial frame, eqn. (4.12a) can be rewritten:

$$\oint G_1 \mathrm{d}x^1 + G_2 \mathrm{d}x^2 + G_3 \mathrm{d}x^3 = 0 \qquad (4.12b)$$

Equation (4.12b) obtains if and only if there is a scalar field $p \mapsto \Phi(p)$ such that

$$G_1 = -\frac{\partial \Phi}{\partial x^1} \qquad G_2 = -\frac{\partial \Phi}{\partial x^2} \qquad G_3 = -\frac{\partial \Phi}{\partial x^3} \qquad (4.13b)$$

or – again in vector notation – if and only if

$$G = -\nabla\Phi \qquad\qquad (4.13a)^{32}$$

In other words, due to the fact – expressed in eqn. (4.12) – that gravitational forces are conservative (see §2.5.3, after eqn. (2.14)), the information content of the vector field G is also encoded more simply in the scalar field Φ, which is known as the *gravitational potential*. The *electrostatic potential* φ is defined in the same way: Substitute E and E for G and G in the foregoing discussion.[33]

Maxwell's theory of electricity and magnetism was first adumbrated in his paper "On Faraday's lines of force" (1855/56), in which he sought to work out Faraday's conception of magnetic lines of force in a mathematically precise and mechanically viable manner. I mentioned in §2.5.2 the discoveries of Oersted and Ampère, as well as the latter's purported explanation of them by means of attractive and repulsive central forces. Faraday did not go along with Ampère's account. He felt no sympathy for instantaneous action at a distance, and his own discoveries led him to see the true protagonist of electromagnetic action, not in the matter-borne electric charges, but in the – possibly empty – space between them.[34] The reader has probably observed how iron filings strewn among magnets spontaneously align themselves along smooth curves that join the magnetic poles and appear to continue inside the magnets. Faraday took this as an indication that such curves – which he termed *magnetic lines of force* – are the site of natural powers that are ready to act on a susceptible material as soon as it is placed on them. Because these lines are *curved*, Faraday felt that he could not conceive them "without the conditions of a physical existence in that intermediate space".[35] He became convinced of the phys-

---

[32] The minus sign in eqns. (4.12) is just a convenient – and well-entrenched – convention. Obviously, if there is a scalar field Φ satisfying these equations, there necessarily is another one for which the equations hold with the minus sign deleted, viz., $-\Phi$.

[33] Need I mention that the potentials Φ and φ are defined in each case only up to an arbitrary constant? Since $\partial k/\partial x^i = 0$ for any constant k, if Φ is a solution of eqns. (4.12b), $\Phi + k$ is another. This is not important in the present context, but it gives an inkling of the conventional element in physico-mathematical representations.

[34] In November 1845, Faraday introduced the word 'field' as a name for just such a site of potential electromagnetic action (*Diary*, §7979; cf. EREM, §2247). In 1850, he dropped the word after defining it in terms of lines of force (EREM, §2806), but it had already gained the favor of younger physicists like Kelvin. See references in Gooding (1981, footnote 35), and in *OED*, s.v. 'field', 17a.

[35] Faraday (EREM, §3258); cf. §3263: "To acknowledge the action in curved lines, seems to me to imply at once that the lines have a physical existence. It may be a

ical reality of magnetic lines of force through his work on the effects of magnetism on light (after 1845); he could, however, find supporting evidence in the phenomena of electromagnetic induction, which was perhaps his most striking – and historically significant – discovery (c. 1831). Faraday had shown that an electric current appears in a closed wire loop in three situations that, using the idea of lines of force, can be described as follows: (i) The loop moves across magnetic lines of force (issuing from a magnet or surrounding a current-carrying wire, as in Oersted's experiment); (ii) the loop rests and a magnet or current-carrying wire is moved in its neighborhood in such a way that the lines of force issuing from and presumably dragged by the latter successively cut the loop; and (iii) an electric current is started (interrupted) near the loop, so that, while that current is increasing (decreasing) the loop is swept by an expanding (collapsing) system of lines of force. Faraday devised a way of quantifying the lines of force that issue from a magnet or a current of known intensity and showed that the electromotive force developed in a loop in the said three cases is always proportional to the number of lines per second intersected by the loop. Faraday also extended this idea of lines of force to electrostatics and gravitation but would grant that gravitational lines are imaginary, because he thought that they are always straight.[36]

Maxwell (1855/56) treated Faraday's electric (and magnetic) lines of force as a simple geometrical consequence of the disposition, present at each point p in the surroundings of an electrified conductor (or, respectively, of a magnetic body), to urge in a definite direction a small charged body (or one of the poles of a small magnet) that is placed at p.

> When a body is electrified in any manner, a small body charged with positive electricity, and placed in any given position, will experience a force urging it in a certain direction. If the small body be now negatively

vibration of the hypothetical aether or a state or tension of that aether equivalent to either a dynamic or a static condition; or it may be some *other state*, which though difficult to conceive, may be equally distinct from the supposed non-existence of the lines of gravitational force [. . .]."

[36] See the quotation in note 35. Maxwell, the mathematician, disabused him in a letter that he wrote to Faraday in 1857: "The lines of Force from the Sun spread out from him and when they come near a planet *curve out from it* so that every planet diverts a number depending on its mass from their course and substitutes a system of its own so as to become something like a comet, *if lines of force were visible*" (Campbell and Garnett 1884, p. 203; quoted in Harman 1982b, p. 88).

electrified, it will be urged by an equal force in a direction exactly opposite.

The same relations hold between a magnetic body and the north or south poles of a small magnet. If the north pole is urged in one direction, the south pole is urged in the opposite direction.

In this way we might find a line passing through any point of space, such that it represents the direction of the force acting on a positively electrified particle, or on an elementary north pole, and the reverse direction of the force on a negatively electrified particle or an elementary south pole. Since at every point of space such a direction may be found, if we commence at any point and draw a line so that, as we go along it, its direction at any point shall always coincide with that of the resultant force at that point, this curve will indicate the direction of that force for every point through which it passes, and might be called on that account a line of force. We might in the same way draw other lines of force, till we had filled all space with curves indicating by their direction that of the force at any assigned point.

(Maxwell 1890, vol. I, p. 158)

Thus, in effect, Maxwell conceived the electric or magnetic force that would be exerted on a test object at a given point as a *vector* at that point, and Faraday's lines of force as the paths in space to which such vectors are tangent. Maxwell remarks that Faraday's method of representing a force field by lines of force will "tell us the *direction* of the force, but we should still require some method of indicating the *intensity* of the force at any point" (Ibid.). He proposes to consider Faraday's lines "as fine tubes of variable section carrying an incompressible fluid". Then, "since the velocity of the fluid is inversely as the section of the tube, we may make the velocity vary according to any given law, by regulating the section of the tube, and in this way we might represent the intensity of the force as well as its direction by the motion of the fluid in these tubes" (Maxwell 1890, vol. I, pp. 158–59). Maxwell emphasizes that he does not mean to postulate a hypothetical fluid to explain actual phenomena. The fluid in question here "is merely a collection of imaginary properties which may be employed for establishing certain theorems in pure mathematics in a way more intelligible to many minds and more applicable to physical problems than that in which algebraic symbols alone are used" (p. 160). Indeed, it is not "assumed to possess any of the properties of ordinary fluids" except those required for the present exercise, viz., "freedom of motion and resistance to compression". If I understand him well, Maxwell is saying that he resorts to his fictitious fluid only as a more familiar alternative realization of the mathematical structure that he submits is also embodied in electromagnetic phenomena. The algebraic symbolism adopted by mathematicians for describing this structure would, strictly speaking, also be sufficient for the physicist's purpose, which is to grasp electromagnetism by means of it; but the hydrodynamic analogy is nevertheless introduced because, in Maxwell's judgment, it is both didactically enticing and heuristically fruitful.

In his next paper on the subject, Maxwell (1861/62) produced a mechanical model of electromagnetic action of a very different character. While in the earlier paper he used "mechanical illustrations to assist the imagination, but not to account for the phenomena", he now proposed "to examine magnetic phenomena from a mechanical point of view, and to determine what tensions in, or motions of, a medium are capable of producing the mechanical phenomena observed" (1890, vol. I, p. 452). Maxwell invites the reader to

> suppose that the phenomena of magnetism depend on the existence of a tension in the direction of the lines of force, combined with a hydrostatic pressure; or in other words, a pressure greater in the equatorial than in the axial direction: the next question is, what mechanical explanation can we give of this inequality of pressures in a fluid or mobile medium? The explanation which most readily occurs to the mind is that the excess of pressure in the equatorial direction arises from the centrifugal force of vortices or eddies in the medium having their axes in directions parallel to the lines of force.
>
> (Maxwell 1890, vol. I, p. 455)

To ensure that neighboring vortices can rotate in the same sense, Maxwell inserts between them rows of little idle wheels. I cannot go further into this quaint and yet admirable construction. Suffice it to say that by pursuing it far enough Maxwell reached the conclusion that light is an electromagnetic process. His judicious choice of mechanical analogues for known electromagnetic quantities enabled him to calculate – "from the electro-magnetic experiments of MM. Kohlrausch and Weber" (p. 500) – the speed of propagation of transverse waves in his hypothetical medium by a straightforward application of a formula from continuum mechanics.[37] He found it to be 310,740 kilometers per second "in air or vacuum" (p. 499), a value so close to the velocity of

---

[37] Viz. $V = \sqrt{m/\rho}$, where V is the said speed, m is the coefficient of transverse elasticity and ρ is the density.

light in air as determined by Fizeau (314,858 kilometers per second), "that we can scarcely avoid the inference that *light consists in the transverse undulations of the same medium which is the cause of electric and magnetic phenomena*" (p. 500; Maxwell's italics). This identification led Maxwell henceforth to designate this medium by the same name 'aether' employed for it in the wave theory of light.[38]

Maxwell did not abide by this "rotatory theory of magnetism", in which he perceived a tendency "towards the to me inconceivable and ∴ no doubt to the misty".[39] In his *Treatise on Electricity and Magnetism* (1873), he noted that this theory should be taken only for "a demonstration that mechanism may be imagined capable of producing a connexion mechanically equivalent to the actual connexion of the parts of the electromagnetic field". However, "the problem of determining the mechanism required to establish a given species of connexion between the motions of the parts of a system always admits of an infinite number of solutions",[40] and he had no evidence that the particular solution described in the rotatory theory was the true one. Therefore Maxwell, although he continued to believe that electromagnetic phenomena are the manifestation of a mechanical process, governed by Newton's Laws of Motion, did not again put forward a hypothetical account of its operation. The third and last of his great papers on electrodynamics, "A Dynamic Theory of the Electromagnetic Field" (1864), turns on the idea

that there is an aethereal medium pervading all bodies, and modified only in degree by their presence; that the parts of this medium are capable of being set in motion by electric currents and magnets; that this motion is communicated from one part of the medium to another by forces arising from the connexions of those parts; that under the action of these forces there is a certain yielding depending on the elasticity of these connexions; and that therefore energy[41] in two different forms may

[38] The word 'aether' is the Latin transcription of Greek αἰθήρ = 'heaven'. Aristotle used the word for his fifth element, of which the heavens are made. Descartes used it for the more subtle form of matter, filling the interstices between grosser, ponderable particles. In this acceptance, 'aether' was an obvious candidate for the name of the medium whose vibrations constitute light according to the wave theories of Huygens, Young, Fresnel, and others.

[39] Maxwell to William Thomson, 15 October 1864; Glasgow University Library, Kelvin Papers M 17 (quoted in Harman 1987, p. 277).

[40] Maxwell (1873, vol. II, p. 417; 1891, vol. II, p. 470).

[41] As noted in Chapter One, note 30, 'energy' – meaning 'capacity to do work' – gained currency in the 1850s. Maxwell has the following to say on his use of it here:

exist in the medium, the one form being the actual energy of motion of
its parts, and the other being the potential energy stored up in the connexions, in virtue of their elasticity.

(Maxwell 1890, vol. I, pp. 532–33)

We are thus led "to the conception of a complicated mechanism
capable of a vast variety of motion, but at the same time so connected
that the motion of one part depends, according to definite relations,
on the motion of other parts, these motions being communicated by
forces arising from the relative displacement of the connected parts, in
virtue of their elasticity. *Such a mechanism must be subject to the
general laws of dynamics*" (p. 533; my italics). But Maxwell no longer
fancies a mechanical contraption that would account for electromagnetic phenomena; he resorts instead to Lagrange's methods of analytical mechanics, which – as I noted in §2.5.3 – were designed to facilitate
the description of a mechanical system whose inner workings are
unknown. Early in the paper, Maxwell introduces the term 'electromagnetic field' for "the space in the neighbourhood of the electric or
magnetic bodies" (p. 527). The connexion between an electric current
and the magnetic forces excited by it in the field is discussed in abstract
dynamical terms, leading to the notions of (*a*) the *electromotive force*,
which "is not ordinary mechanical force, at least we are not as yet able
to measure it as common force", and which does not move "merely
the electricity in the conductor, but something outside the conductor,
and capable of being affected by other conductors in the neighbourhood carrying currents" (p. 539); and (*b*) the *electromagnetic momentum*, "every change of which involves the action of an electromotive
force, just as change of momentum involves the action of mechanical

In speaking of the Energy of the field [. . .] I wish to be understood literally.
All energy is the same as mechanical energy, whether it exists in the form of
motion or in that of elasticity, or in any other form. The energy in electromagnetic phenomena is mechanical energy. The only question is, Where does
it reside? On the old theories it resides in the electrified bodies, conducting circuits, and magnets, in the form of an unknown quality called potential energy,
or the power of producing certain effects at a distance. On our theory it resides
in the electromagnetic field, in the space surrounding the electrified and magnetic bodies, as well as in those bodies themselves, and is in two different
forms, which may be described without hypothesis as magnetic polarization
and electric polarization, or, according to a very probable hypothesis, as the
motion and the strain of one and the same medium.

(Maxwell 1890, vol. I, p. 564)

force" (p. 542). According to Maxwell, the latter is the same quantity that Faraday called "the electrotonic state"; Maxwell later dubbed it "the vector potential" (1873, §405; cf. 1891, vol. II, pp. 29, 187, 232). Like their mechanical prototypes, both quantities are vectors varying smoothly in size and direction from point to point of the field.

After inviting us to pick three orthogonal directions in space as the axes of Cartesian coordinates and "to let all quantities having direction be expressed by their components in these three directions", Maxwell introduces the "General Equations of the Electromagnetic Field", 20 partial differential equations in 20 unknowns, which link the electromagnetic momentum and the electromagnetic force with other electromagnetic quantities (pp. 554–62; the unidentified quotations in this paragraph and in note 42 are all from these pages). Translated into the vector notation that was subsequently developed by his friend Tait and his follower Heaviside, and which so naturally fits Maxwell's thinking, the 20 equations reduce to 8, involving 6 vectorial and 2 scalar variables. Besides the two vectors already named, there are two that were already familiar in theories of electricity, viz., (c) the "Current due to true Conduction", that is, the quantity of electricity transmitted from one part of a body to another in unit time per unit area; and (d) the "Magnetic Intensity", or "force acting on a unit magnetic pole placed at the given point"; plus two more that were first introduced by Maxwell, viz., (e) "Electric Displacement", which he described as "the opposite electrification of the sides of a molecule or particle of a body which may or may not be accompanied with transmission through the body"; and (f) "Total Current (including variation of displacement)", which is defined by Maxwell's equations (A) as the vector sum of the conduction current and the time derivative of the electric displacement. The scalars are (g) "the quantity of free positive electricity" per unit of volume, and (h) the electrostatic potential.

A discussion of Maxwell's equations would be quite out of place here.[42] They contain or entail the laws of electricity that were formerly

---

[42] Some readers will wish to know how the 20 equations of Maxwell 1864 relate to the four streamlined vector equations that go under Maxwell's name in college textbooks such as Feynman et al. (1964; see Table 18-1 in vol. II). The answer is that the latter can be obtained from the former with some rewriting (in modern vector notation), some reinterpretation, and – in the case of two of them – some straightforward mathematical reasoning. I denote electromagnetic momentum by **A**, magnetic intensity by **H**, electromotive force by **F**, conduction current by **j**, electric displacement by **D**,

secured by Coulomb, Ampère, and Faraday but go well beyond them. Maxwell's innovations, notably the inclusion of "variation of displacement" in the total current, drew support mainly from the fact that the theory could, by Lagrange's methods, be shown to be dynamically viable. The theory's most striking novelty concerns the existence of electromagnetic waves: The equations admit solutions consisting of mutually induced electric and magnetic undulatory disturbances propagating in free space with the speed of light. But, apart from Maxwell's bold conjecture that such disturbances are the stuff that light itself is made of, there was not a shred of evidence for them. The theory was resisted on the continent, particularly by Helmholtz, and was looked askance in Britain by Maxwell's older friend, Lord Kelvin (born

quantity of free electricity by e, and electrostatic potential by Ψ (e and Ψ are used by Maxwell in this sense). I use the nabla operator

$$\nabla = \left\langle \frac{\partial}{\partial x}, \frac{\partial}{\partial y}, \frac{\partial}{\partial z} \right\rangle$$

Let me also put $\mathbf{E}$ for the electromotive force on a conductor at rest, as defined in Maxwell's eqns. (35):

$$\mathbf{E} = -d\mathbf{A}/dt - \nabla\Psi \qquad (35)$$

Maxwell's eqn. (G) is

$$e + \nabla\mathbf{D} = 0 \qquad (G)$$

which – restricted to free space and other regions where $\mathbf{D} = \varepsilon_0\mathbf{E}$ and e equals the charge density ρ – agrees, up to a conventional choice of sign, with Feynman's "Maxwell equation I":

$$\nabla \cdot \mathbf{E} = \rho/\varepsilon_0 \qquad (I)$$

Putting, as is now usual, $\mathbf{B} = \mu\mathbf{H}$, where μ is "the ratio of the magnetic induction in a given medium to that in air under an equal magnetizing force", eqns. (B) in Maxwell (1864) can be written:

$$\mathbf{B} = -\nabla\mathbf{A} \qquad (B)$$

Together with eqn. (35) and the mathematical theorem "$\nabla \times \nabla\sigma = 0$ for every scalar σ", this yields "Maxwell equation II", viz.,

$$\nabla \times \mathbf{E} = -d\mathbf{B}/dt \qquad (II)$$

Since, for any vector $\mathbf{V}$, $\nabla \cdot (\nabla \times \mathbf{V}) = 0$, eqn. (B) also yields Feynman's "Maxwell equation III":

$$-\nabla\mathbf{B} = 0 \qquad (III)$$

William Thomson), but it won acceptance among the younger electricians in the United Kingdom – Lodge, Heaviside, FitzGerald – who made it the groundwork of their research. The Maxwell equations became the core of electrical science only after Heinrich Hertz – a former assistant of Helmholtz – showed in 1888 that he could generate and receive electromagnetic waves with electrical laboratory equipment. However, the program of conceiving a working mechanical model of the electromagnetic aether never took root on the continent,[43]

Maxwell's eqns. (C) can be written as

$$\nabla \times \mathbf{B} = 4\pi(\mathbf{j} + d\mathbf{D}/dt) \tag{C}$$

which, under the restriction stipulated above for eqn. (G), translates, in SI units, into Feynman's "Maxwell equation IV":

$$c^2 \nabla \times \mathbf{B} = \mathbf{j}\varepsilon_0^{-1} + d\mathbf{E}/dt \tag{IV}$$

Maxwell's "complete equations of electromotive force on a moving conductor" – labeled (D) in his paper – are also worth looking at. By writing $\mathbf{v}$ for the conductor's velocity, they become, in my notation,

$$\mathbf{F} = (\mathbf{v} \times \mathbf{B}) + \mathbf{E} \tag{D}$$

which reads, of course, just like the standard formula for the so-called Lorentz force – i.e., the "ordinary mechanical force" – exerted by the magnetic field $\mathbf{B}$ and the electric field $\mathbf{E}$ on a body carrying a unit of charge and moving with velocity $\mathbf{v}$. See eqn. (7.3) and the text surrounding it on page 427.

[43] The program, however, lived on in the British Isles. We hear, for instance, of one W. M. Hicks, at St. John's College, Cambridge, who "since graduating 7th wrangler in the Cambridge Mathematical Tripos of 1873 [and until the early years of the twentieth century] devoted the greater part of his research effort to the mathematical analysis of vortex motion in hydrodynamical ether" (Warwick 1995, p. 308). Hunt (1987) gives an enthralling report on G. F. FitzGerald's aether models and their heuristic utility. FitzGerald warned against mistaking his models – "made up of wheels and india-rubber bands, [or] of paddle-wheels, with connecting canals" – for a likeness of reality, but he still believed that "what physicists ought to look for is such a mode of motion in space as will confer upon it the properties required in order that it may exhibit electromagnetic phenomena" (FitzGerald 1885, p. 162; quoted in Hunt 1987). As late as 1904, Larmor, reviewing Kelvin's modeling exercise in his Baltimore Lectures (delivered in 1884 but first published that year), wrote: "We are at the parting of the ways. Is it incumbent on us to treat the aether as strictly akin to the material bodies around us? or may we assign to it a constitution of its own, to be tested by its success in comprehending the complex of known relations of physical systems?" (Larmor 1904, quoted in Wise and Smith 1987, p. 323). Larmor's questions were rhetorical, for he had de facto answered 'No' to the first and 'Yes' to the second in his book, Aether and Matter (1900).

where indeed the very idea that electromagnetic phenomena rest on some form of mechanical interaction – subject to Newton's Laws of Motion – between ponderable matter and the aether fell into abeyance. H. A. Lorentz, surely the most significant contributor to electrodynamics at the turn of the century, postulated a completely motionless aether.[44] So with him the electromagnetic waves could not be mechanical vibrations – involving elastic displacements, however small – but only oscillatory changes of length and direction of the electric and magnetic field vectors at each point. From here to the modern view of geometric object fields pinned directly on space (or on spacetime), without a ghostly material medium to carry them, it was but a small step (though one that only a truly free spirit like Einstein could have taken).[45]

4.3 Heat and Chance

Maxwell is best known for his equations of the electromagnetic field (§4.2). But a no less decisive contribution to physics was his use of the probability calculus in the theory of heat. Through the study of thermal phenomena in gases he sought definitive evidence for the thesis that heat is motion – specifically, that the energy contents of steam and other hot gases consists, for the most part, in the kinetic energy of its wildly agitated particles. To cope with the immense number of molecules in any observable volume of gas Maxwell took his cue from Quêtelet, a former astronomer turned social scientist who successfully employed the probability calculus to handle demographical statistics. In this section I shall try to ascertain the meaning of physical probability as it was understood by Maxwell and his successors. But first let us deal briefly – in §§4.3.1 and 4.3.2 – with the development of thermal physics leading to his move.

---

[44] Contrary to what one sometimes reads, Lorentz's motionless aether does not in any way presuppose the conception of an absolute space. "When I say for brevity's sake that the aether is at rest, I mean only that no part of this medium is displaced with respect to its other parts and that all perceptible motions of the heavenly bodies are motions relative to the aether" (Lorentz 1895, §1, in CP, vol. 5, p. 4).

[45] For a concise and illuminating sketch of the development of the field concept from Faraday and Maxwell, through Lorentz, to Einstein, see Nersessian (1984, Part II). The standard reference on the history of electrodynamics in the late nineteenth century is Whittaker (1951/53, vol. I, Ch. IX–XIII). For a richer and deeper view, see Buchwald (1985, 1994).

4.3.1   Heat as Motion

At the beginning of the nineteenth century the received scientific view was that there exists a weightless, elastic, fluid substance – known as *caloric* – responsible for all thermal phenomena.[46] Parts of caloric were supposed to repel each other and to be attracted by other material substances. The absorption of caloric furnished a seemingly obvious explanation of the expansion of bodies with heat. Temperature gradients reflected the tendency of caloric to flow from one place to another. Local temperature, however, depended not only on the local abundance of caloric but also on the capacity of local matter for storing it. (It was well known that to warm up a pound of water by one degree one must burn more fuel – and so, presumably, release more caloric – than to do the same to a pound of mercury.) The conception of heat as motion favored by Bacon and Boyle in the seventeenth century was generally regarded as obsolete. Indeed, Rumford did his best to gather empirical support for it as he supervised the manufacture of ordnance for the Bavarian army. He observed that the rate at which heat is produced while boring a cannon remains steady and does not gradually diminish as it should if heat were a substance stored in and extracted from the whole metal body. Nor could the heat released proceed exclusively from the thin layer of metal being reduced to chips, for its total quantity was out of all proportion with the mass of the chips obtained.[47] So he concluded that the heat observed was not a substance previously stored in the cannon but only a display of the metal's growing internal agitation. The scientific establishment, however, remained unconvinced[48] until the

---

[46] Caloric, like aether, was the subject of much respectable scientific work before being pronounced nonexistent. My short remarks cannot do justice to caloric physics, its diversity, and its applications. For a detailed report, see Fox (1971).

[47] While the borer detached 54 g of "metallic dust, or, rather, scaly matter" from the bottom of the cylinder, the heat developed in the cannon – apart from any heat loss to the environment – was sufficient to bring to the boil some 2,300 g of ice-cold water. Rumford not only judged it improbable that such a small metal mass should contain that much caloric, but, by "experiments, made for the express purpose of ascertaining that fact", he showed that "the capacity for Heat of the metal of which great guns are cast *is not sensibly changed* by being reduced to the form of metallic chips in the operation of boring cannon" (Rumford 1798, in CW, I, 10).

[48] One notable exception was young Humphry Davy, who, shortly after Rumford communicated his results to the Royal Society of London, published his own friction experiments with blocks of ice, which – he believed – showed that "heat cannot be

1840s, when Joule measured the mechanical equivalent of heat and Helmholtz and others took this as proof that the capacity for doing mechanical work – or *energy*, as it would soon be called – is conserved in nature and that heat is just one of its transitory guises.

In about 1838 Joule began experimenting on what we would now call energy conversion, with the aim of improving the electric motor and eventually replacing with it the fairly inefficient steam engines of the time. He was soon disappointed in his hopes, but he had shown by the way that, since an electric current would produce heat and mechanical work in fixed amounts depending on the current's intensity, the mechanical and heating powers of a current stand in a fixed ratio, which he reckoned as 838 foot-pounds per BTU.[49] Emboldened by the fact that "the magnetic engine [i.e. the electric generator] enables us to convert mechanical power into heat by means of the electric currents which are induced by it", he ventured to think "that by interposing an electro-magnetic engine [i.e., an electric motor] in the circuit of a battery a diminution of heat evolved per equivalent of chemical change would be the consequence, and this in proportion to the mechanical power obtained" (Joule 1843, in Joule 1884, p. 120). This was hard to prove, for heat loss to the environment can easily escape detection,[50] but Joule was able to show by a clever experiment that heat literally *converts* into mechanical work. He compressed air to 22 atmospheres inside a vessel placed in a water tank and released the air into a second vessel in the same tank. If the second vessel was kept at atmospheric pressure, the inflowing air had to do work against it and the water in the tank grew colder, but no temperature change was observed if the second vessel was exhausted and the air entered it without overcoming any resistance. As Joule noted, these results are baffling if heat is a substance, but they "are such as might have been

considered as matter" but must be regarded as a "peculiar motion, probably a vibration of the corpuscles of bodies" (quoted in Wolf 1939, p. 198). However, the effects observed by Davy were probably due to the conduction of heat from the environment to the ice.

[49] One foot-pound is the quantity of work required to raise one pound to the height of one foot (in London, I presume). One BTU (or British Thermal Unit) is currently defined as the quantity of heat required to increase the temperature of one pound of water from 63°F to 64°F; Joule's unit was slightly different; see the quotation in the main text, at the end of the paragraph.

[50] In 1862 it was shown experimentally by Hirn that "when heat does work in an engine a portion of the heat disappears" (Maxwell 1883, p. 147).

deduced a priori from any theory in which heat is regarded as a state of motion among the constituent particles of bodies" (Joule 1844, in Joule 1884, p. 186). Joule developed diverse methods for measuring the mechanical equivalent of heat with increasing precision. In 1849 (Joule 1884, p. 328), he announced that "the quantity of heat capable of increasing the temperature of a pound of water (weighed in vacuo, and taken at between 55° and 60°) by 1° Fahr, requires for its evolution the expenditure of a mechanical force represented by the fall of 772 lbs through the space of one foot", a result that is remarkably close to the modern value of 778.[51]

Joule's results did not meet immediate acceptance. However, Helmholtz cited them as evidence for the conservation of energy in his pioneering paper of 1847. By that time, the idea of energy conservation was cropping up in independent, often obscure, publications – Kuhn names "twelve men who, within a short period of time, grasped for themselves essential parts of the concept of energy and its conservation" (1959, p. 321) – but Helmholtz was the first one to state it with full clarity and precision as a fundamental principle of physics.[52] He puts it forward as "a physical presupposition (*Voraussetzung*)" the implications of which he develops and compares with "the empirical laws of natural phenomena in the different branches of physics" (1847, p. 1). This development and comparison fill most of the paper, but here we need only consider the general remarks contained in the Introduction. Helmholtz notes that his theses can be reached from two propositions that, he contends, are logically equivalent (*identisch*), namely, (i) "the statement that it is not possible to obtain limitless energy (*Arbeitskraft*, i.e., 'force for doing work' – R. T.) from the effects of any combination of natural bodies on one another", and (ii) "the assumption that all effects in nature must be referred to attractive and

---

[51] The said value was obtained by Joule through friction experiments in water. In the same paper, he reported a value of 776.045, obtained through friction against cast iron. However, in Joule's opinion it was "highly probable that the equivalent from cast iron was somewhat increased by the abrasion of particles of the metal during friction, which could not occur without the absorption of a certain quantity of force in overcoming the attraction of cohesion" (1884, p. 328).

[52] Helmholtz (1847) speaks of the conservation of *Kraft*, which literally means *force*. In his text, *Kraft* sometimes designates Newtonian force, but more often the capacity for doing mechanical work. Of course, it is only in the latter sense that *Kraft* is a conserved quantity according to Helmholtz. In a footnote added in 1881, Helmholtz uses 'Energie' for 'Kraft' in this sense (WA, I, 29).

repulsive forces (*Kräfte*) whose intensity depends only on the distance of the interacting points" (1847, p. 1). Proposition (i) amounts to the impossibility of perpetual motion, which we already saw Leibniz invoke against Descartes (§1.5.2). Helmholtz has no difficulty in showing that proposition (i) follows from (ii), that is, that physical systems under the exclusive rule of central forces are conservative (in the sense explained in §2.5.3). But he also argues that proposition (ii) follows from (i), so that a perpetual motion should be feasible in a physical system that is governed by forces other than central:

> If the natural bodies also exhibit forces (*Kräfte*) which depend on time and speed or act in directions other than the straight lines joining each pair of acting material points – e.g. rotatory ones – , then systems of such bodies would be possible in which energy (*Kraft*) is either lost or gained *ad infinitum*.
>
> (Helmholtz 1847, pp. 19f.)

Helmholtz retracted these words in 1881, noting that they hold only if Newton's Third Law of Motion is generally valid (WA, I, 71; cf. the footnotes to WA, I, 20, 21). Even this restriction may, however, be insufficient to rescue Helmholtz's claim, for the equality of action and reaction can hold – in a sense – together with energy conservation in a classical electromagnetic system in which the force on a moving charge depends on the latter's velocity, provided that energy and momentum can be stored in the field.

The Introduction to Helmholtz's paper of 1847 also contains a philosophical argument for energy conservation that deserves our attention. While the "experimental part of our sciences" seeks for the laws that will enable one to bring particular processes in nature under general rules (Helmholtz's examples are the law of the refraction and reflexion of light, and Mariotte and Gay-Lussac's law regarding the volumes of gases), the theoretical part endeavors "to find the unknown causes of the processes from their visible effects". This endeavor is justified and indeed necessitated by "the principle that every change in nature *must* have a sufficient cause". Some of the causes we discover are changeable, so we must look for *their* causes in turn. Thus, "the final goal of the theoretical natural sciences is to discover the ultimate unchangeable causes of natural processes" (p. 2). In its quest science uses two "abstractions", namely, *Materie* ('matter'), that is, the sheer existence of things apart from their effect on other things and on our sense organs, and *Kraft* ('force'? 'energy'?), that is, the capacity of things to produce effects. The problem of referring natural phenomena to ultimate unchanging causes therefore takes this form: "as ultimate causes [. . .] one ought to find unchanging *Kräfte*" (p. 4).[53] Helmholtz takes for granted the analysis of nature into chemical elements, that is, "matters with unchanging *Kräfte* (undestructible qualities)". Under this analysis, the only possible changes in the universe are spatial, that is, motions; "so the forces can only be moving forces, dependent in their action upon spatial relations" (Ibid.). A few additional reflections lead to the concept of central forces and to the following conclusion:

> The problem of the physical sciences is finally determined thus: To refer natural phenomena to unchanging attractive and repulsive forces, whose intensity depends on distance. The solvability of this problem is at the same time the condition of the complete intelligibility of nature.
>
> (Helmholtz 1847, p. 6)

In 1881 Helmholtz reconsidered his distinction between the experimental search for laws and the theoretical search for causes. He says that the philosophical discussion I have just summarized was more strongly influenced by Kant than he would still judge proper. Remarkably, Helmholtz's emended view comes even closer to Kant's philosophy (cf. §3.4):

> Only later have I understood that in fact the principle of causality is nothing but the assumption that all natural phenomena accord with law. Law acknowledged as objective power we call *Kraft*. By dint of its etymological meaning, *cause* (*Ursache*)[54] is the unalterably permanent or existent behind the changes of phenomena, namely Matter (*Stoff*), and the law of its action is *Kraft*.
>
> (Helmholtz WA, I, 68)

Soon after the publication of Helmholtz's paper the English-speaking physicists who shared his views began using 'energy' as a general term for his conserved *Kraft* (cf. Chapter One, note 30). According to this school of thought – to which Kelvin, Rankine, Maxwell, and, of

---

[53] 'Kräfte' is the plural of 'Kraft'. The reader may decide when to render it as 'forces' and when as 'energies'.

[54] 'Ursache', the German word for 'cause', literally means 'primal thing' or 'thing-at-the-origin'.

course, Helmholtz himself belonged – energy is present in bodies in two main forms only: (i) as energy of motion or *kinetic energy*, equal to one-half the body's mass times its squared velocity, and (ii) as energy of position or *potential energy*, equal to the work that must be done against ambient forces to carry the body from a position of (conventionally) zero potential energy to its present place. The received inverse-square laws of gravity, electrostatics, and magnetostatics naturally led to this concept of energy of position, and our physicists looked forward to explaining chemical and electrodynamic energy in some such way. Latent heat – that is, the heat absorbed without temperature change by melting solids and boiling liquids – was readily understood in similar terms, as the work done to break a solid's rigidity or a liquid's cohesion, which could later be recovered by freezing or condensation. But manifest heat – showing up through temperature increases – was regarded by these physicists as kinetic energy. In his textbook on the theory of heat, Maxwell concedes that "the evidence for a state of motion, the velocity of which must far surpass that of a railway train, existing in bodies which we can place under the strongest microscope, and in which we can detect nothing but the most perfect repose, must be of a very cogent nature before we can admit that heat is essentially motion" (1883, pp. 302f.). The alternative is "that the energy of a hot body is potential energy". Now, this form of energy "depends essentially on the relative position of the parts of the system in which it exists", so that motion is necessarily involved "in every transformation of potential energy" (p. 303). Therefore heat transfer, such as is bound to occur wherever there is a temperature gradient, must involve motion, even though none is visible. This proves that every hot body contains invisibly moving parts and that at least part of its thermal energy arises from their motion.[55] The study of gases – says Maxwell – provides good evidence that "a very considerable part of the energy of a hot body" (p. 304) is in fact the kinetic energy of its submicroscopic parts.

---

[55] The proof, of course, depends on the premise that all nonkinetic energy is energy of position. Physicists like Mach, Ostwald, and Duhem, who rejected the analysis of matter into molecules as an unnecessary metaphysical hypothesis, did not subscribe to that premise but simply viewed energy as a fundamental physical quantity that is transformed in fixed ratios from one to another of its many guises: mechanical, thermal, electromagnetic, chemical, and so on. This school of thought presumably inspired Sigmund Freud with his notion of psychic energy, although he never bothered to propose an experiment for ascertaining its mechanical equivalent.

4.3.2 The Concept of Entropy

If heat is energy and energy is conserved, why cannot we satisfy our energy requirements through a scheme like the following? (i) Extract heat from the environment and convert it into kinetic energy; (ii) use the kinetic energy thus obtained to run the transportation system and to generate electricity; and (iii) as wheels chafe the ground and resistors warm up in the electric network, let the heat return to the environment and recycle it. Sadi Carnot (1824) explained within the theory of caloric why this ecophile's dream is unfeasible. Just as the substance water yields mechanical work as it falls from a higher to a lower level, so the substance caloric yields mechanical work as it falls from a higher to a lower temperature. Therefore, to obtain useful work from a hot source heat must be transferred to a *colder* sink. Moreover, Carnot showed by a masterly piece of reasoning that the efficiency of a heat engine – that is, the amount of work that it can obtain from a unit of heat as this falls to a lower temperature – has an upper bound that depends not on the machine's design, nor on the working substance employed in it (e.g., steam, gasified petrol, etc.), but only on the temperatures between which the engine operates.

In view of Carnot's achievement and its well-corroborated predictions and applications, physicists like Kelvin were reluctant to give up the caloric theory.[56] When the kinetic theory prevailed *c.* 1850, the new science of heat founded by Kelvin and Clausius in the wake of Helmholtz and Joule rested on two principles, viz., the Conservation of Energy and the so-called Second Principle of Thermodynamics, embodying the gist of Carnot's idea without the water-level metaphor. Kelvin and Clausius gave two different versions of this principle, which, in fact, entail each other. Let me quote them in Enrico Fermi's words:

---

[56] This may have been partly due to careless reading, for Carnot did not conceal his doubts with regard to the prevailing theory of heat (1824, p. 37n1). He also asked rhetorically: "Can one conceive the phenomena of heat and electricity as due to anything else than certain motions of bodies?" (p. 21n1). Indeed, as Maxwell (1883, pp. 146f.) showed, Carnot's analysis is inconsistent with the caloric theory. To see this, suppose that an amount of kinetic energy W is obtained while heat is transferred from a warmer to a colder body. Let Q be the quantity of heat drawn from the source. W can in turn be converted into a quantity of heat q, for example, by causing it to move a water paddle. So, unless a part of Q is destroyed to compensate for the generation of W – which is of course impossible if heat is caloric – the process will create a net increase of heat q – which again is impossible if heat is caloric.

A transformation whose only final result is to transform into work heat extracted from a source which is at the same temperature throughout is impossible (Postulate of Lord Kelvin).

A transformation whose only final result is to transfer heat from a body at a given temperature to a body at a higher temperature is impossible (Postulate of Clausius).

(Fermi 1937, p. 30)

Building on Carnot's work, Kelvin developed the absolute scale of temperature and Clausius the concept of entropy. Entropy soon became entangled in philosophically significant debates, so I must stop to explain it. The long-winded road leading to its definition is in itself of interest for philosophers, as an egregious instance of creative concept formation in mathematical physics.[57] Ultimately, I do not think that entropy is any more contrived than force or momentum, or the Newtonian concepts of space and time. But these have managed, despite their novelty and artificiality, to pass for mere refinements of familiar ideas, whereas entropy – perhaps because it was carefully thought out from the beginning and was given a newly coined name[58] – has always appeared to be foreign to common experience. In this it anticipates many quantities in twentieth-century physics, such as spin, strangeness, and spacetime curvature, for which no folk prototype can be cited.

I begin with Carnot's theorem on the efficiency of heat engines. Such an engine typically produces work W by taking a quantity of heat Q_1 from a body B_1 at temperature T_1 and surrendering a quantity of heat Q_2 to a body B_2 at temperature $T_2 < T_1$. If all forms of energy are expressed in the same unit (say, joules), obviously $W = Q_1 - Q_2$. The engine's *efficiency* is naturally expressed by

$$\eta = \frac{W}{Q_1} = 1 - \frac{Q_2}{Q_1} \qquad (4.14)$$

[57] I confine my exposition to the original definition in the 1850s. For a philosophical discussion of the subsequent development, generalization and reinterpretation of 'entropy' in physics, see Bartels (1994, pp. 135–219).

[58] By Clausius. Rankine called it "thermodynamic function". Structurally, the Greek word ἐντροπία stands to τροπή ('turn, turning') in the same relation as ἐνέργεια stands to ἔργον ('work'). Ἐνέργεια, meaning 'activity', 'actuality', 'vigor of style', was used abundantly by Greek prose writers since Aristotle. On the other hand, ἐντροπία is extremely rare in extant literature; it occurs once in the Homeric Hymn to Hermes (245), meaning 'trick, wile', and once in Hippocrates (*De decente habitu*, 2), meaning 'modesty'.

To prove his theorem, Carnot constructed the concept of an ideal engine, known as a Carnot engine, that runs cyclically – periodically returning the working substance to its initial state[59] – and reversibly – so that it can also run backwards, converting heat Q_2 at temperature T_2 into heat Q_1 at temperature T_1 with a net expenditure W of work supplied by external sources.[60] Because they are reversible, *all Carnot engines working between a given pair of temperatures are equally efficient,* and *no heat engine working cyclically between the same temperatures can be more efficient than them.* Let $\mathfrak{C}$ denote a Carnot engine that produces work W by taking a quantity of heat Q_1 from a source at temperature T_1 and surrendering a quantity of heat Q_2 to a sink at the lower temperature T_2, and let $\mathfrak{C}^*$ be any other engine – reversible or not – that obtains work W^* by extracting heat Q^* at temperature T_1 and surrendering heat $Q^* - W^*$ at the lower temperature T_2. Without significant loss of generality we may assume that Q_1/Q^* is rational, so that there are integers m and n, relatively prime, such that $Q_1/Q^* = m/n$.[61] We can imagine a heat engine $\mathfrak{C}$, each cycle of which consists of n cycles of $\mathfrak{C}$ running backwards followed by m forward cycles of $\mathfrak{C}^*$. If $\mathfrak{C}^*$ were more efficient than $\mathfrak{D}$, that is, if $(W^*/Q^*) - (W/Q_1) = \Delta\eta > 0$, each cycle of $\mathfrak{D}$ would obtain the quantity of work $\Delta W = mQ^*\Delta\eta > 0$ by taking heat nQ_2 from a source at temperature T_2 and returning to it heat $m(Q^* - W^*)$ at the same temperature, without causing any other net change in the world, in violation of Kelvin's postulate. (And ΔW could, of course, be used for heating up any body already at a temperature higher than T_2, in violation of Clausius's postulate.) Therefore, $\mathfrak{C}$ is at least as efficient as $\mathfrak{C}^*$. In particular, if $\mathfrak{C}^*$ is another Carnot engine, it must be exactly as efficient as $\mathfrak{C}$, for we can then imagine a heat engine $\mathfrak{D}^*$, each cycle of which consists of m cycles of $\mathfrak{C}^*$ running backwards followed by n forward cycles of $\mathfrak{C}$, and show by the above argument that $\mathfrak{D}^*$ would violate Kelvin's postulate if $\mathfrak{C}$ were more efficient than $\mathfrak{C}^*$.

If all Carnot engines operating between two given temperatures have the same efficiency, the ratio Q_2/Q_1 between the heat Q_2 delivered by

---

[59] Thanks to this feature, the working substance in a Carnot engine contains the same amount of energy at the beginning and at the end of each cycle and one need not inquire how much energy is needed to bring it from its initial to its final state.

[60] For a description of a Carnot engine, see the appendix at the end of §4.3.

[61] If Q_1/Q^* is irrational, then, for every positive real number ε, no matter how small, there is a pair of integers m_ε and n_ε, relatively prime, such that $|n_\varepsilon Q_1 - m_\varepsilon Q^*| < \varepsilon$.

such an engine to a sink at temperature T_2 and the heat Q_1 it draws from a source at temperature T_2 can depend only on those temperatures (cf. eqn. (4.14)). We can therefore put

$$\frac{Q_2}{Q_1} = \varphi(T_1, T_2) \qquad (4.15)$$

where φ is a real-valued function of temperature pairs, independent of engine design. The function φ has the following property:[62] For any three temperatures T_0, T_1, and T_2,

$$\varphi(T_1, T_2) = \frac{\varphi(T_0, T_2)}{\varphi(T_0, T_1)} \qquad (4.16)$$

So a function of temperature θ can be defined, relative to an arbitrary temperature T_0, by the condition

$$\theta(T) = k\varphi(T_0, T) \qquad (4.17)$$

where k is a positive, real-valued constant. By substituting from eqn. (4.15) into (4.17) we have that, for any choice of k,

$$\frac{Q_1}{Q_2} = \frac{\theta(T_1)}{\theta(T_2)} \qquad (4.18)$$

(Note that k is the only conventional element in the definition of θ, for if we define θ' relative to T_0' as in eqn. (4.17), θ' satisfies eqn. (4.18) just like θ, so θ and θ' differ at most by a constant factor.)

$T_1 > T_2$ entails that $\theta(T_1) > \theta(T_2)$. Otherwise, a Carnot engine running in reverse between these temperatures could draw an amount of heat Q_2 at temperature T_2 and deliver an equal or smaller amount of heat Q_1 at the higher temperature T_1 without adding any energy from external sources, in violation of Clausius's postulate. So there is a one–one correspondence between the values of θ and temperature states, as labeled by any scale, and it is possible to define a scale of temperature such that $\theta(T) = T$ throughout. This is Kelvin's *absolute scale*. In contrast with traditional scales, which are based on the

---

[62] To prove it, consider two Carnot engines, $\mathfrak{C}_1$ and $\mathfrak{C}_2$, operating between temperatures T_1 and T_0, and T_2 and T_0, respectively. Let $\mathfrak{C}_1$ absorb a quantity of heat Q_1 at temperature T_1 and deliver Q_0 at T_0, while $\mathfrak{C}_2$ absorbs a quantity of heat Q_2 at temperature T_2 and also delivers Q_0 at T_0. The engine $\mathfrak{C}$ whose cycle consists of one forward cycle of $\mathfrak{C}_2$ followed by one backward cycle of $\mathfrak{C}_1$ is also a Carnot engine, for a sequence of two reversible cycles is itself reversible. $\mathfrak{C}$ absorbs Q_1 at temperature T_1 and delivers Q_2 at T_2. Obviously, $\varphi(T_1, T_2) = Q_2/Q_1 = (Q_2/Q_0)/(Q_1/Q_0) = \varphi(T_0, T_2)/\varphi(T_0, T_1)$.

thermal properties of this or that thermometric substance, the Kelvin scale rests on the universal properties of Carnot engines. To set up a temperature scale one normally picks a unit and a zero. In this case, however, the latter choice is not up to us. By substituting from eqn. (4.18) into (4.14) we see that the efficiency η_{12} of a Carnot engine operating between temperatures T_1 and T_2 is given by

$$\eta_{12} = 1 - \frac{\theta(T_2)}{\theta(T_1)} = \frac{\theta(T_1) - \theta(T_2)}{\theta(T_1)} \tag{4.19}$$

Therefore, the temperature represented by $\theta = 0$ must be such that every Carnot engine approaches 100 percent efficiency as its sink approaches that temperature. This limit temperature is the same for all Carnot engines, regardless of the temperature of their heat sources.[63] On the other hand, we are free to define the unit of the Kelvin scale (known as 1 kelvin, or 1 K). This amounts in effect to fixing the multiplicative constant k in eqn. (4.17). To make the kelvin as close as possible to the familiar Celsius degree, we agree that there will be a difference of 100 K between the freezing and the boiling points of water at atmospheric pressure. The efficiency η of a Carnot engine operating between 100°C and 0°C must then satisfy the relation

$$\eta = \frac{(x + 100)\mathrm{K} - x\mathrm{K}}{(x + 100)\mathrm{K}} = \frac{100\mathrm{K}}{(x + 100)\mathrm{K}} \tag{4.20}$$

Equation (4.19) could be solved for x if one could accurately measure η. This would give us the Kelvin temperature of the freezing point of water (at atmospheric pressure). Metrologists, however, have found it more expedient to assign 273.16 K to the triple point of water, an easily reproducible state in which water, ice, and vapor coexist without noticeable changes.

We are finally in a position to define *entropy*. Henceforth all temperatures are assumed to be given in the Kelvin scale. By using eqns. (4.14) and (4.19) we can write

$$W = \frac{Q_1}{T_1}(T_1 - T_2) \tag{4.21}$$

---

[63] Suppose that, for an arbitrarily small positive real number ε, a Carnot engine $\mathfrak{C}_1$, drawing heat from a source at temperature T_1, has the efficiency $(\theta(T_1) - \varepsilon)/\theta(T_1)$, and a Carnot engine $\mathfrak{C}_2$, drawing heat from a source at temperature T_2, has the efficiency $(\theta(T_2) - \varepsilon)/\theta(T_2)$. The sink of both machines must then be at the same temperature $\theta^{-1}(\varepsilon)$.

for the amount of work that a Carnot engine operating between temperatures T_1 (high) and T_2 (low) extracts from a quantity of heat Q_1. As we know, W is an upper bound on the work produced by real heat engines, operating irreversibly in the said conditions. By using eqn. (4.18) we verify that, in the case of Carnot engines, the factor Q_1/T_1 equals Q_2/T_2. In other words, *in a reversible thermodynamic process the ratio Q/T between the quantity of heat exchanged at a fixed temperature at any point of the process and the temperature at which the exchange takes place is conserved*. A physical quantity with this property is worth looking into.

Following Clausius's lead, we consider a cyclic process in the course of which a system $\mathfrak{S}$ exchanges heat with n bodies at temperatures T_1, . . . , T_n. Let Q_k denote the quantity of heat exchanged at temperature T_k. We take Q_k positive if it is heat absorbed by $\mathfrak{S}$ and negative if it is delivered by $\mathfrak{C}$. We shall prove that

$$\sum_{k=1}^{n} \frac{Q_k}{T_k} \leq 0 \qquad (4.22)$$

the equality sign holding if and only if the process is reversible. To do so, we introduce an additional body at an arbitrary temperature $T_0 > 0$ and n Carnot engines $\mathfrak{C}_1, \ldots , \mathfrak{C}_n$, such that $\mathfrak{C}_k$ operates between T_k and T_0 exchanging at T_k the quantity of heat $-Q_k$.[64] So, by eqn. (4.18), the quantity of heat Q_0^k exchanged by $\mathfrak{C}_k$ at T_0 must be

$$Q_0^k = Q_k \frac{T_0}{T_k} \qquad (4.23)$$

Consider now the complex process consisting of one cycle of $\mathfrak{S}$ followed by one cycle of each of the Carnot engines $\mathfrak{C}_1, \ldots , \mathfrak{C}_n$. The net exchange of heat at temperature T_k amounts to $Q_k - Q_k = 0$ ($1 \leq k \leq n$), while the net exchange of heat at temperature T_0 is equal to

$$Q_0 = \sum_{k=1}^{n} Q_0^k = T_0 \sum_{k=1}^{n} \frac{Q_k}{T_k} \qquad (4.24)$$

Now, if $Q_0 > 0$, our complex process will absorb a positive amount of heat at temperature T_0 while surrendering no heat at all at tempera-

---

[64] In other words, $\mathfrak{C}_k$ surrenders $|Q_k|$ at T_k if $\mathfrak{S}$ absorbs $|Q_k|$ at T_k, and $\mathfrak{C}_k$ absorbs $|Q_k|$ at T_k if $\mathfrak{S}$ surrenders $|Q_k|$ at T_k. Therefore, $\mathfrak{C}_k$ absorbs (surrenders) $(T_0/T_k)|Q_k|$ at T_0 if $\mathfrak{S}$ absorbs (surrenders) $|Q_k|$ at T_k; this accounts for the identity of sign in both sides of eqn. (4.22). For different values of k, the reversible engine $\mathfrak{C}_k$ will produce work or consume it, as the circumstances require.

tures $T_1, \ldots, T_n$. So the only net effect of the process is to convert heat Q_0 into work. Since Q_0 is obtained from a source at uniform temperature, this violates Kelvin's postulate. Consequently, $Q_0 \le 0$. Since $T_0 > 0$, inequality (4.22) follows. If the cycle performed by $\mathfrak{S}$ is reversible, our complex process can be run in the opposite sense, in which case the heat quantities $Q_1, \ldots, Q_n$ change signs. By the same argument that has just led us to inequality (4.22) we conclude that, in this case,

$$\sum_{k=1}^{n} -\frac{Q_k}{T_k} \le 0 \qquad (4.25)$$

So, if the process performed by $\mathfrak{S}$ is reversible, relations (4.22) and (4.25) must hold simultaneously. This can happen if and only if both are equalities. Q. E. D.

We now change the conditions on $\mathfrak{S}$. Instead of exchanging finite amounts of heat with a finite number of bodies, $\mathfrak{S}$ will exchange infinitesimal quantities of heat with a continuous spread of sources. Let dQ denote the heat exchanged by $\mathfrak{S}$ with a source at temperature T.[65] In the general case,

$$\oint_{\mathfrak{S}} \frac{dQ}{T} \le 0 \qquad (4.26)$$

the integral being taken over a complete cycle of $\mathfrak{S}$. If the cycle is reversible,

$$\oint_{\mathfrak{S}} \frac{dQ}{T} = 0 \qquad (4.27)$$

Equation (4.27) entails that for all reversible processes that take our system from a state A to a state B, the integral $\int_A^B dQ/T$ has the same value (for if one such process follows the reverse of another, they jointly complete a cycle, the integral over which is equal to 0), dependent only on the initial and final states, not on the course of the process. Therefore, given a conventional reference state O, we can define a property of the thermal state A by

$$S(A) = \int_O^A \frac{dQ}{T} \qquad (4.28)$$

---

[65] We take dQ to be positive if $\mathfrak{S}$ absorbs heat from the source and negative if the source absorbs heat from $\mathfrak{S}$. Hence, if $dQ > 0$, the temperature of $\mathfrak{S}$ is not greater than T, nor is it smaller than T if $dQ < 0$. Therefore, if the process is reversible, the temperature of $\mathfrak{S}$ in each heat exchange must be equal to that of the relevant source.

Clausius called $S(A)$ the *entropy* of A. Evidently,

$$\int_A^B \frac{dQ}{T} = \int_A^O \frac{dQ}{T} + \int_O^B \frac{dQ}{T} = \int_O^B \frac{dQ}{T} - \int_O^A \frac{dQ}{T} = S(B) - S(A \qquad (4.29)$$

Consequently, for any state A, the entropy $S'(A)$ defined as in eqn. (4.28) with respect to another reference state O' differs from $S(A)$ only by an additive constant (equal to $S(O')$). This relativity in the concept of entropy was overcome in the twentieth century by Walther Nernst, who showed that the entropy of a system at 0 K is independent of every macroscopic particularity of the system and so should be regarded as a universal constant that one can put equal to 0.

The definition of entropy as a property of state in eqn. (4.28) presupposes that the integral on the right-hand side depends only on O and A. The integral must therefore be taken over a continuous succession of *reversible* heat exchanges. The same condition holds for the integral $\int_A^B dQ/T$ on the left-hand side of eqn. (4.29). But once entropy has been defined as a property of thermodynamic systems, one may well compare the entropy difference on the right-hand side with the said integral taken over a continuum of *arbitrary* heat exchanges. Consider a cycle formed by an arbitrary process from A to B combined with a reversible one from B to A. Then, by eqn. (4.26), $0 \geq \oint_{ABA} dQ/T = \int_A^B dQ/T + \int_B^A dQ/T$; hence, by eqn. (4.29), $0 \geq \int_A^B dQ/T + S(A) - S(B)$, so that, in the general case,

$$S(B) = S(A) \geq \int_A^B \frac{dQ}{T} \qquad (4.30)$$

If the process under consideration occurs in a completely isolated system, $dQ = 0$; therefore, the entropy of the final state B is always equal to or greater than that of the initial state A. Thus, *in a closed thermodynamic system the entropy can never decrease*. This proposition, which we have derived from the postulates of Kelvin and Clausius, also entails them and is therefore often employed to state the Second Principle of Thermodynamics in a concise and pleasantly esoteric way.

Philosophers became interested in entropy because it was assumed that the universe is a closed system possessing a definite energy and entropy. Since the energy cannot change and the entropy can only increase, the conclusion seemed inevitable that the universe would one day reach the maximum entropy compatible with its energy contents, after which it could no longer change in any way. Many mature

thinkers, who had at most a few decades to live, were deeply impressed
– and depressed – by the anyway distant prospect of this so-called *heat
death* of the universe. The subject was debated in the context of the
mechanical conception of heat that we shall deal with in the next two
subsections, so I shall come back to it in §4.3.4.

4.3.3   Molecular Chances

In his paper "On the Kind of Motion We Call Heat" (1857), Clausius
combined the kinetic conception of heat with a molecular view of
matter to form what became known as the kinetic theory of gases.[66]
He assumes that every body consists of very small particles or *molecules*, roughly similar in size, which are bound to one another in solids
(firmly) and liquids (pliably) but which in the gaseous state move freely
in straight lines and rebound under the action of repulsive forces when
they run against other molecules. Clausius says that he was moved to
publish his ideas by Krönig's "Groundlines of a Theory of Gases"
(1856), which anticipated some of them. Both Krönig and Clausius
appeal to probability at some point in their arguments. Krönig considers a gas consisting of equal, perfectly elastic spheres that move,
without interacting, inside a cubic container; each sphere travels perpendicularly to two of the six container walls and rebounds elastically
in the opposite direction when it collides with either of them. Krönig
admits that, at the scale of gas molecules, even the smoothest wall is
very rough.

> The path of each gas atom must therefore be very irregular, so that it
> eludes calculation. However, *according to the laws of the probability cal-
> culus* one may assume, instead of this perfect irregularity, a perfect reg-
> ularity.
>
>   (Krönig 1856, p. 316; quoted by Schneider 1988, p. 300; my
>   italics)

Clausius (1857) bestowed more freedom on his molecules but retained
the mutual independence of molecule motions, which entails, of course,

---

[66] The theory can be traced back to Daniel Bernoulli's *Hydrodynamica*, Sect. 10 (1738).
   Versions of it had been argued for, with little success, by Herapath (1821) and by
   Waterston (1846). The name 'kinetic theory of gases' is not quite adequate, for a
   theory that conceived gases as continuous fluids could – and surely would – be kinetic
   too. 'Molecular-kinetic' is a better name.

that they never meet each other.[67] Regarding the collisions with the container walls, Clausius notes that if the latter are perfectly smooth and the bouncing is perfectly elastic, the speed and angle of incidence are equal to the speed and angle of reflexion, but that this need not be so in a more general case.

> Still, *by the rules of probability*, one can assume that there are just as many molecules whose angles of reflexion lie within a certain interval, e.g. between 60° and 61°, as there are whose angles of incidence lie in that interval, and also that, on the whole, the speed of the molecules is not changed by the wall. Thus it will make no difference in the final result if one assumes that for each molecule the angle and speed of reflexion are equal to those of incidence.
>
>   (Clausius 1857, §15; my italics)

Neither Krönig nor Clausius – in 1857 – carried their use of probability any further, but in the paper of 1858 mentioned in note 67 Clausius figured out the mean free path of a molecule by straightforward probability calculations. A partial look at these calculations will help us fix the sense in which the concept of probability made its first appearance in thermal physics (and see whether it retained this same sense later). Clausius bids us consider a space – which I shall denote by $\mathcal{S}$ – containing a great many molecules *at rest*, distributed without any regular order but with uniform density. Each molecule has a "sphere of action" of radius ρ, inside which other molecules are repelled. There is, on average, one molecule in each cube of volume λ^3. Clausius calculates the probability W that a molecule of the same sort, moving in a straight line through $\mathcal{S}$, will travel a distance x undeterred by the repulsive forces of the static molecules. The calculation consists of two parts: (i) a formal calculation leading to the result

$$W = \exp(-\alpha x) \qquad (4.31)^{68}$$

and (ii) a calculation of the value of α from the above physical assumptions. Part (i) was redone much more elegantly and concisely by

---

[67] If this condition were even approximately realistic, gases would mix together much faster than they actually do, and one would rarely get to see puffs of smoke. Reminded of this fact by Buijs-Ballot (1858), Clausius (1858) made allowance for intermolecular collisions and calculated the mean free path of a gas molecule, i.e., the average distance it travels between two such collisions. The figure he got was largely compatible with phenomena.

[68] I write $\exp(x)$ for e^x (e the basis of natural logarithms). I reserve e for the electron charge.

Maxwell (1860, Prop. X), whom I shall follow. Anyway, only part (ii) involves Clausius's understanding of physical probabilities. Maxwell writes αdx for the probability of a particle striking other particles while traveling among them through a distance dx. In other words, if N such particles traveling through $\mathcal{S}$ independently – perhaps on separate occasions – reach a distance x, $N\alpha dx$ of them would be stopped before getting to the distance $x + dx$. In other words,

$$\frac{dN}{dx} = -N\alpha \qquad \text{or} \qquad N = C\exp(-\alpha x) \qquad (4.32)$$

"Putting $N = 1$ when $x = 0$, we find $\exp(-\alpha x)$ for the probability of a particle not striking another before it reaches a distance x" (Maxwell 1890, vol. I, p. 386; my notation). This is precisely the relation expressed by eqn. (4.31). To calculate the value of α we reason – after Clausius – as follows. We divide the space $\mathcal{S}$ into slabs perpendicular to the direction of motion of our moving molecule. Let an average slab of thickness λ contain n static molecules. The slab's area is therefore $n\lambda^2$. The section of each molecule's sphere of action is $\pi\rho^2$, so the total area across which passage is blocked is $n\pi\rho^2$. The probability that the moving molecule will be stopped as it traverses a typical slab of thickness λ is the ratio between the blocked area and the total section of the slab: $\pi\rho^2/\lambda^2$. To get the probability αdx that the moving molecule will strike a static molecule as it travels the distance dx, we multiply this ratio by dx/λ. So

$$\alpha dx = \frac{\pi\rho^2}{\lambda^3} dx \qquad (4.33)$$

whence, from eqn. (4.31)

$$W = \exp(-\alpha x) = \exp(-x\pi\rho^2\lambda^{-3}) \qquad (4.34)$$

Clausius's argument evidently treats the molecule-studded space $\mathcal{S}$ as the three-dimensional field of a new-fangled pinball game, into which the moving molecule is thrown in a particular direction *at random*, that is, in such a way that it has precisely *the same chance* of taking *any* of the parallel paths available in that direction. There is no essential difference between the derivation of eqn. (4.33) and the standard calculation of the probability of outcomes, say, in a game of roulette. Here we divide the number of holes corresponding to the outcome in question (e.g., *Red*, or *Third dozen*) by the total number of holes available. The result obtained is correct if the ball has precisely the same chance of falling into any hole (which, one assumes, is the case if the initial angular velocity of the wheel and the initial position and velocity of the ball are picked at random from among all the viable alternatives).[69] Of course, in Clausius's calculation one cannot compare the *number* of blocked paths with the total *number* of paths because there are uncountably many of them. So we take a *measure* over each set of paths. This is naturally given by the area of a section of $\mathscr{S}$, each point of which is perpendicularly intersected by one of the paths in question. We may therefore conclude that the concept of probability was first introduced into thermal physics with the same sense it possessed on its emergence in the late medieval and early modern discussions of games of chance: It quantifies the comparative *ease* with which a definite physical chance setup produces this or that outcome of interest.[70]

To calculate the mean free path of a molecule, Clausius assumes that the molecules in $\mathscr{S}$ are no longer static but move in straight lines in every direction with the same constant speed. This assumption is still too unrealistic and was dropped by Maxwell in his first paper on the subject (1860), in which, among other things, he calculated the likeliest *distribution of velocities* among the molecules of a (streamlined) gas. Prompted by Clausius's work, Maxwell sought here to "lay the foundation" of the kinetic theory of gases "on strict mechanical principles" by demonstrating "the laws of motion of an indefinite number of small, hard, and perfectly elastic spheres acting on one another only during impact" (Maxwell 1890, vol. I, p. 377).

> If the properties of such a system of bodies are found to correspond to those of gases, *an important physical analogy* will be established, which *may lead to more accurate knowledge* of the properties of matter. If experiments on gases are inconsistent with the hypothesis of these propositions, then our theory, though consistent with itself, is proved to be incapable of explaining the phenomena of gases. In either case it is necessary to follow out the consequences of the hypothesis.
>
> (Maxwell 1890, vol. I, p. 378; my italics)

---

[69] Suppose that the roulette game is governed by classical mechanics, so that any given set of initial conditions I determines one and only one outcome O. We conceive I as a point in the system's phase space S. If the roulette game is fair, there will be, within every neighborhood of I in S, no matter how small, sets of initial conditions leading to every possible outcome. Thus, the slightest variation in the conditions is apt to radically change the outcome.

[70] On chance setups see Hacking (1965), or my own presentation in Torretti (1990, §4.3).

Thus, for Maxwell, the aim of a physico-mathematical theory like the one he proposes in this paper is *heuristic*, not *dogmatic*. If the theory is successful, that is, if the logical consequences derived from its ground assumptions roughly agree with the phenomena it is meant to explain, it may be serviceable for scientific inquiry by raising questions, motivating experiments and suggesting theoretical refinements whose consequences agree even better with phenomena. It would be foolish, however, to assert the hypothesis of a successful theory as a dogma concerning the nature of things, for its agreement with phenomena can only be approximate and there might always be another, very different hypothesis that agrees with them equally well.

Our concern here is not with the physics or the mathematics of Maxwell's theory of gases, but with the probability arguments that Maxwell puts forward at some critical junctures. I presume that the probability *assumptions* on which those arguments rest also fall within the scope of the preceding remarks. The first such assumption (Prop. II) concerns the motion after impact of two spheres that strike each other while moving in opposite directions with velocities inversely as their masses. Maxwell has just shown that each ball moves with the same speed before and after impact, and that the directions before and after impact lie in the same plane with the line of centers and make equal angles with it (Prop. I). He goes on to ask for the probability that the direction after impact lies between given limits. He answers as follows:

> In order that a collision may take place, the line of motion of one of the balls must pass the centre of the other at a distance less than the sum of their radii; that is, it must pass through a circle whose centre is that of the other ball, and radius (s) the sum of the radii of the balls. *Within this circle every position is equally probable*, and therefore the probability of the distance from the centre being between r and $r + dr$ is $2rs^{-2}dr$. Now let ϕ be the angle [. . .] between the original direction and the direction after impact, then [the angle between the original direction and the line of centres] $= \phi/2$, and $r = s\sin(\phi/2)$, and the probability becomes $\frac{1}{2}\sin\phi\, d\phi$. The area of a spherical zone between the angles of polar distance ϕ and $\phi + d\phi$ is $2\pi\sin\phi\, d\phi$; therefore if ω be any small area on the surface of a sphere, radius unity, the probability of the direction of rebound passing through this area is $\omega/4\pi$; so that the probability is independent of ϕ, that is, all directions of rebound are equally likely.
>
> (Maxwell 1890, vol. I, p. 379; my italics)

The assumption I have italicized comes out of the blue. There is no thought here of a physical process that might be regarded as a chance setup apt to yield with the same probability every position within the said circle. What we have is a general concept – viz., 'to be found within the said circle' – coupled with a probability distribution over the set of its alternative specifications. In contrast with earlier uses of the concept of probability, this distribution does not express the relative ease with which one or the other alternative is produced by a definite physical process whose outcomes are instances of the general concept. I cannot see that such a distribution has any meaning apart from the statistics by which it would be confirmed.

To establish a semantic connection between probability statements and statistical data was surely the main purpose of the interpretation of probability in terms of relative frequency introduced in the 1840s by several authors, including the Cambridge mathematician R. L. Ellis (1849). In its more sophisticated versions, the probability of finding a feature A among objects of a class B is defined as the *limit* of the *relative frequency* of A's in a *random sequence* of B's.[71] Although the probability calculus was originally developed around the concept of relative ease of occurrence, it admits this interpretation because some of its theorems provide a close link between probability and relative frequency. Thus, the so-called Strong Law of Large Numbers implies that if (i) B is the class of outcomes of a chance setup, (ii) p is the probability of obtaining an A in a single trial, and (iii) P is the probability that the relative frequency of A's in an infinite sequence of independent trials[72] converges to the limit p, then $P = 1$.[73] This is, of course, trivially true if probability is *defined*, as above, by the frequency limit. Note, however, that the tie between frequency and probability that flows from the probability calculus is itself probabilistic, not factual.

[71] By 'random sequence' I mean an infinite sequence such that, given any positive integer n, the list of the first n terms in the sequence provides no hint whatsoever concerning the $(n + 1)$-th term.

[72] Two or more events are said to be (statistically) independent if the probability of each is in no way affected by the occurrence or nonoccurrence of the others. Thus, a sequence of independent trials on a chance setup is always random in the sense of note 71. However, when we resort to the concept of a random sequence in the definition of probability we cannot use independence as a criterion of randomness, for, as we have just seen, independence is itself defined in terms of probability.

[73] Compare the so-called Weak Law of Large Numbers: Take B and p as above, and let P_k denote the probability that the proportion of A's falls within $1/k$ of p in a run of k independent trials: Then the sequence $P_1, \ldots, P_n, \ldots$ converges to the limit 1 as n increases beyond all bounds (Jakob Bernoulli 1713, pp. 225f.).

Moreover, by the very nature of infinite sequences, *any* frequency limit is compatible with *every* finite set of statistics (for let σ be an infinite sequence of B's in which the relative frequency of A's converges to the limit p; let τ be a finite list of B's – no matter how long – in which the relative frequency of A's is $q \neq p$; let σ* denote the sequence formed by τ followed by σ; then the relative frequency of A's also converges to p in σ*).[74]

Maxwell resorts again to probabilistic ideas in the derivation of the distribution of velocities among the molecules in his gas model. The task is "to find the average number of particles whose velocities lie between given limits, after a great number of collisions among a great number of equal particles" (1860, Prop. IV). Maxwell analyzes the velocity v of a typical particle into three mutually perpendicular components that I shall denote by v_x, v_y, and v_z.[75] Following him, I write $Nf(v_i)dv_i$ for the number of particles whose ith velocity component lies between v_i and $v_i + dv_i$ ($i = x, y, z$), where N is the total number of particles and f is a function to be determined (on $\mathbb{R}$). Maxwell says that each velocity component does not in any way affect the other two, "since these are all at right angles to each other and independent" (1890, vol. I, p. 380). So, he concludes, the number of particles whose velocity lies (i) between v_x and $v_x + dv_x$, and also (ii) between v_y and $v_y + dv_y$, and also (iii) between v_z and $v_z + dv_z$, is

$$Nf(v_x)f(v_y)f(v_z)dv_x dv_y dv_z \qquad (4.35)$$

To understand the argument leading to eqn. (4.35) we should note first that if $Nf(v_i)dv_i$ has the above meaning, then $f(v_x)dv_x$, $f(v_y)dv_y$ and

---

[74] According to the great frequentist statistician, Richard von Mises, this obvious fact may be safely ignored because the hypothetical random sequences into which statistical data are embedded are built on the "silent assumption" that "*in certain known fields of application of probability theory* (games of chance, physics, biology, insurance, etc.) *the frequency limits are approached comparatively rapidly* (the rate of approach being different for different problems)" (1964, p. 108). Therefore, the assigned probabilities should lie close to the relative frequencies observed in long, though finite, series of observations. "*This assumption has nothing to do with the axioms of the probability calculus* [. . .] and is not explained by any results of theoretical statistics," for the latter, in fact, depend upon it (Ibid.). However, "the whole body of our experience in applications of probability theory seems to prove that rapid convergence indeed prevails – at least in such domains – as a physical fact, confirmed by an enormous number of observations. [. . .] In such domains, and only in them, statistics can be used as a tool of research" (pp. 109, 110).

[75] Maxwell's notation is x for v_x, y for v_y, and z for v_z.

$f(v_z)dv_z$ stand, respectively, for the relative frequency of the particles meeting conditions (i), (ii), and (iii) among N particles. If conditions (i), (ii) and (iii) are *statistically* independent (in the sense explained in note 72), the probability that one particle satisfies them all is calculated by multiplying the probability assigned to one condition by the probabilities assigned to the others. So, if one equates probability with relative frequency, as one can do with little risk of error if N is very large, $f(v_x)f(v_y)f(v_z)dv_xdv_ydv_z$ is the probability that a single particle meets conditions (i), (ii), and (iii) and therefore the relative frequency of such particles among the total N. Result (4.35) follows at once, and the function f is readily determined from it.[76]

Maxwell (1866) dealt again with the collective behavior of gas molecules, considered this time "not as elastic spheres of definite radius, but as small bodies or groups of smaller molecules repelling one another with a force whose direction always passes very nearly through the centres of gravity of the molecules, and whose magnitude is represented very nearly by some function of the distance of the centres of gravity" (1890, vol. II, p. 29). This change of model was motivated, he says, by his experiments on the viscosity of air at different temperatures. When tackling the distribution of velocities he did not assume the mutual independence of a particle's orthogonal velocity components for "this assumption may appear precarious" (p. 43). So he derived "the Final Distribution of Velocity among the Molecules of Two Systems acting on one another according to any Law of Force" from a consideration of velocity changes in molecular collisions. Mol-

[76] Maxwell reasons thus: If we suppose the N particles to start from the origin at the same instant, eqn. (4.34) is the number in the element of volume $dv_xdv_ydv_z$ after unit time, so the number referred to unit volume is $Nf(v_x)f(v_y)f(v_z)$. "But the directions of the coordinates are perfectly arbitrary, and therefore this number must depend on the distance from the origin alone" (1890, vol. I, p. 381), that is, $f(v_x)f(v_y)f(v_z) = \varphi(v_x^2 + v_y^2 + v_z^2)$. Solving this functional equation, we find

$$f(v_x) = C\exp(Av_x^2) \quad \text{and} \quad \varphi(r^2) = C^2\exp(Ar^2)$$

If A were positive, the number of particles would increase with the velocity, and their total number would be infinite. So we make A negative and equal to $-\alpha^{-2}$. Then the number of particles whose first velocity component lies between v_x and $v_x + dv_x$ is $NC\exp(-v_x^2\alpha^{-2})dv_x$. Integrating over $\mathbb{R}$ we find the total number of particles $N = NC\alpha\sqrt{\pi}$; so $C = (\alpha\sqrt{\pi})^{-1}$. Therefore

$$f(v_x) = \frac{1}{\alpha\sqrt{\pi}}\exp\left(-\frac{v_x^2}{\alpha^2}\right)$$

ecule velocities are represented "in direction and magnitude" by lines drawn from a common origin O (i.e., by vectors). "The extremities of these lines will be distributed over space in such a way that if an element of volume dV be taken anywhere, the number of such lines which will terminate within dV will be $f(r)$dV, where r is the distance of dV from O" (p. 43). Let dV(X) stand for all the velocities whose representative lines terminate inside a sphere of volume dV centered at X. Consider the set $\mathscr{C}$ of pairwise colliding molecules such that the initial velocity of a molecule in each pair lies in dV(A) and its final velocity lies in dV(A'), while the initial and final velocities of the other molecule lie, respectively, in dV(B) and in dV(B'). Let $\mathscr{C}^*$ be defined by reversing all the velocities of the molecules in $\mathscr{C}$. Thus $\mathscr{C}^*$ is the set of pairwise colliding molecules whose initial velocities lie in dV(A') and in dV(B'), while the final velocities lie, respectively, in dV(A) and in dV(B). When the number of molecules in $\mathscr{C}$ equals the number of molecules in $\mathscr{C}^*$, "then the final distribution of velocity will be obtained, which will not be altered by subsequent exchanges". This remark yields at once "a possible form of the final distribution of velocities" (p. 45).[77] According to Maxwell, it is the only form, for if there were another, the exchange between velocities represented by dV(A) and dV(A') would not be equal. Thus, Maxwell's result rests on the unspoken proposition that in a molecule population governed by the laws of classical mechanics any particular configuration $\mathscr{C}$ of colliding pairs is, in the long run, exactly as probable as the configuration defined by reversing the velocity of each molecule in $\mathscr{C}$. The final distribution will, of course, be reached only after a great number of collisions, that is, in a well-shuffled gas. However, "the great rapidity with which the encounters succeed each other is such that in all motions and changes of the gaseous system except the most violent, the form of the distribution of velocity is only slightly changed" (p. 46).

Another way of understanding and using probability in physics,

---

[77] Let a lowercase a (b, etc.) stand for the distance from O to A (B, etc.) and therefore also for the magnitude of the vector OA (OB, etc.). $\mathscr{S}$ and $\mathscr{S}^*$ contain the same number of molecules if and only if $f(a)f(b) = f(a')f(b')$, where f is the function defined in the above quotation from p. 43. Maxwell assumes that all molecules in $\mathscr{S} \cap$ dV(A) have the same mass M_1 and that all molecules in $\mathscr{S} \cap$ dV(B) have mass M_2. Energy conservation entails that $M_1a^2 + M_2b^2 = M_1a'^2 + M_2b'^2$. So $f(a) = C_1\exp(-a^2/\alpha^2)$ and $f(b) = C_2\exp(-b^2/\beta^2)$, where $M_1\alpha^2 = M_2\beta^2$. The constants C_1 and C_2 are readily computed as in 1860 (see note 76).

which is central in the work of Gibbs (1902) and Einstein (1902, 1903, 1904), was adumbrated in Maxwell's derivation of the mean free path (1860, Prop. X). I shall explain it briefly in its mature form and then return to Maxwell's argument. We consider a classical mechanical system formed by N dimensionless particles, so that its mechanical state is fully characterized by three position and three momentum coordinates. Obviously, if N is very large, one can never know the $6N$ coordinates that specify the state of the system. Typically, one can only know a few of its properties, which are then compatible with a large family of alternative states, corresponding to many different sets of additional properties. So we represent every conceivable state of our system by a different point in a $6N$-dimensional Euclidian space, the system's *phase space* (§2.5.3, after eqn. (2.37)). The observable or *macroscopic* state of the system, specified by the few properties we know, is then represented by a proper part of the phase space, formed by all the points whose $6N$ coordinates are compatible with that macroscopic state. The classical physicist assumes, of course, that the actual – *microphysical* – state of the system is represented by a single point in phase space. However, he does not know which one it is. He is therefore content to study an *ensemble* (Gibbs's term) of *possible* systems, any one of which *might be* the real one, for they all share with it the said macroscopic state. This enables him to estimate the *probability* that the system under consideration possesses this or that hidden property of interest and will therefore evolve in this or that way. To do so he *assumes* that the real system has *an equal chance* of being *any* of the possible systems and calculates *what fraction* of the possible systems shares the property of interest. This fraction is the probability that the real system has the said property. Probability is thus equated with relative frequency but, mark you, with frequency in an *imaginary* ensemble, not among actual things or events. Just as Galileo and Pascal calculated the probability of making a certain number with a pair of dice by counting all the alternative equipossible dice throws and computing what fraction of them yields that number, so the statistical physicist estimates the probability that a gas will take a particular course by measuring the set of points representing its alternative equipossible microphysical states and ascertaining what fraction of the set represents states that put the gas on that course. In either case probability is quantified possibility. There is one big difference, however, that dissociates Gibbs and Einstein from Galileo and Pascal: In the ensemble approach probability is in no way tied to a physical chance setup of which the actual dynamical state of the real system is an outcome. This agrees, of course, with what I said earlier about Maxwell's handling of the direction in which a molecule moves after striking another one (1860, Prop. II). It is fairly well known that, 30 years before Gibbs and Einstein, the ensemble approach had been used by Boltzmann (1871a, 1872 in WA, I, 261f., 401f.). I submit that the idea was already at work in Maxwell's argument for eqns. (4.32): The N particles that are supposed to travel *independently* through the given space $\mathcal{S}$ do not constitute a physical system – for what would then prevent them from occasionally interacting with one another? – but an *ensemble* of unimolecular systems.

4.3.4 Time-Reversible Laws for Time-Directed Phenomena?

Maxwell's molecular-kinetic theory of gases was extended and perfected by Ludwig Boltzmann, who made the subject his life work. The theory successfully accounted for many phenomena, at least approximately, but often gave the wrong numbers for specific heats. It won acceptance in the British Isles but was resisted on the continent by Ernst Mach and other prominent physicists (including Max Planck). They criticized it on methodological grounds, as a purely speculative hypothesis that was not justified by experience and was unnecessary for the advancement of science. But they also adduced the theory's wrong predictions[78] and its conceptual difficulties. Among the latter there is one of considerable philosophical interest that I shall now discuss.

To see the difficulty one must bear in mind that the equations of

---

[78] The molecular-kinetic hypothesis, combined with Newtonian mechanics, also entails the absurd Rayleigh–Jeans formula for black-body radiation, according to which the total energy of a hot body that absorbs and emits radiant energy in all frequencies is given by

$$\frac{8\pi kT}{c^3} \int_0^\infty \upsilon^2 d\upsilon = \infty$$

where υ stands for the frequency, T stands for the temperature, k is Boltzmann's constant, and c is the speed of light. Einstein, who independently discovered this formula (1905i), cited its inconsistency as evidence for the emergent quantum theory. That same year Einstein also gave a molecular-kinetic explanation of Brownian motion (1905k) that was subsequently confirmed by Perrin's admirable experimental work and finally persuaded the Continental scientific establishment of the existence of atoms and molecules (see Perrin 1913; for a brilliant philosophical analysis of Perrin's statistical arguments, see Mayo 1996, pp. 220–50).

classical mechanics for a conservative dynamical system are *invariant under time reversal*. In other words, if you replace, say, in the Lagrange equations (2.33) every occurrence of the time variable t with $-t$, the changes cancel out in each term, so you end up with exactly the same set of equations.[79] Therefore, for every solution of the Lagrange equations there is another solution like the former in which $-t$ is substituted for t. In a trivial sense, such a substitution can be taken as a mere relabeling of successive instants; as time goes by, the time coordinate decreases instead of growing. But there is another, physically more significant interpretation: If we read the time coordinate in the standard way (as monotonically increasing with the lapse of time), the substitution of $-t$ for t amounts to a general reversal of velocities. Consider a mechanical system $\mathcal{S}$ moving according to the Lagrange equations during the interval $(-1,1)$. Let $\mathbf{q}(t) = \langle q_1(t), \ldots, q_n(t) \rangle$ be the list of generalized coordinates representing $\mathcal{S}$ at a particular time $t \in (-1,1)$. The mapping $\mathbf{q}:(-1,1) \to \mathbb{R}^n$ by $t \mapsto \mathbf{q}(t)$ is a curve in the configuration space of $\mathcal{S}$ that traces the evolution of $\mathcal{S}$ during $(-1,1)$. The generalized velocities of $\mathcal{S}$ at t are represented by a vector in the tangent space at $\mathbf{q}(t)$, viz., $\dot{\mathbf{q}}(t) = \langle \dot{q}_1, \ldots, \dot{q}_n \rangle$. The pair $\langle \mathbf{q}(t), \dot{\mathbf{q}}(t) \rangle$ fully represents the state of $\mathcal{S}$ at t. Consider now the mapping $\mathbf{r}:(-1,1) \to \mathbb{R}^n$ by $t \mapsto \mathbf{r}(t) = \mathbf{q}(-t)$. $\mathbf{r}$ traces the evolution of a system $\mathcal{S}^*$ whose state at a particular time $t \in (-1,1)$ is represented by $\langle \mathbf{r}(t), \dot{\mathbf{r}}(t) \rangle = \langle \mathbf{q}(-t), -\dot{\mathbf{q}}(-t) \rangle$. $\mathcal{S}^*$ plays back, so to speak, in reverse time order the evolution of $\mathcal{S}$. Time-reversal invariance entails that the motion of $\mathcal{S}^*$ also satisfies the Lagrange equations.

The molecular-kinetic theory purportedly accounts for the macroscopic behavior of gases by conceiving it as the manifestation of molecular motions governed by the Lagrange equations. But familiar instances of gas behavior are not matched in nature by similar series of events in reverse time order. Thus, the pressurized gas in a punctured balloon spurts out of it until the inside and outside pressures are

---

[79] The Lagrangian L equals $T - V$. The potential V depends on the positions but not on t, and the kinetic energy T is proportional to the squared velocities and therefore retains its value when t is multiplied by -1. So time reversal does not affect the second term $\partial L/\partial q_i$ in eqns. (2.33). Nor does it affect the first, $d/dt(\partial L/\partial \dot{q}_i)$, for, as I explain in the text, if $-t$ is substituted for t, $\dot{q}_i$ must be replaced with $-\dot{q}_i$ and, plainly,

$$\frac{d}{dt}\left(\frac{\partial L}{\partial \dot{q}_i}\right) = \frac{d}{d(-t)}\left(\frac{\partial L}{\partial(-\dot{q}_i)}\right)$$

equalized; but we have yet to see air enter of itself into a flat tire until it reaches the pressure prescribed by the manufacturer. If you open the door connecting a cold room with a warm one, the air in both will mix; but the air in a room at even temperature will not separate spontaneously into a hot part that goes to one corner and a cold one that goes to another. The notorious time-directedness of thermal phenomena finds expression in Fourier's equation of heat conduction, which is *not* invariant under time reversal, and, of course, in the Second Principle of Thermodynamics as stated in eqn. (4.30). The molecular-kinetic theory therefore had to face the seemingly impossible task of deriving such laws, and the irreversible natural processes that suggested them, from the time-reversal invariant laws of mechanics.

Boltzmann tackled this task head on. In 1872 he introduced a quantity, which (in 1896)[80] he designated by H and defined by

$$H = \int_\omega f \log f \, d\omega \tag{4.36}$$

Here (i) log is the natural logarithm function; (ii) $f d\omega$ abbreviates $f(v_x, v_y, v_z, t) dv_x dv_y dv_z$, which stands for the number of particles whose velocity components relative to the Cartesian coordinates x, y, and z lie, at time t, between v_x and $v_x + dv_x$, between v_y and $v_y + dv_y$, and between v_z and $v_z + dv_z$, respectively; and (iii) it is assumed that the velocity cell $d\omega$ "can be infinitesimal yet still contain many molecules" (Boltzmann 1896a, §15; LGT, p. 111). Boltzmann showed that in a closed system $dH/dt \le 0$, with equality holding if and only if f happens to be Boltzmann's improved version of Maxwell's velocity distribution

---

[80] In 1872 Boltzmann defined the quantity of interest in terms of the distribution of *kinetic energies* among the molecules of the gas and denoted it by E (for entropy). The original definition, which can be found in Boltzmann (1872, eqn. 17, in WA, I, p. 335), reads as follows:

$$E = \int_0^\infty f(x,t) \left\{ \log\left[\frac{f(x,t)}{\sqrt{x}} \right] - 1 \right\} dx$$

Here f does not stand for the velocity distribution function as in eqn. (4.35), but it is defined thus: Let r be a region of the space occupied by the gas, of arbitrary shape and unit volume, large enough to contain many molecules; $f(x,t)dx$ denotes the number of molecules in r whose kinetic energy at time t lies between x and $x + dx$ (WA, I, 321). The letter H, introduced by Burbury (1894), is in all likelihood not the eighth letter of the English alphabet, but a Greek eta (Gibbs used a lowercase eta as a symbol for entropy).

function. In other words, H can only decrease, unless the distribution of velocities has reached its final equilibrium form, in which case H is constant. This proposition, known as Boltzmann's H-Theorem, evokes at once the law (4.30), by which entropy can only *increase*, except in the limiting case of reversible processes, in which it remains unchanged. And indeed Boltzmann shows that in all cases in which the entropy can be defined as in §4.3.2, it turns out to be, in the molecular-kinetic interpretation, equal to $-H$ (up to a positive constant factor).[81]

Boltzmann (1872) openly suggested that his H-Theorem follows from the laws of classical mechanics.[82] This would imply that these laws, though invariant under time reversal, impose a time-directed evolution on large molecular populations. The absurdity of this implication was perceived by Kelvin (1874), and also by Loschmidt (1876), himself a noted contributor to the molecular-kinetic theory.[83] They made the point that I explained above, namely, that for every dynamically possible evolution $\mathscr{E}$ of a classical system of molecules in motion, the reverse evolution $\mathscr{E}^*$ is equally possible. Specifically, suppose that $\mathscr{E}$ comprises a continuum of states, a chronologically ordered finite subset of which we number from 0 to $2m$ ($m > 0$). Consider now the dynamical evolution $\mathscr{E}^*$ of a copy of the former system, such that in a

---

[81] Of course H also has a meaning in situations for which the Clausius entropy is not defined. Boltzmann therefore boasted that he had generalized the entropy principle "in that we have been able to define the entropy in a gas that is not in a stationary state" (1896, §8; in Boltzmann LGT, p. 75).

[82] After deriving the form of f (in the sense of note 77) at the terminal stage of evolution, Boltzmann writes: "Thus it is *rigorously* proved that, no matter what the distribution of kinetic energy at the beginning of time, after a very long time it must *always necessarily* approach the distribution discovered by Maxwell" (1872 in WA, I, 345). Cf. the title of Boltzmann (1871b): "Analytical proof of the Second Principle of the Mechanical Theory of Heat from the Theorems on the Equilibrium of Kinetic Energy" (WA, I, 288). On the other hand, the introductory section of Boltzmann (1872) stresses the fundamental role of probability in the molecular-kinetic theory and warns the reader against confusing "an incompletely proven theorem, the correctness of which is therefore problematic, with a completely demonstrated theorem of the probability calculus: the latter – like the result of any other calculus – is a necessary consequence of certain premises, and if these premises are correct, it is confirmed by experience if sufficiently numerous cases are subject to observation, as is always the case in the theory of heat, given the enormous number of molecules in a body" (WA, I, 317).

[83] Indeed Boltzmann (1905, p. 236) gave his friend and senior colleague Loschmidt credit for setting him on the path of molecular-kinetic research.

definite halfway state m^* the molecules hold the same relative positions as in state m of $\mathscr{E}$, but have their respective velocities reversed. Since $\mathscr{E}^*$ before and after m^* merely repeats $\mathscr{E}$ in the reverse time order, it must successively go through $2m + 1$ states, which we shall label 0^* to $2m^*$, such that – if the integer k ranges from $-m$ to m – the molecules hold the same relative positions in state $(m - k)^*$ as in state $(m + k)$ of $\mathscr{E}$, but have their respective velocities reversed. Evidently, the distribution function f, which depends on the *squared* velocities (cf. note 76), is the same at $(m + k)$ and at $(m - k)^*$; so, if H_i and H_{i^*} denote the value of H for either system in the ith and i^*th states, respectively, it is clear that

$$H_{(m+k)} = H_{(m-k)^*} \qquad (-m \leq k \leq m) \tag{4.37}$$

Therefore, if in agreement with Boltzmann's H-Theorem the values of H for the said stages of $\mathscr{E}$ satisfy the inequality

$$H_0 \geq H_2 \geq \ldots \geq H_m \geq H_{m+1} \geq \ldots \geq H_{2m} \tag{4.38}$$

the values of H for the said stages of $\mathscr{E}^*$ must satisfy the inequality

$$H_{0^*} \leq H_{2^*} \leq \ldots \leq H_{m^*} \leq H_{(m+1)^*} \leq \ldots \leq H_{2m^*} \tag{4.39}$$

This clashes openly with the H-Theorem, except in situations in which the Maxwell–Boltzmann velocity distribution holds (so H is constant).

Countering Loschmidt's criticism Boltzmann stressed that the H-Theorem does not follow from the laws of mechanics alone, but from these laws together with the probability calculus and suitable probability assumptions. He formulated those assumptions in several not altogether clear ways. Applied to the above example the gist of them may be stated thus: In the arbitrarily chosen dynamical evolution $\mathscr{E}$ the initial molecular positions and velocities should be totally uncorrelated. Since $\mathscr{E}^*$ is expressly defined in terms of a set of previously specified positions and velocities, that are reached in $\mathscr{E}$ after numerous collisions, the initial positions and velocities in $\mathscr{E}^*$ are not uncorrelated, so it is no wonder that the H-Theorem does not apply to it. Therefore, the H-Theorem is not meant to be universally true, but only overwhelmingly probable. The same holds, according to the molecular-kinetic theory, for the Second Principle of Thermodynamics. In this theory it is not *impossible* that gin and tonic spontaneously separate after being thoroughly mixed in a glass, but only *extremely improbable* – more so, indeed, than that every inhabitant of a large country commits suicide, purely by accident, on the same day, or that every building burns down at the same time.[84] Still, Loschmidt's objection should also be understood statistically. It is not a matter of choosing at random a typical evolution $\mathscr{E}$ and then, after it is picked, reversing the molecular velocities in one of its states. Think rather of the ensemble of dynamically possible evolutions of any given collection of molecules. For every member $\mathscr{E}$ of the ensemble in which H never increases, there must be a matching member $\mathscr{E}^*$ in which H never decreases. Boltzmann's mature reading of the H-Theorem, developed in answer to criticism by Zermelo, implicitly vindicates Loschmidt's objection in this ensemble form.

To fight the molecular-kinetic theory, Planck's assistant, Ernst Zermelo, recalled that, by a theorem proved by Poincaré, every closed mechanical system must eventually return to a state as similar as you wish to its present state.[85] "Therefore, in such a system *irreversible processes are impossible,* for (apart from singular initial states) no single-valued continuous function of the state variables, such as entropy, can increase continuously; if there is a finite increase, there must be a corresponding decrease when the initial state recurs" (Zermelo 1896, p. 485). Boltzmann (1896b, 1897) did not deny Zermelo's claim, but was content to show that the (microphysical) states of a gas in which H is at or very near its absolute minimum are overwhelmingly more probable than those in which it is far above it. From this he concluded that:

(I)   the gas spends most of the time in or near the equilibrium condition characterized by the Maxwell–Boltzmann velocity distribution, in which H is as small as it can be and most likely to remain constant, and

(II) if the gas departs from that condition by any chance, then, at any particular moment in which H is not minimal, H will most probably decrease, indeed all the more probably the greater the departure from minimality.[86]

---

[84] The last two examples are given by Boltzmann, who wryly remarks: "Yet the insurance companies get along quite well by ignoring the possibility of such events" (LGT, p. 444).

[85] Represent the different states of the system by points in its phase space. Let P represent the system's initial state. Take any neighborhood $N(P)$ of P. Then, by Poincaré's theorem, the system's trajectory in phase space will enter $N(P)$ again and again in the course of everlasting time.

[86] Paul and Tatiana Ehrenfest (1906) constructed a beautiful example of a function of time that is always more likely to decrease than to increase – unless it stands at its absolute minimum – although it depends on a process that is invariant under time reversal.

Thus, the long-term evolution of a closed molecular system certainly includes periods of increasing H in which the Second Principle of Thermodynamics is transgressed; but most transgressions are small and short-lived, so it is no wonder that we do not usually perceive them.[87] Still, they can be seen under the microscope in the guise of Brownian motion, the ceaseless agitation of particles suspended in fluids, about one micron in diameter, which was first accurately described by the botanist Robert Brown (1828), and which Einstein (1905k, 1906b) successfully explained as a consequence of the frequent collisions of such particles with the much smaller molecules that surround them.[88] The Brownian particles perform uninterrupted mechanical work at the expense of heat extracted from the fluid even if the latter is at the same temperature throughout. As Perrin eloquently puts it:

> We have only to follow, in water in thermal equilibrium, a particle denser than water, to notice that at certain instants it rises spontaneously, thus transforming a part of the heat of the medium into work. If we were no bigger than bacteria, we should be able at such moments to fix the dust particle at the level reached in this way, without going to the trouble of lifting it and to build a house, for instance, without having to pay for the raising of the materials.
>
> (Perrin, *Atoms*, p. 87)

[87] Of course, if the universe is a Boltzmann molecular-kinetic system spending most of its time near equilibrium, it must have gone through an enormous entropy decrease when our sun and the other stars were formed. Boltzmann's theory does not preclude such a stroke of luck, without which, in fact, Boltzmann himself could not have been born. Moreover, if the universe were *infinite* and *eternal* – as Boltzmann apparently thought – , the occurrence at some time and place of a low-entropy bubble, less than a trillion light-years in diameter, and returning to heat death in a few dozen billion years, would be practically certain. Cf. the long quotation from Boltzmann (1898, §90) given in the main text, near the end of this section.

[88] A molecular-kinetic explanation of Brownian motion had been suggested by Gouy (1888), among others. But it ran against the following difficulty: the observed speed of Brownian particles was very much less than the speed calculated from the molecular-kinetic theory. It took Einstein's peerless imagination to realize that we can at no time properly observe the momentary velocity of a particle that moves to and fro propelled by some 10^{22} collisions per second, and that the particle's trajectory from one second to the next is a polygonal line that is very much richer in turnabouts and therefore very much longer than what it looks to us. Therefore, the particle's true speed is a good deal greater than what it seems. To come anywhere near perceiving the Brownian motion's actual complexity we would need, besides the microscope that enables us to discern particles about 10^{-18} cubic meters in size, a device enabling us to discern events lasting, say, less than 10^{-18} second.

In reaching conclusion (II), Boltzmann took for granted that a physical system is more likely to evolve from a less probable to a more probable state than vice versa. To infer conclusion (I) he had to assume that the time average of a physical quantity throughout the evolution of a closed physical system equals its so-called phase average, that is, its present average over an ensemble of similar closed systems. Both assumptions are questionable, but this is not the place to discuss them.[89] Before changing the subject I must, however, refer to a remark by Boltzmann that, although irrelevant to his physics and unworthy of his genius, has had a lasting effect on lesser minds. It is contained in one of the final sections of his *Lectures on Gas Theory*, entitled "Application to the universe". I reproduce it in context:

> One can think of the world as a mechanical system of an enormously large number of constituents, and of an immensely long period of time, so that the dimensions of that part containing our own "fixed stars" are minute compared to the extension of the universe; and times that we call eons are likewise minute compared to such a period. Then in the universe, which is in thermal equilibrium throughout and therefore dead, there will occur here and there relatively small regions of the same size as our galaxy (we call them single worlds) which, during the relative short time of eons, fluctuate noticeably from thermal equilibrium, and indeed the state probability in such cases will be equally likely to increase or decrease. For the universe, the two directions of time are indistinguishable, just as in space there is no up or down. However, just as at a particular place on the earth's surface we call "down" the direction toward the center of the earth, *so will a living being in a particular time interval of such a single world distinguish the direction of time toward the less probable state from the opposite direction (the former toward the past, the latter toward the future)*. By virtue of this terminology, such small isolated regions of the universe will always find themselves "initially" in an improbable state. This method seems to me the only way in which one can understand the second law – the heat death of each single world – without a unidirectional change of the entire universe from a definite initial state to a final state.

(Boltzmann 1898, §90; LGT, p. 447; my italics)

The whole passage is interesting for the light it throws on Boltzmann's final reading of his *H*-Theorem. Clearly, Loschmidt's objection

---

[89] The reader who wishes to pursue the fascinating subject of heat and chance will find in Sklar (1993) a trustworthy and illuminating guide, with abundant references to the scientific and philosophical literature.

has been fully assimilated. I have set the remark in question in italics. By virtue of it, Boltzmann's controversial probabilistic assumption about the transition from states of lesser to states of greater probability is upgraded to a truth of grammar. No scientist could ever observe a global entropy decrease in his world region, because advance toward a more probable state, that is, one of maximal entropy, is *by definition* the mark of the time direction from past to future. Boltzmann's maneuver is of a type that – although common in philosophy – is quite unusual in science. He needs, of course, to retain the everyday meaning of 'past' and 'future' and of the temporal direction from one to the other, for otherwise his gambit would be no more than wordplay. Yet he produces a definition of those terms that is not warranted by ordinary usage. I have just seen two swallows fly past my window, head forward, from left to right. In perceiving their flight in this time order I took no cognizance of the global thermal state of my "single world" – extending to the furthest quasars – , let alone of its progress toward a more probable condition. Would I see the birds' heads trail their bodies if I lived in a "single world" in which entropy is decreasing? To be specific, imagine that the system of observable galaxies, which is currently expanding, begins to recontract. (Such a turnabout would be bound to occur if – as seems likely – General Relativity is approximately right and – contrary to current data – the average density of nongravitational energy were greater than the so-called critical density.) If the concept of entropy is applicable to the system of galaxies, their coming together would involve a sustained, very large entropy decrease. Would this affect our perception of temporal order? Would we *see* birds fly backwards? Or would birds living in a contracting "single world" fly tail first, so that people in that "single world" – who identify the future with the time direction of increasing entropy – see them fly in the same direction as we do? What about the galaxies? Shouldn't they too look as if they moved backwards, so that even after they are closing in on us they still seem to move away from us? If Boltzmann's criterion of time order makes any sense, "the direction of time" depends on the global evolution of entropy over our entire region of the universe,[90] and it must therefore be reversed in our minds as soon as the distant galaxies begin to approach us. Yet at that moment the

[90] If the sense of time direction depended on local entropy change, then, by Boltzmann's criterion, the observer who follows the Brownian motion of a particle would experience the future direction toggle back and forth.

light signals we are receiving from them are still old signals, issued many millions of years earlier, while their distance from us was growing. Will that light still appear redshifted after the sources turn around? Or will it appear blueshifted? And how will the *new* light appear when it reaches us from the nearest, now incoming galaxies? Does the change in the direction of time change the way the galactic spectra are shifted? Or do scientists who live in a "single world" in which entropy is decreasing, prompted by their particular experience of time order, replace the Doppler shift formula with one in which blueshift indicates recession? Such questions are silly, indeed, but no more so than the remark that motivates them.[91]

Appendix

A Carnot engine can be built from the following pieces: a piston P in a cylinder that contains the working substance G; a stand S; and two bodies B_1 and B_2, at temperatures T_1 and T_2, respectively, with $T_1 > T_2$ (see Fig. 11). It is assumed that the piston, the cylinder walls, and the stand are perfect nonconductors of heat, but that the bottom of the cylinder is a perfect conductor whose capacity for storing heat is negligible. On the other hand, B_2 and B_1 have such a large heat capacity that their temperature does not vary significantly when they exchange heat with G in the manner described below. A cycle of the Carnot engine consists of four stages:

I. With G at temperature T_2, the cylinder is placed on the stand S. No heat can go in or out. The piston P is forced down very slowly until the temperature of G rises to T_1. Let W_I denote the amount of work done by P on G in this stage.

II. With G at temperature T_1, the cylinder is placed on B_1, which, as we may recall, is at that same temperature. The piston P is allowed to rise. Since the bottom of the cylinder is a perfect conductor, heat from B_1 flows into G and prevents it from being cooled by the expansion. Temperature remains constant. Let Q_1 be the quantity of heat that G absorbs from B_1 and W_{II} the amount of work done by G on P in this stage.

[91] It may be objected that such questions only *look* silly to me because I do not take an Archimedean standpoint outside time from which to view them (cf. Price 1996). Such a feat, however, is beyond my capacity. Indeed, in my irrevocably timebound situation, the very idea of standing outside time only compounds the silliness.

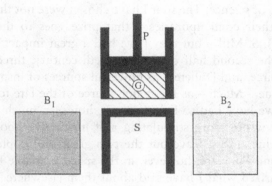

Figure 11

III. With G still at temperature T_1, the cylinder is once again placed on S. P is allowed to rise. Since no heat can go in or out of the cylinder, the temperature of G falls as it expands. The process is stopped when the falling temperature is back at T_2. Let W_{III} be the amount of work done by G on P in this stage.

IV. With G at temperature T_2, the cylinder is placed on B_2, which, as we may recall, is at that same temperature. The piston P is forced down very slowly until it returns to the position it had at the beginning of stage I. This time, compression does not cause a rise in temperature, for the bottom of the cylinder is a perfect conductor and the extra heat flows to B_2 while the temperature remains constant. Let Q_2 be the quantity of heat that G surrenders to B_2 and W_{IV} the amount of work done by P on G in this stage.

At the end of stage IV the working substance G has recovered the temperature and the volume, and thus also the pressure, it had at the beginning of stage I and the engine is ready to begin a new cycle. The net amount of work extracted during one cycle is $W = -W_I + W_{II} + W_{III} - W_{IV}$.

4.4 Philosophers

Nineteenth-century philosophy is overwhelming in variety, density, and volume. Much of it concerns the natural sciences, which were increasingly regarded as the most rewarding form of intellectual endeavor. I can deal here only with a very small sample of nineteenth-century philosophers of science. The four I have chosen were not the ones most revered by their contemporaries – that prize goes to the positivists Comte and J. S. Mill – but they have had a great impact on my generation, in the second half of the twentieth century, three of them – Whewell, Peirce, and Duhem – as powerful sources of inspiration, and the fourth one – Mach – as the major source of the late form of positivism that we have had to fight (he was, however, much more open-minded and wrote more stimulating and instructive books than his epigons). Reluctantly I leave out the two great philosopher-scientists Helmholtz and Poincaré; however, in the space available I could only have summarized what I have said about them elsewhere.[92]

4.4.1   William Whewell (1794–1866)

To a casual reader it may seem that Whewell's philosophy of human knowledge is only a watered-down version of Kant's. Like Kant, Whewell insists on the active role of the knowing mind, through which alone knowledge achieves universality and necessity. However, disregarding Kant's meticulous classification of the mental ingredients of knowledge into "forms of sense", "categories of the understanding", and "ideas of reason", he brings every contribution of the mind under the single term 'ideas', which, besides *space*, *time*, and *cause*, also covers *number*, *resemblance*, *force*, *polarity*, *medium* (for the transmission of signals), and so on. Whewell maintains that ideas – in his sense – are in one way or another conditions of the possibility of experience, and repeats Kant's "transcendental" vindication of space as the foundation of geometric truths (1847, I, 85f.). But he does not try to show – as Kant did in the "proofs" we examined in §3.4 – that the ideas of cause, etc., are prerequisites of objectivity and *therefore* justified in their ordinary scientific use. Now, Kant's conclusions concerning the limits of a priori knowledge depend on his precise grounding of it. By cavalierly forgoing this grounding, Whewell is able to avoid those conclusions and cheerfully lists "natural theology" among the sciences, right after linguistics and ethnography (1847, II, 117).

To a closer look, however, Whewell's writings offer some new and

---

[92] On Poincaré, see Torretti (1978, pp. 320–58; 1983, pp. 83–87). On pages 155–71 of the earlier book, I comment on Helmholtz's philosophy of geometry. The discussion of extensive quantities in Torretti (1990, pp. 58–65), is ultimately based on Helmholtz (1887).

illuminating thoughts on the natural sciences, which I shall epitomize and illustrate by a series of quotations.[93] Let me first note that Whewell's portmanteau description of the mental principles of knowledge as 'ideas' is anything but careless. Such looseness is required to make room for the historical development of knowledge. Although Whewellian ideas are supposed to express the permanent nature of the human mind, they do so progressively, as human experience evolves. For Whewell

> Ideas are not Objects of Thought, but rather Laws of Thought. Ideas are not synonymous with Notions; they are Principles which give to our Notions whatever they contain of truth.
>
> (Whewell 1847, I, 29)

And yet, according to him, ideas are *modified* by their *employment* in experience.

> Ideas cannot exist where Sensation has not been [. . .]. Hence, at whatever period we consider our Ideas, we must consider them as having been already engaged in connecting our Sensations, and as having been modified by this employment. By being so employed, our Ideas are unfolded and defined; and such development and definition cannot be separated from the Ideas themselves.
>
> (Whewell 1847, I, 43–44)

Therefore, the philosopher may not claim to have now – or at any other time – a definite and definitive awareness of the mind's contribution to experience.[94] This contribution does not consist of innate factors whose inventory and mutual relations one could establish once and for all.

> Our fundamental ideas are necessary conditions of knowledge, universal forms of intuition, inherent types of mental development; they may even be termed, if any one chooses, results of connate intellectual tendencies; but we cannot term them *innate* ideas [. . .]. For innate ideas were considered as capable of composition, but by no means of simplification: as most perfect in their original condition; as to be found, if any where, in

---

[93] Whewell's main books, the *History of the Inductive Sciences* (1837) and the *Philosophy of the Inductive Sciences* (1840, 2nd much revised ed. 1847), are very readable, but most students will probably be put off by their size. They can profitably use the excellent selection in Butts (1968).

[94] Kant failed to see that this follows from his assertion that *all* knowledge begins with experience and *none* precedes it (1787, p. 1), and went on to prolixly set forth his tripartite architecture of reason.

the most uneducated and most uncultivated minds; as the same in all ages, nations, and stages of intellectual culture; as capable of being referred to at once, and made the basis of our reasonings, without any special acuteness or effort: in all which circumstances the Fundamental Ideas of which we have spoken, are opposed to Innate ideas so understood.

(Whewell 1851, in Whewell 1860, pp. 530f.)

Such "fundamental ideas" are not necessarily ultimate elements of knowledge; "*they are the results of our analysis so far as we have yet prosecuted it; but they may themselves subsequently be analysed*" (p. 531; my italics).

Indeed, for Whewell, 'ideas and sensations' form just one of several pairs of terms by which English speakers denote the "fundamental antithesis of philosophy". Its "simplest and most idiomatic expression" is the opposition between 'thoughts and things'; others are 'necessary and experiential truths', 'subjective and objective', 'form and matter', and 'theories and facts'.

The fundamental antithesis of philosophy is an antithesis of inseparable elements. Not only cannot these elements be separately exhibited, but they cannot be separately conceived and described. The description of them must always imply their relation; and the names by which they are denoted will consequently always bear a relative significance. And thus *the terms which denote the fundamental antithesis of philosophy cannot be applied absolutely and exclusively in any case.*

(Whewell 1844, in Whewell 1847, II, 651–52)

The revolution of the stars about the poles, the rotation and translation of the earth, and the mutual attraction of the sun and planets are described by some as *theories* and as *facts* by others.

In these cases we cannot apply absolutely and exclusively either of the terms, Fact or Theory. Theory and Fact are the elements which correspond to our Ideas and our Senses. The Facts are Facts so far as the Ideas have been combined with the sensations and absorbed in them: the Theories are Theories so far as the Ideas are kept distinct from the sensations, and so far as it is considered as still a question whether they can be made to agree with them. A true Theory is a fact, a Fact is a familiar theory.

(Whewell 1844, in Whewell 1847, II, 652)

In a Fact, the Ideas are applied so readily and familiarly, and incorporated with the sensations so entirely, that we do not see *them*, we see *through them.* A person who carefully notes the motion of a star all night, sees the circle which it describes, as he sees the star, though the circle is, in fact, a result of his own Ideas.

(Whewell 1847, I, 40)

In Whewell's vocabulary, 'ideas' are "certain comprehensive forms of thought"; while "the special modifications of these ideas which are exemplified in particular facts" are termed 'conceptions' (1847, II, 5–6). The construction of science comprises two main processes, viz., the *Explication of Conceptions*, by which conceptions are "carefully unfolded" and "made more clear in themselves", and the *Colligation of Facts*, "by which the conceptions more strictly bind together the facts" (II, 5). These two processes must go hand in hand. The "right definition of a Term may be a useful step in the explication of our conceptions", but only when we contemplate "some Proposition in which the Term is employed" (II, 12). For "that which alone makes it worth while" to clarify a conception "is the opportunity of using it in the expression of Truth" (II, 13). So "the question really is, how the Conception shall be understood and defined in order that the Proposition may be true". Now, as Whewell emphatically recalls, "the establishment of a Proposition requires an attention to observed Facts, and can never be rightly derived from our Conceptions alone" (II, 12). However, "facts cannot be observed as Facts, except in virtue of the Conceptions which the observer himself unconsciously supplies" (II, 23). Therefore, the two processes by which science is constructed, "the Explication of the Conceptions of our own minds, and the Colligation of observed Facts by the aid of such Conceptions", are *inseparably connected with each other* (II, 46). When jointly employed in collecting knowledge "they constitute the mental process of *Induction*; which is usually and justly spoken of as the genuine source of all our *real general knowledge* respecting the external world" (II, 46–47). The vulgar conceive of induction as a process "by which we collect a *General Proposition* from a number of *Particular Cases*: and it appears to be frequently imagined that the general proposition results from a mere juxta-position of the cases, or at most, from merely conjoining or extending them" (II, 48). Not so, says Whewell.

In each inference made by Induction, there is introduced some General Conception, which is given, not by the phenomena, but by the mind. The conclusion is not contained in the premises, but includes them by the introduction of a New Generality. In order to obtain our inference, we travel beyond the cases which we have before us; we consider them as mere exemplifications of some Ideal Case in which the relations are complete and intelligible. We take a Standard, and measure the facts by it; and this Standard is constructed by us, not offered by Nature. We assert, for example, that a body left to itself will move on with unaltered velocity; not because our sense ever disclosed to us a body doing this, but because (taking this as our Ideal Case) we find that all actual cases are intelligible and explicable by means of the Conception of *Forces*, causing change and motion, and exerted by surrounding bodies.

(Whewell 1847, II, 49)

So in every induction "there is some Conception *superinduced* upon the facts", and Whewell proposes that we henceforth regard this as "the peculiar import of the term *Induction*" (II, 50). This decisive step in the formation of science, "the Invention of a new Conception in every inductive inference", had rarely been noticed by Whewell's predecessors. The reason for this is not hard to see: Such acts of invention soon slip out of notice.

Although we bind together facts by superinducing upon them a new Conception, this Conception, once introduced and applied, is looked upon as inseparably connected with the facts, and necessarily implied in them. Having once had the phenomena bound together in their minds in virtue of the Conception, men can no longer easily restore them back to the detached and incoherent condition in which they were before they were thus combined. [. . .] As soon as the leading term of a new theory has been pronounced and understood, all the phenomena change their aspect. There is a standard to which we cannot help referring them.

(Whewell 1847, II, 51–53)

Although Whewell occasionally refers to a "logic of induction" – perhaps as a concession to his contemporaries – and devotes several chapters to "methods of induction" (II, 395–425), he saw and said clearly – a full century before Feyerabend (1970) – that scientific discovery is not subject to any rules.

Scientific discovery must ever depend upon some happy thought, of which we cannot trace the origin; – some fortunate cast of intellect, rising above all rules. No maxims can be given which inevitably lead to discovery.

(Whewell 1847, II, 20f.)

The Conceptions by which Facts are bound together, are suggested by the sagacity of discoverers. This sagacity cannot be taught. It commonly succeeds by guessing; and this success seems to consist in framing several *tentative hypotheses* and selecting the right one. But a supply of appropriate hypotheses cannot be constructed by rule, nor without inventive talent.

(Whewell 1847, II, 467f.)

Talent can certainly benefit from instruction. This must be sought, however, not in abstract methodological prescriptions, but in actual instances of scientific thinking and scientific practice. So Whewell filled page after page of his philosophical books with historical examples and wrote the three volumes of his *History* (1837).

Twentieth-century positivists would maintain, of course, that the rules of inductive logic are not meant to preside over the process of discovery, but to control the validity of its findings. Whewell expressly asks how we are to find what new conception succeeds in binding together the facts. How is the trial to be made? What is meant here by 'success'? (II, 45). Compared with the convoluted contraptions of modern confirmation theory, his approach to the testing of hypotheses may well seem casual (see II, 60–95). He repeats the usual bland generalities about the comparison of hypotheses with the facts and stresses the force of conviction issuing from the prediction of novel phenomena (as opposed to the mere explanation of those already familiar). But he makes no move to quantify degrees of confirmation or the weight and relevance of evidence. In view of the dismal failure of all attempts to achieve cogency in this matter, his relaxed informality seems wise. Only in cases of a peculiar and indeed quite extraordinary sort did he claim finality for the conclusions of induction. He coined for them the term 'consilience of inductions' (II, 65). This "takes place when an Induction, obtained from one class of facts, coincides with an Induction, obtained from another different class" (II, 469). Consilience – says Whewell – "is a test of the truth of the Theory in which it occurs" (Ibid.).

> The evidence in favor of our induction is of a much higher and more forcible character when it enables us to explain and determine cases of a *kind different* from those which were contemplated in the formation of our hypothesis. The instances in which this has occurred, indeed, impress us with a conviction that the truth of our hypothesis is certain. No accident could give rise to such an extraordinary coincidence. No false supposition could, after being adjusted to one class of phenomena, exactly represent a different class, when the agreement was unforeseen and uncontemplated. That rules springing from remote and unconnected

quarters should thus leap to the same point, can only arise from *that* being the point where truth resides.

(Whewell 1847, II, 65)

Whewell's paramount example of consilience is Newton's discovery that the three Keplerian laws of planetary motion follow from his Law of Gravity although no link between them had been noticeable before. Moreover, Newton's Law also accounted for such dissimilar phenomena as the tides and the precession of the equinoxes (II, 66). Now, despite its enormous predictive fruitfulness, Newtonian instantaneous attraction at a distance is no longer thought to be the best, let alone the true, explanation of the phenomena that we nevertheless continue to call 'gravitational' (see §5.4). So Newton's theory is also a preeminent instance of inconclusive consilience, and Whewell has been rightly taken to task for thinking otherwise. Still, there is no question that consilience, when it occurs, impresses us with a conviction of certainty. Moreover, there is one respect in which the consilience of inductions seems indeed to be definitive: The unifications that it brings about appear to be irreversible, at least within a given scientific tradition. Thus, no one today would dream of placing different subsets of the so-called gravitational phenomena under disparate conceptions. We naturally expect every purported successor to Newton's theory of gravity to account for the fall of heavy bodies *and* the circulation of the planets, not forgetting the tides and the precession of the equinoxes, just as any substitute for the wave theory of light – Whewell's second example of consilience – must countenance some sort of waves, be it only waves of probability.

4.4.2   Charles Sanders Peirce (1839–1914)

Peirce once characterized philosophy as "the attempt to form a general informed conception of the *All*" (CP 7.579). Since we cannot reasonably hope to attain individually "the ultimate philosophy which we pursue", it must be a goal "for the *community* of philosophers" (CP 5.265), a public work, "meant for the whole people and [...] erected by the exertions of an army representative of the whole people", like a cathedral (CP 1.176; cf. 6.315). Early "a passionate devotee of Kant",[95] Peirce retained a lifelong preference for triads (CP 4.3,

---

[95] "I believed more implicitly in the two tables of the Functions of Judgment and the Categories than if they had been brought down from Sinai" (CP 4.2).

6.321ff., 1.369, 6.202, 1.471ff., etc.; cf. 1.568) and fondly cited Kant's teaching about the "architectonic" construction of philosophy (CP 6.9, 5.5, 1.176; cf. Kant 1781, pp. 831ff.). And yet, were it not for these reminders, you would hardly think of philosophical system building as you navigate the ocean of his *Collected Papers*. Gathered by "pitch-forking" Peirce's published articles and massive *Nachlass* (CP 1.179), they glaringly flaunt their variegated origin, despite the editors' Procrustean endeavor to fit them into a systematic plan.[96] Moreover, it would seem that Peirce, a profoundly original thinker of incisive intelligence, had a hyperkinetic mind, which made it difficult for him to keep a steady course even within a single paper or a connected series of papers. So his writings resemble anything but a masterpiece of architecture. In a way this is all for the best. As we zigzag among Peirce's startling insights and clever arguments, the *Collected Papers* come out as immensely more attractive and instructive than, say, the fossil treatises of Schopenhauer or Spencer.

One can, indeed, by dexterous hermeneutics elicit one or more systems of philosophy[97] from Peirce's "odds and ends of commentary" (CP 6.150). Maybe this is the proper way to handle the heritage of a man who relished coining "ism" words to bear on his standard. But I see no gain in corralling a thinker I admire into a philosophical "position" that anyway is bound to prove untenable. So, instead of trying to weave Peirce's thoughts on science into the systematic context to which they presumably belong, I shall consider only three topics relevant to physics on which he threw much light. This approach, I think, comes closer than would a rational reconstruction of Peirce's system or systems to the "ism" for which he is best remembered, of which he said:

> *Pragmatism* is not a *Weltanschauung* but is a method of reflexion having for its purpose to render ideas clear.[98]

The topics I have chosen for consideration are: (i) the maxim of pragmatism, (ii) inference, and (iii) determinism and chance.

---

[96] A new edition of Peirce's writings in chronological order and satisfying modern standards of scholarship is being published by Indiana University Press since 1982. As I write this, only five volumes have appeared, covering the period from 1857 to 1886. Peirce remained a very prolific writer until two or three years before his death in 1914.

[97] M. G. Murphey distinguishes four such systems in his article on Peirce in Edwards's *Encyclopedia of Philosophy*.

[98] CP 5.13n1; c. 1902. From Peirce's personal interleaved copy of the *Century Dictionary*.

(i) In the first of six papers on "the logic of science" published in 1877–1878, Peirce defines 'inquiry' as the struggle to overcome the irritation of doubt by reaching a state of belief (CP 5.374). "Hence" – he concluded – "the sole object of inquiry is the settlement of opinion" (5.375). He considers four methods of fixing belief. The first, the method of *tenacity*, consists in "taking any answer to a question which we may fancy, and constantly reiterating it to ourselves" (5.377). The second, the method of *authority*, confers this function on an institution whose sole purpose is "to keep correct doctrines before the attention of the people, to reiterate them perpetually, and to teach them to the young; having at the same time power to prevent contrary doctrines from being taught, advocated, or expressed" (5.379). The failings of tenacity and authority have led to the development of a third method, which Peirce describes as follows: "Let the action of natural preferences be unimpeded, then, and under their influence let men, conversing together and regarding matters in different lights, gradually develop beliefs in harmony with natural causes" (5.382). However, this method, well exemplified by the history of metaphysical philosophy, has also failed to produce a general agreement and to stabilize belief. "To satisfy our doubts, therefore, it is necessary that a method should be found by which our beliefs may be caused by nothing human, but by some external permanency – by something upon which our thinking has no effect" (5.384). This must be something "which affects, or might affect, every man. And, though these affections are necessarily as various as are individual conditions, yet the method must be such that the ultimate conclusion of every man shall be the same. Such is the method of science" (Ibid.).

The next paper in the series examines the first step of the scientific method, namely, *how to make our ideas clear*. Peirce assumes that the sole aim of thought is to produce belief (5.396),[99] and that all belief

---

[99] Did Peirce abide by this view of thought? Twenty years later, in the Cambridge lectures of 1898 on "Reasoning and the Logic of Things", he asserted that "what is properly and usually called *belief*, that is, the adoption of a proposition as a κτῆμα ἐς αἰεί [a possession forever] has no place in science at all" (CP 1.635). However, in the 1878 paper, Peirce did not equate *belief* with something everlasting. "Belief" – he wrote then – "is a rule for action, the application of which involves further doubt and further thought", so, "at the same time that it is a stopping-place, it is also a new starting-place for thought" (5.397). Thus, Peirce's views of 1878 and 1898 might

"involves the establishment in our nature of a rule of action, or, say for short, a *habit*" (5.397). If two beliefs "appease the same doubt by producing the same rule of action, then no mere differences in the manner of consciousness of them can make them different beliefs" (5.398). Hence the root of every real distinction of thought, no matter how subtle, lies in what is tangible and conceivably practical: "There is no distinction of meaning so fine as to consist in anything but a possible difference of practice" (5.400). Therefore, to make our ideas clear we must follow this maxim:

> Consider what effects, that might conceivably have practical bearings, we conceive the object of our conception to have. Then, our conception of these effects is the whole of our conception of the object.
>
> (Peirce CP 5.402)

In Baldwin's *Dictionary* (1902), Peirce defined 'pragmatism' as "the opinion that metaphysics is to be largely cleared up by the application of the [above] maxim", which he quoted verbatim (CP 5.2). He added that, "after many years of trial", he still thought the maxim to be "of great utility"; but he explicitly distanced himself from William James, who in his *Will to Believe* (1897) and elsewhere had "pushed this method to such extremes as must tend to give us pause" (5.3). Taken literally the maxim would indeed sweep away the difference between rational and irrational numbers, for actual measurements can only yield rational values. This was a consequence that Peirce, as a trained mathematical physicist, could not stomach (cf. 5.32–33). He mentioned it in the said *Dictionary* article, and went on to add that the doctrine of pragmatism "appears to assume that the end of man is action – a stoical axiom which to the present writer at the age of sixty, does not recommend itself so forcibly as it did at thirty" (5.3). If action wants an end and that end must be something of a general description, then "the spirit of the maxim itself, which is that we must look to the upshot of our concepts in order rightly to apprehend them, would direct us

---

not be as irreconcilable as they seem. Still, one notices at least a change of emphasis from his early claim that the "sole motive, idea and function" of thought is "to produce belief" (5.396), to the subsequent one that "there is . . . no proposition at all in science which answers to the conception of belief" (1.635). Such change probably has to do with his later view that the end of man is not simply "action" – as he may have thought at age 30 – but "the development of concrete reasonableness" (5.3; cf. 5.433, 1.615).

towards something different from practical facts, namely, to general ideas, as the true interpreters of our thought" (Ibid.).

(ii) Peirce distinguished three kinds of inference: deduction, induction, and abduction.

Deduction comprises every inference in which the premises are so related to the conclusion that it is impossible that the latter be false while the former are true (CP 2.778). It is worth remembering that in the 1870s Peirce himself and Gottlob Frege, unbeknownst to one another, revolutionized the theory of deductive inference (which – as Kant 1787, p. viii, noted – had not progressed since Antiquity). They founded what Peirce called 'the logic of relatives', involving predicates assertible of two or more objects. They also introduced the now familiar analysis of general statements as quantified compound statements.[100] Thanks to their momentous work they were both able to identify deduction with mathematical proof. "Every science has a mathematical part, a branch of work that the mathematician is called to do. We say: 'Here, mathematician, suppose such and such to be the case. Never you mind whether it is really so or not; but tell us, supposing it be so, what will be the consequence.'" (CP 1.133). This the mathematician does by deduction, which, of course, "does not lead to any positive knowledge at all, but only traces out the ideal consequences of hypotheses" (7.207).[101] Peirce's work on deduction persuaded him that every deductive inference can be represented by a diagram from which one intuitively gathers that the conclusion necessarily follows from the premises. Deduction – he wrote to Calderoni *c.* 1905 – "consists in constructing an image or diagram in accordance

---

[100] Readers unacquainted with modern formal logic may benefit from the following. Denote arbitrary objects by italic lowercase letters. The expression 'is greater than' – as in 'x is greater than y' – is a binary predicate; 'gives' – as in 'x gives y to z' – is a ternary one. Aristotle classified general statements into universal statements of the form 'All Ps are Qs' and particular statements of the form 'Some Ps are Qs'. For Peirce and Frege, 'All Ps are Qs' is just a short way of saying that 'For any object x (in the universe of discourse), if x is P, then x is Q', whereas 'Some Ps are Qs' says that 'In the universe of discourse there is at least one object x such that x is P and x is Q'. The generalizing operators 'For any x' and 'There is an x' are called *quantifiers.*

[101] "It is impossible to reason necessarily concerning anything else than a pure hypothesis. Of course, I do not mean that if such pure hypothesis happened to be true of an actual state of things, the reasoning would thereby cease to be necessary. Only, it never would be known apodictically to be true of an actual state of things" (Peirce CP 4.232).

with a general precept, in observing in that image certain relations of parts not explicitly laid down in the precept, and in convincing oneself that the same relations will always occur when that precept is followed out" (8.209; cf. 1.66).

In 1878 Peirce defined *induction* as synonymous with statistical inference, viz., "the inference that a previously designated character has nearly the same frequency of occurrence in the whole of a class that it has in a sample drawn at random out of that class" (6.409). He gave a similar definition in 1883 (2.702), but in an important manuscript, "On the Logic of Drawing History from Ancient Documents Especially from Testimonies", written perhaps in 1901 (CP 7.163–255), he tentatively proposed a distinction between three kinds of induction, such that the definition of 1878 applies only to the first.[102] All three kinds share the property that Peirce regards as "essential" and "intrinsic" (5.579) to this method of inference, viz., that it is self-correcting:

> Induction is the experimental testing of a theory. The justification of it is that, although the conclusion at any stage of the investigation may be more or less erroneous, yet the further application of the same method must correct the error. The only thing that induction accomplishes is to determine the value of a quantity. It sets out with a theory and it measures the degree of concordance of that theory with fact.
>
> (Peirce CP 5.145)

> The process of testing it will consist, not in examining the facts, in order to see how well they accord with the hypothesis, but on the contrary in examining such of the probable consequences of the hypothesis as would be capable of direct verification, especially those consequences which would be very unlikely or surprising in case the hypothesis were not true.
>
> (CP 7.231)

---

[102] Strictly speaking, to a subspecies of the first. But this is because in the paper of c. 1901 Peirce gives a definition of randomness by virtue of which no random samples can be drawn from infinite collections. Consequently, the definition of 1878 does not cover the case in which we sample a countable collection "in order to ascertain the proportionate frequency with which its members have a certain character designated in advance of the examination" (7.212), nor that in which we find a countable series "in an objective order of succession, and wish to know what the law of occurrence of a certain character among its members is, without at the outset so much as knowing whether it has any definite frequency in the long run or not" (7.213). These are, then, the other subspecies of the first kind of induction.

When we adopt a certain hypothesis, it is not alone because it will explain the observed facts, but also because the contrary hypothesis would probably lead to results contrary to those observed. So, when we make an induction, it is drawn not only because it explains the distribution of characters in the sample, but also because a different rule would probably have led to the sample being other than it is.

(CP 2.628)

Readers acquainted with modern statistics will perceive that Peirce anticipates here the approach subsequently developed by Neyman and Pearson.[103] He explicitly renounced the opposite statistical school, purportedly based on Bayes's Theorem, long before it became fashionable:

The theory here proposed does not assign any probability to the inductive or hypothetic conclusion, in the sense of undertaking to say how frequently *that conclusion* would be found true. It does not propose to look through all the possible universes, and say in what proportion of them a certain uniformity occurs; such a proceeding, were it possible, would be quite idle. The theory here presented only says how frequently, in this universe, the special form of induction or hypothesis would lead us right. The probability given by this theory is in every way different – in meaning, numerical value, and form – from that of those who would apply to ampliative inference the doctrine of inverse chances.

(Peirce CP 2.748)

Throughout his life, Peirce always insisted in the existence of a third type of inference besides deduction and induction. He called it by different names – viz., 'abduction', 'retroduction', 'hypothesis', 'presumption' – and described it in mildly different ways while consistently regarding it as the mainstay of scientific thinking, the sole source of new ideas in science (5.145) and indeed of new truths (7.219). Here is how Peirce characterized *abduction* in Baldwin's *Dictionary* (s.v. 'Reasoning'):

Upon finding himself confronted with a phenomenon unlike what he would have expected under the circumstances, [the reasoner] looks over its features and notices some remarkable character or relation among them, which he at once recognizes as being characteristic of some conception with which his mind is already stored, so that a theory is

---

[103] Also Deborah Mayo's "error statistics", as she points out in her illuminating analysis of Peircean error correction (Mayo 1996, pp. 412-41). I owe to Mayo the four Peirce extracts on induction.

suggested which would *explain* (that is, render necessary) that which is
surprising in the phenomena.

(Peirce CP 2.776)

There is no warrant for this inference. The hypothesis inferred is often
"utterly wrong" and "even the method need not ever lead to the truth;
for it may be that the features of the phenomena which it aims to explain
have no rational explanation at all". Nevertheless, abduction is justified
insofar as it is "the only way in which there can be any hope of attaining a rational explanation" (2.777). Or, as Peirce concisely put it in his
1903 lectures on pragmatism, "Abduction consists in studying facts and
devising a theory to explain them. Its only justification is that if we are
ever to understand things at all, it must be in that way." (5.145).

Peirce once proposed some trivial rules by following which "the
process of making an hypothesis should lead to a probable result"
(2.634). Clearly he was paying lip service to the contemporary prejudice in favor of rules of method; for in this matter of theory building
– as Paul Feyerabend (1970) said after Cole Porter – "anything goes".
Peirce also comments on "a rule of abduction much insisted upon by
Auguste Comte, to the effect that metaphysical hypotheses should be
excluded" (7.203). By 'metaphysical' Comte means a hypothesis that
has no experiential consequences. Peirce agrees that

> an explanatory hypothesis, that is to say, a conception which does not
> limit its purpose to enabling the mind to grasp into one a variety of facts,
> but which seeks to connect those facts with our general conceptions of
> the universe, ought, in one sense to be *verifiable*; that is to say, it ought
> to be little more than a ligament of numberless possible predictions con-
> cerning future experience, so that if they fail, it fails.

(Peirce CP 5.597)

However, "Comte's own notion of a *verifiable* hypothesis was that it
must not suppose anything that you are not able directly to observe",
and he ought therefore "to forbid us to suppose that a fossil skeleton
had ever belonged to a living ichtyosaurus" and indeed "to believe in
our memory of what happened at dinnertime today" (5.597). Thus, in
Comte's own narrow sense of 'verifiable', his rule is perverse. On the
other hand, in Peirce's sense, the rule is sound but superfluous, for a
hypothesis that is unverifiable in this wider sense cannot even be constructed. A proposition cannot "refer to anything with which experience does not connect us". Indeed, "the entire meaning of a hypothesis

lies in its conditional experiential predictions" and "if all its predictions are true, the hypothesis is wholly true" (7.203).

Although "abduction is, after all, nothing but guessing" (7.219), it is apparently guided by some kind of instinct (1.630, 5.604) or natural affinity for truth (7.220), for otherwise it would be well-nigh impossible for us ever to guess right. "Think of what trillions of trillions of hypotheses might be made of which one only is true; and yet after two or three or at the very most a dozen guesses, the physicist hits pretty nearly on the correct hypothesis. By chance he would not have been likely to do so in the whole time that has elapsed since the earth was solidified" (5.172). Peirce repeatedly identified this human talent with the "natural light" (*lume naturale*) to which Galileo appealed "at the most critical stages of his reasoning" (1.80; cf. 6.477, 1.630, 5.604). He was fond of equating it with the innate abilities of animals and to trace it somehow to natural selection:

> If man had not had the gift, which every other animal has, of a mind adapted to his requirements, he not only could not have acquired any knowledge, but he could not have maintained his existence for a single generation.
>
> (Peirce CP 5.603)

Yet he must have realized that the instinctive understanding of natural relations that is required to preserve our species could well be confined to the human scale and fail utterly at the frontiers of physical inquiry, for he referred to correct abductive guessing as a *hope* implicit in scientific practice (much in the way that Kant talked of life after death as a hope alive in ordinary moral practice):

> We are therefore bound to hope that, although the possible explanations of our facts may be strictly innumerable, yet our mind will be able, in some finite number of guesses, to guess the sole true explanation of them. *That* we are bound to assume, independently of any evidence that it is true. Animated by that hope, we are to proceed to the construction of a hypothesis.
>
> (Peirce, CP 7.219)

(iii) Brilliantly anticipating twentieth-century physics, Peirce (1892) rejected "the doctrine of necessity", that is, the thesis of thoroughgoing physical determinism adopted by so many of his contemporaries, and subjected it to damaging criticism. Instead of it, he professed "tychism", that is, the doctrine that ultimately nature is the realm of

4.4 Philosophers
231

chance. This, he said, is subsidiary to "synechism", the metaphysics of continuity that he embraced in the 1890s.[104]

The thesis of determinism is "that the state of things existing at any time, together with certain immutable laws, completely determine the state of things at every other time" (CP 6.37). Some "thinking men" who professed it told Peirce that it is a presupposition or "postulate" of scientific reasoning. However, "this does not make it true, nor so much as afford the slightest rational motive for yielding it any credence" (6.39). Moreover, the very idea that scientific reasoning, and in particular induction, may require such a postulate should, according to Peirce, be dismissed in the light of our present understanding of inference. On the other hand, there is no hope that induction might itself provide a ground for determinism. Every purported application of this view requires "that certain continuous quantities have certain exact values". To someone who is familiar with experimental work "the idea of mathematical exactitude being demonstrated in the laboratory will appear simply ridiculous" (6.44).

> Those observations which are generally adduced in favor of mechanical causation simply prove that there is an element of regularity in nature, and have no bearing whatever upon the question of whether such regularity is exact and universal or not. Nay, in regard to this *exactitude*, all observation is directly *opposed* to it; and the most that can be said is that a good deal of this observation can be explained away. Try to verify any law of nature, and you will find that the more precise your observations, the more certain they will be to show irregular departures from the law. We are accustomed to ascribe these, and I do not say wrongly, to errors of observation; yet we cannot usually account for such errors in any antecedently probable way. Trace their causes back far enough and you will be forced to admit they are always due to arbitrary determination, or chance.
>
> (Peirce CP 6.46)[105]

---

[104] Peirce (CP 6.202). 'Tychism' is formed from the Greek τύχη = 'luck, chance'; 'synechism', from συνεχής = 'continuous'.

[105] Here, Peirce, like almost everybody else, associates physical determinism with causation. But elsewhere he forcefully criticizes this association, much on the same grounds I adduced in §3.4.3. For him, causation is a "confused notion" (6.600) that men in different stages of scientific culture have conceived in entirely different and inconsistent ways. In particular, the common view "that our conception of cause is that of the Aristotelian efficient cause will hardly bear examination. The efficient cause was, in the first place, generally a thing, not an event; then, something which

For Peirce, "every throw of sixes with a pair of dice is a manifest instance of chance" (6.53). He debates this issue with an imaginary determinist who recalls that "each die moves under the influence of precise mechanical laws". Yet, counters Peirce, "the laws act just the same when other throws come up". True, the determinist concedes, but "the diversity is due to the diverse circumstances under which the laws act: the dice lie differently in the box, and the motion given to the box is different"; the "number of numbers, which expresses the amount of diversity of the system, remains the same at all times" (6.55–56). Peirce clinches the issue thus:

> You think all the arbitrary specifications of the universe were introduced in one dose, in the beginning, if there was a beginning, and that the variety and complication of nature has always been just as much as it is now. But I, for my part, think that the diversification, the specification, has been continually taking place.
>
> . . . .
>
> By thus admitting pure spontaneity or life as a character of the universe, acting always and everywhere though restrained within narrow bounds by law, producing infinitesimal departures from law continually, and great ones with infinite infrequency, I account for all the variety and diversity of the universe, in the only sense in which the really *sui generis* and new can be said to be accounted for.
>
> (Peirce, CP 6.57, 6.59)

I am not sure that I understand how tychism follows from synechism according to Peirce. It certainly has to do with his unorthodox con-

---

need not do anything; its mere existence might be sufficient. Neither did the effect always necessarily follow. True when it did follow it was said to be compelled. But it was not necessary in our modern sense. That is, it was not invariable" (6.66). After poking fun at those who admire John Stuart Mill for regarding "the cause as the aggregate of all the circumstances under which an event occurs", Peirce notes that this – otherwise commonplace view – rests on a misconception. "So far as the conception of cause has any validity – that is [. . .], in a limited domain – the cause and its effect are two *facts*"; however, contrary to what Mill "thoughtlessly" assumed, "the objective history of the universe for a short time, in its objective state of existence in itself" is not what a fact is; "a fact is an abstracted element of that; a fact is so much of the reality as is represented in a single proposition" (6.67). Nor does modern physical determinism illustrate Kant's law of causality (the Second Analogy of Experience). "There is no mechanical truth in saying that the past determines the future, rather than the future the past" (6.600). Moreover, "Kant's 'Analogy' ignores that continuity which is the life blood of mathematical thought" (Ibid.).

ception of the continuum. Peirce studied and partially embraced Cantor's ideas about infinite sets and transfinite cardinalities, but he firmly rejected Cantor's view that continua are equinumerous with 2^ω, that is, the set of all mappings from the set of natural numbers into {0,1}. A collection that meets this condition must consist of "absolutely distinct individual objects" and therefore should be called a *pseudo-continuum* (6.176), for "continuity is fluidity, the merging of part into part" (1.164). According to Peirce, mathematics calls for continuity in this sense. To show it, he proves, after Cantor, that no collection of distinct individuals "can have a multitude [i.e., cardinality – R. T.] as great as that of the collection of possible collections of its individual members" (Peirce RLT, p. 158). He then bids us consider "a collection containing an individual for every individual of a collection of collections comprising a collection of every abnumeral [i.e., uncountable – R. T.] multitude".

> This collection shall consist of all finite multitudes together with all possible collections of those multitudes, together with all possible collections of collections of those multitudes, together with all possible collections of collections of collections of those multitudes, and so on *ad infinitum*. This collection is evidently of a multitude as great as that of all possible collections of its members. But we have just seen that this cannot be true of any collection whose individuals are distinct from one another. We, therefore, find that we have now reached a multitude so vast that the individuals of such a collection melt into one another and lose their distinct identities. Such a collection is *continuous*.
>
> (Peirce RLT, p. 159)[106]

In other words, a continuum is what Peirce elsewhere calls a "collection too great to be discrete" (CP 4.180). More exactly, "a true continuum is something whose possibilities of determination no multitude of individuals can exhaust" (1.170). Synechism regards all distinct actualities as issuing from some such primeval continuum of potentialities. "It must be by a contraction of the vagueness of that poten-

---

[106] In this way, if Peirce's conception of the continuum makes any sense, it provides an instantaneous solution to Cantor's antinomy of the set of all sets. Peirce's argument for the reality of such continua is no less original and worth looking into: "The argument which seems to me to prove, not only that there is such a conception of continuity as I contend for, but that it is realized in the universe, is that if it were not so, nobody could have any memory. If time, as many have thought, consists of discrete instants, all but the feeling of the present instant would be utterly nonexistent." (CP, 4.641; cf. 1.167ff.)

tiality of everything in general, but of nothing in particular, that the world of forms comes about" (6.196). So "the very first and most fundamental element that we have to assume is a Freedom, or Chance, or Spontaneity, by virtue of which the general vague nothing-in-particularness that preceded the chaos took a thousand definite qualities" (6.200). At this point, I presume, tychism joins synechism.

4.4.3   Ernst Mach (1838–1916)

Mach saw himself as a physicist, not as a philosopher.[107] However, it is in the latter capacity that he is best known. In our eyes, his most valuable work is contained in his "historico-critical" studies on the energy principle (1872), the science of mechanics (1883), the theory of heat (1896), and physical optics (1921). His criticism of what he considered superfluous metaphysical ingredients in Newton's conceptual frame impressed young Einstein, encouraged his defiance of established physics, and guided his first steps toward a new theory of gravity. Mach is also remembered for his teachings about the aim of science and the common ground of physics and psychology, mainly because of the influence they exercised on the Vienna Circle, and through it on twentieth-century positivism. The following notes focus on these teachings – beginning with the second point – and on his criticism of Newton.

(i) Mach felt that mind–body dualism, especially in the clear-cut version inherited from Descartes, is a needless, unwarranted obstacle to the unity of science. A smooth transit from physics (and physiology) to psychology and vice versa can be secured if we acknowledge that we have to do with two *perspectives*, two different ways of articulating and combining the same set of elements. Such elements are "colors, sounds, temperatures, pressures, spaces, times, and so forth, [. . .] connected with one another in manifold ways; with them are associated

---

[107] Cf. Mach (AS, p. 47): "I am a scientist and not a philosopher" (Mach AS, p. 47). Ibid., p. 368: "Once more, there is no such thing as 'the philosophy of Ernst Mach'." See also pp. xxxvi, 30 and Mach (EI, pp. 15–16, 143). Through his experimental work, Mach made significant contributions to various fields of physics. Several quantities are named after him in fluid dynamics; for example, the *Mach number* is the ratio of the local velocity of flow to the velocity of sound in a compressible fluid.

dispositions of mind, feelings, and volitions" (Mach AS, p. 2). Certain "complexes of colors, sounds, pressures, and so forth, functionally connected in time and space" are comparatively stable. They are called *bodies*. Physics studies their properties and mutual relations. Among such bodies there is one, whose elements Mach denotes by $K\,L\,M\ldots$, that appears to be more intimately connected than the other bodies (denoted by $A\,B\,C\ldots$) with "the complex of volitions, memory-images, and the rest", which he denotes by $\alpha\,\beta\,\gamma\ldots$ "Usually, now, the complex $\alpha\,\beta\,\gamma\ldots K\,L\,M\ldots$, as making up the ego, is opposed to the complex $A\,B\,C\ldots$, as making up the world of physical objects; sometimes also $\alpha\,\beta\,\gamma\ldots$ is viewed as ego, and $K\,L\,M\ldots A\,B\,C\ldots$ as world of physical objects" (AS, p. 9). Psychology studies all elements in their relation to and interdependence with the body $K\,L\,M\ldots$ and its associated elements $\alpha\,\beta\,\gamma\ldots$ Considered from its standpoint, the elements $A\,B\,C\ldots K\,L\,M\ldots$ are called *sensations*; while in the perspective of physics they are physical objects.[108]

> Any one who has in mind the gathering up of the sciences into a single whole, has to look for a conception to which he can hold in every department of science. Now if we resolve the whole material world into elements which at the same time are also elements of the psychical world and, as such, are commonly called sensations; if, further, we regard it as the sole task of science to inquire into the connexion and combination of these elements, which are of the same nature in all departments, and into their mutual dependence on one another; we may then reasonably expect to build a unified monistic structure upon this conception, and thus to get rid of the distressing confusions of dualism.
>
> (Mach AS, p. 312)

At first blush Mach's solution may seem incredibly amateurish. It will appear to be less so if one considers how broadly he uses the term 'sensation'. To prove that there exist definite, specific sensations of time, he proposes the example of two bars of music, in which the sequence of the notes is quite different but which, when played consecutively, will at once be recognized as rhythmically identical.[109] He describes them as "two tonal entities which, acoustically, are differ-

---

[108] "Phenomena may be subdivided into elements, which, in so far as they are connected with certain processes of bodies, and can be regarded as conditioned by these processes, we call sensations" (Mach 1875, p. 54; quoted by him in AS, p. 17n.).

[109] The notes in the first bar are, in this order, C = ♩, E = ♪, F = ♪, G = ♩, followed by ♩. The notes in the second bar are, in this order, B = ♩, G = ♪, F = ♪, E = ♩, followed by ♩.

ently colored, but possess the same temporal form" (AS, p. 248). By 'temporal form' Mach means a certain relational system or structure embodied in both sound sequences. Now, if the concept of 'sensation' in Mach's parlance is wide enough to include *structures*, his approach is a good deal more promising than if the term had its usual meaning and referred only to the notorious sense-data – viz., sounds, colors, odors – of the empiricist tradition. Mach's intention probably was to count only very simple structures among "the elements of the world" (AS, p. 32), perhaps only such as can be apprehended at one fell swoop. But there is no way of drawing a limit here, and, of course, if none is drawn, Mach's "elements $A\ B\ C\ldots K\ L\ M\ldots$", the common building-ground of the sciences, comprise not just the sensory contents suggested by Mach's manner of speech, but every contentual and formal aspect discernible in experience. If such is the case, few will dispute Mach's contention that "every physical concept means nothing but a certain definite kind of connexion of the [...] elements which I have denoted by $A\ B\ C\ldots$" (AS, p. 42).

My vindication of Mach is perhaps disingenuous. Thus, in the last quotation I have brazenly substituted '[...]' for the word 'sensory'. I have also disregarded Mach's more obnoxious utterances, for instance, that "the whole inner and outer world are put together, in combinations of varying evanescence and permanence, out of a small number of homogeneous elements" (AS, p. 22). Indeed, the very term 'element' – which I have not tried to conceal – connotes a degree of finality and irreducibility that the referents of current physical concepts plainly cannot lay claim to. However, one can find plausible answers to these objections. Mach, for one, stressed that his elements are *provisional* (*vorläufig* – twice in italics in EI, p. 12), viz., they are that beyond which we cannot go *for the time being*. Science is concerned with "the *functional dependence* (in the mathematical sense) of these elements *from one another*" (EI, p. 11; cf. AS, p. 35); usually, indeed, with the smooth variation of elements of one kind as the elements of some other kind – or some ordered n-tuple of kinds – vary smoothly. This alone entails that the elements cannot all be homogeneous, so that Mach's contrary indication must be a slip of the pen. Moreover, nontrivial functional dependence and, in particular, *smooth* dependence presuppose that the elements in question belong to structures apart from which they are nothing or, at any rate, not what the scientists take them to be.

(ii) Mach insisted that no science would be necessary, and none would ever have arisen, "if every particular fact, every particular phenomenon were immediately accessible as soon as we require to know it" (Mach 1872, in 1909, p. 31). The function of science is, therefore, "to serve as a substitute for experience"; its task is "to represent the facts as completely as possible with the *least expenditure of thought*" (Mach M, p. 465). We should be content to know the distance s traversed by a freely falling body in each time interval t. "But what enormous memory would be necessary to keep in our heads the relevant table of s and t. Instead of it, we retain the formula $s = 1/2gt^2$, [. . .] by which we find the s pertaining to a given t [. . .]. This formula, this 'law' does not in the least possess a greater objective value than all the particular facts together. Its value lies merely in the ease with which it is used. It has an economic value" (1909, p. 31; cf. M, p. 461, on Snell's law). Besides gathering and filing a maximum of facts in a manageable shape, science has a second task, namely, to analyze the more complex facts into a minimal number of maximally simple ones. "This we call *explaining*. The simple facts to which we reduce the more complex ones are in themselves always *incomprehensible*, i.e. not further analyzable, e.g. that a mass bestows acceleration on another mass. It is again a question of *economy*, on the one side, and of *taste*, on the other, at which incomprehensibles one chooses to stop" (1909, p. 31; my italics). "No basic fact is more understandable than another. The choice of basic facts is a matter of comfort, of history and of habit" (p. 33).

The pursuit of such economic aims is naturally led by interest. "When we portray facts in thoughts, we never portray the facts *as a whole*, but only the aspect which is *important* to us; we have here a goal which has issued mediately or immediately from a practical interest. Our portrayals are always abstractions" (M, p. 458). Thus, "when we speak of cause and effect we arbitrarily highlight those aspects to whose connection we ought to *pay attention* as we portray a fact in the respect which is important to us. In nature there is no cause and no effect. Nature is simply there, just *once*." (M, p. 459). Likewise, when we bestow a name on a "thing", that is, a fairly stable complex of elements, we look away from its surroundings and the many little changes that such a complex is continually going through. "The thing is an abstraction, its name a symbol for a *complex* of elements, whose changes we ignore" (Ibid.). "*Bodies* are nothing but *bundles of reactions connected by law*. The same holds for *processes* of every sort which we classify and name to satisfy our need for overview. Be it waterwaves [. . .] or soundwaves in the air [. . .] or an electric current [. . .], that which is constant is always the *connection* of reactions according to law, and this *alone*. This is the *critically purified concept of substance* which, scientifically, ought to replace the vulgar one" (EI, p. 148).

Summing up: "Our concepts arise from the sensations, by way of their connections; the aim of concepts is to lead us in every given case by the shortest and most comfortable ways to the sensory representations which best agree with the sensations" (EI, p. 144).[110] Of course, the agreement need not be better than is required by the momentary interests and circumstances under which it takes place. Since such interests and circumstances vary from case to case, the intellectual portrayals of phenomena do not exactly agree with each other. According to Mach, "biological interest impels the mutual correction of different results of portrayal, towards the best possible, most advantageous balancing of differences" (EI, p. 164). Thus, our concepts need not only adjust themselves to facts, but also to one another. "The adjustment of *thoughts* to *facts* [. . .] we call *observation*; the adjustment of *thoughts* to *each other* we call *theory*. Observation and theory cannot be sharply separated, for almost every observation is already influenced by theory and, if sufficiently important, will react upon theory" (pp. 164f.).

Scientific knowledge is concerned with the connection of phenomena. Whatever we might make out as standing behind phenomena "exists *only* in our understanding and has for us just the value of an aid to memory or formula, whose form, being arbitrary and indifferent, can very easily change with our cultural standpoint" (Mach 1872, in 1909, pp. 25f.). We are therefore free to devise transphenomenal objects in any way we think useful for connecting the phenomena. We are under no constraint to impose on them the familiar conditions of sensory experience. In particular, there is no need to think of them as spatial, that is, as sporting the same kind of relations as the visible and the tangible, "any more than it is necessary to think of them as having a definite pitch of sound" (1909, p. 27). As we shall see in Chapters Five and Six, early twentieth-century physicists took advantage of this freedom to a wholly unprecedented degree.

---

[110] The passage continues with this remarkable metaphor: "Our true mental *workers* are the sensory representations, while concepts are their supervisors and *regulators*, which put the multitude of the former into place and assign them their jobs. In simple affairs the intellect deals immediately with the workers, but in larger enterprises it has to do with the leading engineers, which, however, would be of no use to it if it had not also furnished them with reliable workers."

(iii) Mach's "critical purification" of Newtonian mechanics concentrates on the concept of mass and the principle of inertia (Newton's First Law).

According to Mach, Newton's definition of mass as the product of density and volume is useless, for density, in turn, can only be evaluated in terms of mass. In the wake of Euler's *Mechanica* (1736), most writers on the subject opted for defining the mass of a particle as the ratio between the net external force acting on the particle and the particle's acceleration.[111] This approach, however, could not satisfy Mach, who sought to "demythologize" the notion of force by defining it as the product of mass and acceleration. So he tackles the problem in quite another way. He considers two isolated bodies A and B, interacting with one another. Initially he assumes that A consists of m and B consists of m' equal bodies a, where m and m' are integers. Let φ and φ' denote the accelerations experienced by A and B, respectively, due to their interaction. Then, says Mach, "if we take into consideration the sign of the acceleration", we have that $m/m' = -\varphi'/\varphi$. If $|\varphi| = |\varphi'|$, we say that A and B have the same mass. This linguistic convention would be highly inconvenient if there existed three bodies, A, B, and C, such that A and B had the same mass, and B and C had the same mass, but C and A had different masses. Mach argues, however, that if this could happen, it would lead to situations blatantly incompatible with the experiences summarized by the principle of energy conservation. Let A, B, and C be three perfectly elastic bodies constrained to move around a fixed, circular, frictionless ring. Suppose that, under our convention, A has the same mass as B and B has the same mass as C, but that, in an interaction between A and C, A experiences a greater acceleration than C. Then, if the three bodies are initially at rest on the ring, and we impress on A a speed v toward B, A will eventually transmit this speed to B which will in turn transmit it to C; but when C reaches A the latter will acquire a speed $v' > v$. The process will repeat itself, and new kinetic energy will accrue to the three-body system with each cycle, thus generating a *perpetuum mobile*. This thought experiment shows, in the light of experience, that sameness of mass as defined

---

[111] Paul Appell gave this definition in his *Traité de mécanique rationelle*: "The mass of a particle is the constant ratio that exists between the intensity of a constant force and the acceleration impressed by it on the particle" (1893, p. 87; quoted in Jammer 1961, p. 89).

above is a transitive relation. Therefore a definite mass can be consistently assigned to every body, at least ideally, by the following procedure: Pick any body A as the standard of mass. Any body with the same mass as A is assigned 1 unit of mass. The mass of any other body B is put equal to $|\varphi'/\varphi|$ units, where φ' is the acceleration experienced by B and φ is the acceleration experienced by a body C of mass 1, when B and C interact in isolation. Note that $|\varphi'/\varphi|$ can be any real number, not necessarily a rational one.

Mach discussed Newton's First Law in a course of lectures in the summer of 1868. The gist of his comments was printed in a note at the end of his booklet on the energy principle (1872; in Mach 1909, pp. 46–50). A longer and better known version of his criticism appeared in *Die Mechanik* (1883) and was amplified in later editions (see M, pp. 216–71). Mach complains that the First Law – quoted in §2.1 – is ambiguous. It speaks of "uniform motion straight ahead" but says nothing of the bodies to which "the direction and speed of the moving body" are referred. Newton, of course, thought that the direction should be taken in absolute space and the speed in absolute space and time. Mach regarded this as sheer nonsense.

> A motion can be uniform with respect to another motion. The question whether a motion is uniform *in itself* has *no sense at all*. Likewise, we cannot speak of an "absolute time" (independent of every change). This absolute time cannot be measured against any motion, so it has no practical and also no scientific value; no one is entitled to say that he knows something about it, it is an idle, "metaphysical" concept.
>
> (Mach M, pp. 217)

> No one can say anything about absolute space and absolute motion; they are mere figments of thought (*Gedankendinge*) which cannot be pointed out in experience. All our principles of mechanics [. . .] are experiences about the *relative* positions and motions of bodies. Before testing them it was neither possible nor permissible to adopt them for the regions in which they are now considered valid. No one is entitled to extend these principles beyond the limits of experience. Such an extension is indeed senseless, for no one would know how to apply it.
>
> (Mach M, pp. 222–23)

Mach was indeed aware of the experiment through which, according to Newton, the presence of absolute space is made manifest: A bucket is suspended from a long cord and turned about until the cord is strongly twisted. The bucket is then filled with water and held at rest.

If it is suddenly impelled in the opposite direction, and continues to move for some time while the cord untwists, "the surface of the water will at first be flat, as before the bucket began to move; but after that, the bucket, by gradually communicating its motion to the water, will make it begin sensibly to revolve, and recede little and little from the center, and ascend to the sides of the bucket, forming itself into a concave figure (as I have experienced), and the swifter the motion becomes, the higher will the water rise, till at last, performing its revolutions in the same times with the vessel, it becomes relatively at rest in it" (Newton 1726, p. 10). According to Newton, the water's endeavor to climb the bucket's walls and recede from its axis bears witness to its real rotation in absolute space, for the water's surface remained initially unchanged while it rotated only in appearance with respect to the adjacent bucket. Mach proposes an altogether different reading of these phenomena:

> Newton's experiment with the rotating water bucket teaches us only that the rotation of the water relative to the *bucket walls* does not stir any noticeable centrifugal forces; these are prompted, however, by its rotation relative to the mass of the earth and the other celestial bodies. Nobody can say how the experiment would turn out, both quantitatively and qualitatively, if the bucket walls became increasingly thicker and more massive – eventually several miles thick.
>
> (Mach M, p. 226)

According to this view the steady speed and direction of inertial motion depends on and should be referred to, not the chimera of absolute space, but the actual distribution of matter in the entire world (pp. 227, 228). As we shall see in §5.4, Einstein believed for some time that his General Theory of Relativity embodied this view of inertia.

The following passage, highlighted by Mach, contains what he describes as "the most important result" of the preceding reflections:

> *The seemingly simplest propositions of mechanics are quite complicated by nature. They rest on experiences which are still incomplete and indeed can never be fully completed. Practically, they are sufficiently assured to furnish a basis for mathematical deduction with a view to the needed stability of our environment; but they should by no means be regarded as mathematically final truths, but rather as propositions which admit and indeed require a continual experimental control.*
>
> (Mach M, p. 231)

I do not doubt that words like these encouraged Einstein and his younger contemporaries to proceed with their reform of physics.

4.4.4   Pierre Duhem (1861–1916)

Like Whewell and Mach, Duhem was a practicing scientist who devoted an important part of his adult life to the history and philosophy of physics. With his studies on the origins of statics (1905–1906), on the mechanical tradition linked to Leonardo da Vinci (1906–1913), and on cosmology from Plato to Copernicus (1913–1958), he founded single-handed the history of medieval physics.[112] His philosophy is contained in *La théorie physique: son objet, sa structure* (1906), which may well be, to this day, the best overall book on the subject. Its main theses, although quite novel when first put forward, have in the meantime become commonplace, so I shall review them summarily without detailed argument, just to associate them with his name. But first I ought to say that neither in the first nor in the second (1914) edition of his book did Duhem take into account – or even so much as mention – the deep changes that were then taking place in physics. It is mainly for this reason that I classify him as a nineteenth-century philosopher.[113] Still, the subsequent success and current entrenchment of Duhem's ideas are due above all to their remarkable agreement with – and to the light they throw on – the practice of mathematical physics in the twentieth century.

In the first part of *La théorie physique* Duhem contrasts two opinions concerning the aim of a physical theory. For some authors, it ought to furnish "the *explanation* of a set of experimentally established laws", while for others it is "an abstract system whose aim is to *summarize* and *logically classify* a set of experimental laws, without pretending to explain these laws" (Duhem 1914, p. 3). Duhem resolutely sides with

---

[112] When I was young, Duhem's scholarship used to be invidiously compared with Anneliese Maier's, whose admirable "Studies on the Natural Philosophy of Late Scholastics" were being published just then. I suppose that, by now, one may commend Duhem's pioneer work with equanimity.

[113] In a review of his own work that he submitted to the Académie des Sciences when he was proposed for membership, Duhem (1913) professed support for Energetics. This was the nineteenth-century school that opposed the analysis of matter into atoms and molecules (see note 55). By 1913, Einstein's theoretical work on Brownian motion and Jean Perrin's experimental research had already persuaded the more vocal energetists that their position was untenable.

the latter. His rejection of the former rests on his understanding of 'explanation' ('explication' in French), which he expresses as follows: "To explain, *explicare*,[114] is to divest *reality* from the *appearances* which enfold it like veils, in order to see that reality face to face" (pp. 3–4). Authors in the first group expect from physics the true vision of things-in-themselves that religious myth and philosophical speculation have hitherto been unable to supply. Their expectation makes no sense unless (i) there is, "beneath the sense appearances revealed to us by our perceptions, [. . .] a reality different from these appearances" and (ii) we know "the nature of the elements which constitute" that reality (p. 7). Thus, physical theory cannot explain – in the stated sense – the laws established by experiment unless it depends on metaphysics and thus remains subject to the interminable disputes of metaphysicians. Worse still, the teachings of no metaphysical school are sufficiently detailed and precise to account for all the elements of a physical theory (p. 18). Duhem instead assigns to physical theories a more modest but autonomous and readily attainable aim:

> A physical theory is not an explanation. It is a system of mathematical propositions, derived from a small number of principles, whose purpose is to represent a set of experimental laws as simply, as completely and as exactly as possible.
>
> (Duhem 1914, p. 24)

Duhem's book is a long gloss on this definition. At one point, it may seem trite: Physics advances by abstraction and generalization, from facts to laws and from laws to theories. First, an enormous variety of particular, complex facts is analyzed to find what they have in common. This is summarized "in a *law*, that is, a general proposition combining abstract notions". Then a whole set of laws is considered. "A very small number of extremely general statements, about some very abstract ideas" is substituted for them. The physicist "chooses these *primary properties*, formulates these *fundamental hypotheses* in such a way that every law belonging to the set under study can be drawn from them by a possibly very long but very secure deduction. This system of hypotheses and the consequences that flow from them [. . .] constitute the physical theory as defined by us" (p. 77). However, Duhem gives a peculiar twist to this seemingly facile view: Physical

---

[114] In Latin in the original. From 'plico' = 'I fold'; so 'explico' = 'I bring out of its folds, I unfold'.

laws are *experimental* and physical theories are *mathematical*; this combination of mathematics and experiment can only be achieved by putting theories to work in the design of experiments and the gathering of facts under laws.

According to Duhem, there are three operations involved in the constitution of any physical theory (pp. 197f.). In the first place, a few properties are chosen among the many revealed by observation; they are regarded as primary and are represented by algebraic or geometric symbols. Certain relations are postulated between these symbols; they are the principles or "fundamental hypotheses" of the theory. The third operation is the mathematical development of the theory. "Its purpose is to teach us that, by virtue of the fundamental hypotheses [. . .], the combination of such and such circumstances will bring about such and such consequences; to announce to us, for example, that by virtue of the hypotheses of thermodynamics, if we subject an ice block to such-and-such compression, the block will melt when the thermometer will indicate such-and-such a degree" (p. 198).[115] The mathematical development must bond (*se souder*) with observable facts at both ends. This can be achieved only by a double translation: "To introduce the circumstances of an experiment into the calculations, a translation must replace the language of concrete observation with the language of numbers; to check the outcome predicted by the theory for this experiment, a numerical value must be transformed into an indication stated in the language of experience" (p. 199). According to Duhem, both translations involve a measure of indeterminacy (or, as he bluntly says, of betrayal). He bases this on the presumption that algebraic variables must stand for numbers (p. 158). Such a view of algebra is surely too narrow, but it is not one that can be easily countered with examples drawn from the literature of physics. Anyway, to vindicate Duhem's allegation it is enough to consider that every term occurring in a mathematical argument is precise and unambiguous in a way in which "the language of concrete observation" is not. "The facts of experience, in all their native brutality, cannot be used in mathematical reasoning. To

---

[115] Duhem's description of the mathematical development of a theory agrees fairly well with what Hempel and Oppenheim (1948), Braithwaite (1953), and others call the *explanation* of laws by theories. They chose to retain this prestigious name for the aim of science, although they conceived it very much like Duhem. Their decision gave rise to a vast and mostly tedious literature about the true meaning of 'explanation'. Whoever has had to wade through it is bound to admire the wisdom of Duhem, who gave up the term without fuss.

feed such reasoning they must be transformed and put into symbolic form" (p. 298).

The indeterminacy of translation follows at once from Duhem's distinction between "practical" and "theoretical" facts. A practical fact is a concrete fact described in ordinary language just as one happens to observe it. A "theoretical fact", instead, is the "set of mathematical data which replaces a concrete fact in the reasonings and calculations of the theoretician" (p. 199). In such a fact nothing is vague or undecided. If it concerns the temperature distribution in a body, the latter is geometrically defined, its edges are widthless lines, its points are dimensionless, lengths and angles are exactly assigned, and each point of the body has a definite temperature, a number that cannot be confounded with any other number. A practical fact is a far cry from this, and it cannot be described "without attenuating by means of the words *more or less* (*à peu près*) whatever is too determined in each statement" (p. 200). Because of it, "an infinity of different theoretical facts can be taken as the translation of the same practical fact" (p. 201), and vice versa (pp. 225, 229, 230).

Of course, Duhem uses 'translation' as a metaphor.[116] While the rules of genuine translation, for example, between Greek and Latin, are not a part of either language but can be established by comparison from the outside (say, by a German philologist), the rules of Duhemian translation are an integral part of the physical theory to and from which the translation is effected. This affects the very essence of physical experiments.

> A *physical experiment is the precise observation of a group of phenomena together with the* INTERPRETATION *of these phenomena; this interpretation substitutes, for the concrete data actually collected, abstract and symbolic representations which correspond to them by virtue of the theories accepted by the observer.*
>
> (Duhem 1914, p. 222; Duhem's emphasis)

One cannot use the instruments found in physical laboratories if one does not substitute for the concrete objects which compose those instru-

---

[116] The metaphor was popularized in the English-speaking world by Campbell (1920). I am not sure that everyone who used it realized that it was only a metaphor, and that, for instance, every sentence in a paper by Rutherford is written in one and the same language, viz., English. (Indeed, the mathematical and chemical formulas are ideograms that can be read in any civilized language, just as Chinese characters can be read in Mandarin and Cantonese.)

ments an abstract and schematic representation which provides a handle
for mathematical reasoning, if one does not subject this combination of
abstractions to deductions and calculations which involve the acceptance
of theories.

(p. 231)

When a physicist makes an experiment two quite different representations of the instrument on which he operates occupy his mind. One is
the image of the concrete instrument he actually handles; the other is a
schematic model (*type*) of that same instrument, constructed by means
of symbols furnished by the theories. And, mark you, it is about this
ideal and symbolic instrument that he reasons, applying the laws and
formulas of physics to *it*.

(p. 235)

Since the laws of physics are based on the results of physical experiments, the character of the latter affects the nature and scope of the
former. "A physical law is a symbolic relation whose application to
concrete reality requires that one knows and accepts a whole collection of theories" (p. 254). Such a law is always provisional and relative. It is provisional, not because it is true today and will be false
tomorrow, but because it represents the facts to which it is applicable
with an approximation that physicists now consider satisfactory, yet
will eventually judge insufficient. It is relative, not because it is true for
one physicist and false for another, but because the approximation
involved in it is good enough for one physicist's use of it and not for
the other's (p. 260). Indeed, "one can see the same physical law simultaneously adopted and rejected by the same physicist in the course of
the same work" (p. 262).[117]
Yet physicists show, as a matter of fact, "an irresistible aspiration"
toward a physical theory that would represent all experimental laws
by means of a single, logically consistent theory (p. 449). This aspiration has been present and has exerted a powerful influence throughout
the history of physics. According to Duhem, neither physical knowledge nor the philosophical analysis of the structure of physical theory

---

[117] An outstanding example of this procedure was provided soon thereafter by Einstein
(1915h) when he computed the anomalous perihelion advance of Mercury using his
new law of gravity, while at the same time accepting the value computed by Newton's
law of the – 12 times larger – perihelion advance attributable to the presence of the
other planets. See §§5.4, 5.5, and 7.2.

can justify this aspiration; so the physicist who yields to it – and which does not? – implicitly endorses a metaphysical creed:

Which is this metaphysical proposition that the physicist will affirm, as if by force, despite the restraints imposed by the method he habitually uses? He will aver that, beneath the sensible data solely accessible to his methods of research, realities are hidden whose essence cannot be grasped by those same methods; that these realities are disposed in a certain order which physical theory cannot contemplate directly; but that physical theory, by its successive perfectionments, tends to arrange its experimental laws in an order which is ever more similar to the transcendent order by which realities are classed; so that, by virtue of this, physical theory gradually approaches its limiting form as a *natural classification*; finally, that logical unity is a character without which physical theory cannot claim the status of a natural classification.

(Duhem 1914, p. 450)

These are the central tenets of a "believer's physics",[118] which no reason can justify, but without which "it would be unreasonable to work in the progress of physical theory" (1908, p. 19). It often happens that a hitherto unsuspected physical law is derived from a physical theory and subsequently corroborated by experiment (1914, p. 450). Events like this "press the physicist to assert that *physical theory, as it progresses, becomes more similar to a natural classification, which is its ideal and its goal*" (p. 452). However, Duhem's strong and disciplined intellect would not let him claim that this assertion is *proved* by such events.

At any rate, it is clear that physics cannot proceed by induction alone. For, as we have seen, no generalization from experiment can be of any use to physics "unless it undergoes an interpretation which transforms it into a symbolic law; and this interpretation involves the acceptance of a whole set of theories" (p. 303). Many "symbolic translations" being equally admissible, the physicist must choose among them "the one which will provide a fruitful hypothesis to the theory, without his choice being guided by experience in any way" (Ibid.). Since "the execution and interpretation of any physical experiment involve the acceptance of a whole ensemble of theoretical propositions", theoretical physics can only be tested as a corporate body.

---

[118] "Physique de croyant" is the title of the article from which the last quotation is taken; it appeared in *Annales de Philosophie chrétienne* in the fall of 1905 and was later included by Duhem in the second edition of *La théorie physique*.

The physicist can never subject an isolated hypothesis to experimental control, but only a whole ensemble of hypotheses; when experience disagrees with his previsions it teaches him that at least one of the hypotheses in this ensemble is unacceptable and should be modified; but it does not indicate to him which one should be changed.

. . . .

Physics is not a machine one can take apart; one cannot try each piece in isolation and wait, to adjust it, until its solidity has been minutely checked. Physical science is a system that must be taken as a whole. It is an organism no part of which can be made to function without the remotest parts coming into play, some more, some less, but all in some degree.

                                        (Duhem 1914, pp. 284–85)

In every epoch physicists agree that certain elements in their system are beyond question, so that any modification required by experimental evidence must bear on other elements. However, the privileged status of this hard core of physics does not rest on any logical necessity. So a physicist can always refuse to reconcile the theoretical scheme with the facts by invoking causes of error and introducing corrections; instead, "by resolutely bringing reform to the propositions declared intangible by common consent, [he or she might] fulfill the work of a genius who opens theory to a new career" (pp. 321f.; cf. p. 328). According to Duhem, the choice of physical hypotheses is subject only to these restrictions: Each must be consistent with itself and with the rest, and they must form a system such that, "from *their ensemble*, mathematical deduction can draw consequences that represent, to a sufficient approximation, *the ensemble* of experimental laws" (p. 335). Otherwise, theoreticians are free to lay down the foundations of their systems in any way they think fit. When Duhem died in 1916, Einstein's dazzling display of freedom in physics had been going on for over a decade.

CHAPTER FIVE

✦

Relativity

'Relativity theory' or simply 'Relativity' is the standard name of two quite different, yet subtly related theories put forward by Albert Einstein in 1905 and 1915, respectively. The first, Special Relativity (SR), rescued the Maxwell equations from seemingly catastrophic experimental results by making deep changes in the basic concepts and laws of Newtonian mechanics. The second, General Relativity (GR), solved the problem of reconciling SR with Newton's theory of gravity by transcending them both.[1] For those of us who cherish physico-mathematical theories more for their inherent beauty than for their transient accuracy, SR and GR remain unmatched. Moreover, to this day, they have enjoyed tremendous empirical success. SR is corroborated daily in every high-energy lab. GR accounts for all the phenomena Newton classified as gravitational just as well or even better than his theory. Moreover, it provides an amazing gravitational explanation of other phenomena – such as the systematic shift in the spectrum of light from distant galaxies and the pervasive microwave background radiation –, which nobody even suspected c. 1910 and which would not easily fit in a Newtonian framework.

This is not the place to deal even superficially with the many fruitful applications of SR and GR. Our attention must go to their con-

---

[1] The difference and the relation between SR and GR will, I hope, be made clear in this chapter. For the time being let me just remark that SR can hold in a world governed by GR only if that world is completely lacking in gravitational sources. Still, in a world like ours, GR agrees excellently with SR in a small, freely falling lab, over short periods of time. It is therefore not altogether unjustified to describe SR as a special version and GR as a general version of one and the same theory. On the other hand, this description conceals the drastic change of meaning and scope that SR suffers when embedded in GR.

ceptual problems and their philosophical significance. The latter has been judged differently by different authors. In my view, it lies chiefly in the fact that both theories are exemplary cases of far-reaching conceptual change in fundamental physics, firmly rooted in the tradition they go beyond. As such, they illustrate with exceptional clarity the way in which rupture and continuity are combined in the history of physics. Conceptual innovation was also the main source of the so-called philosophical problems of Relativity, which generally stem from misunderstandings and were proposed by physicists and philosophers who resisted Einstein's reckless assault on their encrusted modes of thought.

5.1   Einstein's Physics of Principles

Let me first explain the problem to which Einstein responded with SR. By virtue of Newton's principle of relativity, any inertial frame of reference may be looked on as a frame at rest to which all motions are referred (see Chapter Two, note 17, and the text linked to it). The validity of this principle remained unquestioned while physics dealt with interactions that cause acceleration and depend on the mutual distances – but not on the relative velocities – of the interacting bodies, that is, until the 1850s (see the quotations from Helmholtz 1847, p. 6, on page 185). This changed with the advent of Maxwell's electrodynamics. The constant speed c with which light travels *in vacuo* takes pride of place in the Maxwell equations. Moreover, the velocity $\mathbf{v}$ with which a charged body moves through an electromagnetic field determines both the direction and the magnitude of the force that the field exerts on it. Now, if $\mathbf{v}$ is the momentary velocity of an object B relative to an inertial frame $\mathcal{F}$, and $\mathcal{F}$ in turn moves with constant velocity $\mathbf{w}$ relative to another frame $\mathcal{G}$, then – one presumed – the same object's velocity relative to $\mathcal{G}$ is surely $\mathbf{v} + \mathbf{w}$. So, if F stands for the force per unit charge exerted on our object by the electric and magnetic fields $\mathbf{E}$ and $\mathbf{B}$, the expression $\mathbf{F} = (\mathbf{v} \times \mathbf{B}) + \mathbf{E}$ (eqn. (7.3)) must be referred to a particular such frame.[2] One assumed, rather naturally, that this is the frame in which the electromagnetic aether, the site of the stresses represented by $\mathbf{E}$ and $\mathbf{B}$, is permanently at rest. It was – one concluded – relative

---

[2] Otherwise, an object of unit mass that experiences acceleration $(\mathbf{v} \times \mathbf{B}) + \mathbf{E}$ relative to frame $\mathcal{F}$ must experience acceleration $((\mathbf{v} + \mathbf{w}) \times \mathbf{B}) + \mathbf{E}$ relative to frame $\mathcal{G}$, which is absurd.

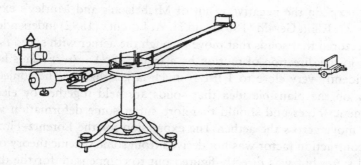

Figure 12

to this frame, and to this frame only, that light travels *in vacuo* in any direction with speed *c*.

Of course, the earth does not rest in the aether, at least not the whole year round.[3] So one should, in principle, be able to measure the speed *u* of the earth in the aether by the observation of optical effects. None were detected. It was shown, however, that all first-order effects of the earth's motion – that is, effects that depend on multiples of the ratio u/c, but not on $(u/c)^2$, $(u/c)^3$, or other higher powers of this ratio – cancel out if one assumes with Fresnel that a small fraction of the aether is dragged by matter. On the other hand, the means of detection available until 1880 were insensitive to second-order effects, depending on the much smaller ratio $(u/c)^2$. But Albert Michelson's interferometer, which was first built about that time, was capable of disclosing such effects. It consists of two perpendicular steel arms, one of which is aligned with the motion of the earth (see Fig. 12). Light issuing from a common source is divided into two rays that travel along each arm, are reflected by mirrors at either end, and then are mixed together to produce a pattern of interference fringes that are observable through a microscope. The pattern should vary when the instrument is rotated by 90° (so that the arm that was formerly aligned with the motion of the earth is now perpendicular to it, and vice versa). However, neither Michelson (1881) nor Michelson and Morley (1887) – using an improved interferometer – could measure any significant change in the interference pattern.

---

[3] The motion of the earth in the aether follows at once from the Copernican hypothesis. Empirical proof was provided by Bradley's discovery of the aberration of starlight (1728).

To explain the negative result of Michelson's and Morley's experiment, G. F. FitzGerald (1889) and H. A. Lorentz (1892) independently conjectured that solids that move through the aether with speed v contract in the direction of motion by a factor of $\sqrt{1 - (v/c)^2}$. This factor is evidently very close to 1 if c is much larger than v. The conjecture drew on the plausible idea that solids are held together by electromagnetic forces and should therefore suffer some deformation when they move across the aether. The expression for the Lorentz–FitzGerald contraction factor was not derived from some specific theory of the solid state, but was directly figured out to compensate for the difference in the speed with which light was supposed to travel along one and the other arm of the interferometer. Lorentz realized later that, to obliterate all the predictable – yet not forthcoming – effects of motion in the aether, he would have to tamper not only with lengths but also with durations. He introduced the enigmatic notion of a *local time*, measured on moving objects and $1/\sqrt{1 - (v/c)^2}$ times longer than the real time elapsed between the same events. In this way, the discussion of space and time coordinates and of coordinate substitution eventually took center stage in Lorentz's study of the motion of bodies in the quiescent aether. It should be noted, though, that the coordinate substitutions found in Lorentz's papers as late as 1904 are not quite the same thing as the exact transformations that now bear his name (courtesy of Poincaré 1906), but are more in the nature of coordinate adjustments, made necessary by the physical alteration of moving clocks and rods, and avowedly valid only to a specified approximation.[4]

The task of describing electromagnetic phenomena relative to moving bodies was handled in a totally different spirit by Einstein (1905r). His work on radiation (1905i; see §6.1) persuaded him that contemporary views concerning the microphysical basis of electrodynamic phenomena needed thorough revision. Time was not ripe yet for

---

[4] Lorentz (1904, §4, eqns. (4) and (5); my notation) introduces the auxiliary substitution $x' = kl\xi$, $y' = l\eta$, $z' = l\zeta$, $t' = lk^{-1}\tau - klv\xi c^{-2}$, where $k = 1/(1 - (v/c)^2)$ and "the coefficient l . . . is to be considered as a function of v whose value is 1 for $v = 0$, and which, for small values of v, differs from unity no more than by a quantity of the second order". Suppose now that ξ, η, ζ, and τ, are related to x, y, z, and t by the Galilean transformation $\xi = x - vt$, $\eta = y$, $\zeta = z$, $\tau = t$. Then the coordinate systems (x,y,z,t) and (x',y',z',t') will be related by a Lorentz transformation – eqns. (5.5) on page 257 – if and only if $l = \text{const.} = 1$. Lorentz argued for this equality from physical considerations. Poincaré showed that unless $l = 1$, the coordinate substitutions introduced by Lorentz do not constitute a group.

a new construction founded on hypotheses about the structure of matter. So Einstein took his cue from classical thermodynamics, which accounts for a vast and ubiquitous class of phenomena, without making any assumptions about their deep underlying structure, by means of two universal principles (cf. §4.3.2). He also proposed two universal principles, which can be paraphrased as follows:

Relativity Principle. The laws by which the states of physical systems change do not depend on the particular inertial frame of reference to which those changes of state are referred.
Light Principle. Every light signal travels *in vacuo* relatively to a given inertial frame of reference with constant velocity c, no matter what the state of motion of its source.

The meaning of these statements will be clarified below; but there is no doubt that, *prima facie*, they agree with experience. Indeed, Lorentz and others strained their imaginations to explain why they did so, although, on purely conceptual grounds, they ought not to. Instead of postulating new microphysical interactions to account for the unexpected agreement, Einstein embarked in conceptual criticism. He showed that one can very well conceive that the same light signal travels with the same speed relative to two different frames moving with constant velocity relative to each other; and that the common presumption that this is logically impossible rests on the uncritical acceptance of a specific, not logically necessary, scheme for the description of motion. As it turns out, this scheme is admissible if and only if neither light nor any other physical disturbance propagates with the same speed in all inertial frames; and so its adoption by classical physics unwittingly begs the question at issue.

To simplify matters I shall compare two inertial frames $\mathcal{F}$ and $\mathcal{F}'$, endowed with right-handed systems of Cartesian coordinates x, y, z and x', y', z', such that on a particular instant the three axes of the primed Cartesian coordinate system lie, respectively, along the homonymous axes of the unprimed system.[5] The frames are also endowed with universal time coordinates t and t' – on which I shall have more to say –, which take the same value 0 at the origins of the

[5] A system of Cartesian coordinates x, y, z is right-handed if one can place the right hand at the origin in such a way that the thumb points along the x-axis in the direction of increasing x, the forefinger points along the y-axis in the direction of increasing y, and the middle finger points along the z-axis in the direction of increasing z.

Cartesian systems when both origins coincide. All coordinates of the same kind are expressed in the same measurement units (e.g, meters and seconds, or light-years and years).[6] I assume moreover that $\mathcal{F}'$ moves relatively to $\mathcal{F}$ parallel to the x-axis, with constant speed v, in the direction of increasing x. Before Einstein, mathematical physicists took for granted that, in such a case, the primed and unprimed coordinates described above are related by the following so-called *Galileian transformation*:[7]

$$x' = x - vt$$
$$y' = y$$
$$z' = z \tag{5.1}$$
$$t' = t$$

If this is so, then, necessarily, a photon traveling in $\mathcal{F}$ along the x-axis, with constant speed $c = dx/dt$, travels in $\mathcal{F}'$ with constant speed

$$c' = \frac{dx'}{dt'} = \frac{d(x - vt)}{dt}\frac{dt}{dt'} = \frac{dx}{dt} - v = c - v \neq c \tag{5.2}$$

(For brevity's sake – here and in the rest of this chapter – I say 'photon' instead of 'light signal *in vacuo*'.[8]) Obviously, Einstein's two principles

[6] The general case can be reduced to this special case by performing a few, presumably unproblematic operations, viz., (i) transform whatever space coordinates each frame is endowed with into right-handed Cartesian coordinates; (ii) choose an instant of time and apply to one of the Cartesian systems a translation that takes its origin to the point which lies at that instant on the origin of the other Cartesian system; (iii) apply to one Cartesian system a rotation that brings its axes to point in the same directions as the axes of the other; (iv) add or subtract a constant to each value of the time coordinates so that both take the value 0 at the common origin of the Cartesian systems on the chosen instant; and (v) convert all coordinates to common measurement units.

[7] A Galileian transformation is any coordinate transformation that results from applying – in any order – one or more coordinate transformations chosen from the following sets: (i) the Galileian *boosts*, defined by eqns. (5.1); (ii) the space translations such as the one mentioned in (ii) of note 6; (iii) the space rotations such as that mentioned in (iii) of note 6; and (iv) the time translations described in (iv) of note 6. The Galileian transformations form a group, of which each one of the sets (i)–(iv) is a subgroup. Of course, Galileo Galilei never wrote down a Galileian transformation, and presumably did not even dream of them.

[8] The idea that electromagnetic radiation is emitted and absorbed in discrete quanta of energy – subsequently called 'photons' – is due to Einstein (1905i); but he did not

cannot be jointly true if the right scheme for the description of motion includes (5.1).

The transformation (5.1) involves essentially the quantity v, equal to the distance in $\mathcal{F}$ that any point of $\mathcal{F}'$ traverses in unit time. Consider the origin A′ of the primed Cartesian coordinate system. At time $t = 0$, it lies on the origin A of the unprimed system. So, if A′ passes by the point B in $\mathcal{F}$ at time $t = 1$, then v equals the distance from A to B. This statement makes no sense unless the time coordinate t associated with the unprimed frame $\mathcal{F}$ is defined at both A and B. Since B is arbitrary, there must be a definite way of setting up t throughout $\mathcal{F}$. Since $\mathcal{F}$ is, by definition, an inertial frame of reference, t must be such that a force-free particle traverses equal distances (in $\mathcal{F}$) in equal t-intervals. But this requirement leaves a wide latitude of choice. Suppose that it is satisfied by a time coordinate function t. Then, evidently, it is also satisfied by the coordinate function t^* given by

$$t^* = ax + by + cz + dt + k \qquad (5.3)$$

where a, b, c, d, and k are arbitrary constants.

The problem of defining a time coordinate for an inertial frame of reference is dealt with by Einstein in 1905r, §1.[9] He discusses it in picturesque, seemingly down-to-earth terms, as a matter of synchronizing distant clocks; but, of course, anyone trying to practice the "operational" method of definition that he proposes would fail to reach the *perfect* accuracy that he subsequently takes for granted (when the time coordinate thus set up is used in the equations of physics). An obvious method of distant clock synchronization is implicit in eqns. (5.1). If $t = t'$, clocks placed along the x-axis of $\mathcal{F}$ can be synchronized by comparing them with the clock at A′ as it passes by them; clocks placed along other axes through A can be set by a clock at the origin of other suitably moving frames, whose time coordinate also agrees with t by the appropriate analogues of eqns. (5.1). Since (5.1) precludes the joint validity of Einstein's principles, this method of defining the time coordinate function on the inertial frame $\mathcal{F}$ – known as *synchronization of distant clocks by clock transport* – begs the question. The

---

mingle it with his work on the Relativity Principle. I therefore use 'photon' in this chapter not as a term of art, but only as a useful abbreviation.

[9] As far as I know, James Thomson (1884) was the first to mention this problem in writing, but he glossed over it; cf. Chapter Two, note 16, and the text linked to it. It was later discussed by Poincaré (1898).

method proposed by Einstein is, at first blush, no less obvious than the former; indeed, the big surprise is that, as a matter of fact, they do not agree with each other (unless clock transport takes place at infinitely slow speed – cf. Eddington 1923, §§4, 11). The time coordinate t defined at point A of the inertial frame $\mathscr{F}$ is diffused through all space by means of rebounding photons. Let B be a point on $\mathscr{F}$ outside A. A photon is sent from A to B at time t_1. If the photon, after rebounding at B, returns to A at time t_3, the time t_2 at which it reaches B is, *by definition*,

$$t_2 = t_1 + \frac{1}{2}(t_3 - t_1) \qquad (5.4)$$

Thus the photon takes the same time to go from A to B and from B to A. Since B is arbitrary and ranges over all $\mathscr{F}$, the photons employed in setting up a time coordinate function in this way satisfy the Light Principle *by definition* – if indeed the procedure can be consistently performed. A coordinate function t defined by this method will henceforth be referred to as *Einstein time*. It must be understood that the references to time implicit in the Relativity Principle – when it talks about laws of *change* – and in the Light Principle – when it mentions light *velocity* – are references to Einstein time.

The blatantly conventional character of the definition of Einstein time took most physicists and philosophers by surprise and became the source of endless debate. I shall deal with some of it in §5.3.2. But before proceeding any further it is important to realize that, although the definition of Einstein time rests on an agreement, this could not work as expected if some things happened differently. As Einstein says, one assumes that such a definition "is possible, without contradiction, for any number of points" (1905r, p. 894), that is, that no inconsistencies will arise as points A and B range freely over space. Specifically, (i) Einstein time defined from A at B should agree with Einstein time defined from B at A; and (ii) if Einstein time is defined from A at B and C, and then again from B at C, both time coordinates should agree at C. Einstein does not add that (iii) Einstein times defined from A by sending out photons at different moments must agree with each another – but he presumably took it for granted. Evidently, such agreements cannot be secured by convention. I shall refer to them as 'the consistency conditions of Einstein time'. They can hold only if *all* photons obey the Light Principle, not just the one by means of which Einstein time is defined from a particular point at a particular moment.

Suppose then that the coordinate functions t and t' are Einstein times defined, respectively, for the inertial frames $\mathscr{F}$ and $\mathscr{F}'$, and that (x,y,z) and (x',y',z') are Cartesian systems for each frame. Einstein was able to prove, from his two principles, that the primed and unprimed coordinates are then related by the following transformation, called a *Lorentz boost*:[10]

$$x' = \frac{x - vt}{\sqrt{1 - \dfrac{v^2}{c^2}}}$$

$$y' = y$$

$$z' = z \qquad\qquad (5.5)$$

$$t' = \frac{t - \dfrac{vx}{c^2}}{\sqrt{1 - \dfrac{v^2}{c^2}}}$$

In his proof, Einstein assumes from the outset that the sought for transformation must be linear "because of the properties of homogeneity which we attribute to space and time" (1905r, p. 898). In fact, however, linearity follows from homogeneity only if the latter is a feature of the four-dimensional "space" compounded of space and time (i.e., the 4-manifold charted by the coordinate systems (x,y,z,t) and (x',y',z',t') – cf. §4.1.3), not just of space on one hand and of time on the other. This is interesting because it shows that Minkowski's reading of SR (§5.2) was already implicit in Einstein's first steps. Yet at this point he could still have done without it, for the linearity of the transformation

---

[10] A Lorentz transformation is any coordinate transformation that results from applying – in any order – one or more coordinate transformations chosen from the following two sets: (i) the Lorentz boosts, defined by eqns. (5.5), and (ii) the space rotations such as that mentioned in (iii) of note 6. The Lorentz transformations form a group, of which the Lorentz boosts are a subgroup. The Lorentz group, in turn, is a subgroup of the Poincaré group. A Poincaré transformation is any compound of Lorentz transformations and space and time translations, such as those described in (ii) and (iv) of note 6. The designation does justice to Poincaré's independent discovery of the transformation (5.5) and the major group to which it belongs (see note 12).

$(x,y,z,t) \mapsto (x',y',z',t')$ can also be inferred from Einstein's two principles (cf. Torretti 1983, pp. 75f.). Ignatowsky (1910, 1911) tried to
derive eqns. (5.5) from the Relativity Principle alone. However, to dis-
pense with the Light Principle he had to use the following

Principle of Reciprocity. If the inertial frame $\mathscr{F}'$ moves with velocity v
relatively to the inertial frame $\mathscr{F}$, the inertial frame $\mathscr{F}$ moves with velocity $-$v relatively to the inertial frame $\mathscr{F}'$.

Contrary to Ignatowsky's belief, the Principle of Reciprocity does not
follow from the Relativity Principle.[11]
    It follows at once from eqns. (5.5) that two events that happen at the
same time t cannot happen at the same time t' unless their x-
coordinate is the same. In other words, two events simultaneous in an
inertial frame of reference are also simultaneous in another frame
moving uniformly with respect to the former only if they take place on
a plane perpendicular to the direction of motion. Such *relativity of
simultaneity* flies in the face of ingrained commonsense representations
and is the source of a galling philosophical problem (see §5.3.4). Yet
without it the Light Principle and the Relativity Principle cannot be
jointly held. Consider a light wave emitted from A in every direction at
the one instant in which A coincides with A'. By the Light Principle, the
wave front takes up in $\mathscr{F}$ and $\mathscr{F}'$, at each subsequent instant, a sphere
centered at A and A', respectively. This would be absurd if each instantaneous location of the wave front in $\mathscr{F}$ were also an instantaneous location of the wave front in $\mathscr{F}'$. It is, however, perfectly possible if, due to
the frame dependence of simultaneity, the events that constitute the
wave front at a given time in $\mathscr{F}$ are not simultaneous in $\mathscr{F}'$ and hence do
not constitute the wave front at any time in $\mathscr{F}'$. The sphere in $\mathscr{F}$ centered
at A, which the wave front covers at some instant of Einstein time t, is

---

[11] Ignatowsky (1911, p. 5) argues thus for the Principle of Reciprocity: Let AB and $A'B'$
be rods of unit length at rest, respectively, in the inertial frames $\mathscr{F}$ and $\mathscr{F}'$. As $\mathscr{F}'$
moves, $A'B'$ slides along AB, so that B' passes A and B, respectively, at times t_0 and
t_1 of the unprimed system. Let t'_0 and t'_1 be the times of the primed system at which
A passes B' and A', respectively. Then, according to Ignatowsky, the Relativity Prin-
ciple requires that $\Delta t = t_1 - t_0 = t'_1 - t'_0 = \Delta t'$. Consequently, the speed $1/\Delta t$ of $\mathscr{F}'$ in
$\mathscr{F}$ equals the speed $1/\Delta t'$ of $\mathscr{F}$ in $\mathscr{F}'$, and the velocities of each frame relative to the
other can only differ in sign. But Ignatowsky overestimates the strength of the Rela-
tivity Principle. Relativity entails that Δt is *the same function* of $\Delta t'$ as $\Delta t'$ is of Δt,
not that this function is the *identity*.

certainly not a sphere in $\mathscr{F}'$ centered at A', but neither is it the shape of the wave front at any instant of Einstein time t'.

Equations (5.5) also entail the relativity (frame-dependence) of three physical quantities hitherto regarded as fundamental, viz., length, duration and inertial mass. This caused astonishment and much discussion. I deal with these issues in §§5.3.1, 5.3.3, and 5.3.5.

Equations (5.5) lead to a rule for the transformation of velocities – from one coordinate system to another – quite different from the simple classical law of addition used in eqn. (5.2). An object moving in $\mathscr{F}$ along the x-axis with velocity $u = dx/dt$, moves in $\mathscr{F}'$ along the x'-axis with velocity $u' = dx'/dt'$. But u' is not equal to $u - v$. By substituting from eqns. (5.5), we obtain after a short calculation:

$$u' = \frac{u - v}{1 - (uv/c^2)} \tag{5.6}$$

Note that, in consonance with the assumption that photons travel with the same speed in all inertial frames, eqn. (5.6) implies that if $u = c$, then $u' = c$ as well.

The Relativity Principle, as stated above, is a straightforward generalization to all laws of physics of Newton's principle of relativity embodied in Corollary V to his Laws of Motion. However, in combination with the Light Principle and with Einstein's introduction of a physically significant, frame-dependent, universal time coordinate, the Relativity Principle yields eqns. (5.5), which are incompatible with Newton's Laws of Motion. In this new context, the Relativity Principle translates into the following

Principle of the Lorentz Invariance of the Laws of Physics. The laws by which the states of physical systems change, expressed in terms of Cartesian coordinates and Einstein time for an inertial frame of reference, are invariant under Lorentz transformations.

Now, the Maxwell equations of electrodynamics automatically meet this requirement, as if by magic, if the electric and magnetic field vectors are transformed in a certain way.[12] But Newton's Laws of

---

[12] Einstein (1905r, §6) showed this for the Maxwell equations as formulated by Heinrich Hertz. The transformation of the electric and magnetic field components by the Lorentz boost (5.5) is given in eqns. (5.21). Continuing with notes 4 and 10, one might wish to say that what Poincaré (1905, 1906) and Einstein (1905r) discovered in the wake of Lorentz's groping efforts (1895, 1899, 1904) was the symmetry group

Motion – and his Law of Gravity – do not. So Einstein's principles brought about at once a deep revision of classical mechanics (and soon moved him to seek for a new account of gravitational phenomena).[13] Still, if the relative speed v of the inertial frames being considered is much smaller than the speed of light c, the Lorentz transformation (5.5) does not differ significantly from the Galilei transformation (5.1), under which Newton's laws are invariant. And the rule of transformation of velocities (5.6) practically agrees with the rule of addition u' $= u \pm v$, if both u and v are much less than c. So, whatever support accrued to Newtonian physics from experiments involving speeds much less than that of light stood ready to corroborate the new physics based on Einstein's principles.[14]

5.2   Minkowski's Spacetime

In lectures delivered on 5 November 1907 (Minkowski 1915) and 21 September 1908 (Minkowski 1909), the mathematician Hermann Minkowski proposed a cogent reading of Einstein's new physics that clarified its baffling features. He showed how to substitute for the *three*-dimensional space continuum and the *one*-dimensional time continuum with their separate metrics, presupposed by Newtonian physics, a *four*-dimensional continuum metrically structured in consonance with the physical behavior of photons and free material particles. Einstein's principles follow at once from this new geometry. Therefore, the experimental results that confirm these principles also support Minkowski's declaration that, "from now on, space for itself

---

of Maxwell electrodynamics, which is now usually called the Poincaré group. (Still, Minkowski 1909, p. 106n1, traces this discovery back to Voigt 1887.) But, whereas Einstein boldly proceeded to make this group into the symmetry group of nature, Poincaré squandered the glory of his discovery in a hopeless inquiry into the deformation of electrons in motion.

[13] Of course, the Maxwell equations cannot remain unscathed if mechanics is reformed, for, in the classical understanding of them, they concern the relations between electric and magnetic forces (in the Newtonian sense of 'force'). So, even if in SR the Maxwell equations retain their shape, they must change their meaning. I touch on this matter in passing in §5.3.5; see also Torretti (1983, pp. 108ff.).

[14] I should also mention that Einstein derived from his principles new, more accurate formulas for the optical effects hitherto explained by a partial aether drag. Indeed, after Einstein, twentieth-century physicists discarded the aether as lightheartedly as their nineteenth-century predecessors had jettisoned caloric.

and time for itself should completely reduce to shadows, and only a sort of union of both ought to retain autonomy" (1909, p. 104). For this union of space and time Minkowski used the term 'world', but today we normally call it 'spacetime'.

Before proceeding any further with the explanation of spacetime, I should emphasize that the idea of a four-dimensional continuum in which we live, move, and have our being comes a good deal closer to our experience than the traditional ideas of time separate from space and hence from motion, and of space separate from time and hence from all forms of change. Think of traffic on a motorway. One might, with considerable effort, imagine the different geometrical figures that, at successive instants, the vehicles cut in space; but the phenomenon one perceives is a single coherent spacetime flow. Still, Minkowski's idea of spacetime does not stem from a desire to grasp life's course less abstractly, but from a consideration of what is actually involved in the physicist's practice of describing the phenomena of motion by means of four coordinate functions, especially if these are related to inertial frames and to one another in the manner proposed by Einstein. The familiar Cartesian coordinate system (x,y) used in plane geometry is a mapping of the Euclidian plane onto the (structured) set of real number pairs $\mathbb{R}^2$. The coordinate systems (t,x,y,z) used in Newtonian and SR kinematics – combining a universal time coordinate with Cartesian space coordinates – are mappings onto the (structured) set of real number quadruples $\mathbb{R}^4$. But what are they mappings *of*? What is the domain of such kinematic coordinate systems? We can only think of it as a (structured) set of points, each one of which is the possible location of an instantaneous pointlike physical event. Minkowski showed that by adopting Einstein's principles and using his coordinates one automatically bestows a specific structure on this point set. A set of points endowed with this specific structure is called *Minkowski spacetime*. Following the standard practice, I refer to spacetime points as *events*.

The structure of Minkowski spacetime is a geometry in Klein's sense (§4.1.2) and also in Riemann's (§4.1.3). Klein's approach is more economical and in this case more fitting,[15] but only Riemann's can enable us to understand the transition from SR to GR and the conceptual relationship between these theories. I shall therefore base my presentation

[15] Klein (1911) greeted SR as a most welcome confirmation of his insight.

on both. Let me first show that spacetime is a 4-manifold. To see
this it is enough to recall the definition of n-manifold in §4.1.3 and to
note that any coordinate system that – like the primed and the
unprimed systems of §5.1 – combines Cartesian coordinates and
Einstein time tied to an inertial frame of reference is in effect a one–
one mapping of all spacetime onto $\mathbb{R}^4$. Therefore, a single coordinate
system of this type constitutes, all by itself, an *atlas* for spacetime
and determines – in the manner explained in Chapter Four, note 17 –
a maximal atlas, which in turn bestows on spacetime a definite manifold structure.

In agreement with the literature on differential geometry, I shall
henceforth use lowercase Roman letters such as x, y, z to designate different spacetime *charts*, that is, one–one mappings of an open region
of spacetime into or onto $\mathbb{R}^4$. If x is such a chart, it labels each event
P with four real numbers or coordinates that I shall designate by $x^0(P)$,
$x^1(P)$, $x^2(P)$, and $x^3(P)$. In this way, each coordinate function is denoted
by x^k (with the index – not exponent! – k ranging from 0 to 3). The
index 0 corresponds to the time coordinate, and the other indices to
the spatial coordinates.

In the remainder of this section I shall only use spacetime charts
of a special kind, which I call *Lorentz charts*. Just as Cartesian
coordinate systems are especially useful in Euclidian geometry, so
Lorentz charts are particularly well suited for describing relations in
Minkowski spacetime. A Lorentz chart combines Cartesian coordinates
and Einstein time, defined as in §5.1 for an inertial frame to which the
chart is said to be *adapted*. As is usual in classical physics, the coordinate values measure distances along the coordinate axes; but the measurement units are chosen in such a way that the speed of light c is one
unit of length per unit of time.[16] For this purpose we agree that all
Lorentz charts meet the following condition: The time unit is our
second (i.e., the duration of 9,192,631,770 periods of the radiation
corresponding to the transition between the two hyperfine levels of the
ground state of caesium-133), but the unit of length is the light-second
(i.e., the distance *in vacuo* traversed by a photon in one second), not
the all too human meter (defined since 1983 as 1/299,792,458 of the
said distance). Consider now the collection of *all* Lorentz charts x such

---

[16] If space and time are fused, lengths and time intervals must, of course, be regarded
as aspects of a single metrical quantity. Therefore, speeds are pure numbers. But first
we must work our way to the vantage point from which this will be obvious.

that (i) there is a unique event O at the origin of all coordinate axes (in other words, there is an O such that, for all x, $x^0(O) = x^1(O) = x^2(O) = x^3(O) = 0$); (ii) the time coordinate increases as time advances (for all x, and any two events X and Y in the history of a given particle, $x^0(X) > x^0(Y)$ if and only if X occurs later than Y); and (iii) the Cartesian coordinates form a standard, right-oriented system. This collection is obviously an atlas for spacetime. I call it the *standard Lorentz atlas based at* O and denote it by $\mathscr{A}_{\mathscr{L}(O)}$. Note that if the primed and the unprimed system of eqns. (5.5) meet the above condition on measurement units, they both belong to $\mathscr{A}_{\mathscr{L}(O)}$.

Let x and y belong to $\mathscr{A}_{\mathscr{L}(O)}$. x and y are one-one mappings of all spacetime onto $\mathbb{R}^4$, so they have inverses, x^{-1} and y^{-1}, that map $\mathbb{R}^4$ onto spacetime. The composite mapping $y \circ x^{-1}$ is therefore a one–one mapping of $\mathbb{R}^4$ onto $\mathbb{R}^4$, that is, a one–one correspondence between number quadruples. Consider the x-coordinates of an arbitrary event P. They form the number quadruple $\langle x^0(P), x^1(P), x^2(P), x^3(P) \rangle = x(P)$. The composite mapping $y \circ x^{-1}$ assigns to $x(P)$ the number quadruple $y(P) = \langle y^0(P), y^1(P), y^2(P), y^3(P) \rangle$ formed by the y-coordinates of P. Thus, $y \circ x^{-1}$ is the coordinate transformation that substitutes the y-coordinates for the x-coordinates. And, of course, $x \circ y^{-1}$, that is, the transformation that substitutes the x-coordinates for the y-coordinates, is the inverse of $y \circ x^{-1}$. If z is a third chart in $\mathscr{A}_{\mathscr{L}(O)}$, the composite mapping $(z \circ y^{-1}) \circ (y \circ x^{-1})$ is identical with the coordinate transformation $z \circ x^{-1}$. Thus, the set of all transformations $y \circ x^{-1}$, where x and y range over $\mathscr{A}_{\mathscr{L}(O)}$, forms a group, with the trivial transformation $x \circ x^{-1}$ as a neutral element (see §4.1.2). In the light of note 10, it should be clear that this group is a realization of what I called there the Lorentz group.[17] Consider now the composite mapping $y^{-1} \circ x$, where x and y belong to $\mathscr{A}_{\mathscr{L}(O)}$. $y^{-1} \circ x$ is a one–one mapping of spacetime onto itself, which assigns to each event P the one and only event Q whose y-coordinates are identical with the x-coordinates of P.[18] Such a mapping is called a *point transformation* (to distinguish it from coordinate transformations, which map number n-tuples to number n-tuples). The set of point transformations $(y^{-1} \circ x)$, where x and y range over $\mathscr{A}_{\mathscr{L}(O)}$, is of course another

---

[17] Let $\mathscr{A}_{\mathscr{P}}$ be the union of all atlases $\mathscr{A}_{\mathscr{L}(P)}$, where P ranges over all of spacetime. Then the set of all transformations $y \circ x^{-1}$, where x and y range over $\mathscr{A}_{\mathscr{P}}$, is a realization of the Poincaré group, as defined in note 10. One may naturally call $\mathscr{A}_{\mathscr{P}}$ the standard Poincaré atlas of spacetime.

[18] If $Q = y^{-1} \circ x(P)$, $y(Q) = x(P)$; therefore, $y^i(Q) = x^i(P)$ ($i = 0, 1, 2, 3$).

realization of the Lorentz group[19] acting on spacetime itself in the manner contemplated by Klein. In accordance with Klein's Erlangen program, the geometric structure of Minkowski spacetime consists of – or rests on – the properties and relations of events that are Lorentz-invariant.

One can easily verify that neither lengths nor durations remain invariant under the Lorentz boost (5.5). To simplify calculations, let us first rewrite eqns. (5.5) using the new nomenclature (x^0 for t, x^1 for x, etc.) and putting $c = 1$:

$$y^0 = \frac{x^0 - vx^1}{\sqrt{1 - (v)^2}}$$

$$y^1 = \frac{x^1 - vx^0}{\sqrt{1 - (v)^2}} \qquad (5.7)$$

$$y^2 = x^2$$

$$y^3 = x^3$$

(where a superscript right after a letter is an index, but a superscript after a right parenthesis is an exponent).

Consider two events, P and Q, at the origin of the Cartesian system (x^1, x^2, x^3). The time interval between them is given by $|x^0(\mathrm{P}) - x^0(\mathrm{Q})|$ in terms of the x-chart and by $|y^0(\mathrm{P}) - y^0(\mathrm{Q})|$ in terms of the y-chart. These two numbers are not equal, for

$$|y^0(\mathrm{P}) - y^0(\mathrm{Q})| = \frac{|x^0(\mathrm{P}) - x^0(\mathrm{Q}) - v(x^1(\mathrm{P}) - x^1(\mathrm{Q}))|}{\sqrt{1 - (v)^2}}$$

$$= \frac{|x^0(\mathrm{P}) - x^0(\mathrm{Q})|}{\sqrt{1 - (v)^2}} > |x^0(\mathrm{P}) - x^0(\mathrm{Q})| \qquad (5.8)$$

(unless $v = 0$).

Consider now a rod of length λ, at rest on the y-frame. Assume that it lies along the y^1-axis, with one end at $y^1 = 0$ and the other end at $y^1 = \lambda$. So, in the x-frame the rod moves with speed v in the direction along which it lies. We wish to ascertain the rod's length in the x-frame, that is, the distance between two simultaneous positions of the rod's ends on this frame. So let P and Q denote the events at which either end of the rod takes those positions. The distance in question is equal

---

[19] Let x and y range over $\mathcal{A}_{\mathscr{L}(\mathrm{O})}$. The mapping $(y \circ x^{-1}) \mapsto (y^{-1} \circ x)$ is a group isomorphism.

to the square root of $(x^1(Q) - x^1(P))^2 + (x^2(Q) - x^2(P))^2 + (x^3(Q) - x^3(P))^2$. Our assumptions concerning the position of the rod on the y-frame imply that $y^2(P) = y^2(Q)$ and $y^3(P) = y^3(Q)$. Hence, by virtue of eqns. (5.7), the distance on the x-frame between the positions of the rod ends at events P and Q equals $|x^1(Q) - x^1(P)|$. We have that

$$0 = y^1(P) = \frac{x^1(P) - vx^0(P)}{\sqrt{1-(v)^2}} \qquad \lambda = y^1(Q) = \frac{x^1(Q) - vx^0(Q)}{\sqrt{1-(v)^2}} \qquad (5.9)$$

Since $x^0(P) = x^0(Q)$ (for P and Q are simultaneous on the x-frame), on subtracting the first equation from the second we obtain:

$$|x^1(Q) - x^1(P)| = \lambda\sqrt{1-(v)^2} < \lambda \qquad (5.10)$$

(unless $v = 0$).

On the other hand, as the reader can easily verify, if P and Q are any two events, the quantity $\sigma(P,Q)$, expressed below in terms of chart x, is invariant under the Lorentz boost (5.7):[20]

$$
\begin{aligned}
\sigma(P,Q) &= (x^0(P) - x^0(Q))^2 \\
&\quad - \left((x^1(P) - x^1(Q))^2 + (x^2(P) - x^2(Q))^2 + (x^3(P) - x^3(Q))^2\right) \\
&= (x^0(P) - x^0(Q))^2 - \sum_{i=1}^{3}(x^i(P) - x^i(Q))^2
\end{aligned}
$$
$$(5.11)$$

Note that $\sigma(P,Q)$ is also invariant under spatial rotations; it is the difference between $(x^0(P) - x^0(Q))^2$, which is the squared time interval between events P and Q, and $\sum_{i=1}^{3}(x^i(P) - x^i(Q))^2$, which is the squared spatial distance between their respective locations, and spatial rotations preserve both these quantities. Since every Lorentz transformation is a product of boosts and space rotations (see note 10), it follows that $\sigma(P,Q)$ is Lorentz-invariant. I call $\sigma(P,Q)$ the *spacetime interval* between events P and Q and regard it as the fundamental invariant of Minkowski spacetime.

Since the *difference* between the frame-dependent quantities $(x^0(P) - x^0(Q))^2$ and $\sum_{i=1}^{3}(x^i(P) - x^i(Q))^2$ is Lorentz-invariant, so is the following classification of spacetime intervals:

---

[20] Ignore the second and third spatial coordinates and just prove – by using (5.7) – that

$$\left(x^0(P) - x^0(Q)\right)^2 - \left(x^1(P) - x^1(Q)\right)^2 = \left(y^0(P) - y^0(Q)\right)^2 - \left(y^1(P) - y^1(Q)\right)^2$$

(i) $\sigma(P,Q)$ is *timelike* if the temporal separation $|x^0(P) - x^0(Q)|$ prevails over the spatial distance $\sqrt{\Sigma_{i=1}^3 (x^i(P) - x^i(Q))^2}$, and $\sigma(P,Q) > 0$.

(ii) $\sigma(P,Q)$ is *spacelike* if the spatial distance $\sqrt{\Sigma_{i=1}^3 (x^i(P) - x^i(Q))^2}$ prevails over the temporal separation $|x^0(P) - x^0(Q)|$, and $\sigma(P,Q) < 0$.

(iii) $\sigma(P,Q)$ is *null* if the spatial distance $\sqrt{\Sigma_{i=1}^3 (x^i(P) - x^i(Q))^2}$ equals the temporal separation $|x^0(P) - x^0(Q)|$, and $\sigma(P,Q) = 0$.

Note that if $\sigma(P,Q)$ is null, P and Q could be two events in the history of a photon (or any object traveling with the speed of light).[21] Because of this, some authors call zero-valued spacetime intervals 'lightlike' instead of 'null'. On the other hand, if $\sigma(P,Q)$ is timelike, P and Q could be two events in the history of an inertial particle traveling with a velocity less than that of light; and if $\sigma(P,Q)$ is spacelike, P and Q could be two events in the history of an inertial particle traveling with a velocity greater than that of light.[22]

Pick an event P and consider all events X, such that $\sigma(P,X) = 0$. These are the events from which a photon received at P could have been sent, or at which a photon sent from P could be received. The set of events $\{X \mid \sigma(P,X) = 0\}$ is the *null-cone* (or *light-cone*) of P.[23] It comprises three mutually exclusive parts, viz., (*a*) P itself, (*b*) all events X from which a photon could be sent for reception at P, and (*c*) all events X at which a photon could be received from P. Part (*b*) is the *past* null-cone of P, and part (*c*) is the *future* null-cone of P. Parts (*b*) and (*c*) are three-

---

[21] Case (iii) obtains if and only if the ratio

$$\frac{\sqrt{\sum_{i=1}(x^i(P) - x^i(Q))^2}}{|x^0(P) - x^0(Q)|}$$

is equal to 1, which, remember, is the velocity of light in the units we are using.

[22] Faster-than-light particles are called *tachyons* (from the Greek ταχύς, 'fast'). Much has been written about tachyons, although there is not a whit of evidence that they exist. SR is not incompatible with them but imposes some constraints on their behavior. Tachyons could never be slowed down to the speed of an ordinary massive particle, nor ordinary massive particles accelerated to tachyon speed. If real numbers are employed, as usual, to express the inertial mass of ordinary particles, the inertial mass of tachyons must be expressed by multiples of $\sqrt{-1}$. It is generally agreed that, if tachyons existed, they could not be used to transmit information between ordinary particles.

[23] If we discard one spatial dimension and use our intuitive image of it to represent time, the set $\{X \mid \sigma(P,X) = 0\}$ covers a two-sheeted cone with its vertex at P. By analogy, the three-dimensional figure cut by $\{X \mid \sigma(P,X) = 0\}$ in four-dimensional spacetime is also called a *cone* or, more strictly, a *hypercone*.

dimensional submanifolds of four-dimensional spacetime (hypersur-faces, in mathematical jargon). Let $\kappa(P)$ denote the null-cone of P. Let Y and Z be events such that $\sigma(P,Y) > 0$ and $\sigma(P,Z) < 0$. Then, any smooth curve joining Y to Z intersects $\kappa(P)$.[24] Thus the null-cone $\kappa(P)$ separates two mutually exclusive regions of spacetime, constituted, respectively, by events Y such that $\sigma(P,Y)$ is timelike – which, we say, are *inside* $\kappa(P)$ – and by events Z, such that $\sigma(P,Z)$ is spacelike – which lie *outside* $\kappa(P)$.

The function σ assigns real numbers to event-pairs much like the ordinary distance function assigns numbers to point-pairs in space. Indeed, if $\sigma(P,Q)$ is spacelike, the distance between the spatial location of P and Q in an inertial frame in which both events are Einstein-simultaneous is given precisely by $\sqrt{\sigma(P,Q)}$. On the other hand, if $\sigma(P,Q)$ is timelike, and P and Q are events in the history of the same inertial particle, $\sqrt{\sigma(P,Q)}$ equals the time lapse between them in the rest frame of that particle.[25] But, in stark contrast with the ordinary distance func-tion, $\sigma(P,Q)$ – with $P \neq Q$ – can take negative and zero values, so it cannot be used for comparing timelike with spacelike intervals and is worthless as a measure of the separation between events on the same null-cone. Still, we could use the null-cones for distinguishing between spacelike, timelike, and null curves and then resort to $\sigma(P,Q)$ or $\sqrt{\sigma(P,Q)}$ for separately defining the length of timelike and of spacelike curves in Minkowski spacetime, in the way that the ordinary distance function is used for defining the length of curves in Euclidian 3-space. However, the same purpose is achieved in a neater and – with a view to GR – more illuminating manner by introducing the *Minkowski metric* η.

This is defined on the analogy of a Riemannian metric **g** (see §4.1.3, after eqn. (4.5)) but with one important difference: η assigns a bilin-ear function η_P to each event P, but η_P is not required to be positive definite and nondegenerate; therefore, if v is a tangent vector at P, dif-ferent from the zero-vector, it may happen that $\eta_P(v,v)$ is greater than,

---

[24] To prove this statement consider a curve γ such that $\gamma(0) = Y$ and $\gamma(1) = Z$. Let σ_P denote the function that assigns to each event X the real number $\sigma(P,X)$. σ_P is smooth. Therefore, the composite mapping $\sigma_P \circ \gamma$ is a smooth real-valued function of one real variable whose value at Y is $\sigma(P,Y) > 0$ and whose value at 1 is $\sigma(P,Z) < 0$. So there must exist a real number a, between 0 and 1, such that $\sigma_P \circ \gamma(a) = 0$. This implies that $\sigma(P,\gamma(a)) = 0$, so $\gamma(a)$ lies on $\kappa(P)$.

[25] Let P and Q be events in the history of an inertial particle, and let x be a Lorentz chart adapted to the rest frame of that particle. Then, if $P \neq Q$, $x^0(P) \neq x^0(Q)$, but $x^\mu(P) = x^\mu(Q)$ for $\mu = 1, 2, 3$.

equal to, or less than 0. The definition of η is made easy by the following basic fact: The Minkowski spacetime manifold admits global charts (i.e., charts defined on the whole manifold), for example, the Lorentz charts. Take any Lorentz chart x. x determines four curves through each event P, viz., the parametric curves through P of the four coordinate functions x^0, x^1, x^2, x^3.[26] Thus, the mere existence of x defines the assignment

$$P \mapsto \left\langle \left.\frac{\partial}{\partial x^0}\right|_P, \left.\frac{\partial}{\partial x^1}\right|_P, \left.\frac{\partial}{\partial x^2}\right|_P, \left.\frac{\partial}{\partial x^3}\right|_P \right\rangle \tag{5.12}$$

where $\partial/\partial x^i|_P$ denotes the vector tangent at P to the parametric curve of x^i through P ($0 \leq i \leq 3$). The four vectors are linearly independent and so constitute a basis of the tangent space at P, a so-called *tetrad* at P. The mapping (5.12) is a *tetrad field* on the spacetime manifold. By means of it, one can equate directions and orientations at different points of the manifold. Because of this feature, the Minkowski spacetime manifold is said to be *parallelizable*. Since every vector that is tangent to Minkowski spacetime is a linear combination of the four vectors in a local tetrad, the Minkowski metric η is fully determined by defining, for each event P, the value of η_P at each vector pair formed from the tetrad

$$\left\langle \left.\frac{\partial}{\partial x^0}\right|_P, \left.\frac{\partial}{\partial x^1}\right|_P, \left.\frac{\partial}{\partial x^2}\right|_P, \left.\frac{\partial}{\partial x^3}\right|_P \right\rangle$$

The definition is as follows:

$$\eta_P\left(\left.\frac{\partial}{\partial x^h}\right|_P, \left.\frac{\partial}{\partial x^k}\right|_P\right) = \eta_{hk} \tag{5.13}$$

where h and k range over $\{0,1,2,3\}$, $\eta_{00} = 1$, $\eta_{kk} = -1$ if $k > 0$, and $\eta_{hk} = 0$ if $h \neq k$. The metric η assigns a "length" of sorts to each vector v at P, on the analogy of Riemannian metrics. If $v = \Sigma_{i=0}^3 v^i \partial/\partial x^i|_P$.

$$\eta_P(v,v) = \sum_{h=0}^{3} \sum_{k=0}^{3} \eta_{hk} v^h v^k \tag{5.14}$$

---

[26] As I explained in Chapter Four, note 21, given a point P in a manifold $\mathfrak{M}$ and a coordinate function x^k, defined at P and whose values range over an interval I, there is a unique curve $\gamma: I \to \mathfrak{M}$ such that (i) P is the value of γ at $x^k(P)$, and (ii) for every other point Q in the range of γ we have that $Q = \gamma(x^k(Q))$ and $x^h(Q) = x^h(P)$ if $h \neq k$. γ is the parametric curve of x^k through P.

I note for future reference that the Riemann tensor defined from the Minkowski metric η is everywhere 0. Therefore, on the analogy of Euclidean space, the spacetime endowed with η is said to be *flat*.

In contrast with proper Riemannian metrics, the Minkowski metric η is indefinite and degenerate, and it is therefore dubbed *semi-Riemannian*. As noted above, for any given tangent vector v at P it may occur that:

(i) $\eta_P(v,v) > 0$, in which case we say that v is *timelike*; or
(ii) $\eta_P(v,v) < 0$, in which case we say that v is *spacelike*; or
(iii) $\eta_P(v,v) = 0$, in which case we say that v is *null* (or *lightlike*).

Through this tripartite classification of tangent vectors it is possible to classify all spacetime curves into four kinds. (On curves and their paths, see §4.1.3.) A curve γ through P is timelike, spacelike, or null at P if its tangent vector at P is timelike, spacelike, or null, respectively. γ is a *timelike* (*spacelike, null*) curve if it is timelike (spacelike, null) *everywhere*; otherwise, γ is said to be *mixed*. This classification is extended in a natural way to two- and three-dimensional submanifolds of spacetime; for example, a hypersurface is said to be spacelike if every vector tangent to it is spacelike. The distinction between timelike, spacelike, and null curves has a special physical significance. Ordinary massive particles can never attain the speed of light (see §5.3.5), so all events in their lives must lie on the path of a timelike curve (in the case of all known particles) or on that of a spacelike curve (in the case of tachyons, if such particles exist – see note 22). Null curves are reserved for massless objects, traveling with the speed of light. Henceforth, I shall use Minkowski's term 'worldline' for the curve – null or timelike – that tracks the life history of a photon or an ordinary massive point-particle. (Tachyons will be ignored.) With mild abuse of language I shall occasionally speak of the 'worldline' of a larger object, such as a clock or a planet, whose volume is irrelevant in the context.

If the curve γ is defined on the interval (a,b), a measure of its "length" is naturally obtained by substituting the metric η in the integrand of integral (4.3):

$$\int_a^b \eta_{\gamma(u)}(\dot{\gamma}(u), \dot{\gamma}(u)) \qquad (5.15)$$

Note that integral (5.15) is positive if γ is timelike, negative if γ is spacelike, and zero if γ is null. So this measure of "length" yields significant quantitative information only if it is used for comparing a timelike

curve with other timelike curves or a spacelike curve with other spacelike curves. Applied to null curves it yields the uninteresting information that they all have "length" zero. And, of course, it cannot be
applied significantly to mixed curves. Now, once it is clear that a set
of spacetime curves cannot be compared as to "length" unless they are
all timelike, or all spacelike, it is perhaps preferable to measure them
by the following integral, which comes closer to proper Riemannian
length (4.6):

$$\int_a^b \sqrt{|\eta_{\gamma(u)}(\dot{\gamma}(u), \dot{\gamma}(u))|}\, du \qquad\qquad (5.16)$$

Integral (5.16) is, of course, equal to 0 if γ is null, and it conveys no
information if γ is mixed. If γ is timelike, the "length" measured by
integral (5.16) is called *proper time*. It is a remarkable fact of life that
– to the extent that gravitational fields are uniform – atomic clocks
keep proper time along their respective worldlines.[27]

Due to the nature of the Minkowski metric, the classical definition
of geodesic as a curve of extremal length cannot be universally applied
to spacetime curves. However, Levi-Città (1917) showed how to
define geodesics without appealing to a concept of length. By his
definition, a geodesic is a curve whose tangent vectors are all parallel
to each other along it. A geodesic is therefore a curve of constant direction. The definition uses Levi-Città's notion of parallel transport of
vectors along a curve, which I cannot explain here. But no explanation
is necessary in the case of Minkowski spacetime, which, as we saw, is
a parallelizable manifold. Two vectors v and w at different points of
such a manifold are parallel – in Levi-Città's sense – along every line
joining those points if and only if they are parallel absolutely. This will
be the case if and only if they have proportional components relative
to the local values of a global tetrad field. To see this more clearly,
remember the tetrad field (5.12) determined by a Lorentz chart x. Let
$v = \Sigma_{i=0}^3 v^i \partial/\partial x^i|_P$ and $w = \Sigma_{i=0}^3 w^i \partial/\partial x^i|_Q$ be vectors at spacetime points P
and Q, respectively. v and w are parallel if and only if $v^i = kw^i$ for some

---

[27] If gravity's actual lack of uniformity is brought back into the picture, the time measured by an atomic clock along its worldline still agrees splendidly with eqn. (5.16),
provided that the GR metric is substituted for the Minkowski metric (see §5.4).
Neither SR nor GR contains the slightest hint that could help one understand this
agreement; but were it not for it, they would surely be less interesting as theories of
the physical world.

$k > 0$ ($i = 0$, 1, 2, 3). Hence, if v and w are parallel, $\eta_P(v,v) = k\eta_Q(w,w)$. Thus, a spacetime geodesic can be either spacelike, timelike, or null, but not mixed.

Let me note further that every Lorentz chart x maps geodesic paths onto straight lines in $\mathbb{R}^4$, and that its inverse x^{-1} maps straight lines in $\mathbb{R}^4$ onto geodesic paths. This means, in particular, that the worldlines of ordinary inertial particles and photons – which Lorentz charts evidently map onto straight lines – are the paths of timelike and null geodesics, respectively. Here we have spelled out the geometry of Minkowski spacetime – somewhat artificially – in terms of the Lorentz charts. From our present vantage point we realize that the Minkowski structure of null-cones and timelike and null geodesic paths can be built on a proper physical basis, supplied by the behavior of ordinary inertial particles and photons (see Ehlers, Pirani, and Schild 1972).

5.3 Philosophical Problems of Special Relativity

In the central portions of this section I shall discuss some difficulties raised by the seemingly arbitrary definition and the frame dependence of Einstein time (§§5.3.2, 5.3.3, 5.3.4). Then I shall turn to the question of conceptual change in physics, as illustrated by the contrast between Newtonian mass and relativistic mass (§5.3.5). I begin, however, with a problem that is more physical than philosophical (§5.3.1), which will prepare us for dealing with the difficulties of Einstein time.

5.3.1 The Length of a Moving Rod

Consider a rod, 10 meters long, moving with constant speed v along the x-axis of our inertial frame $\mathscr{F}$. To simplify calculations I put $(v)^2 = 0.75(c)^2$. By eqn. (5.10) the length of the rod in $\mathscr{F}$ is equal to $10\sqrt{1/4}$ meters = 5 meters, so it fits comfortably in a rectangular barn that is 10 meters long, standing in $\mathscr{F}$ along the x-axis. On the other hand, in the rod's own rest frame $\mathscr{F}'$, it is the barn that moves with speed v in the opposite direction, so the barn's length is 5 meters, while the rod's length is 10 meters. Thus, from this perspective, there is no way that the rod can fit into the barn. At first blush this is baffling, for one tends to think that a solid object either fits or does not fit into another, as a matter of frame-independent fact. However, we would not say that the rod fits into the barn, or that the barn surrounds the rod, unless there

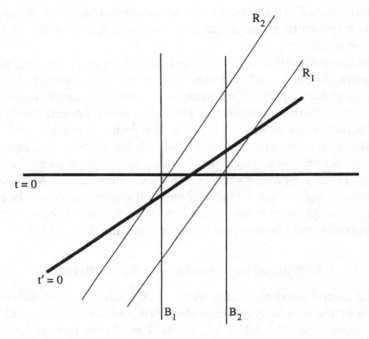

Figure 13

are two simultaneous positions of the barn and the rod such that the former lies within the latter. Now this is certainly the case if simultaneity is defined by Einstein's method in the barn's rest frame $\mathscr{F}$: There is a short interval of Einstein time adapted to $\mathscr{F}$ during which each successive position of the moving rod lies wholly inside the barn. But the situation is very different if simultaneity is defined by Einstein's method in the rod's rest frame $\mathscr{F}'$: There is no instant of Einstein time adapted to $\mathscr{F}'$ at which both ends of the rod are contained within the barn; when the forward end of the rod has already crossed the front and back entrances of the barn and is moving away from it, the rod's rear end still has not reached the barn and is advancing toward it. That this follows necessarily from the principles of SR can be proved by straightforward calculation, by substituting the appropriate data into eqns. (5.5) or (5.7). It can also be shown very easily if we represent the x-t spacetime plane on the page, mark the loci of Einstein simultaneous events on each frame, and draw the worldlines of the two ends of the rod and of each entrance to the barn. This is done in Fig. 13, where the lines R_1 and R_2 represent, respectively, the worldlines of the forward and the rear ends of the rod, while B_1 and B_2 represent, respectively, the worldline of a point at the front and the back entrances of the barn. The thick lines indicate the locus of events at time zero on the primed and the unprimed frames. The relative speed of the rod and the barn is represented by the angle between the B-lines and the R-lines, and also by the angle between the line $t = 0$ and the line $t' = 0$. To keep the line $t' = 0$ well distinguished from the R-lines, Fig. 13 represents a relative speed that is not as high as the one assumed in the text. Still, the drawing shows clearly that R_1 and R_2 cross the line $t = 0$ at points lying *between* the intersections of that line with B_1 and B_2, and they cross the line $t' = 0$ at points placed *outside* the intersections of B_1 and B_2 with this line.

5.3.2 Simultaneity in a Single Frame[28]

Einstein's Light Principle is curiously intertwined with his definition of a time coordinate function adapted to an inertial frame of reference. On the one hand, without *some* such function the Light Principle is meaningless. On the other hand, the Light Principle is true *only* under Einstein time. The situation closely resembles Lange's handling of Newton's First Law (§2.2). Lange used three free particles for defining an inertial frame and an inertial time scale. For these particles the First Law holds *by definition*. However, once the frame and its time scale are fixed, all other free particles bear witness to the First Law's validity. Likewise, Einstein picks a swarm of bouncing photons issuing at a particular instant from a point P in an inertial frame. His definition of time implies that *these* photons move with the same speed in every direction. But once Einstein time is fixed in an inertial frame, *all other* photons bear witness to the validity of the Light Principle in that frame.

To devise a scheme of description through which the universal facts of nature shine splendidly is perhaps the most telling sign of scientific genius. But many philosophers and scientists would rather draw a neat line between hard, dependable facts and soft, whimsical conventions. So some philosophers – notably Reichenbach (1924, §7; 1928, §19) – have come up with the notion that the Light Principle, insofar as it

---

[28] For a recent, lucid, and fair discussion of this subject, with references to the literature, see Redhead (1993).

makes a factual claim, can only refer to what they call the "two-way speed" of light – that is, the average speed with which a radar signal travels *in vacuo* from its source to a reflecting body and back to its source again –, whereas Einstein's application of the Light Principle to the "one-way speed" of light, involving as it does the definition of Einstein time, is purely conventional. Reichenbach treats Einstein time as a particular case in an infinite family of admissible time coordinates, each one of which assigns a different pair of values to the one-way speed of light to and from a given point in space.

Of course, no amount of philosophical argumentation can do away with the fact that the constant c that occurs in the equations of physics designates an ordinary, one-way speed.[29] Disturbed by the idea that this fundamental constant of nature could be fixed by convention, other authors have racked their brains in search for ways of measuring the one-way speed of light that would not presuppose a universal time coordinate. Unfortunately, any such procedure must at some point resort openly or furtively to synchronized distant clocks,[30] so their proposals have about as great a chance of success as the familiar attempts at squaring the circle. To expose the fallacy, proposals of this type must be patiently scrutinized, preferably by the editors of the professional journals to which they are regularly submitted. The following remark might be helpful: As I noted above, one meter is, by definition, 1/299,792,458 of the distance traveled by a photon in one second.[31] Therefore, no conceivable instrument can measure the one-way speed of light as anything else than exactly 299,792,458 meters per second. Under these circumstances, any attempt to measure the one-way speed of light c without relying on synchronized distant clocks would, if successful, do no more than establish experimentally that the distance trav-

---

[29] Indeed, the very idea of a "two-way speed" evinces insensitivity to one of the great achievements of modern mathematical physics, viz., the conception of motion as a state of the moving body, fully actual at any instant (see in §1.3 the quotation from Descartes (1644, II §39), and the remarks that follow).

[30] A good example of this is Rømer's method for measuring the speed of light, which was discussed in §1.5.3. To calculate the one-way speed of light c between Jupiter's satellite Io and the Earth, one needs to use the speed u with which the Earth travels toward Io or away from it between successive eclipses of the latter. Any estimate of u involves a comparison of time at the different places from which the said eclipses are observed.

[31] This convention is not a whim, but the outcome of a vast experience with other standards that proved to be harder to reproduce in a reliably stable way.

eled by a photon in a certain direction is *approximately* equal – within the limit of admissible experimental error – to the distance traveled by it over the same path in the opposite direction. But this is ridiculous, for, by dint of a universal linguistic convention, the distance between two points in space is *exactly* the same, no matter in what direction you travel it.

The definition of the meter ratifies Reichenbach's view of the one-way speed of light as conventional but also undercuts his claim that its two-way speed is factual. With a single clock affixed to a radar transmitter at rest on an inertial frame one can measure the *distance* from it to any radar-reflecting post in the frame, not the *average speed* with which the signal travels to and fro. To be precise, one measures the length of the two-way trip; that this is exactly twice the length of the one-way trip is a consequence of the age-old convention noted above: 'Distance' is a *symmetric* real-valued function on point-pairs. Moreover, if the one-way speed of light is by definition the same everywhere and in every direction, a photon takes exactly the same time in traveling back and forth between two points at rest on an inertial frame. Evidently, Einstein time is the only time coordinate function that agrees with this.

Reichenbach wrote about time coordinates and the speed of light at a time when the meter was identified with the distance between two marks on a metal rod kept at Sèvres. Although his concepts are out of step with today's metrology, they still deserve attention. He concentrates on the definition of simultaneity between distant events. Let a photon be issued from A to B, where it is reflected back toward A. Let E_1 be the emission of the photon and E_3 its final reception at A. What event at A shall we judge simultaneous with the bouncing of the photon at B? Reichenbach says that *any* event E_2 will do, provided that it comes after E_1 and before E_3. This condition is satisfied if

$$t(E_2) = t(E_1) + \varepsilon(t(E_3) - t(E_1)) \tag{5.17}$$

where t assigns to events at A their time as given by a standard clock at that point, and ε is *any* positive real number less than 1. By means of eqn. (5.17), every moment of time t is – in Newton's phrase – "diffused indivisibly throughout all spaces", as point B ranges over them. For some unknown reason, when Reichenbach formally introduced eqn. (5.17) he postulated that "ε is an arbitrary factor which, however, must be equally large for all points B" (1924, §7, Def. 2; my notation). By *Reichenbach time* I mean a universal time coordinate defined in this way. According to Reichenbach, Einstein time is just a special case of Reichenbach time, in which $\varepsilon = \frac{1}{2}$. Although eqn. (5.4) might seem to indicate this, it is not true without further qualification. Einstein's definition presupposes that A and B are points at rest on the same inertial frame, a requirement that Reichenbach does not bother to mention (so one is free to wonder how his definition of simultaneity might work if A and B stood on the rim of a rotating disk). Moreover, Einstein assumes that the universal time coordinate defined by his method is an inertial time scale, so that free material particles traverse equal distances in equal times as measured by it; but if ε takes the same value in every direction issuing from A, the universal time coordinate defined by eqn. (5.17) cannot be an inertial time scale unless $\varepsilon = \frac{1}{2}$.[32] In other words, Einstein time as given by eqn. (5.4) is an instance of Reichenbach time as given by eqn. (5.17) only if Reichenbach time is so constrained that Einstein time is its sole instance.

The relevance of Reichenbach time to SR physics can somehow be rescued if, following Reichenbach's own suggestion (1928, §26), we forget his earlier demand that ε be fixed and define it as a continuous function of direction that (a) takes in each direction about A a value equal to 1 minus the value it takes in the opposite direction, (b) ranges between a maximal value $\varepsilon_{max} < 1$ and a minimal value $\varepsilon_{min} = 1 - \varepsilon_{max} > 0$, and (c) satisfies the condition $\varepsilon = \frac{1}{2}$ in every direction perpendicular to that in which $\varepsilon = \varepsilon_{max}$. Let us say that a universal time coordinate that constitutes an inertial time scale, defined as in eqn. (5.17), with A and B affixed to the same inertial frame and ε subject to (a), (b), and (c), is a *modified Reichenbach time*. Given the facts of nature that secure the consistency conditions of Einstein time, a modified Reichenbach time can also be defined consistently. This is no wonder, for if ε varies with direction in the manner I have just described, the modified Reichenbach time associated by eqn. (5.17) to the inertial frame $\mathcal{F}$ in which A rests agrees precisely with the Einstein time associated by eqn. (5.4) to an inertial frame $\mathcal{F}'$ that moves relative to $\mathcal{F}$, in the direction in which $\varepsilon = \varepsilon_{max}$, with a definite speed v depending on ε_{max}.[33] Thus, except in the case

[32] For a proof of this statement, see Torretti 1983, pp. 224f.
[33] Here is a proof-sketch. To make things clearer I describe any figure F in spacetime by the same geometric terms – for example, 'straight', 'plane', 'hyperplane' – which apply to the image of F by a Lorentz chart. Pick an event P. The condition $0 < \varepsilon < 1$ implies that any other event that is Reichenbach-simultaneous with P lies outside the null-cone $\kappa(P)$. Let y be a spacetime chart combining a modified Reichenbach time

that ε = const. = $\frac{1}{2}$, modified Reichenbach time is, if I may say so, mal-adapted Einstein time.[34]

5.3.3   Twins Who Differ in Age

Perhaps the most picturesque expression of conceptual novelty in SR is the so-called paradox of twins. Jack and Jill are born on earth in quick succession from the same mother. At an early age, Jill is placed on a spaceship in which she tours the galaxy at enormous speed, say, $0.9c$. When she returns, a handsome woman in her thirties, Jack is a decrepit octogenarian. This is not what one would have expected, not, at any rate, before Einstein. But then we have never met any woman traveling at such speeds, so why should her aging process meet our expectations? The worldlines of Jack and Jill are not the same, and proper time as measured by eqn. (5.16) is a good deal shorter along Jill's. So, if the twins' biological clocks run more or less like atomic clocks, Jack must be a good deal older than his sister when she returns.[35]

coordinate y^0 with Cartesian coordinates y^1, y^2, y^3, all adapted to the same inertial frame $\mathscr{F}$. The sets of Reichenbach-simultaneous events $\{X \mid y^0(X) = \text{const.}\}$ are flat 3-manifolds, i.e., hyperplanes. Let us say that a hyperplane in spacetime is spacelike if the spacetime interval σ between any two events on it is spacelike. Then, each hyperplane y^0 = const. is spacelike. Each family of parallel spacelike hyperplanes corresponds precisely to a particular partition of spacetime into classes of Einstein simultaneous events. Consequently, the modified Reichenbach time coordinate y^0 is identical with the Einstein time coordinate x^0 of a particular Lorentz chart x. Of course x is not adapted to $\mathscr{F}$ but to some other inertial frame $\mathscr{F}'$ in which $\mathscr{F}$ moves with velocity v. If r_{max} denotes a unit vector in the direction in which $\varepsilon = \varepsilon_{max}$,

$$v = \tanh(\text{arc tan } |1 - 2\varepsilon_{max}|)r_{max}.$$

[34] Someone who does not care for historical precedence might state this identity the other way around: Each Einstein time coordinate x^0 is none other than the one modified Reichenbach time coordinate that is well adapted to the inertial frame for which it is defined. I say that this time coordinate is "well adapted" because it does not require for its everyday use that one keep the bearings of a fixed direction in space, a practically impossible thing to do given the spatial isotropy of inertial frames.

[35] That atomic clocks behave like our twins was shown by Hafele and Keating (1972). Thirty years earlier, Rossi and Hall (1941) studied short-lived fast particles generated in the upper atmosphere by cosmic rays, and showed that their decay rate, as measured by earthbound clocks, was – in good agreement with SR – less than it would be if the particles were at rest on the earth.

Obviously, this could not be true if biological clocks kept absolute universal Newtonian time, but nobody claims that they do. Still, in the literature the twins' story has taken an air of paradox due to misunderstandings. Thus, it was argued that if Jack ages sooner than Jill when events are referred to his rest frame, Jill must age sooner than Jack if events are referred to hers. After all, SR is a theory of relativity, in which any inertial frame may be considered to be at rest. This argument overlooks that Jack's life is not a mirror image of Jill's. Suppose, for simplicity's sake, that he stays all the time in a single inertial frame; then Jill must sit in at least two such frames as she speeds away and then returns. Others contend that, due to this asymmetry, the twin paradox lies outside SR's ken, for Jill must undergo an acceleration that only GR can describe. But, as we saw in §5.2, Minkowski geometry assigns lengths to timelike curves of any shape, not just geodesics. Since only the latter represent the spacetime tracks of ordinary massive inertial particles, the often heard remark that SR kinematics is unable to cope with accelerated motion is groundless.

However, we can disregard Jill's acceleration if we suppose that she travels successively in two inertial spaceships and jumps instantaneously from one to the other. This violent exercise may cause her to look older than she is, but her chronological age, as measured by the atomic clocks that surround her, will anyway be much less than Jack's when they come together. Let x, y, and z be Lorentz charts adapted, respectively, to Jack's inertial frame, to Jill's as she goes away, and to Jill's as she returns. Let Jack stand all the time at the spatial origin of chart x (so that, for any event E in his life, $x^1(E) = x^2(E) = x^3(E) = 0$). Put $c = 1$ and let Jill travel on the plane $x^2 = x^3 = 0$ with speed 0.9. To make the numbers smaller, I express time in years and distance in light-years. Jack and Jill separate at P and reunite at Q. Let J denote Jill's jump, midway between P and Q, from the outward bound to the inward bound spaceship. For later reference, I designate by G and H the events in Jack's life that are simultaneous with J by chart y and by chart z, respectively (see Fig. 14.)

If Jack lives on earth 80 years while Jill is traveling, $x^0(Q) - x^0(P) = 80$, $x^0(J) = 40$, and $x^1(J) = 36$. Proper time along Jill's worldline amounts to $y^0(J) - y^0(P)$ plus $z^0(Q) - z^0(J)$. These values can be obtained by straightforward calculation after making the appropriate substitutions in eqn. (5.8):

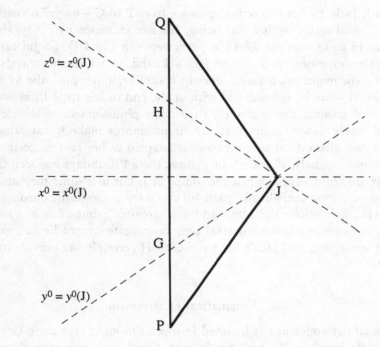

Figure 14

$$y^0(J) - y^0(P) = \frac{x^0(J) - x^0(P) - v(x^1(J) - x^1(P))}{\sqrt{1 - (v)^2}}$$

$$= \frac{40 - 0.9(36)}{\sqrt{1 - (0.9)^2}} = \frac{7.6}{0.43589} = 17.4356$$

(5.18)

$$z^0(Q) - z^0(J) = \frac{x^0(Q) - x^0(J) + v(x^1(Q) - x^1(J))}{\sqrt{1 - (-v)^2}}$$

$$= \frac{40 + 0.9(-36)}{\sqrt{1 - (-0.9)^2}} = \frac{7.6}{0.43589} = 17.4356$$

So $y^0(J) - y^0(P) + z^0(Q) - z^0(J) = 34.8712$, and Jill has lived less than 35 years while her brother lives 80.

Another seeming paradox might irk us if we allow Jill to keep track of Jack's aging in her calendar. In Jill's own rest frame Jack's birthdays occur less frequently than hers. Thus, in the 17.4356 years between

P and J, Jack, by her reckoning, grows – from P to G – only 7.6 years older.[36] And again by her reckoning, his age increases by 7.6 years – from H to Q – in the 17.4356 years between J and Q. So Jill can celebrate no more than 15 of Jack's birthdays while she travels. When – she might ask herself – does he complete, in her time, the 64.8 additional years he is burdened with at the end of her trip? If, as we may well assume, she is cleverer than some philosophers of science, it will surely dawn on her that her circumstances make it advisable to use two methods of time-reckoning, adapted to her two successive rest frames, namely, y^0 and z^0. In y^0-time, the 64 birthdays between G and H are still to come when she jumps at J; but in z^0-time they are, at that very moment, already past. Jill could try a composite chronology, viz., "y^0 before the jump and z^0 thereafter"; but such a non-Einstein, noninertial, nonuniversal time coordinate cannot be defined on all spacetime, and Jack's life from G to H certainly lies outside its domain.[37]

5.3.4   Kinematical Determinism

Universal determinism has haunted European thought ever since Leucippus, the inventor of atoms, proclaimed: "Nothing occurs at random; but everything for a reason and by necessity" (DK 67B2). Some Greek philosophers argued for determinism from logic: Since every declarative sentence is either true or false, every fact – past, present, or future – must be determined once and for all.[38] But in modern times, determinism has normally been based on dynamics (see the end of §2.5.3). It is the privilege of our postmodern age to have produced an argument for universal determinism on purely kinematical grounds, apart from any conception of dynamics that might be required to enforce it.

The argument, based on Einstein's SR scheme for the description of motion, was independently put forward by Rietdijk (1966; cf. his

---

[36] To verify this figure, substitute G for P and P for Q in eqn. (5.8). Remember that $x^1(G) = x^1(P) = 0$. So $x^0(G) - x^0(P) = 17.4356 \times 0.43589 = 7.6$.

[37] My handling of Jill's time-keeping was prompted by Debs and Redhead (1996), but I am not certain that it agrees with their approach.

[38] Consider the sentence: 'On 4 July 2076, abortion will be illegal in the United States'. If this is either true or false, there is a fact of the matter concerning the legality of abortion in the United States in 2076 that is already determined, although unknown to us and to the justices and legislators – still unborn perhaps – who will be responsible for it.

1976) and by Putnam (1967). It rests on the following unquestionable facts of SR: Suppose that P and Q are two events in the worldline of an inertial particle, such that Q lies in the future light cone of P, and let x be a Lorentz chart adapted to the particle's inertial frame; then there is

(i)   an event E, such that $x^0(P) = x^0(E)$, and
(ii) a Lorentz chart y, adapted to a different inertial frame, such that $y^0(E) = y^0(Q)$.[39]

Now if P is present, and therefore fully determinate, in some inertial frame, and E is simultaneous with P in that frame, E is just as present as P in that frame and therefore no less determinate; but then, if Q is simultaneous with E in another frame, it is no less present than E in that frame and therefore quite as determinate. Note that in this argument, as I present it,[40] 'X is determinate' is a frame-independent predicate, as it well may be, but 'X is present' and 'X is simultaneous with Y' are frame-dependent, as required by SR. But is it right to assume that two events must be equally determinate (absolutely) merely because they are simultaneously present (in some frame or another)? This is no doubt so in the natural philosophy of Kant, in which events earn the same place in time through their thoroughgoing mutual determination (§3.4.4). But in SR it is just the opposite: Two events can be simultaneous (in a frame) if and only if (absolutely) they do not influence each other.

The story of Jack and Jill in §5.3.3 can throw some light on the issue at hand. In the frame *from* which Jill jumps at J, her jump is simultaneous with event G in Jack's life; but in the frame *to* which she jumps,

---

[39] This can be verified by putting numbers into eqns. (5.7). Let P and Q occur at the spatial origin of chart x, adapted to frame $\mathcal{F}$. and let E occur at the spatial origin of chart y, adapted to frame $\mathcal{F}'$ moving in $\mathcal{F}$ with speed 0.9 along the x^1-axis, in the direction of decreasing x^1. Put $x^1(E) = 900$ and $x^2(E) = x^3(E) = 0$. Under these conditions, if P and E are x-simultaneous and x and y are related by the Lorentz boost (5.7), $x^0(P) = x^0(E) = -1{,}000$ and $y^0(E) = -190$. But then $y^0(Q) = y^0(E)$, provided that $x^0(Q) = -82.82 > -1{,}000 = x^0(P)$. You can adjust this result to any pair of events P and Q by rescaling the Lorentz charts (i.e., by changing the unit of time).

[40] I give my own presentation, rather than follow Rietdijk or Putnam, because they use more words than are necessary, including some that do not speak to their credit, as when Putnam refers to "the coordinate system of x", where x is an event (1979, p. 200, line 15 from the bottom), or when Rietdijk mentions two observers "experiencing the same 'present'" while they rest at separate points on the same inertial frame (1966, p. 342). See also Stein (1991).

the jump is simultaneous with H. Jill's jump, completed in an instant, is surely determinate there and then. Does this imply that H is determinate when G is? Remember that Jack is 7.6 years old at G and 72.4 years old at H. It is ridiculous to expect that from the purely kinematic information provided Jack can infer at G that he will – or will not – live until H. Neither can Jill infer it at J; indeed, as she jumps she cannot even know that Jack is alive at G.

Kinematic determinists cannot claim that the information required for kinematic description in SR is also sufficient for determining the course of events in full. So their claim must amount to this: SR kinematic description is not possible unless all events are fully determined at all times.[41] This, in my view, groundless claim stems perhaps from a confusion between the ordinary and the relativistic sense of 'event'. In the jargon of Relativity, 'event' is short for 'spacetime point'. These abstract events should not be confused with the events of real life, nor with their idealized version, the pointlike yet concrete events of physics. Spacetime points are, of course, fully determined by the spacetime structure (Minkowskian or otherwise); indeed, were they not fully individuated, they would not be available as arguments for spacetime charts (Lorentzian or otherwise).[42] The abstract events P, Q, J, G, and H in Fig. 14 must be given from the outset or the kinematic relations that we have assumed between them could not be specified. But this says nothing about the concrete events at P and Q, J, G and H, viz., the twins' separation and reunion, Jill's jump and the contemporary happenings in Jack's life. A concrete event is what it is, where, and when it is. Through the laws of physics one can infer at least some of its features from other concrete events on or inside its null-cone. But, according to SR, nothing that happens outside an event's null-cone can be a necessary or a sufficient condition of any concrete aspect of that event. Thus, it is possible to argue – on the strength of some relativis-

---

[41] Full determination would be the job of natural forces. In a similar vein, the Stoic philosopher Chrysippus did not just let logical determinism stand by itself but invoked it as a proof of dynamical determinism. "For Chrysippus argues thus: 'If there is motion without a cause, not every proposition [. . .] will be either true or false, since anything lacking efficient causes will be neither true nor false. But every proposition is either true or false. Therefore, there is no motion without a cause. If this is so, everything that happens happens through antecedent causes. If this is so, everything happens by fate. Hence everything that happens happens by fate'" (Cicero, De fato, 20–21).

[42] On individuation by structure see Newton in Hall and Hall (1962, p. 103) (quoted in §2.2).

tic theory of dynamics – from the determination of an event P to that of another event Q on a worldline through P; but one cannot buttress up such an argument by referring to a third event E that lies outside the null-cones of both P and Q.

5.3.5 The Quantities We Call 'Mass'

Newton's Laws of Motion do not meet the requirement of Lorentz invariance and are therefore incompatible with SR.[43] So Einstein had to develop a new mechanics. However, in stark contrast with Galileo and Newton, he did not try his best to forget the received view and to build from scratch a new physics of motion. Instead he reached for equations of motion that take the Newtonian form at the low speeds to which Newtonian mechanics had hitherto been successfully applied and significantly diverge from it only at speeds closer to the speed of light. This approach was reasonable but not inevitable. Evidently, the new mechanics ought to yield the well-corroborated predictions of the old, but their agreement need only be numerical and concern just the experimental results. The new theory was therefore free to refer to those results with new terms, embedded in new equations that do not collapse to the earlier ones when one deletes every term that tends to zero together with speed. Einstein exercised this kind of freedom a few years later, when he produced a theory of gravitation constrained to agree numerically with Newton's in the weak-field low-speed region but which differs from it drastically even in typography. But in 1905 his new Lorentz invariant mechanics was formulated as a variation of the old. Success has proved him right. Yet by using, say, the same old word 'mass' or writing m (actually, μ) in his equations, he fostered the illusion that he was merely proposing new laws for the familiar quantity that Newton named thus, when he was in fact replacing it with new quantities that were deeply at variance with it.

Einstein (1905r, §10) discusses the motion of a charged particle in an electromagnetic field. Let x, y, z be Cartesian coordinates and t

---

[43] See §5.1, at the end. Of course, the requirement of Lorentz invariance holds only for physical laws referred to a Lorentz chart. Therefore, the requirement cannot properly apply to Newton's laws, for the time variable that occurs in them is not Einstein time. However, the incompatibility of SR with Newton's Laws of Motion can be restated more accurately as follows: In the light of Einstein's criticism, the time variable in Newton's Laws is meaningless; if one charitably replaces it with Einstein time, the Laws thus refurbished are not Lorentz invariant.

Einstein time, adapted to an inertial frame $\mathcal{F}$. I denote by E_x, E_y, and E_z the electric field components relative to this coordinate system, and I denote the magnetic field components by B_x, B_y and B_z. Let m be the mass and q the electric charge of a particle momentarily at rest in $\mathcal{F}$. Then, says Einstein, "in the next small bit of time" ("im nächsten Zeit-teilchen") the particle obeys the equations of motion:

$$m\frac{\mathrm{d}^2 x}{\mathrm{d}t^2} = qE_x \qquad m\frac{\mathrm{d}^2 y}{\mathrm{d}t^2} = qE_y \qquad m\frac{\mathrm{d}^2 z}{\mathrm{d}t^2} = qE_z \qquad (5.19)$$

Equations (5.19) combine Newton's Second Law $m\dot{\mathbf{v}} = \mathbf{f}$ (eqn. (2.1)), with the electrodynamic "Lorentz force" law, $\mathbf{f} = q(\mathbf{E} + (\mathbf{v} \times \mathbf{B}))$ (cf. eqn. (7.3)). This move is admissible, despite all that has been said about the incompatibility of Newtonian mechanics with SR, because "in the next small bit of time" the particle is still moving very slowly in frame $\mathcal{F}$, so the SR-compatible laws of mechanics, whatever they might be, must agree well with Newton's Second Law in this case. Now consider a time in which the particle moves in $\mathcal{F}$ with instantaneous speed v along the x-axis in the direction of increasing x. There is an inertial frame in which the particle's instantaneous speed is 0. Clearly this frame in which our particle is momentarily at rest stands to $\mathcal{F}$ in the same relation as the frame we called $\mathcal{F}'$ in §5.1. Once again we employ the primed letters x', y', z', and t' to denote a system of Cartesian coordinates and Einstein time adapted to $\mathcal{F}'$, which share a common origin and spatial orientation with the unprimed coordinates. Let $E'_{x'}$, $E'_{y'}$, $E'_{z'}$ and $B'_{x'}$, $B'_{y'}$, $B'_{z'}$ denote, respectively, the electric and the magnetic field components relative to the primed system. In the immediately ensuing very short time interval, the particle obeys the equations of motion

$$m\frac{\mathrm{d}^2 x'}{\mathrm{d}t'^2} = qE'_{x'} \qquad m\frac{\mathrm{d}^2 y'}{\mathrm{d}t'^2} = qE'_{y'} \qquad m\frac{\mathrm{d}^2 z'}{\mathrm{d}t'^2} = qE'_{z'} \qquad (5.20)$$

As we know, eqns. (5.20) are not Lorentz invariant. What shape do they take when referred to the unprimed coordinate system? This depends on the way the electric field components transform under the Lorentz boost (5.5). The Lorentz invariance of Hertz's version of the Maxwell equations is secured if the electric and magnetic field components relative to the primed and unprimed coordinate systems are linked to one another as follows:

$$E'_{x'} = E_x \qquad\qquad B'_{x'} = B_x$$

$$E'_{y'} = \frac{E_y - \dfrac{v}{c} B_y}{\sqrt{1 - \left(\dfrac{v}{c}\right)^2}} \qquad B'_{y'} = \frac{B_y + \dfrac{v}{c} E_y}{\sqrt{1 - \left(\dfrac{v}{c}\right)^2}}$$

$$\text{(5.21)}$$

$$E'_{z'} = \frac{E_z + \dfrac{v}{c} B_z}{\sqrt{1 - \left(\dfrac{v}{c}\right)^2}} \qquad B'_{z'} = \frac{B_z - \dfrac{v}{c} E_z}{\sqrt{1 - \left(\dfrac{v}{c}\right)^2}}$$

Substituting from eqns. (5.5) and (5.21) in (5.20) and reshuffling terms, we obtain:

$$\frac{m}{\left(\sqrt{1 - \left(\dfrac{v}{c}\right)^2}\right)^3} \frac{d^2 x}{dt^2} = qE_x = qE'_{x'}$$

$$\frac{m}{1 - \left(\dfrac{v}{c}\right)^2} \frac{d^2 y}{dt^2} = \frac{q\left(E_y - \dfrac{v}{c} B_y\right)}{\sqrt{1 - \left(\dfrac{v}{c}\right)^2}} = qE'_{y'} \qquad \text{(5.22)}$$

$$\frac{m}{1 - \left(\dfrac{v}{c}\right)^2} \frac{d^2 z}{dt^2} = \frac{q\left(E_z + \dfrac{v}{c} B_z\right)}{\sqrt{1 - \left(\dfrac{v}{c}\right)^2}} = qE'_{z'}$$

Einstein (1905r, p. 919) reminds us that $qE'_{x'}$, $qE'_{y'}$, and $qE'_{z'}$ are "the components of the ponderomotive force acting upon" the charged particle, referred to an inertial system that – at the moment considered – moves with the same velocity as the particle. The said components can be measured, for instance, with a spring dynamometer at rest in $\mathscr{F}'$. So, if 'mass' is inertia or resistance to acceleration, and it is therefore equal to the ratio between the magnitude of the acting force and that of the acceleration it produces in the direction in which it acts, we are forced to distinguish between two kinds of mass, the *longitudinal mass* m_λ, or resistance to acceleration in the direction of motion, and the *transversal mass* m_τ, or resistance to acceleration perpendicular to the direction of motion. We gather at once from eqn. (5.22) that these quantities are given by

$$m_\lambda = \frac{m}{\left(\sqrt{1-\left(\dfrac{v}{c}\right)^2}\right)^3} \qquad m_\tau = \frac{m}{1-\left(\dfrac{v}{c}\right)^2} \qquad (5.23)$$

where v is the particle's speed relative to the inertial frame $\mathscr{F}$ from which the accelerations are measured, and m denotes the particle's mass in whatever inertial frame it happens to be at rest.

Einstein uses eqn. (5.22) to calculate the kinetic energy that the particle acquires as its speed relative to $\mathscr{F}$ increases from 0 to v. The work W done by the force exerted on the particle in the direction of motion is given by:

$$W = \int qE_x\,dx = m\int_0^v \frac{v\,dv}{\left(\sqrt{1-\left(\dfrac{v}{c}\right)^2}\right)^3} = mc^2\left(\frac{1}{\left(\sqrt{1-\left(\dfrac{v}{c}\right)^2}\right)}-1\right) \qquad (5.24)$$

Forces perpendicular to the direction of motion do no work on the particle, so the total kinetic energy acquired by the latter is W. The energy obviously grows beyond all bounds as v approaches c. It is therefore impossible for an ordinary massive particle to achieve the speed of light.

Since q in the above arguments can be any arbitrarily small charge, Einstein concludes that eqns. (5.23) also hold for a particle with charge equal to 0. Since every force can be balanced by an electromagnetic force, they all must have the same effect on matter. So m_λ and m_τ measure – in the manner explained – our particle's resistance to acceleration by any type of force. Einstein (1905r, p. 919) emphasizes that "with a different definition of force and acceleration one would naturally obtain for the masses other numerical values" than those in eqn. (5.23). In fact, a different definition eventually did prevail; it was put forward by Planck (1906). While Einstein gave a unique, frame-independent representation of the force accelerating a particle in any inertial frame by equating it formally with the force defined according to Newton's Second Law in the inertial frame in which the particle is momentarily at rest, Planck provided a physically warranted representation of the force peculiar to each frame. This rests on Einstein's discovery that "the mass of a body is a measure of its energy content" (1905s, p. 641).[44] If m denotes the mass

---

[44] On the derivation of mass-energy equivalence in Einstein (1905s), see Stachel and Torretti (1982). Other derivations are found in Einstein (1906e, 1935).

of a particle in its rest frame, its rest energy equals mc^2. The total energy $E(v)$ of a particle moving with speed v is the sum of its rest energy and its kinetic energy. So, by eqn. (5.24),

$$E(v) = mc^2 + W = \frac{mc^2}{\left(\sqrt{1-\left(\dfrac{v}{c}\right)^2}\right)} \tag{5.25}$$

Prompted by this result we define the *relativistic mass* of our particle by

$$m(v) = \frac{E(v)}{c^2} = \frac{m}{\left(\sqrt{1-\left(\dfrac{v}{c}\right)^2}\right)} \tag{5.26}$$

The particle's *relativistic momentum* **p**, in an inertial frame in which it moves with velocity **v** (with $|\mathbf{v}| = v$), is equated with $m(v)\mathbf{v}$, and the *relative force* **k** acting on the particle in that frame is defined, in formal agreement with Newton's Second Law, by

$$\mathbf{k} = \frac{d\mathbf{p}}{dt} = \frac{d}{dt}m(v)\mathbf{v} \tag{5.27}$$

where t is Einstein time adapted to the frame.[45]

Evidently, relativistic mass, a function of speed, cannot be the same physical quantity as Newtonian mass, the absolute measure of matter. In fact, $m(v)$ is not normally the ratio of force to acceleration and thus not even a measure of inertia in the received sense.[46] The quantity that

---

[45] These definitions have some very nice consequences. For instance, (i) given a system of freely colliding but otherwise unrelated particles, such that $\mathbf{v}_i$ is the velocity and $m_i(v_i)$ the relativistic mass of the ith particle at any given moment in the chosen inertial frame of reference, the conservation of energy ($\Sigma_i m_i(v_i)c^2 = \text{const.}$) and the conservation of momentum ($\Sigma_i m_i(v_i)\mathbf{v}_i = \text{const.}$) hold if and only if relativistic mass is defined by eqn. (5.26). Also, (ii) the *power* or rate at which the relative force **k** does work on a particle moving with velocity **v** is

$$\mathbf{v} \cdot \mathbf{k} = \frac{d}{dt}m(v)c^2$$

Moreover, (iii) the relative force exerted by the electric and magnetic fields **E** and **B** on a particle with charge q that moves with velocity **v** is $\mathbf{k} = q(\mathbf{E} + \mathbf{v} \times \mathbf{B})$. In other words, if 'force' is defined by (5.27), the Lorentz force law (cf. eqn. (7.3)) holds in each inertial frame as a corollary of Einstein's Relativity Principle and the Maxwell equations.

[46] By result (ii) in note 45,

we have called m – known in the literature as 'rest mass' or 'proper mass' – is of course constant, frame-independent, and a measure of inertia (in the particle's momentary inertial rest frame), and it is thus a closer analogue of Newtonian mass; but it is still a far cry from it. Newtonian mass is additive in a literal physical sense, and proper mass is not. If several bodies are put together to make a bigger body, the Newtonian mass of the latter equals the sum of the masses of the former. Likewise, if a body explodes and breaks asunder, the Newtonian masses of the splinters add up to the original mass of the body. In stark contrast with this, the proper mass of the bigger body includes, besides the proper masses of the ingredient bodies, the mass equivalent of the energy invested in bringing them together. And the sum of the proper masses of the splinters equals the proper mass of the original body minus the mass equivalent of the energy released in the explosion.

These deep differences in the meaning of a central term of physics, which is common to SR and Newtonian mechanics, fueled the claim made by Kuhn (1962) and others that revolutionary science changes the referents of scientific discourse and therefore, since it no longer speaks about the same things, is "incommensurable" with the system it seeks to displace. Countering this extravagant claim, Putnam promoted the doctrine of reference without sense, which he later abandoned, but which is still endorsed by some philosophers. According to this doctrine, a reference is assigned to a general term of science – or ordinary language – in an original act of name-giving, patterned after *Genesis* 2:19–20, and is subsequently transmitted from generation to generation by word of mouth, independently of any connotations accruing to the term. We can readily imagine Adam pointing with his forefinger at a tall, dappled, long-legged creature with a towering neck as he mutters "giraffe, giraffe". But what could Newton – or the medieval inventor of the term – have pointed at to provide 'quantitas materiae' with a reference? It is hard to believe that by mere gestures and without elucidating its sense he could get anyone to understand what he was talking about. And, had he succeeded, it is not at all likely that the reference thus bestowed on 'quantity of matter' or 'mass' would be transferred by

$$k = \frac{d}{dt} m(v)\mathbf{v} = m(v)\frac{d\mathbf{v}}{dt} + \mathbf{v}(\mathbf{v}\cdot\mathbf{k})c^{-2}$$

So $m(v)$ is the ratio of force to acceleration if and only if $\mathbf{v}(\mathbf{v}\cdot\mathbf{k}) = 0$, that is, if and only if the particle is at rest or the relative force is perpendicular to the direction of motion.

Einstein or Planck to 'longitudinal mass', or 'transversal mass', or 'relativistic mass', or 'proper mass', or to all these terms at once when they defined them in the manner explained above.

I have discussed with some care the introduction of new concepts of mass in SR so that the reader can form a judgment on this question. Apparently no finger-pointing was done to secure the continuity of reference despite the momentous shift in meaning. But the definitions were so contrived by Einstein and his followers that all mass measurements recorded in a Newtonian setting could be received and utilized in SR.[47] Analogous precautions have surrounded the main conceptual changes in twentieth-century physics.

5.4   Gravitation as Geometry

It is clear that Newton's law of gravity does not meet the requirement of Lorentz invariance. The attractive force exerted *now* by a massive particle on another depends on their *present* distance, and as that distance varies the force changes *instantaneously*.[48] Einstein soon persuaded himself that a Lorentz invariant theory of gravitation was not likely to succeed and adopted a wholly new approach that he fondly regarded as an extension or "generalization" of his original Principle of Relativity. Einstein's search lasted many years, from 1907 to the hectic month of November 1915, when he finally struck gold. Throughout that time he remained faithful to one guiding insight that he had in the fall of 1907 and which he later described as his "happiest" – or "luckiest" – idea.[49] The gist of it can be expressed thus: A man falling freely, say, from a tall building, must feel completely weightless, and so will all the other heavy objects – purse, watch, keyholder – feel that fall together with him. A faint-hearted scientist would have viewed this merely as a manifestation of the fact that gravity

---

[47] Indeed, Newtonian measurements made in the so-called Newtonian limit can be employed to test SR only after an SR-meaning has been imposed on all the basic concepts, starting of course with time (cf. note 43). So Kuhn is right in saying that Newton's laws cannot be regarded as a "special case of the laws of relativistic mechanics" unless they "are reinterpreted in a way that would have been impossible until after Einstein's work" (1970, p. 101).

[48] See eqn. (2.3). The independent variable t in this equation is absolute Newtonian time and cannot be subjected to a Lorentz boost.

[49] See Pais (1982, p. 178). Einstein says that the idea came to him while he was working on his (1907j), that is, between 25 September and 4 December 1907.

imparts the same acceleration on all bodies, regardless of the stuff that they are made of.[50] But Einstein, true to the spirit of Newton's Rules of Philosophy, boldly jumped to the conclusion that a frame of reference falling freely in a uniform gravitational field is fully equivalent to an inertial frame. This is Einstein's Equivalence Principle.[51] It implies that an astronaut shut up in a capsule in interstellar space cannot tell, by means of physical experiments, whether the capsule moves inertially or falls freely, unless the gravitational field it traverses varies significantly in space and time. The Equivalence Principle suggests that inertia is just a limiting case of gravity, in agreement with Mach's idea (§4.4.3) that inertial phenomena – for example, the deformation of the liquid surface in Newton's rotating bucket – reflect the presence of distant matter.

Einstein promptly derived two testable consequences of the Equivalence Principle, viz., (i) that gravity bends light rays,[52] and (ii) that the

[50] See Chapter Two, note 34. This is sometimes described as the equality of *inertial* and *gravitational* mass, as if there were two distinct Platonic ideas of *mass* that happen to be always realized in equal amounts in each piece of matter.

[51] Professional literature distinguishes between the *strong* Equivalence Principle that I have just stated and the so-called *weak* Equivalence Principle, which equates inertial with (passive) gravitational mass. Note that the latter does not assert the *equivalence* of certain types of frames and of the physical experiments referred to them, but merely the *equality* of mass in its twofold Newtonian role of resistance to acceleration and susceptibility to gravity (cf. eqns. (2.2*) and (2.3)). Einstein usually formulated his Equivalence Principle from the point of view of observers *at rest* – not *freely falling* – in a uniform gravitational field (cf. 1907j, p. 454). He repeatedly gave the following example: A man inside an elevator completely shut up to the outside world cannot tell by means of physical experiments whether the elevator is at rest in a uniform gravitational field that pulls everything toward the elevator's floor with a force −g per unit mass, or whether the elevator is moving in gravitation-free space with acceleration g in the opposite direction. However, the formulation that I chose above not only exactly expresses the import of the daydream that he described as his most fortunate thought, but prepares us for GR in a way in which the alternative formulation does not.

[52] Einstein (1907j, p. 483). In Einstein (1911h, §2), he showed by means of one of his clever thought-experiments that radiation must gravitate if, in agreement with SR, a body's mass increases when it absorbs radiant energy. If radiation were not affected by gravity and could be lifted at no cost, say, from the ground level A to a point B at a higher gravitational potential, we would be able to create energy by radiating it from A to B and returning it from B to A stored in a falling body. After a full cycle the initially radiated energy E would increase to $E(1 + \gamma l c^{-2})$, where γ is the magnitude of the gravitational acceleration, l is the height of B above A, and c is the speed of light.

frequency of radiation varies with the gravitational potential.[53] Again, someone more timid than Einstein might have conceived this gravitational frequency shift as a distortion that, if not corrected, disqualifies the readings of atomic clocks. But in Einstein's thinking it became tangible evidence that gravity and spacetime geometry are intimately bound together. Atomic clocks measure proper time along their respective worldlines. If they seem to go faster or slower in the presence of different gravitational potentials, it is simply because the length of intervals along timelike spacetime curves depends on local gravity. Gravity does not therefore impair the timekeeping function of natural clocks; it shapes the several strands of time measured by differently placed clocks. In Einstein's eyes this unexpected turn of thought combined well with Mach's approach to inertia. In SR, the motion of free particles is determined by the spacetime geometry, which prescribes which worldlines are timelike or null geodesics. So, in this theory – as Einstein would later complain to Born (Einstein and Born 1969, p. 258) – spacetime, a ghostly presence, acts on matter and radiation, and yet is not acted upon. This asymmetry is overcome if the spacetime geometry, in turn, embodies the gravitational field structure determined by the distribution of matter.

The real gravitational field is, of course, uniform only within an agreed approximation, inside a more or less small neighborhood of each spacetime point. If we lift these restrictions, we must face the fact that existing fields are far from uniform, with the field here and that in the antipodes obviously pointing in opposite directions. Einstein was undoubtedly aware of this when he began to work on gravity. However, we do not know exactly when he first had the idea of piecing together the one highly diversified gravitational field of the universe from all those gravitationally almost uniform small neighborhoods, on the analogy of a curved surface, pieced together from the almost flat neighborhoods of its points. The first published testimony of this idea is the long paper he wrote with Grossmann in 1913. But Einstein declared later that the said analogy occurred to him shortly after arriving in Zürich in August 1912, before Grossmann introduced him to the mathematics with which he

---

[53] Einstein (1907j, pp. 458f.; 1911h, §3). The gravitational frequency shift follows at once from the argument in note 52 combined with the quantum equation $E = h\nu$ (where h is Planck's constant and E and ν are, respectively, the energy and the frequency of a photon). Einstein (1905i) was the first to postulate this equation with full generality, but he never appealed to it in his papers on Relativity (cf. note 8).

worked it out (CP 6, 535, n. 4). And it is hard to imagine that the Equivalence Principle, the link between gravity and proper time, and the familiar fact of the nonuniformity of gravity could live long together in his mind without combining to yield some such insight.[54]

The Einstein–Grossmann paper (1913) presents a theory of gravity that agrees with GR in intent and in many basic features but sports a different set of field equations. The central ideas shared by both theories can be summarily stated thus:

(A) Spacetime is a 4-manifold endowed with a semi-Riemannian metric g, which agrees well with the Minkowski metric η in a small neighborhood of any spacetime point P (in the same way that a proper Riemannian metric agrees with the Euclidean metric – see Chapter Four, text linked to note 25).[55] By virtue of this property, spacetime curves can be classified – just as in SR – as *timelike, spacelike,* or *null* according to the nature of their tangent vectors. Natural clocks measure proper time along their respective worldlines.

(B) The metric field g guides free fall, just as the Minkowski metric η guides inertial motion in SR: A chargeless nonrotating freely falling particle – an ordinary massive particle that has no angular momentum and is not subject to any external influence except that of gravity – describes a timelike geodesic. Photons describe null geodesics.

(C) The metric tensor field g is therefore none other than the gravitational field, so it must depend on the distribution of matter and nongravitational energy in spacetime.

(D) The dependence of g on matter-energy is governed by the gravitational field equations, a set of second-order differential equations in the metric components that are designed to agree, in the limiting case in which the gravitational field is weak and the bodies

---

[54] The delay until 1912 in reaching this insight was probably due to Einstein's initial resistance to Minkowski's approach to SR (cf. Einstein CP 5, 121n12). He was finally persuaded of its usefulness by Sommerfeld's papers on four-dimensional vector algebra and vector analysis (1910a, 1910b; cf. Einstein's letter to Sommerfeld of July 1910 in CP 5, 246). Later he acknowledged that without Minkowski's contribution GR "would probably never have got out of its nappies" (1917a, §17, p. 39).

[55] This implies, among other things, that on a small neighborhood of any spacetime point P one can define a chart relative to which the components of the metric at P satisfy the relations $g_{ij}(P) = \eta_{ij}$. Thus, $\partial g_{ij}/\partial x^k|_P = 0$; therefore, the metric g agrees to first order with η at P. Of course, the second derivatives of the g_{ij} need not vanish at P, so the Riemann tensor is not identically 0; the spacetime characterized by the metric g is generally not flat (or parallelizable).

concerned move slowly, with the Poisson equation of Newtonian theory (viz., $\nabla^2 \Phi = 4\pi\rho$, where Φ is the gravitational potential introduced in eqns. (4.13) and ρ is the density).

Points (A)–(D) require further elucidation. Note first that the metric **g** is not known a priori. Hence it is generally impossible to use coordinates that are metrically significant, like the Lorentz coordinates of SR. From now on, coordinates are just numerical labels with no quantitative meaning.[56] Physical quantities must therefore be represented, not by real-valued functions of the coordinates, as in earlier physics, but by tensor fields or other geometric objects that can be specified in terms of arbitrary coordinate systems, but are themselves coordinate-independent. Equations between such objects are therefore generally covariant, that is, they hold in every coordinate system. As explained in §4.1.3, a Riemannian (or semi-Riemannian) metric is a tensor field, and so is the Riemann tensor, which measures the metric's departure from flatness. Generally speaking, a tensor field **A** of rank n on spacetime assigns to each spacetime point P a tensor $\mathbf{A}_P$ of rank n at P, that is, an n-linear function on the tangent space at P (if **A** and hence $\mathbf{A}_P$ are *covariant*), or on its dual, the cotangent space at P (if **A** and hence $\mathbf{A}_P$ are *contravariant*).[57] Because of linearity, a tensor is fully specified by its action on a basis of the vector space on which it acts. The local tetrad

$$\left\langle \left.\frac{\partial}{\partial x^0}\right|_P, \left.\frac{\partial}{\partial x^1}\right|_P, \left.\frac{\partial}{\partial x^2}\right|_P, \left.\frac{\partial}{\partial x^3}\right|_P \right\rangle$$

---

[56] Of course, this condition does not hold in those special cases in which some features of the metric are known or postulated. Thus, in the spherically symmetric Schwarzschild solution of the Einstein field equations applied in the study of planetary motion, one can define a radial space coordinate that measures the distance from the planet to the central body. In Friedmann solutions in which there is a singularity in the past of every worldline of matter (see §5.5), one can define a universal time coordinate that measures the proper time elapsed since the singularity along each worldline of matter.

[57] Multilinear functions on vector spaces are explained in Supplement I.5. The above description can be made more precise as follows: A (p,q)-tensor field **A** on spacetime assigns to each spacetime point P a (p,q)-tensor $\mathbf{A}_P$ at P. This is a $(p + q)$-linear function acting on $(p + q)$-tuples of vectors, q of which are drawn from the tangent space at P and p of which are drawn from the cotangent space at P. If $p = 0$, **A** and $\mathbf{A}_P$ are said to be *covariant* of rank q. If $q = 0$, they are said to be *contravariant* of rank p. If both p and q differ from 0, **A** and $\mathbf{A}_P$ are said to be *mixed*, p times contravariant and q times covariant, of rank $(p + q)$. Of course, the definition of a mixed tensor must specify the position of the p tangent vectors and the q cotangent vectors in the

defined at P by an arbitrary chart x is a basis of the tangent space at P (see above, after eqn. (5.12)). Likewise, the corresponding quadruple of coordinate differentials $\langle dx^0, dx^1, dx^2, dx^3 \rangle$ is a basis of the cotangent space at P. Suppose then that A is a tensor field of rank 2. The real-valued function that assigns to each point P in the domain of x the number $A_P(\partial/\partial x^i, \partial/\partial x^j)$ – if A is covariant – or the number $A_P(dx^i, dx^j)$ – if A is contravariant – is the (i,j)th component of A *relative* to the chart x. When there can be no confusion as to the chart in question, this function is denoted by A_{ij} if A is covariant and by A^{ij} if A is contravariant. The set of functions A_{ij} (respectively, A^{ij}) fully specify the tensor field A on the domain of x.[58] But it is not enough to use tensor fields or other coordinate-free geometric objects for representing physical quantities. The physicist also needs a coordinate-free representation of the way these quantities vary continuously over spacetime. Partial derivatives with respect to the coordinates will not do if the coordinates are physically meaningless. So Grossmann drew Einstein's attention to a paper by Ricci and Levi-Cività (1900) that presented an absolute – that is, coordinate-independent – differential calculus for tensor fields (now known as the *tensor* or *Ricci* calculus). Given a tensor field A of rank n on a Riemannian manifold, the Ricci calculus constructs from it a new tensor field of rank $n + 1$, the covariant derivative of A, which suitably reflects A's change along any curve in the manifold.

The geodesic law of motion (B) was introduced by Einstein as a postulate. However, Einstein and Grossmann (1913, p 10) proved it as a theorem for matter distributed in a particularly simple way (as pres-

---

$(p + q)$-tuples that are its arguments. One speaks then, for example, of a tensor of rank 6, covariant in the first and third, and contravariant in the second, fourth, fifth, and sixth *indices*. The origin of the terms 'covariant' and 'contravariant' is too contorted to deserve explanation. The term 'index' comes from the habit of representing tensors by their components relative to a chart and labeling those components with indices ordered like the vectors in each argument of the tensor.

[58] Let y be another chart defined on the same spacetime region as x. Then, the vector fields $\partial/\partial y^j$ (resp. dy^j) are linear combinations of the $\partial/\partial x^i$ (resp. the dx^i), so the components of A relative to y can be readily computed from its components relative to A through formulas involving the partial derivatives $\partial x^i/\partial y^j$ (resp. $\partial y^j/\partial x^i$). This led to the popular description of tensor fields as "arrays of functions that transform according to such-and-such rules" (the transformation formulas being given). Einstein warned Besso against this unperspicuous description on 31 October 1916: "Definition of tensors: not 'things that transform thus and thus'. But: things that can be described, with respect to an (arbitrary) coordinate system by an array of quantities $(A_{\mu\nu})$ obeying a definite transformation law" (Einstein and Besso 1972, p. 85).

sureless dust). In the context of GR it was readily shown that photons describe null geodesics (von Laue 1920; Whittaker 1928). Finally, in the late 1930s, Einstein, in collaboration with several young mathematicians, inferred the geodesic law of motion for massive particles from the field equations of GR by a difficult and controversial argument.

Point (C) raises the following question: How shall the distribution of matter and nongravitational energy be represented in a manner that is appropriate for conveying the metric tensor field's dependence on it? In the context of SR, Minkowski had devised a tensorial representation for the energy of the electromagnetic field, whose components relative to a Lorentz chart were familiar classical quantities: the charge density, the components of the Poynting vector (representing energy flux), and the components of Maxwell's stress tensor (initially conceived for representing tensions in the aether). Inspired by this example, von Laue (1911) constructed, again in the context of SR, the mechanical energy–momentum tensor, a symmetric tensor field representing the dynamical state of a material continuum whose components relative to a Lorentz chart are equal to or built from classical quantities: the mass-energy density, the components of momentum density, and the components of the classical stress tensor.[59] Von Laue argued that in each domain of physics there is a tensor field whose components have a physical meaning corresponding to the above. As these diverse domains interact in a region of spacetime, their respective tensor fields should be added to define a single symmetric tensor field of rank 2, representing the presence of matter-energy in that region. Einstein seized on this idea. In his geometric theories of gravity, which culminated with GR, the distribution of matter and nongravitational energy is represented by just such a general, generally unspecified, energy–momentum tensor, which I shall denote by T. Late in life Einstein declared that "all attempts to represent matter by an energy–momentum tensor are unsatisfactory" (Einstein and Infeld 1949, p. 209), for all such tensor fields "must be regarded as purely temporary and more or less phenomenological devices for representing the structure of matter, and their entry into the equations makes it impossible to determine how far the results obtained are independent of the particular assumption made concerning the con-

---

[59] It is perhaps worth noting that the classical stress tensor – employed since the nineteenth century by structural engineers for representing *tensions*, that is, strains and stresses, inside a beam, column, etc., of a building – motivated the adoption of the name 'tensor' for the geometric objects under discussion here.

stitution of matter" (Einstein, Infeld, and Hoffmann 1938, p. 65). Still, as we shall see in §5.5, simple, highly idealized models of T play an essential role in relativistic cosmology.

Finally we come to point (D), the field equations of the new theory of gravity. One would expect that, once the gravitational field and the distribution of gravitational sources are represented, respectively, by the tensors g and T, the field equations would equate geometric objects constructed from these tensors. This would ensure that the equations are generally covariant and thus physically meaningful even if the coordinate systems are not. However, Einstein and Grossmann, while insisting on the value of a generally covariant formulation of the laws of physics (1913, pp. 7, 10, 18), make an exception with the equations of gravity, which, in the form proposed by them, hold only for a certain ample but anyway restricted family of coordinate systems. Einstein was distressed by it, but he soon found consolation in an argument by which he purported to prove that a generally covariant system of gravitational field equations would be incapable of determining the course of events under its sway. The argument – later dubbed "the hole argument" – runs as follows: Suppose that T is identically zero on a compact, open, simply connected spacetime region $\mathcal{H}$, a veritable hole in the fullness of matter. Let γ be a timelike curve that crosses $\mathcal{H}$, joining two points P and Q on $\mathcal{H}$'s border. Let f be a smooth one–one mapping of spacetime onto itself that agrees outside $\mathcal{H}$ with the identity mapping but differs from it on $\mathcal{H}$ ($f(X) = X$ if $X \notin \mathcal{H}$; $f(X) \neq X$ if $X \in \mathcal{H}$). Then, if the field equations are generally covariant and γ is the worldline of a freely falling test particle, so is $f \circ \gamma$. Thus a generally covariant system of gravitational field equations cannot determine the trajectory that, given the matter distribution T, a freely falling particle would follow from P to Q, but leaves more than one way open to it. To see the force of this argument one must bear in mind that a smooth one–one mapping f of an n-manifold onto itself "drags along" all the geometric objects defined on the manifold; in other words, if A is such an object, f defines an object $f_* A$ of the same type that behaves at $f(X)$ just like A at X.[60] If the field equations are generally covariant and the metric g is a solution for the matter distribution T, then, under the conditions prescribed above for T and f, the metric $f_* g$ is also a solution. And, of course, if γ is a geodesic of the metric g, $f \circ \gamma$ is a geodesic of the metric $f_* g$. So we have two equally admissible worldlines for our

[60] To be specific, let A be a covariant tensor field of rank 2. Then, if V and W are two vector fields and P is any point in the manifold, $f_* A_{f(P)}(f_* V_{f(P)}, f_* W_{f(P)}) = A_P(V_P, W_P)$.

freely falling test particles starting from the same event P (indeed infinitely many, since f is arbitrary on $\mathcal{H}$). So far, so good. But the argument forgets the fact, so clearly set forth by Newton, that points in a structured manifold have no individuality apart from their structural relations.[61] Let X be a point in the hole $\mathcal{H}$ that is equidistant by the metric g from five points A, B, C, D, and E outside $\mathcal{H}$. Then $f(X)$ is equidistant, by the metric f_*g, from A, B, C, D, and E. In what sense can $f(X)$ stand in the manifold $\langle\mathcal{H},f_*g\rangle$ for a different physical event than the one X stands for in the manifold $\langle\mathcal{H},g\rangle$? An answer eludes me. Nor can a physicist conceive of any difference between a test particle aptly represented by the geodesic γ in $\langle\mathcal{H}\cup\{P,Q\},g\rangle$ and one aptly represented by the geodesic $f\circ\gamma$ in $\langle\mathcal{H}\cup\{P,Q\},f_*g\rangle$.[62] After Einstein discovered tensorial equations for gravity in the fall of 1915, the hole argument was no longer necessary. He never publicly retracted it, but in letters to Ehrenfest (26 December 1915) and Besso (3 January 1916) he explained that physical reality consists of spacetime coincidences, which are preserved by all point transformations; thus, "to demand that the laws do not determine anything beyond the aggregate of spacetime coincidences is the most natural thing in the world".[63]

Einstein submitted to the Prussian Academy a system of generally covariant equations for the gravitational field on 11 November 1915 (1915g). They equate the Ricci tensor – a symmetric tensor of rank 2 built from the metric tensor and its first and second derivatives – with the energy–momentum tensor T multiplied by a constant:

$$R_{jk} = -\kappa T_{jk} \qquad\qquad (5.28)$$

These equations responded to a very questionable physical assumption and Einstein very soon discarded them, but not without first having derived from them – in a paper submitted on 18 November (1915h) –

---

[61] See the long quotation from Hall and Hall (1962, p. 103), in §2.2. I learned this in my view decisive objection to Einstein's hole argument from John Stachel, before either of us became aware of Newton's text. Of course, in 1914 Einstein did not have access to it.

[62] Of course, *each* geodesic can represent more than just one thing. Mathematical models fit many applications, to such-and-such a degree of accuracy, at this or that level of abstraction. But whatever is well represented by γ in $\langle\mathcal{H}\cup\{P,Q\},g\rangle$ is equally well represented by $f\circ\gamma$ in $\langle\mathcal{H}\cup\{P,Q\},f_*g\rangle$.

[63] Einstein and Besso (1972, p. 39). I quote the relevant passage of the letter to Ehrenfest in English translation in Torretti (1983, p. 166). At the time of writing neither letter had yet been printed in Einstein CP. Einstein's hole argument attracted the attention of philosophers after Earman and Norton (1987) reformulated it as a refutation of absolutist theories of spacetime.

a correct prediction of the observed motion of Mercury, which Newtonian celestial mechanics did not fully account for.[64] On 25 November, Einstein (1915i) proposed to the Academy another system of
generally covariant equations. These are the *Einstein field equations*
(EFE), the core of GR. They agree with eqns. (5.28) in free space –
where T is identically zero – so they also support Einstein's solution of
the anomaly of Mercury. In the form in which they were first published,
the EFE equate the Ricci tensor with a tensor field constructed by Einstein from the energy–momentum tensor T:

$$R_{jk} = -\kappa \left(T_{jk} - \frac{1}{2} g_{jk} \sum\nolimits_{i=0}^{3} T_i^i \right) \qquad (5.29)$$

Equations (5.29) are algebraically equivalent to the following system,
in which T multiplied by the gravitational constant –κ stands alone on
the right-hand side and is equated with the so-called Einstein tensor –
built from the Ricci tensor – on the left:

$$R_{jk} - \frac{1}{2} g_{jk} \sum\nolimits_{i=0}^{3} R_i^i = -\kappa T_{jk} \qquad (5.30)$$

The Einstein tensor has a mathematical property that T must have, if
energy is conserved. This is a strong motive for equating these two
tensors fields.[65] Due to the symmetry of the tensors involved ($R_{jk} = R_{kj}$,
etc.), the EFE are only ten distinct equations, with ten unknowns (the
metric components g_{jk}). However, due to the said mathematical property, there are no more than six independent equations. Ten independent equations with ten unknowns would unduly constrain the choice

---

[64] The perihelion of Mercury – the point where the planet comes closest to the Sun –
advances each year on the celestial sphere some 56″ (seconds of arc). Almost 90%
of this advance merely reflects the slow precession of the Earth's axis. Over 9% can
be attributed, in Newtonian celestial mechanics, to the perturbing action of the other
planets. But there remains a perihelion advance of some 43″ per century for which
no satisfactory Newtonian explanation was ever found. (For more precise figures see
§7.2.) Einstein expressed his interest in this anomaly in connection with his research
on gravity in the letter that he wrote to Habicht on Christmas eve, 1907 (CP 5, p.
82). His disappointment with the Einstein–Grossmann field equations stemmed in
part from the fact that they gave the wrong number for the unexplained part of
Mercury's perihelion advance (letter to Sommerfeld of 28 November 1915; Einstein
and Sommerfeld 1968, p. 32). Einstein's work on the perihelion of Mercury is discussed again in §7.2.

[65] For a stricter explanation of this point, see Torretti (1983, p. 323n34, and the text
(on p. 175) to which this note is appended).

of a coordinate system. But this is not the right place for discussing the mathematics of the EFE. In the following section I shall touch on some of their wholly unexpected physical implications.

5.5   Relativistic Cosmology

Less than two months after the publication of system (5.29), Schwarzschild (1916a, b) produced an exact solution. It provided a model of the motion of a small planet or satellite in the gravitational field of a large central body. Schwarzschild assumed a field static in time and spherically symmetric in space that converges to the flat Minkowski metric at spatial infinity and is free of matter except at most on a small neighborhood of the axis of symmetry. The Schwarzschild solution ratifies Einstein's (1915h) calculation of Mercury's perihelion advance through the approximate solution of system (5.28). To obtain this particular application one assumes that the Sun is alone in the universe, on the symmetry axis of the Schwarzschild field. The constant of integration that occurs in the solution is put equal to two solar masses[66] and a test particle – that is, a nonrotating object of negligible mass – is imagined circulating around the Sun, under the sole influence of the Schwarzschild field. Then, if the test particle is initially placed at Mercury's distance from the Sun and given Mercury's velocity, it will go around the Sun in 80 days, tracing an ellipse in space with the Sun at one focus, and a perihelion advancing slightly more than 43″ per century.

The Minkowski metric at spatial infinity implicit in the Schwarzschild solution made good sense in applications to the Sun, moving through empty space at great distance from comparable bodies. The GR spacetime curvature on a sphere drawn about the Sun at one-half the distance from α Centauri is surely insignificant compared to that on Mercury's orbit. On the other hand, flatness at *true* spatial infinity is incompatible with Einstein's Machian view of inertia, for it prescribes a definite worldline – in agreement with SR – to any free particle infinitely far from other matter. Now, in accordance to the Machian view, the worldline of a free particle can only be prescribed by the matter outside it. It is all right for *very* distant matter to guide the free particle along a (nearly) Minkowski geodesic, but *infinitely* distant matter can have no such effect. A free particle that has no matter-

---

[66] Bartels (1994, §2.3), "The Interpretation of the Schwarzschild Mass", provides a lucid, philosophically motivated analysis of this move.

dependent gravitational field to guide it should not know what to do. Einstein expressed his Machian standpoint as follows:

> In a consistent theory of relativity there can be no inertia *relative to* "*space*", but only an inertia of masses *relative to each other*. Hence if I take a mass sufficiently far away from all other masses in the world its inertia must fall down to zero.
>
> (Einstein 1917b, p. 145)

He therefore explored together with Jakob Grommer solutions of eqn. (5.29) that are (i) static, (ii) spherically symmetric in space, and (iii) such that the metric components g_{ij} relative to a suitable chart do not converge at spatial infinity to the values η_{ij} but degenerate in a way that entails the vanishing of inertial masses. They concluded that these requirements could not be reconciled with the fairly slow motion that astronomers attributed to stars.[67]

Confronted with this difficulty Einstein realized, in a stroke of genius, that in the context of GR the problem he was trying to solve could be simply done away with. A Riemannian metric of positive curvature can be defined on a boundless yet finite space in which no lump of matter is infinitely far from the others and the question of the behavior of free particles and the metric at infinity does not even arise. Assuming that matter in the large is evenly distributed and can therefore be passably represented by a homogeneous and isotropic fluid, Einstein searched for a solution of the EFE such that spacetime is uniformly filled with matter and can be partitioned into finite spacelike slices of the same constant curvature. When he saw that eqns. (5.29) admit no such solution, he added a new term to the left-hand side, to make them read as follows:

$$R_{jk} - \lambda g_{jk} = -\kappa \left(T_{jk} - \frac{1}{2} g_{jk} \sum_{i=0}^{3} T_i^i \right) \qquad (5.31)$$

The factor λ is a constant, known as the *cosmological constant*, which in the light of observation must be so small as to be negligible.[68] Still

---

[67] In fact they were pursuing a will-o'-the-wisp, for – as G. D. Birkhoff would prove shortly (1923) – a solution of the EFE that is spherically symmetric in space *must* converge to flatness at spatial infinity (besides being *necessarily* static).

[68] Current data imply that $|\lambda| \lesssim 3 \times 10^{-52}\,\mathrm{m}^{-2}$ (Ciufolini and Wheeler 1995, p. 209). In what other branch of physics would an adjustable parameter with such a tiny upper bound be pronounced different from 0?

eqns. (5.31), with $\lambda > 0$, do admit the spatially finite static solution that Einstein was looking for.

Einstein's static world model marks the beginning of modern cosmology. The notion that the world fills only a finite space, even though it has no limits, resolved in an unexpected, intellectually most satisfactory way one of the antinomies that – according to Kant (§3.5) – banned cosmology from the realm of rational inquiry. On the other hand, the modified EFE did not serve Einstein's Machian motivation very well. First, a nonzero cosmological constant constitutes precisely the kind of sourceless universal field acting on matter and not acted upon that the Machian approach was supposed to avoid. Second, de Sitter (1917a, 1917b, 1917c) soon found a nonflat solution of eqns. (5.31) that is totally exempt of matter. As originally presented, the solution is static, but a handful of test particles scattered in de Sitter space will fly away from each other. De Sitter, Weyl, and others noted the agreement of this surprising prediction with the no less surprising data on the recession of galaxies that were coming in from California.

The next momentous step in the creation of modern cosmology was hardly noticed at the time. The Russian meteorologist Alexander Friedmann (1922, 1924) discovered a family of nonstatic, spatially homogeneous and isotropic solutions of the EFE, of which the static solutions of Einstein (1917b) and de Sitter (1917c) were particular – and, as it turned out, unstable – cases. Friedmann's first paper appeared in the *Zeitschrift für Physik* followed by a note in which Einstein (1922) asserted, without any further explanation, that Friedmann's work rested on a mathematical error. That the paper was printed despite this authoritative disclaimer speaks well of German science in the 1920s. Einstein (1923) soon conceded that the error was his, but Friedmann earned little or no credit for his world models until long after Lemaître (1927) had independently rediscovered them.

The Friedmann world models are solutions of the modified EFE – eqns. (5.31) – admitting any value for the cosmological constant λ, including 0. Like all particular solutions of a system of partial differential equations, they involve certain special assumptions. These concern the geometric structure of spacetime and the state and motion of matter. As to the former, Friedmann assumed (a) that spacetime admits a global time coordinate x^0; (b) that the parametric curves of x^0 are everywhere at right angles to the spacelike hypersurfaces $x^0 =$ const.; and (c) that on each hypersurface $x^0 =$ const. the spacetime metric amounts to a proper, time-dependent, Riemannian metric of constant curvature.[69] With regard to matter, Friedmann assumed (d) that it can be represented as a pressureless fluid whose worldlines agree with the paths of the parametric curves of x^0. Assumption (c) can be seen as a geometric statement of the spatial homogeneity and isotropy of matter in the large. On the other hand, according to Friedmann, assumption (b) is to be adopted only because it simplifies calculations. Now, assumption (b) is a strong requirement that severely restricts the available solutions. However, Robertson showed that, if the worldlines of matter are geodesics – a natural assumption in GR, given assumption (d) –, then assumptions (b) and (c) follow from the philosophical demand that no observer shall be able to distinguish between any two directions about him by some intrinsic property of spacetime, or to "detect any difference between his observations and those of any contemporary observer" (1929, p. 823 – this is known in the literature as "the Cosmological Principle").

Except for the limiting case in which spacetime is static, the Friedmann world models share a remarkable feature: The worldlines of matter either diverge from one another, or converge toward one another, or first diverge and then converge. Suppose that we live in a Friedmann universe in which the matter worldlines are currently diverging, either forever or until such time as they will begin to converge. Suppose that we mentally go backwards in time along one such worldline γ. Then, as we delve into the past, second by second, year by year, all the other worldlines will converge towards γ, so that matter in every spatial neighborhood of γ will be denser and denser, its average density increasing beyond all bounds as the time we have gone backwards approaches a certain value (the currently fashionable figure is 12 to 15 billion years). Of course, the density of matter cannot be actually infinite at a given spacetime point, since neither the energy–momentum tensor nor the metric would be defined at such a point. On the other hand, in a universe of this type, if the time coordinate x^0 is defined so as to measure the temporal distance between events along each worldline of matter, the range of x^0 obviously cannot be the entire real number field $\mathbb{R}$, but only a finite connected open interval in $\mathbb{R}$. Thus, if you choose a time unit, assign time 0 to the present, and use the negative reals to label the past, the entire past of each worldline of

[69] Let $K(t)$ denote the constant curvature of the hypersurface $x^0 = t$. Friedmann (1922) considers only the case $K(t) > 0$; the case $K(t) < 0$ is the subject of Friedmann (1924). He did not consider the case $K(t) = 0$ countenanced by Einstein and de Sitter (1932).

matter will be mapped onto some interval $(-T, 0)$. You may then, with Friedmann, call T "the time since the creation of the world" (*die Zeit seit der Erschaffung der Welt*; Friedman 1922, p. 384n.). But you would be wrong to think that you have found the date of creation. $-T$ is the greatest lower bound of the range of your time coordinate. But it is not the date of any event, for events can only occur on matter worldlines, and therefore they must carry a date greater than $-T$. Indeed, in a Friedmann universe every process whose beginning can be dated is preceded by some length of time.

Relativistic cosmology achieved somewhat greater generality at the hands of Lemaître (1927), who proceeded from the weaker assumption that matter is a perfect fluid (with constant, isotropic pressure), not a presureless one. Einstein and de Sitter (1932) proposed an expanding universe with flat (Euclidean) spacelike slices orthogonal to the worldlines of matter. Robertson (1935) and Walker (1935, 1937) derived the most general expression for the metric of a spatially homogeneous and isotropic spacetime, which is independent of GR (not necessarily a solution of the EFE) but agrees with SR on each tangent space (is "locally Minkowskian"). Spacetimes with this metric are termed FRW worlds (for Friedmann, Robertson, and Walker), or FLRW worlds (if one is mindful of Lemaître), and constitute the so-called *standard model* of current cosmology. The success of a world picture so inimical to traditional myth and science is mainly due to the accumulation of favorable empirical evidence. By the end of the 1920s, Hubble and his colleagues at the Mt. Wilson observatory had established that the Galaxy, with its estimated hundred million stars, is only a speck in the universe, and that most of the observable nebulae are in fact other such star groupings, at enormous distances from each other. The systematic study of the light received from these sister galaxies showed that, except in the case of a few nearby ones, they are receding from us at speeds that – according to Hubble (1929) – are proportional to their distance. Lemaître (1927) proposed his version of the Friedmann solutions as a theoretical explanation of Hubble's empirical results and was the first professional scientist to speak of the dense, hot beginning of the universe (the poet Edgar Allan Poe had done so in 1848).

For 20 years, the GR theory of the expanding universe ran against one grave obstacle. According to Hubble's calculations, the time elapsed since the start of cosmic expansion was about 2 billion years, a good deal less than the age of some rocks, as calculated by geolo-

gists. To overcome this difficulty, Milne proposed that there are two
types of natural clocks, embodied in different processes and running at
different rates, that yield two seemingly inconsistent series of chronological data. The steady-state cosmology put forward in two different
versions, by Bondi and Gold (1948) and by Hoyle (1948), simply
denied the finite age and evolution of the universe, and compensated
for its manifest expansion by postulating the constant creation of
matter everywhere at a rate too low to be detected by us.[70] The
cosmological and the geological time scales were reconciled when
Baade (1956) showed that intergalactic distances had been underesti-
mated due to an error concerning the luminosity of Cepheid variable
stars. Baade's recalibration lengthened the age of distant light signals
by a factor of 2, and subsequent findings have further increased this
factor.

Still, the great majority of physicists did not take GR cosmology very
seriously until quantum physics – or rather, a factual discovery that
they understood in the light of it – persuaded them that the world we
live in is well portrayed, in the large, by one or the other of the expanding FLRW models. I think that this turn of events is highly instructive,
for GR, which provides the framework for understanding the expansion of the universe, is in fact incompatible with quantum physics. But
let me just tell the story very briefly here and leave the philosophical
morals for Chapter Seven. In the 1940s Gamow and others speculated
on the origin of the elements.[71] The relative abundance and uniform
distribution of helium could be readily understood if it was generated
by thermonuclear reactions when the universe was very dense and hot.
If that was the case, matter and radiation must have once been coupled
together in continual mutual exchange, and a relic of radiation as it
stood at the time of decoupling should still be observable. This radiation will now be cooled down to almost 0 K due to expansion, but it

[70] Bondi and Gold (1948, §4.1) put the average rate of creation at 10^{-43} grams per cubic
centimeter per second. The steady-state theory was not in itself a response to the time-
scale problem. It issued from a purely methodological consideration. "The universe
is postulated to be homogeneous and stationary in its large-scale appearance as well
as in its physical laws" not because the authors claim that this so-called *Perfect Cosmological Principle* must be true, but because "if it does not hold, one's choice of the
variability of the physical laws becomes so wide that cosmology is no longer a science;
one can then no longer use laboratory physics without relying on some arbitrary principle for their extrapolation" (1948, §1.2).

[71] Alpher, Bethe, and Gamow (1948), and Gamow (1948). For a lively and highly
instructive report on these developments, see Kragh (1996, Ch. 3).

will retain its isotropy and universal presence, as well as its thermal character (i.e., the distribution of energy flux at each frequency will agree with Planck's law). Alpher and Hermann (1948) predicted a current temperature of 5 K. Little was made of this until a decade and a half later, when the idea was revived just in time to furnish a cosmological explanation of the pervasive unremitting isotropic apparently thermal radiation at c. 2.7 K discovered by Penzias and Wilson (1965; cf. Dicke et al. 1965) while they investigated background microwave noise in radio communications. Very accurate and detailed measurements have since confirmed that this radiation is indeed thermal and almost perfectly isotropical.[72] This is generally regarded as overwhelming evidence for the hot dense beginning of the universe and its subsequent cooling by expansion (of course, there are dissenters).

In the standard model all worldlines of matter proceed as if from a common point at which, if it existed, the energy density would be infinite and the spacetime metric would be undefined. In the early days of modern cosmology this "initial singularity" was glossed over as an idealization that would be automatically avoided in a more realistic, not quite perfectly isotropic and homogeneous world model. For example, de Sitter wrote that "the conception of a universe shrinking to a mathematical point at one particular moment of time [. . .] must [. . .] be replaced by that of a near approach of all galaxies during a short interval of time" (1933, p. 631). But since after 1965 there has been a tendency to take the singularity literally, due to the almost perfect isotropy of the cosmic background radiation and to the singularity theorems proved by Penrose, Hawking, and Geroch.[73] Some believe that this raises a problem. If matter has existed only for a finite

---

[72] A small but systematic anisotropy is attributable to the motion of the Solar System across this ocean of radiation. When that is subtracted there remain anisotropies of up to 1 part in 10,000, short of which the articulation of matter into galaxies and stars would perhaps have been impossible.

[73] In a series of papers culminating with Hawking and Penrose (1970). For a textbook exposition, see Wald (1984, Ch. 9). For a philosophically minded discussion, see Earman (1995, Ch. 2). The singularity theorems establish the existence of incomplete timelike geodesics in every GR spacetime that meets certain plausible physical conditions, when such geodesics converge to one another as in FLRW worlds. An incomplete geodesic is an inextensible geodesic defined on an interval that has a greatest lower bound, or a least upper bound, or both. A geodesic γ defined on an interval I is said to be inextensible unless there is another geodesic γ' defined on an interval I' such that I is a proper part of I' and γ' takes the same values as γ on I.

time in an FLRW world, some parts of it have never had an opportunity to interact with others. For example, the most distant galaxies that we barely manage to observe at opposite sides of the sky cannot yet be observed from each other. How could they get to be so similar? In particular, when could the cosmic background radiation in those non-interacting regions of the world reach the state of thermal equilibrium that we ascribe to it? Guth (1982) thought that he could give an unexpected solution to this question – and a few others – by boldly assuming the EFE to work in a cosmic setting subject to a "Grand Unified" quantum field theory of nuclear forces. Under his assumptions the universe expands exponentially during a very short "inflationary" period (less than 10^{-30} second), after which the standard FLRW expansion takes over. Before the latter begins, even the more distant parts of the currently observable universe have been sufficiently close to mix together by normal thermal processes. However, Guth's original model leads to gross inhomogeneities, that render it untenable (Blau and Guth 1987, p. 550). To overcome this difficulty other inflationary schemes have been proposed, the most remarkable of which is due to Linde (1983, 1986). Here "the *local* structure of the universe is determined by inflation," governed by the EFE; however, "its *global* structure is determined by *quantum* effects" (Linde 1987, p. 607).

> It proves that the large-scale quantum fluctuations of the scalar field ϕ generated in the chaotic inflation scenario lead to an infinite process of creation and self-reproduction of inflationary parts of the universe. In this scenario the evolution of the [. . .] universe has no end and may have no beginning. As a result, the universe becomes divided into many different domains (mini-universes) of exponentially large size, inside which all possible (metastable) vacuum states are realized.
>
> (Ibid.)

Such mini-universes can possess very different basic physical properties (e.g., different dimension number). One of them evolves through inflation into the four-dimensional roughly FLRW world in which we live, which, in stark contrast with its siblings, sports the very unlikely combination of features required for human life.[74]

---

[74] Guth (1997) explains inflationary cosmology to nonspecialists, with ample references to the original literature. For a less sanguine appraisal, see Earman and Mosterín (1999).

CHAPTER SIX

✦

Quantum Mechanics

The word 'quantum' was used in Latin as a relative and interrogative adjective, adverb, or pronoun, to mean 'how much' or 'how many'. Before the twentieth century it was occasionally used in English, as in German,[1] to mean a definite amount of something. It was in this sense that Planck (1900, in PAV I, 706) spoke of "the elementary quantum of electricity e", meaning "the electric charge of a positive univalent ion or an electron". In that paper Planck derived the law of thermal (or "black body") radiation that now bears his name. He used as a model a collection of many linearly oscillating monochromatic resonators enclosed in a cavity with reflecting walls,[2] and he assumed that the energy of the resonators oscillating with a particular frequency ν was an integral multiple of the quantity

$$E_\nu = h\nu \tag{6.1}$$

where h is a constant of nature with the dimension of action (= energy × time), subsequently known as *Planck's constant*. It is unlikely that Planck meant this assumption as a general physical hypothesis and not just as a prop for his argument, a peculiarity of his fictitious model. At any rate, he did not then call h an 'elementary quantum of action' nor E_ν one of energy. Five years later, Einstein – building on Planck's work – conjectured that all electromagnetic radiation consists of "energy quanta (*Energiequanten*) localized at points in space, which move undi-

---

[1] See, for example, Kant (1787, p. 224), quoted at the beginning of §3.4.2.

[2] He could profitably adopt this far-fetched and seemingly arbitrary model because in such a cavity – as Kirchhoff (1860) proved from the conservation of energy principle – the distribution of energy over frequency in thermal equilibrium does not depend on the nature of the radiating bodies.

vided, and can only be absorbed or generated as wholes" satisfying the condition (6.1) (1905i, p. 133). Thereafter, the word 'quantum' was increasingly used as a noun to designate the minimum amount in which some physical quantity is found in nature and by multiples of which it increases or decreases, as well as an epithet for hypotheses, theories, and the like, that imply the reality of such quanta and, in particular, the "quantization" of energy in accordance with eqn. (6.1). Following standard usage, I speak of 'quantum physics' in this general sense and reserve the name Quantum Mechanics (QM) for the theory formed in the late 1920s by the conflation of Heisenberg's "matrix mechanics" of 1925 and Schrödinger's "wave mechanics" of 1926. QM involves a wholly new way of understanding the purpose and the basic concepts of physics, which has become the subject of endless philosophical debate. This state of affairs is all the more irksome in view of the theory's unblemished record of experimental success. The equations of QM are not Lorentz invariant and therefore can hold well only in situations in which relative speeds are small, but most authors agree that its philosophical difficulties are inherited by the Lorentz invariant quantum theories developed in its wake. The philosophical problems of QM are no doubt by far the chief source of current interest in the philosophy of physics, especially since the – mostly imaginary – philosophical problems of Relativity have been tamed.

In the space available here I can only introduce the reader to the main disputed questions and opposing theses. I begin with a summary of events leading to the birth of QM (§6.1). In §6.2 I refer to matrix and wave mechanics, the equivalence of both theories, Born's probabilistic interpretation of Schrödinger's ψ-function, and Heisenberg's indeterminacy relations; before explaining the latter I sketch the QM formalism of vectors and operators in Hilbert space (§6.2.5). In §6.3 I discuss the two chief philosophical problems of QM, viz., the EPR paradox (thus called after Einstein, Podolski, and Rosen 1935) and the problem of measurement. In §6.4 I consider a few of the philosophical proposals concerning the right meaning of the theory. I close the chapter with a few remarks on relativistic quantum physics (§6.5).[3]

---

[3] §§ 6.1 and 6.2 owe much to Jammer (1966), Darrigol (1992), and the editor's introduction to Van der Waerden (1967). Readers who are unsatisfied with my sketches should turn to these works. For greater detail and abundant references, see Mehra and Rechenberg (1982–88).

6.1   Background

6.1.1   The Old Quantum Theory

Quantum physics before QM is loosely referred to as the Old Quantum Theory. It was the theoreticians' response to the surprising news about the fine structure of matter that began to flow from physical laboratories at the end of the nineteenth century. In 1895 Röntgen discovered X-rays, which, after long discussions, were finally classified as high-frequency electromagnetic radiation when von Laue and his associates succeeded in diffracting them in 1912. In 1896 Becquerel discovered radioactivity. In 1898 Rutherford analyzed radioactive output into α-rays (to be identified many years later as twice ionized Helium atoms) and β-rays (soon assimilated to cathode rays), to which in 1900 Villard added γ-rays (ultra-high-frequency radiation). At about the same time, Rutherford established that a given amount of a radioactive element loses one-half of its radioactivity in a period of time that depends solely on its nature and is not affected by changes in the surrounding circumstances. In 1899 J. J. Thomson proved that cathode rays − first observed by Faraday in the 1830s and instrumental to Röntgen's discovery − are beams of extremely light particles, which he dubbed 'electrons' and were thereafter regarded as the ultimate free carriers of negative electric charge.[4] The use of these powerful effects as laboratory tools produced a flood of information. Side by side with them, the spectroscopists supplied increasingly accurate measurements of the spectral lines characteristic of the several chemical elements, which displayed enticingly simple yet utterly incomprehensible regularities.

Philosophers who reflect on quantum physics should not lose sight of these events. In stark contrast with the birth of mathematical physics in the seventeenth century, which consisted for the most part in reconceiving familiar facts under new standards of rigor, the "quantum revolution" was forced on a fairly conservative scientific establishment by a surfeit of unexpected phenomena. By stressing this I do not mean to

---

[4] For the date of J. J. Thomson's discovery of the electron, see Pais (1986, pp. 78ff.). Other authors give other dates, not because of any disagreement about laboratory records, but because they have different notions of what it takes to discover an elementary particle (for an instructive discussion, see Arabatzis 1996). The term 'electron' had been introduced in 1891 by Stoney, who since 1874 had argued for the existence of an indivisible unit of electric charge.

endorse thoughtless empiricism. To be sure, experiments were designed and results gleaned from them in terms of some preconceived theoretical scheme. Even Becquerel's photographic plates, which were blackened by radioactivity when accidentally exposed to uranium salts while they lay idle in a drawer, could only convey their startling revelation to someone versed in the carefully articulated late nineteenth-century view of nature. Democritus or Aristotle would probably have been content with a more casual explanation of a much lesser consequence. But late nineteenth-century physics, the classical system that some had come to think would thenceforth progress only by making better measurements of well-defined, well-understood quantities, proved unable to cope with the new phenomena, and the efforts of the early quantum theorists were driven by their perception of this somehow paradoxical state of affairs. This explains their partial yet persistent reliance on classical mechanics and electrodynamics, whose concepts and principles were, after all, built into the experimental and computational procedures by which the new information was obtained. It also accounts for their readiness to combine seemingly inconsistent notions, to try out obscure algebraic relations in lengthy and laborious calculations, in short, to follow every lead that might take physics out of the morass in which it was caught.

The Old Quantum Theory can be traced back to the work by Planck (1900) and Einstein (1905i) mentioned above, but it began in earnest with Bohr's three-part series, "On the Constitution of Atoms and Molecules" (1913). By that time it was generally agreed that matter consisted of atoms of a number of (the order of 10^2) different kinds. Despite the etymology of their name (< 'a-' privative + 'tomos' = 'slice'), atoms were supposed to have parts. These were required in view of the complex structure of atomic spectra and because atoms, which under ordinary conditions are electrically neutral, acquire a positive or negative charge through ionization. After Thomson's discovery of the electron, the opinion prevailed that every atom contains such particles, that ionization is due to the atom's catching or shedding one or more of them, and that the spectral lines characteristic of each element reflect the periodic motion that is proper to the electrons in its atoms. Two alternatives were contemplated: Either the extremely light, negatively charged electrons circulate around a massive, positively charged nucleus, like planets around the sun, or they are buried, like raisins in a cake, inside a positively charged sphere. The latter arrangement had greater hope of being proved stable according to classical electrodynamics, but it foundered on some stunning experimental results. Geiger and Marsden (1909), working under Rutherford in Manchester, bombarded thin sheets of metal foil with α-rays, or α-particles, as they were already called by then. Most of them went through, but gold foil 6 × 10^{-5} centimeters thick deflected about 1 particle in 8,000 by an angle ≥90°. Years later Rutherford described his reaction as follows: "It was quite the most incredible event that has ever happened to me in my life. It was almost as incredible as if you fired a 15-inch shell at a piece of tissue paper and it came back and hit you" (quoted by Pais 1986, p. 189). He concluded that almost all the gold mass was concentrated in pointlike positively charged nuclei separated by comparatively enormous distances; this explained both the overall transparency of the foil to the α-particles and the energetic rebound of the few that hit the mark. Rutherford therefore embraced the planetary model of the atom. However, according to classical electrodynamics, an electron moving circularly, and thus acceleratedly, around the atomic nucleus would continually emit radiation, thereby losing energy, so the time should soon arrive when the electron, no longer able to resist the electric attraction of the nucleus, would plunge into it. Even if this process, as some conjectured, was infinitely slow, radiation from a body of continuously decreasing energy could hardly be responsible for the neatly defined spectral lines characteristic of each element; nor could one reasonably hope to explain the stability of such lines from classical premises if, as all agreed, every atom in an incandescent gas is being struck by others at the rate of 100 million collisions per second.

This led, in Bohr's words, to "a general acknowledgment of the inadequacy of the classical electrodynamics in describing the behavior of systems of atomic size" (1913, p. 1). It was "necessary to introduce in the laws in question a quantity foreign to the classical electrodynamics, i.e. Planck's constant, or as often it is called the elementary quantum of action" (Ibid.). In Bohr's new scheme of things an atom of a given element can exist in different "stationary states" in which each electron circulates around the nucleus at a characteristic energy level. The electrons absorb or emit radiation only during the brief transition from one stationary state to another. Bohr assumed that:

(i) classical mechanics can be of help in discussing the dynamics of the stationary states, but it does not apply to the transition from one such state to another; and

(ii) the transition of an electron from a higher energy level E_n to a lower energy level E_m is accompanied by the emission of monochromatic radiation whose frequency $\nu(n,m)$ is related to the energy difference $E_n - E_m$ by

$$E_n - E_m = h\nu(n,m) \qquad (6.2)$$

Equation (6.2) agreed beautifully with the Ritz Combination Principle. According to this rule, which epitomized the experience of spectroscopists, every frequency of the observable spectrum can be expressed as the difference between two terms, and any difference between two such terms corresponds to a possibly observable difference. Thus, given spectral lines of frequency $\nu(n,k) = T_n - T_k$ and $\nu(k,m) = T_k - T_m$, one may expect a line of frequency

$$\nu(n,m) = \nu(n,k) + \nu(k,m) \qquad (6.3)$$

Bohr's scheme secured the atom's stability. Its straightforward application to the single-electron hydrogen atom enabled him to calculate the frequencies of the well-known Balmer series of spectral lines. At first Bohr ignored the fact, already established by Michelson in 1891, that most of the lines in the hydrogen spectrum can be resolved into multiplets, that is, sets of two or more lines with slightly different frequencies. However, Sommerfeld (1915a, 1915b, 1916) was able to account for them. To do so, he assumed that electrons move, like planets, on elliptic orbits; he took into consideration the relativistic variation of inertia with speed, and, most importantly, he postulated the famous *quantum condition* reminiscent of Pythagoric arithmology. By virtue of it, the stationary states of a (multiply) periodic mechanical system with r degrees of freedom satisfy the equations

$$\oint p_k dq_k = n_k h \qquad (1 \le k \le r) \qquad (6.4)$$

where the q_k and p_k are the generalized position and momentum coordinates employed for describing the system (§2.5.3), each integral is taken over a period of the respective q_k, and the n_k are nonnegative integers, sometimes referred to as "quantum numbers".[5]

Still, Bohr's atom theory had great difficulty in dealing with the

---

[5] The quantum condition implies that, among the continuously many mathematically admissible solutions of the classical equations of motion for a system of bound electrons orbiting under Coulomb's Law around an atomic nucleus, only those that satisfy eqns. (6.4) are physically possible. Wilson (1915) introduced these equations a few months before Sommerfeld and unbeknownst to him, but he did not calculate spectral lines from them.

observed effects of magnetic and electric fields on atomic spectra, and it never yielded accurate values for the spectrum of helium – element number 2 – not to mention anything more complex. But it did give a promising overall qualitative understanding of atomic structure, valence, and the periodic system. So Bohr and his followers persisted in trying to extend the scope of their quantitative predictions. The perturbation methods of classical celestial mechanics were skillfully applied by Sommerfeld and Born to the calculation of electron orbits in many-electron atoms, while Bohr conducted with admirable tact the guesswork prompted by his Correspondence Principle. As explained by him, this is "a law of the quantum theory" (1924, p. 22n), which asserts "a far-reaching *correspondence* between the various types of possible transitions between the stationary states on the one hand and the various harmonic components of the motion on the other hand" (1922, pp. 23f.; cf. Bohr 1934, p. 37). Its sole motivation (or justification) was to secure that, in contexts in which Planck's constant h is insignificant, the quantum theory would agree with classical electrodynamics, "according to which the nature of the radiation emitted by an atom is directly related to the harmonic components occurring in the motion of the system" (Ibid.). But Bohr and his collaborators wielded the Correspondence Principle, not without success, like a flexible magic wand, also in situations where the finite value of h makes a difference, as a guide to quantitative predictions and in selecting the stationary states of an atom that are physically possible from among the continuously many that are mechanically conceivable.[6]

6.1.2   Einstein on the Absorption and Emission of Radiation

In 1916 Einstein, who was long busy with his theory of gravity, again entered the lists of quantum physics. On 11 August he wrote to Besso: "A splendid idea on the absorption and emission of radiation has dawned on me". Its fruits were two papers (1916j, 1916n) in which Einstein derives Planck's Law by a statistical argument that does not depend on the premises from classical mechanics and electrodynamics that Planck had used. The argument concerns a gas consisting of equal molecules in statistical equilibrium with thermal radiation. Mindful of Kirchhoff's theorem on thermal radiation (see note 2), Einstein does

---

[6] For a fairly clear illustration of the latter use of the Correspondence Principle, see Bohr (1918) in Van der Waerden (1967, pp. 110f.).

not assume anything about the constitution of such molecules, except that each will go from a state Z_m with energy E_m to a state Z_n with energy E_n by emitting or absorbing radiation of a definite frequency v_{nm}. The argument yields not only a formula for the density of thermal radiation equivalent to Planck's, but also a *proof* that the difference $|E_m - E_n|$ between the energy levels must be *proportional* to the frequency v_{nm}, the proportionality constant filling precisely the place of h in Planck's formula. What mainly matters for our present story is Einstein's innovative approach to energy emission, in case $E_m > E_n$. He recalls that an oscillating Planck resonator would radiate energy whether or not it is excited by an external field. Correspondingly – he assumes – each of his molecules may go from a state Z_m to a state Z_n emitting radiation energy $E_m - E_n$ with frequency v_{nm} without excitation from external causes. "One can hardly think of it otherwise than on the analogy of radioactivity" (1916j, p. 321); this refers to Rutherford's findings on radioactive decay mentioned at the beginning of §6.1.1. Let M be a mass of some radioactive element L. After time t_L – the so-called half-life of L – one-half of the atoms in M will yield their radioactive output and transmute into another element; t_L is characteristic of L and does not depend at all on external circumstances. Since all atoms of L are equal, one is bound to think of each as a chance setup with a 0.5 probability of decaying within time t_L. Based on this analogy Einstein equates the probability dW that the transition from Z_m to Z_n occurs within time dt with $A_m^n dt$, where A_m^n is a constant characteristic of the index combination (m,n). Turning now to energy absorption, he naturally assumes that it depends on the surrounding thermal radiation, which will do work on our molecule in proportion to the radiation density of the relevant frequency. The work done on a Planck resonator by the surrounding classical electromagnetic field can be either positive or negative, depending on their respective phases. Correspondingly, Einstein introduces "the following quantum-theoretical hypotheses" (1916n, p. 51): Under the action of radiation density ρ_{nm} of frequency v_{nm} a molecule can either pass from state Z_n to state Z_m by absorbing energy $E_m - E_n$ or pass from Z_m to state Z_n by releasing energy $E_m - E_n$, the probability that transition $Z_n \rightarrow Z_m$ occurs in time dt being $B_n^m \rho_{mn} dt$ and the probability that transition $Z_m \rightarrow Z_n$ occurs in time dt being $B_m^n \rho_{mn} dt$ (where B_n^m and B_m^n are constants characteristic of the respective index combinations). Now, in thermal equilibrium, the number of transitions from Z_m to Z_n (due to negative absorption or spontaneous emission) must balance the number of transitions from Z_n to Z_m (due to absorption). So, denoting by N_j the number of molecules in state Z_j, we have that

$$(A_m^n + B_m^n \rho_{mn})N_m = B_n^m \rho_{mn} N_n \qquad (6.5)$$

Let the probability W_j that a molecule finds itself in the state Z_j be given, as in classical statistical physics, by

$$W_j = p_j \exp(-E_j/kT) \qquad (6.6)$$

where p_j is "a constant, characteristic of the molecule's quantum state Z_j and independent of the gas temperature".[7] From eqns. (6.5) and (6.6), Planck's Law and the relation $E_m - E_n = h\nu_{nm}$ follow – as Einstein announced to Besso – by "bafflingly simple" algebra.

In the second paper Einstein also investigates the motion of his molecules under the influence of radiation. The brilliant discussion leads to the following conclusions:

If a radiation bundle strikes a molecule causing it through an elementary process to receive or release the quantity of energy $h\nu$ in the form of radiation, a momentum $h\nu/c$ is always transferred to the molecule, in the direction of propagation of the bundle if energy is received, and in the opposite direction if energy is released. If the molecule is acted upon by several directed bundles of radiation, there is always only one of them taking part in a particular elementary process of irradiation; this bundle alone determines then the direction of the momentum transferred to the molecule.

If the molecule undergoes without external excitation a loss in energy of magnitude $h\nu$ by emitting this energy in the form of radiation, this process is also a *directed* one. There is no outgoing radiation in spherical waves. In the elementary process of emission the molecule experiences a recoil of magnitude $h\nu/c$ in a direction which, in the present state of the theory, is only determined by "chance".

(Einstein 1916n, p. 61)

Einstein expressly regrets that the time and direction of emission are being left to "chance" ("*Zufall*" – a word he pointedly surrounds with scare quotes). Nevertheless, in the light of the above conclusions, the establishment of a "proper quantum theory of radiation" – in accordance with his earlier conjecture (1905i; see above) – "appears almost

---

[7] Einstein (1916j, p. 320), where he stresses that eqn. (6.4) can also be obtained "from thermodynamical considerations" (thus without assuming the validity of classical mechanics).

inevitable". On the other hand, he admits that he is not a bit closer to harmonizing this approach with the wave theory of radiation, based on undeniable facts of radiation interference.

6.1.3 Virtual Oscillators

It was this latter, apparently insurmountable difficulty that moved the leading scientists to resist Einstein's proposal on light quanta. Planck himself, when recommending Einstein's incorporation to the Prussian Academy, noted it as a forgivable blemish in an otherwise extraordinary creative career.[8] And Bohr – who continued to spurn a corpuscular view of radiation even after Compton arranged X-rays to ricochet from electrons like tiny bullets[9] – is said to have replied, as late as 1924, to critical remarks by Einstein that if Einstein telegraphed him that he had found an irrevocable proof of the existence of light quanta, "the telegram could only reach me by radio on account of the waves which are there".[10] So one had to devise some way of linking the discontinuous emission and absorption of discrete amounts of radiation energy by atoms with the supposedly continuous propagation of radiation through space. In the paper that motivated the said exchange with Einstein (Bohr, Kramers, and Slater 1924), Bohr appeared ready to sacrifice the conservation of energy and momentum,[11] which would henceforth hold only in the statistical average but not for every par-

---

[8] "That he may sometimes have missed the target in his speculations as, for example, in his hypothesis of light quanta, cannot really be held against him. For without taking a risk no innovation can really result in the most exact natural science" (from a request to the Prussian Ministry of Education, dated 12 June 1913, handwritten by Planck and signed by him, W. Nernst, H. Rubens, and E. Warburg; quoted in Seelig 1957, p. 145).

[9] Compton showed that X-rays scattered at less than a right angle by a block of paraffin have a smaller frequency than the incident radiation. This result is readily explained on the hypothesis of light quanta: if a light quantum strikes an electron, it will transmit to it a quantity of energy ΔE, so its frequency will decrease by $\Delta \nu = \Delta E/h$.

[10] Quoted in Jammer (1966, p. 187), from an interview with Werner Heisenberg on 15 February 1963.

[11] Bohr had been toying with this idea for some time. Here is a passage from an early draft of Bohr (1918) that was not included in the final version: "It would seem that any theory capable of an explanation of the photoelectric effect as well as the interference phenomena must involve a departure from the ordinary theorem of conservation of energy as regards the interaction between radiation and matter" (quoted by Darrigol 1992, p. 214).

ticular energy–momentum transfer. This seemingly desperate expedient was combined with the novel idea of "virtual" oscillators creating "virtual" fields, contributed by Slater.[12]

> We will assume that a given atom in a certain stationary state will communicate continually with other atoms through a time-spatial mechanism which is virtually equivalent with the field of radiation which on the classical theory would originate from the virtual harmonic oscillators corresponding with the various possible transitions to other stationary states. Further, we will assume that the occurrence of transition processes for the given atom itself, as well as for the other atoms with which it is in mutual communication, is connected with this mechanism by probability laws which are analogous to those which in Einstein's theory [described in §6.1.2 – R. T.] hold for the induced transitions between stationary states when illuminated by radiation. On the one hand, the transitions which in this theory are designated as spontaneous are, on our view, considered as induced by the virtual field of radiation which is connected with the virtual harmonic oscillators conjugated with the motion of the atom itself. On the other hand, the induced transitions of Einstein's theory occur in consequence of the virtual radiation in the surrounding space due to other atoms.
>
> . . . .
>
> We shall assume an independence of the individual transition processes, which stands in striking contrast to the classical claim of conservation of energy and momentum. Thus we assume that an induced transition in an atom is not directly caused by a transition in a distant atom for which the energy difference between the initial and the final stationary state is the same.
>
> (Bohr, Kramers, and Slater 1924, in Van der Waerden 1967, pp. 164–65, 166)

The authors grant that "at present there is unfortunately no experimental evidence at hand which allows to test these ideas," but they insist that the independence that they ascribe to the transition processes "would seem the only consistent way of describing the interaction between radiation and atoms by a theory involving probability considerations" (Ibid., in Van der Waerden 1967, pp. 166–67). This BKS theory – as it is known in the literature – was forcefully criticized by

---

[12] On Slater's original idea, a new conception of light with "both the waves and the particles", see Darrigol (1992, pp. 218f.). When Slater went to Copenhagen, Kramers and Bohr apparently persuaded him to get rid of light quanta by sacrificing energy–momentum conservation.

Einstein and Pauli and finally wrecked by tests of energy conservation in individual Compton scattering events performed by Geiger and Bothe and by Compton himself. Bohr graciously accepted defeat. The BKS scheme, he wrote to Geiger on 21 April 1925, "was more an expression of an endeavor to attain the greatest possible applicability of the classical concepts than a completed theory" (CW, 5, 353). Thereafter, Bohr rejected all "the space-time pictures previously used in the quantum theory: electronic orbits in stationary states, trajectories in collision processes, radiation fields, *and* corpuscular light-quanta" (Darrigol 1992, p. 252).

Nevertheless, Slater's virtual oscillators still played a role in the theory of dispersion put forward by Kramers in two communications to *Nature* (1924a, 1924b) and a longer article coauthored by Heisenberg (1925). According to Kramers, such oscillators were introduced not as an "additional hypothetical mechanism" but "only as a *terminology* suitable to characterise certain main features of the connexion between the description of optical phenomena and the theoretical interpretation of spectra" (Kramers 1924b, p. 311; my italics) – in other words, they were just a manner of speaking. Kramers and Heisenberg's work on dispersion was only partially successful. The "translation" of classical into quantum terms and relations pursuant to Bohr's Correspondence Principle attained here new levels of virtuosity. Finite difference quotients systematically replace the derivatives in the classical formulas. A similar "transition from differential to difference equations" was advocated by Born in "Über Quantenmechanik" (1924), the first paper to use this name in print.[13] Heisenberg (1925) carried such translation methods one long step forward, with amazing success.

6.1.4   On Spin, Statistics, and the Exclusion Principle

I shall finish this sketch of the immediate background of QM with a short note on three novelties that made their appearance more or less

---

[13] Born described this work as "a first step towards a quantum theory of coupling" (1924, p. 379). He endorsed Kramers's handling of dispersion caused by the interaction between atomic electrons and radiation and sought to extend it to the interaction between the several electrons in a single atom (heavier than hydrogen). Born's method of substituting finite differences for classical differentials is concisely explained by Tian Yu Cao (1997, p. 134) in a manner that clearly brings out its affinity with Heisenberg's (1925).

at the same time as the new theory, namely, Pauli's Exclusion Principle, the "spin" of the electron, and the new approach to physical statistics initiated by Bose. None of them played a role in the foundation of QM, but they enabled it to cope with experimental data that would otherwise have eluded it, so they were received within it as welcome additions. Later they found in the context of relativistic quantum theory a rational justification that bound them together.

Pauli lighted on the Exclusion Principle in his endeavor to understand the so-called anomalous Zeeman effect. Spectral lines are split when the source of light is placed in a magnetic field (Zeeman 1897). A threefold split could be readily accounted by classical electrodynamics (Lorentz 1897) and also by Bohr's atom theory. But experiments with heavier elements and better instruments disclosed further splits, which were dubbed "anomalous" because they did not fit the available explanations. When Pauli tackled the problem in the 1920s the developing Old Quantum Theory characterized the stationary state of a bound electron by three quantum numbers, which were thought to reflect, respectively, the size of the electron's orbit, its shape, and its inclination with respect to the external magnetic field. Pauli (1925a, p. 385) concluded that the doublet structure of alkali spectra is due to "a peculiar kind of duplicity (*Zweideutigkeit*) in the quantum-theoretic properties of the optical electron, which cannot be described in classical terms". Shortly thereafter, in a second, brilliant paper, Pauli added a fourth, two-valued quantum number to the former three and postulated his Exclusion Principle:

> *There can never be two or more equivalent electrons in an atom for which the values of all* [four] *quantum numbers coincide in strong fields. If there is an electron in an atom for which these quantum numbers take certain values (in the external field), this state is "occupied".*
>
> (Pauli 1925b, p. 776)

With this powerful and very original assumption Pauli was able to explain both the "anomalous" Zeeman multiplets and the number of electrons allowed in each group or "shell" surrounding the atomic nucleus.

In his two papers Pauli carefully avoided any suggestion that might help to visualize the physical meaning of the quantum numbers by associating them with the area, excentricity, tilt, etc., of an electron orbit. He was probably convinced that – as he put it in the survey of quantum theory that he wrote in the same year, 1925 – "one must renounce the

practice of attributing to electrons in the stationary states trajectories
that are uniquely defined in the sense of ordinary kinematics" (1926a,
p. 167). This may have motivated his initial rejection of the solution
of the Zeeman anomalies put forward by Uhlenbeck and Goudschmidt
(1925), which bestowed a definite – classical – physical meaning on
Pauli's fourth quantum number. From a classical standpoint, the atom's
behavior in a magnetic field depends on its angular momentum. By eqn.
(6.4), the angular momentum due to the electron's orbital motion must
be an integral multiple of $h/2\pi$ (henceforth abbreviated $\hbar$). Uhlenbeck
and Goudschmidt assumed that the electron also has an intrinsic
angular momentum or *spin*, which can be conceived classically as due
to the electron's rotation about a stable axis and takes the value $+\hbar/2$
or $-\hbar/2$ (depending on the sense of rotation). By taking spin into
account they obtained the number of distinct atom states required to
match the observed Zeeman multiplets. The spin hypothesis accounts
precisely for the peculiar duplicity or twofoldness in the quantum-
theoretic properties of the electron that Pauli had noted (1925a, p. 385;
quoted above). However, the conception of the electron as a finite
rotating sphere is fraught with difficulties, for not only is it hard to
understand what keeps its charge together, but its equatorial velocity
would have to exceed the speed of light to yield spin $= \pm\hbar/2$. After the
image of the stationary atom as a classical mechanical system was scut-
tled by Heisenberg (§6.2.1) it was no longer necessary or even possible to understand the spin of the electron as a manifestation of rotation.
Just as Pauli anticipated, spin is now conceived as an irreducible
quantum property of matter, but the name 'spin' has stuck.

Finally, I turn to the matter of quantum statistics. Bose (1924)
showed that Planck's Black-Body Radiation Law could be established
by a purely statistical argument, without appealing to the classical electrodynamic assumptions invoked by Planck (1900), if one regarded
photons of a given frequency as utterly undistinguishable for statistical purposes. This has the effect of reducing the number of distinct
states available to a photon gas. A simple, abstract example will show
why this is so. Consider a system consisting of two objects, each of
which can be in one of three possible states. If each object has an individuality of its own, we may denote them by proper names, say a and
b. The system can be in one of nine different states, viz., $[a]$-$[b]$-$[\ ]$, $[b]$-
$[a]$-$[\ ]$, $[\ ]$-$[a]$-$[b]$, $[\ ]$-$[b]$-$[a]$, $[a]$-$[\ ]$-$[b]$, $[b]$-$[\ ]$-$[a]$, $[ab]$-$[\ ]$-$[\ ]$, $[\ ]$-$[ab]$-
$[\ ]$, and $[\ ]$-$[\ ]$-$[ab]$, where $[a]$-$[b]$-$[\ ]$ is the systemic state in which a is
in the first object-state and b is in the second, and so on. (This approach

is characteristic of the classical, Maxwell–Boltzmann statistics examined in §4.3.) But if the objects are indistinguishable and are denoted, say, by o and o, there are no more than six different states in which the system can find itself, viz., [o]-[o]-[ ], [ ]-[o]-[o], [o]-[ ]-[o], [oo]-[ ]-[ ], [ ]-[oo]-[ ], and [ ]-[ ]-[oo]. Bose's approach was generalized and fruitfully applied to several pending problems by Einstein (1924, 1925a, 1925b) and is commonly known as Bose–Einstein statistics. Note, however, that the last three states mentioned are not possible if the objects obey Pauli's Exclusion Principle. In this case, the available systemic states reduce to three, viz., [o]-[o]-[ ], [ ]-[o]-[o], and [o]-[ ]-[o]. This statistical approach was introduced by Dirac (1926a) and Fermi (1926) and is known as Fermi–Dirac statistics. The physical particles are currently classified into fermions (such as the electron and the proton), which obey the Exclusion Principle, and bosons (such as the photon), which do not. Fermi–Dirac statistics apply to the former, and Bose–Einstein statistics to the latter. In QM one postulates that the spin of a boson is an integral multiple and the spin of a fermion a half-integral multiple of $\hbar$. In relativistic quantum theory this connection between spin, statistics, and the Exclusion Principle is proved as a theorem (Pauli 1940).

6.2   The Constitution of Quantum Mechanics

6.2.1   Matrix Mechanics

According to classical electrodynamics, the frequency of radiation emitted or absorbed by an atom is directly related to the frequency with which its electrons vibrate or circulate. In Bohr's theory, this connection is suppressed. The measurement of spectral lines can now yield no information at all about the periodic motion of electrons. Bohr and his followers continued to speculate about electron orbits to calculate energy levels. By 1925, however, it was clear that this activity was yielding quickly diminishing returns. Born and Jordan sternly warned that the "quantities that enter the true laws of nature" must all be "observable and ascertainable in principle" (1925a, p. 493), while Kramers and Heisenberg boasted that the formulas of their dispersion theory "contain only the frequencies and amplitudes which are characteristic for the transitions [between stationary states of the atom], while all those symbols which refer to the mathematical theory of periodic systems will have disappeared" (1925, in Van der Waerden p. 234). In line with this trend, Heisenberg resolved "to give up altogether the hope of observing the hitherto unobservable quantities (such as the electron's position and period)" and "to try to develop a quantum-theoretical mechanics, analogous to classical mechanics, in which only relations between observable quantities occur" (1925, p. 880).[14]

He secured the analogy between the new and the old mechanics by what he called a "reinterpretation of kinematic and mechanical relations". The classical equations of motion were preserved, but the letters that occur in them were made to stand for new-fangled mathematical objects, each of them representing not a single real number but an infinite array of complex numbers. A reinterpretation of the mathematical operations at play was therefore inevitable: One continued to talk of multiplication or differentiation and to use the familiar symbols, but they now meant something quite different. To say that kinematic and dynamic relations were merely being "reinterpreted" is perhaps an understatement. At any rate, the equations of motion did not apply now to something one could call "motion" in a standard sense. The name "mechanics" could be retained, insofar as the new theory was intended, like Newton's, as a basic framework for describing and explaining all forms of physical change; but physical change was no longer to be explained by the displacement of point-masses in three-dimensional space. A failure to accept this, or perhaps even to realize it, may account for some of the difficulty in understanding QM.

One could not say that Heisenberg had a rational justification for proceeding as he did, but here are some hints as to the intellectual motives that guided him. Any periodic function corresponding to a time-dependent physical quantity can be represented by a Fourier series. In the Old Quantum Theory one considered time-dependent

[14] Heisenberg's stance on this question should be taken with a pinch of salt. On the one hand, his "observables" include the very amplitudes that Born and Jordan (1925a) left aside as unobservable; on the other, although Heisenberg discarded electron positions as unobservable, in mature QM the three Cartesian coordinates of a particle are considered observables. Many years later, Heisenberg (1969, p. 95) approvingly recalled some remarks on observability that he heard from Einstein in 1926: Observation presupposes a known, unambiguous relation between the phenomenon to be observed and our sense perceptions. We can be certain of this relation only if we know the laws that determine it. If the laws of nature are in doubt – as they obviously were for Heisenberg in 1925 – the concept of observation has no clear meaning. The theory must then by itself determine what is observable. (Indeed, it was Bohr's atom *theory* that made electron orbits *unobservable*; cf. Bohr 1934, pp. 12, 36.)

quantities pertaining to each stationary state of an atom. Thus, a single electron's displacement $x_n(t)$ in the nth state could be expressed as:

$$x_n(t) = \sum_{k=-\infty}^{\infty} x_{nk} \exp(2\pi i v_n kt) \qquad (6.7)^{15}$$

Now, "in quantum theory it has not been possible to associate the electron with a point of space, considered as a function of time, by means of observable quantities; however, even in quantum theory it is possible to ascribe to an electron the emission of radiation" (Heisenberg 1925, p. 881). Radiation is emitted (or absorbed) in the transition between two stationary states and is related by eqn. (6.2) to the difference between the respective energy levels; thus,

$$v(n, n-k) = \frac{1}{h}(E_n - E_{n-k}) \qquad (6.8)$$

Bohr and his school took the frequencies $v(n,n-k)$ as the quantum-theoretic analogue of the classical frequencies v_n. Heisenberg introduces quantum-theoretic analogues $x(n,n-k)$ for the classical displacement components x_{nk}. So each Fourier term $x_{nk}\exp(2\pi i v_n kt)$ in eqn. (6.7) is to be replaced in the quantum mechanics by $x(n,n-k)$ $\exp(2\pi i v(n,n-k)t)$. Due to the symmetric role of the indices n and $(n-k)$, these new terms cannot be meaningfully gathered into an infinite series such as (6.7). So, Heisenberg says, one ought to regard the whole array of complex amplitudes $x(n,n-k)\exp(2\pi i v(n,n-k)t)$ – with $n = 1, 2, \ldots$, and with k ranging over all the integers – as the quantum-theoretic analogue of the classical quantity $x(t)$. This is the backbone of his new "kinematics". He then asks: "What will represent the quantity $[x(t)]^2$?" (Ibid., p. 882). The answer to this question will yield the quantum-theoretic analogue of $[x(t)]^r$ for every integer r and hence of any function $f[x(t)]$ that can be expanded in a power series. Now, classically, one got the square of the left-hand side of eqn. (6.7) by multiplying term by term the Fourier series on the right-hand side by itself. This can be written as

$$[x_n(t)]^2 = \sum_{k=-\infty}^{\infty} c_{nk} \exp(2\pi i v_n kt) \qquad (6.9)$$

where

$$c_{nk} \exp(2\pi i v_n kt) = \sum_{j=-\infty}^{\infty} x_{nj} x_{n(k-j)} \exp(2\pi i v_n (j+k-j)t) \qquad (6.10)$$

---

[15] Remember that $\exp(i\alpha) = \cos\alpha + i\sin\alpha$.

According to Heisenberg, in quantum theory, instead of eqn. (6.10), one must write:

$$c(n,n-k)\exp(2\pi i v(n,n-k)t)$$
$$= \sum_{j=-\infty}^{\infty} x(n,n-j)x(n-j,n-k)\exp(2\pi i v(n,n-k)t) \qquad (6.11)^{16}$$

Indeed, according to him, the Ritz Combination Principle leads to "this form of composition almost compulsorily" (Ibid., p. 883). Generally speaking, the product xy of two quantities x and y, represented in quantum theory by arrays with typical elements $x(n,k)$ and $y(m,j)$, respectively, should be represented by an array z whose typical element is

$$z(n,m) = \sum_{j} x(n,j)y(j,m) \qquad (6.12)^{17}$$

Equation (6.12) evidently implies that $xy \neq yx$, except in special cases. In contrast with ordinary multiplication, the product of two quantum quantities is not commutative. Heisenberg initially had some difficulty in accepting this. Equation (6.12) states the rule of matrix multiplication, which is nowadays familiar to countless professionals who use linear algebra in their work (Supplement I.6). But in 1925 Heisenberg had no idea what a matrix was. It was Born who, after pondering Heisenberg's manuscript for a whole day and night, remembered a course that he had taken in his youth with the mathematician Rosanes and recognized Heisenberg's complex number arrays as matrices.

Heisenberg tried his translation rules on a simple mechanical problem: the anharmonic oscillator. From the matrix analogue of the classical equation

$$\ddot{x} + \omega_0^2 x + \lambda x^2 = 0 \qquad (6.13)$$

he obtained good values for the energy levels. Papers by Born and Jordan (1925b) and by Born, Heisenberg, and Jordan (1926 – the "three-men paper") developed Heisenberg's scheme into a system of matrix mechanics. The authors gave plausible rules for symbolic matrix differentiation – a different one in each paper – and adopted the following precise measure of the noncommutatitivity of the matrices q

---

[16] By eqn. (6.8), $v(n,n-j) + v(n-j,n-k) = v(n,n-k)$. So

$$\exp(2\pi i v(n,n-j)t)\exp(2\pi i v(n-j,n-k)t) = \exp(2\pi i v(n,n-k)t)$$

[17] That eqn. (6.11) is a special case of eqn. (6.12) can be seen by substituting j for $n-j$ and m for $n-k$.

and p, which replace the classical position and momentum coordinates q_k and p_k:

$$qp - pq = i\hbar 1 \qquad (6.14)$$

(1 denotes the unit matrix, with diagonal elements equal to 1 and all other elements equal to 0). Equation (6.14) was discovered in rapt meditation by Born, who dubbed it the "sharpened quantum condition", because it replaces in the new theory the Sommerfeld quantum condition (6.4). On the analogy of the Hamilton equations (2.41), the quantum-mechanical equations of motion take the following canonical form (where the dot signifies matrix differentiation with respect to time and H is the matrix representing the classical Hamiltonian function (2.38)):

$$\dot{q} = \frac{\partial H}{\partial p} \qquad \dot{p} = \frac{\partial H}{\partial q} \qquad (6.15)$$

The new mechanics won general approval when Pauli (1926b) calculated from it the hydrogen spectrum in agreement with experience both in conditions already covered by the Old Quantum Theory and in the presence of crossed electric and magnetic fields, in which the Old Quantum Theory failed.

6.2.2   Wave Mechanics

In the same year Schrödinger published the four installments of a long paper, "Quantisation as a problem of proper values" (1926a, 1926b, 1926d, 1926e),[18] in which he tackled the unsolved problems of atomic physics in a way that seemed completely different from and almost opposite to that of matrix mechanics. Schrödinger drew his inspiration from the bold ideas put forward by Louis de Broglie in two communications to the Paris Academy (1923a, 1923b) and in his doctoral thesis (1925). Rather than seek, like Bohr, to reconcile by means of an

---

[18] 'Proper values' of linear operators are defined in Supplement I.6. The following example illustrates the use of the term in the present context. Let D stand for the differential operator d/dx and let f be a unknown function of x. Consider the equation

$$D^2 f - \lambda f = 0 \qquad (*)$$

It can be readily verified that for each positive integer n the function $f_n = \exp(nx)$ satisfies eqn. (*), with $\lambda = n^2$. The functions $f_1, f_2, f_3, \ldots$ are proper functions of the operator D^2, and $1^2, 2^2, 3^2, \ldots$ are the corresponding proper values.

opportunistic hypothesis the new evidence for light quanta with the well-established fact of radiation interference, de Broglie extended the difficulty to ordinary massive matter, thus pressing the need for a radical solution. He recalled that Hamilton had been led to his equations of motion (eqns. (2.41)) by a formal analogy between geometrical optics and analytical mechanics, and proposed that the new mechanics, required by atomic phenomena, should stand to the classical mechanics of corpuscular trajectories in a relation similar to that of the wave theory of light to the geometrical optics of light rays. Special Relativity (SR) established that 'matter' and 'energy' are synonymous expressions that denote the same physical reality. Since quantum physics associates to every isolated portion of energy E a certain frequency $v = E/h$, it is plausible to think that every material particle of rest mass m_0 is associated with a periodic phenomenon of frequency $v_0 = m_0 c^2/h$. De Broglie proved from SR that an observer relative to whom a particle moves uniformly with velocity v must assign to the said wave a "phase velocity" (velocity of propagation) $V = c/\sqrt{1-(v/c)^2} > c$. On the other hand, if two or more such waves are superposed, the said observer will see the maximum beat formed by them move with a "group velocity" that is precisely the same as the classical velocity of the particle. De Broglie therefore suggested that so-called material particles are simply the traveling beats formed by very narrow packets of superposed waves. Since energy and momentum stand to one another in SR as the timelike and the spacelike parts of a single spatio-temporal reality, the same relation that holds between the energy E and the frequency v – which is the number of wave cycles per unit of time – must hold between the momentum p and the reciprocal value of the wavelength λ – which is the number of cycles per unit of length –, viz., $p = h/\lambda$. De Broglie's examiners applauded his brilliant mathematics but met his physical suggestions with scepticism. Einstein (1925a, p. 9) was more receptive and contributed to their diffusion. Two years later, Davisson and Germer published their discovery of electron diffraction, that is, the interference of electron beams.[19]

---

[19] Born says that the discovery by Stern and associates, in 1932, that molecular beams of hydrogen and helium also show diffraction phenomena when reflected by crystals was particularly impressive. "De Broglie's equation $[p = h/\lambda]$ was confirmed for these particles with an accuracy of about 1 per cent. Here, surely, we are dealing with material particles, which must be regarded as the elementary constituents not only of gases but also of liquids and solids" (1962, p. 96).

Schrödinger seized on de Broglie's idea before its spectacular experimental confirmation. It furnished him a welcome means of reconciling the continuity of nature with the presence of integers in quantum physics, instead of linking these to "jumps" or some other unpalatable discontinuity. It is a well-known fact that a vibrating string with both ends fixed displays a definite number of motionless points or *nodes*. Likewise, if bound electrons are conceived as standing waves around the atomic nucleus, the differential equation governing the wave motion could yield the integers characteristic of their state (cf. eqns. (6.4)).[20] So Schrödinger set out to find the equation of the matter waves. He met great difficulties in setting up a Lorentz invariant equation, so, despite, the fact that de Broglie's arguments were firmly rooted in SR, Schrödinger decided to work provisionally on a wave equation valid in the nonrelativistic approximation. To find it he resorted to familiar classical methods. It is not necessary to explain them here. I give, however, a superficial description of Schrödinger's procedure to show the strong roots of QM in analytical mechanics. This may help dispel some common misrepresentations concerning revolutions in physics.

Schrödinger replaces the action S in the classical Hamilton–Jacobi equation (2.43) with $K \log \psi$, where ψ is an unknown function and K is a constant with the dimensions of action. If we ignore the relativistic change of mass with velocity, the equation thus obtained can then be reformulated as $Q(\psi) = 0$, where $Q(\psi)$ is a quadratic form of ψ and its first derivatives. Schrödinger assumes that the ψ-functions of interest to physics are twice continuously differentiable finite functions defined on the entire configuration space of the mechanical system under study.[21] In the tradition of classical variational principles (see remark (iii) after eqns. (2.33)), Schrödinger postulated the following requirement: The integral of $Q(\psi)$ over the entire configuration space is stationary (takes either a minimum or a maximum value). The ψ-

---

[20] "It is hardly necessary to emphasize how much more congenial it would be to imagine that at a quantum transition the energy changes over from one form of vibration to another, than to think of a jumping electron. The changing of the vibration form can take place continuously in space and time, and it can readily last as long as the emission process lasts empirically" (Schrödinger 1926a, in WM, pp. 10–11).

[21] Schrödinger (1926a) also required that the ψ-functions be real-valued (WM, p. 2); this requirement was omitted in 1926b (WM, p. 28) and explicitly left aside in 1926e (WM, p. 104), where the Schrödinger equation is said to allow complex-valued solutions.

functions that satisfy this requirement are the solutions of a certain differential equation, the *time-independent* form of the famous Schrödinger equation that holds center stage in QM.[22] Its solutions represent waves in configuration space, with $|\psi(q)|^2$ the wave intensity at point q. The mysterious integers of quantum theory make their appearance in a natural way in the equation's proper values. So Schrödinger could justifiably regard his variational approach as a substitute for – and a welcome improvement on – the arbitrarily postulated "quantum condition" (6.4).

Schrödinger (1926a) applied the procedure sketched above to the motion of a single electron in the Coulomb field of a much heavier nucleus with the opposite charge. He showed that ψ-functions satisfying the variational postulate exist for every conceivable positive energy value but only for a discrete set of negative energy values.[23] If we put

[22] When one speaks nowadays of the Schrödinger equation, without further qualification, one usually refers to the time-dependent equation first introduced in Schrödinger (1926b, eqn. 18). This can be schematically written:

$$H\psi(t) = i\hbar \frac{d\psi(t)}{dt} \qquad \text{(ES1)}$$

where $\psi(t)$ denotes the state of the system at time t and H stands for the Schrödinger Hamiltonian operator, obtained from the classical Hamiltonian $H(q_k, p_k)$ of the mechanical problem under study by substituting $-i\hbar\partial/\partial q_k$ for p_k. The time-independent equation of Schrödinger (1926a) can be summarily written:

$$H\psi = w\psi \qquad \text{(ES2)}$$

where ψ denotes a stationary state and w is a real number interpreted as a proper value of energy. Nowadays one views eqn. (ES2) as a special case of eqn. (ES1) – thus Mainzer (1995, p. 430). However, Schrödinger originally derived eqn. (ES1) from eqn. (ES2) as is explained in note 23.

[23] They are the solutions of the time-independent Schrödinger equation

$$\nabla^2\psi + \frac{2m}{K^2}(E - V)\psi = 0 \qquad \text{(ES2*)}$$

with the potential $V = -e^2/r$. (As Falkenburg 1995, p. 154 n. 20, pointedly remarks: "Quantum Mechanics is on the whole somewhat of a hybrid theory, which describes a *quantum object* in a *classical potential*.") Here is how Schrödinger (1928, §6) derived from eqn. (ES2*) the corresponding form of his time-dependent equation (cf. note 22). The time dependence of the wave function is expressed by $\psi \sim \exp(i\omega t)$, where $\omega = 2\pi\nu = E/\hbar$. Therefore, $\partial\psi/\partial t = (E/\hbar)i\psi$ and $E\psi = (\hbar/i)\partial\psi/\partial t$. Substituting for $E\psi$ in (ES2*) and putting $\hbar$ for K (as indicated in the main text), one obtains

$$\nabla^2\psi + \frac{2mi}{\hbar}\frac{\partial\psi}{\partial t} - \frac{2mV}{\hbar^2}\psi = 0 \qquad \text{(ES1*)}$$

the constant $K = \hbar$, the said negative values agree numerically with the Bohr energy levels corresponding to the Balmer terms of the hydrogen spectrum. The continuous spectrum of positive energy levels suits the hyperbolic orbits of unbound electrons that pass by the nucleus, from infinity to infinity, like comets go past the sun. In the next installments of his paper (1926b, 1926d) Schrödinger solved other simple examples and developed a perturbation theory, which he applied to the effect of an electric field on the Balmer lines (Stark effect). All results agreed admirably with experimental data, in the limit in which SR effects are negligible; they also tallied with the predictions of matrix mechanics, even where these differed from those of the Old Quantum Theory. This was very surprising, for, as Schrödinger noted, due to "the extraordinary difference between the starting-points and between the methods" of Heisenberg's program and his own, one might expect them "to supplement one another" (1926b, in WM, p. 30) but not to coincide.

6.2.3   The Equivalence of Matrix and Wave Mechanics

The mysterious coincidence was explained by Schrödinger himself in a paper, "On the Relation Between the Heisenberg–Born–Jordan Quantum Mechanics and Mine" (1926c), which he published after the second part of his longer work and before the third. He argued in it that the theories are built around isomorphic mathematical structures, so that "from the formal mathematical standpoint one might well speak of the *identity* of the two theories" (WM, p. 46). The matter was later explained with much greater clarity and precision by the mathematician John von Neumann (1932), whose exposition I summarize.[24]

Faced with a quantum-mechanical problem, matrix mechanics first

---

[24] According to Muller (1997) the equivalence was first *proved* by von Neumann (1932), who established it not for wave mechanics and matrix mechanics as they existed in 1926 – which were *not* equivalent – but for theories developed from them in the next year or two, in part under the influence of Schrödinger's purported proof. Strictly speaking, Muller is right, but perhaps one ought not to speak too strictly about historical creatures such as physical theories. While a theory is still growing in the minds of its authors one cannot simply equate it with what has appeared in print. It is nonetheless remarkable that the community of physicists, including the matrix mechanicists in Göttingen, accepted Schrödinger's proof of equivalence, and obligingly strengthened both theories until the conclusion followed. Note also that von Neumann (1927) already presents, quite rigorously, matrix and wave mechanics as a single theory.

solves the canonical equations (6.15) for the Hamiltonian matrix H. The q and p are, as we know, infinite matrices of complex numbers, which satisfy the commutation rule (6.14). H is then transformed to a diagonal matrix, whose diagonal elements are the admissible energy levels of the quantum system under study. To achieve this one must find a matrix S such that the matrix $W = S^{-1}HS$ is diagonal, with real-valued diagonal elements W_{jj}. Thus, one seeks an invertible matrix S such that HS = SW, or, in other words, such that, for each index pair (i,j),

$$\sum_{k=1}^{\infty} H_{ik}S_{kj} = S_{ij}W_{jj} \qquad (6.16)$$

Therefore, each column $S_{1j}, S_{2j}, \ldots$ of S is a solution of the following problem: To find the sequences $x = \langle x_1, x_2, \ldots \rangle$ such that Hx is a multiple of x, say wx, with w a real number. Every such sequence x is said to be a proper sequence of H, the factor w being the corresponding proper value. Equations (6.16) constitute a complete solution of this problem in the following sense: Every proper value of H occurs in the sequence $W_{11}, W_{22}, \ldots$; and if w is a proper value of H, the corresponding proper sequences are the linear combinations of the columns $S_{1j}, S_{2j}, \ldots$ such that $W_{jj} = w$.

Wave mechanics explicitly seeks the proper values of a differential operator. The Schrödinger equation can be summarily written as

$$H\psi = w\psi \qquad (6.17)$$

where w is a real number and H is Schrödinger's Hamiltonian operator (see note 22). Each solution ψ is a proper function of H and w is the corresponding proper value (which may or may not be different for different proper functions). Again, the w's are the energy levels of the system.

The energy levels obtained from eqns. (6.16) for a particular quantum-mechanical problem coincide with the energy levels given for the same problem by eqn. (6.17). This alone cannot justify the confusing use of the same term 'proper value' in two apparently so diverse mathematical contexts. The fact is, however, that from a mathematician's standpoint the two contexts are intimately related. The sequences of complex numbers $x = \langle x_1, x_2, \ldots \rangle$ that we met above can be regarded as vectors in a vector space $\mathfrak{S}$ characterized as follows:[25]

---

[25] General information about vector spaces is provided in Supplement I. The Dirac symbol $\langle x|y \rangle$ for the inner product of vectors x and y is explained in I.5.

($\mathfrak{S}$1) Vector addition is term-by-term addition: $x + y = \langle x_1 + y_1, x_2 + y_2, \ldots \rangle$.

($\mathfrak{S}$2) Scalar multiplication is term-by-term multiplication by a complex number: $ax = \langle ax_1, ax_2, \ldots \rangle$.

($\mathfrak{S}$3) There is an inner product defined by $\langle x|y \rangle = \Sigma_{k=1}^{\infty} x_k{}^* y_k$, where $x_k{}^*$ denotes the complex conjugate of x_k.

($\mathfrak{S}$4) For every sequence x in $\mathfrak{S}$, the series $\Sigma_{k=1}^{\infty} |x_k|^2$ converges to a finite value.

The Hamiltonian matrix H is plainly the matrix of a linear mapping $\mathfrak{S} \to \mathfrak{S}$. On the other hand, the arbitrary complex-valued functions ψ among which the solutions of eqn. (6.17) are to be found, can be regarded as vectors in a vector space $\mathfrak{F}_\Omega$ characterized as follows:

($\mathfrak{F}$0) Each vector in $\mathfrak{F}_\Omega$ is a twice continuously differentiable finite complex-valued function on the configuration space Ω of a specific mechanical system.

($\mathfrak{F}$1) Vector addition is pointwise addition: For each $\mathbf{q}$ in Ω, $(\psi_1 + \psi_2)(\mathbf{q}) = \psi_1(\mathbf{q}) + \psi_2(\mathbf{q})$.

($\mathfrak{F}$2) Scalar multiplication is multiplication by a complex number: For each a in $\mathbb{C}$ and $\mathbf{q}$ in Ω, $(a\psi)(\mathbf{q}) = a\psi(\mathbf{q})$.

($\mathfrak{F}$3) There is an inner product defined by $\langle \psi|\phi \rangle = \int_\Omega \psi^*(\mathbf{q})\varphi(\mathbf{q})d\mathbf{q}$, where ψ^* denotes the complex conjugate of ψ.[26]

($\mathfrak{F}$4) Every function ψ in $\mathfrak{F}_\Omega$ is square integrable: The integral $\int_\Omega |\psi(\mathbf{q})|^2 d\mathbf{q}$ is well defined.

The vector spaces $\mathfrak{S}$ and $\mathfrak{F}_\Omega$ are isomorphic: There are one-one mappings of each onto the other that preserve vector addition, scalar multiplication, and the inner product. This was proved first in 1906 by Hilbert for $\mathfrak{S}$ and a certain part of $\mathfrak{F}_\Omega$, and shortly thereafter by Riesz and Fischer for the whole of $\mathfrak{F}_\Omega$. The abstract mathematical structure that is common to these vector spaces is called *Hilbert space*. Since both matrix and wave mechanics have isomorphic Hilbert spaces at their core, it is no wonder that both theories make precisely the same quantitative predictions.

6.2.4   Interpretation

One often hears that matrix and wave mechanics, although *formally* equivalent, differed profoundly in physical contents. Two theories can

---

[26] In other words, for every $\mathbf{q}$ in Ω, if $\psi(\mathbf{q}) = a + bi$, $\psi^*(\mathbf{q}) = a - bi$.

certainly share the same mathematical structure and yet have each an empirical meaning of its own. However, if the structures underlying two theories are isomorphic – and so, formally indistinguishable – any physical interpretation bestowed on one of them is, *via* the isomorphism, applicable to the other. Indeed, in the case in point, the surmise that both theories were closely related was prompted by the fact that they assigned the same values to certain physical quantities. This could happen only if some identical numbers obtained by calculation from either theory were being interpreted in the same way, for example, as the frequencies and intensities of some definite spectral lines. Of course, such shared meanings involve only the points of contact of matrix and wave mechanics with laboratory data, and the philosophers who see a big semantic gap between both theories must have something deeper in mind. As we know, Heisenberg scorned the pretension of reaching with his theory anything beyond what is actually observable, while Schrödinger was deliberately trying to conceive and describe continuous (wavelike) processes responsible for the quantum integers and the appearance of "jumps". To this extent, the philosophers are clearly right.

However, Schrödinger's early attempts at a physical interpretation of his ψ-function ended in failure.[27] He had expected that the solutions of his equation would describe the undulatory processes envisioned by de Broglie, which, superposed in narrow stable packets, would be the carriers of particle-like phenomena. But he had to admit that, except in the simplest cases, all wave packets spread out and are therefore incapable of simulating particles.

Moreover, the ψ-function is defined on the configuration space of the system under study, which possesses as many dimensions as the latter has degrees of freedom. This in itself is neither new nor baffling: In §2.5.3 we saw how Lagrange represented the evolution of a mechanical system by the trajectory of a point in the system's configuration space. However, if the system consisted of n particles one could always – at least in principle – calculate their n trajectories in Newtonian 3-space from the trajectory of its representative in $3n$-space. In wave mechanics things are quite different. If the system consists of two or more interacting subsystems, they are usually so entangled that it is impossible to assign a ψ-function to each. Even if the ψ-function of the

---

[27] Another, more promising interpretation advanced in his Dublin seminars of the 1950s never attained a final, sufficiently definite form. See Schrödinger (1995).

system is initially constructed from ψ-functions describing the subsystems right before the interaction, its subsequent evolution under
Schrödinger's equation precludes its analysis into distinct subsystemic
ψ-functions. Thus, the wavelike process constituted by ψ in configuration space cannot normally be translated into an equivalent, equally
continuous process in ordinary space.

Nor could Schrödinger successfully carry through the "heuristic
hypothesis on the electrodynamical significance" of ψ (1926e, in CW,
p. 108) that he subsequently proposed. This consisted in the generalization to any number of particles of an interpretation he found to
work for a single electron, viz., that the squared amplitude $|\psi|^2$ of the
complex wave function ψ represents the electrical density as a function
of the space coordinates and the time.

In July 1926, Schrödinger met Bohr and Heisenberg in Copenhagen.
Bohr argued forcefully to persuade him to give up every hope of
understanding quantum physics in classical terms. At one point
Schrödinger shouted in despair: "If all this damned quantum jumping
were really here to stay then I should be sorry I ever got involved with
quantum theory" (quoted from Heisenberg's recollections in Pais 1991,
p. 299).

The standard interpretation of the ψ-function was proposed by
Born (1926a, 1926b), in full agreement with the spirit of Heisenberg's
approach. According to it, the squared amplitude $|\psi|^2$ measures
the *probability* that certain observable quantities might take certain
values at the location represented by the ψ-function's argument. The
individual process, the "quantum jump", is a chance event whose
antecedent probability is determined by Schrödinger's equation. "The
jump therefore spans a considerable abyss; what happens during the
jump [. . .] perhaps cannot be described at all in a language which
suggests pictures to our visualizing faculty" (Born 1927a, p. 172).
In this way, the conception of atoms and their component particles
as chance setups, which was foreshadowed in Rutherford's law of
radioactive decay and was openly applied by Einstein to the absorption and emission of radiative energy (§6.1.2), now became universal.
In Born's revolutionary view, the paths of particles "are determined
only insofar as they are constrained by the principle of energy and
momentum conservation; apart from this, the value distribution
of the ψ-function determines only the probability that a particle will
follow a particular path. [. . .] The motion of particles obeys the laws
of probability, but probability itself spreads in accordance with the

principle of determinism [embodied in Schrödinger's differential equation – R. T.]."[28]

Born did not impose the said meaning on ψ *by an arbitrary decision* – as one is supposed to do when interpreting an uninterpreted formal system – but *read it off* a particular application of wave mechanics. He noted that the new quantum mechanics, in both the matrix and the wave versions, had only been applied to stationary states, and that many assumed that additional conceptual developments would be necessary to deal with transitions. He himself, "impressed with the closed character of the logical structure of quantum mechanics, conjectured that the theory is complete and must encompass the problem of transitions" (1926a, p. 863). He was now ready to prove it from the application of wave mechanics to collisions.[29]

Born considers an electron coming from infinity that strikes an atom and rebounds toward infinity. Clearly, both before and after the collision, when the electron is far enough and the coupling is small, a definite state must be specifiable for the atom and a definite rectilinear motion for the electron. Now,

> according to Schrödinger, the atom in its *n*-th quantum state is a vibration of a state function of fixed frequency [. . .] spread over all of space. An electron moving in straight line is [. . .] a vibratory phenomenon corresponding to a plane wave. When two such waves interact, a complicated vibration arises. However, one sees at once that one can determine it through its asymptotic behavior at infinity.
>
> (Born 1926a, p. 864)

One must therefore solve the Schrödinger equation for the atom-plus-electron system subject to this boundary condition: In the particular direction of electron space from which the electron is assumed to come the solution goes over asymptotically into a plane wave propagating from exactly this direction. In the solution thus selected "we are interested mainly in the behavior of the 'scattered' wave at infinity, for it describes the behavior of the system after the collision" (Ibid.).

As usual in physics, the mathematics is introduced not as a mean-

---

[28] Born (1926b, p. 804). I translate Born's term "das Kausalgesetz" as "the principle of determinism". See Chapter Three, note 33, and the main text leading to it.

[29] "Of the different forms of the theory only Schrödinger's has proved suitable for this, and precisely for this reason I would regard it as the deepest formulation of the quantum laws" (Born 1926a, p. 864).

ingless calculus, but as a way of conceiving a specific physical situation. This in turn is considered abstractly, in the simplified and idealized form required for it to be conceived in that way. Let $\psi_1^0(q)$, $\psi_2^0(q)$, ... denote the proper functions of the unperturbed atom. A free electron of mass m_e moving in from infinity with energy E in the direction of the unit vector with components α, β, and γ is assigned the continuous spectrum of proper functions $\sin(2\pi/\lambda)(\alpha x + \beta y + \gamma z + \delta)$, where the wavelength λ satisfies the de Broglie relation $E = p^2/2m_e = h^2/2m_e\lambda^2$ and the phase δ takes all possible values. If the electron comes from the direction $+z$, the atom-plus-electron system, before interaction, should be assigned the proper function:

$$\psi_{n_E}^0(q,z) = \psi_n^0(q)\sin\frac{2\pi}{\lambda}z \qquad (6.18)$$

Born figured out by simple perturbation calculations that, for a given interaction potential, there is a unique solution of the Schrödinger equation which at $+z \rightarrow \infty$ goes over asymptotically into the said function $\psi_n^0(q) \sin(2\pi/\lambda)z$. Further calculation shows that "after collision" the scattered wave goes over asymptotically into this superposition of solutions of the unperturbed process:

$$\psi_{n_E}^{(1)}(q,x,y,z) = \sum_m \iint_{\alpha x+\beta y+\gamma z>0} \psi_{n_E m}(\alpha,\beta,\gamma)\sin\frac{2\pi}{\lambda_{n_E m}}(\alpha x + \beta y + \gamma z + \delta)\psi_m^0(q)d\omega$$

$$(6.19)$$

where $d\omega$ is the element of solid angle in the direction (α,β,γ) and $\psi_{n_E m}(\alpha,\beta,\gamma)$ is what is now called the differential cross section for that direction. Born then draws the momentous conclusion: "If one is to understand this result in corpuscular terms, only one interpretation is possible: $\psi_{n_E m}(\alpha,\beta,\gamma)$ determines the probability that the electron coming from the z-direction be thrown in the direction specified by the angles α, β and γ, with the phase change δ" (1926a, pp. 865f.; my notation).

Two questions are in order here: (i) Why must one reach for a corpuscular understanding? and (ii) why, if one does reach for it, is it necessary to accept Born's interpretation? Question (ii) is fairly easy: If the electron is a corpuscle, it can only move in a single direction at a time; the function $\psi_{n_E m}(\alpha,\beta,\gamma)$ can take nonzero values in continuously many different directions and so reflects a property that is tied not to the electron's actual path to infinity, but rather to the whole set of its possible paths; whence Born's conclusion that $\psi_{n_E m}(\alpha,\beta,\gamma)$ measures the probability that the electron takes the particular direction designated by the respective argument (α,β,γ).[30] Question (i) is harder. The only answer that comes to my mind is that we are dealing with a physical situation that was understood in corpuscular terms to begin with. A plane wave incoming from infinity in a definite direction was associated with the electron precisely because the latter was assumed to be a corpuscle all along. The output of our calculations must of course be interpreted in the same corpuscular terms as the input, and therefore the value of the functions $\psi_{n_E m}(\alpha,\beta,\gamma)$ for each set of angles α, β, and γ must have the said probabilistic meaning.

One may wish to know how Born could be so sure that the electron is a corpuscle. Late in life, he gave this explanation: He worked in Göttingen in the same building as James Franck, so he "was witnessing the fertility of the particle concept every day in Franck's brilliant experiments on atomic and molecular collisions and was convinced that particles could not simply be abolished" (1968, p. 55).

6.2.5   Quantum Mechanics in Hilbert Space

Once it was understood that Heisenberg's approach was equivalent to Schrödinger's, they became fused into the theory that we call QM, which promptly attained its mature, streamlined mathematical form through the efforts of Jordan (1926, 1927), London (1926a, 1926b), Dirac (1926a, 1926b, 1930), and von Neumann (1927, 1932). The following sketch will be useful in the discussion of philosophical problems in the next two sections.[31]

Like all theories of mathematical physics, QM comes to life through its application to physical systems, that is, simple or complex, small or large chunks of the world we live in, which are picked out and con-

---

[30] In a note added in proof, Born observes that the said probability must be proportional to $|\psi_{n_E m}(\alpha,\beta,\gamma)|^2$.

[31] The sketch is rather too abstract, but I expect that, combined with Supplement I at the end of the book, it will be sufficient for our purposes. Good philosophical books on QM, such as Redhead (1987) and Hughes (1989), provide more informative – although scarcely less abstract – summaries of the mathematics. For concreteness one must turn to the standard textbooks, for example, Messiah (1961) or Cohen-Tannoudji et al. (1977); however, philosophy students will probably feel overwhelmed by their length. Sudbery (1986, Ch. 2–5), provides a shorter, clean and clear exposition; Chapter 4 works out some of the theory's simplest applications. The new textbook by Hannabuss (1997) has a manageable size and looks very attractive to me, but at the time of writing I had not yet formed an opinion about it.

ceived by the physicist, under suitable idealization, in terms of the theory.[32] We saw in §5.5 that GR has been fruitfully applied to systems that are drastically simplified models of the whole world. And there are authors who write Ψ, with a capital letter, for the ψ-function of the universe. They might even write down a mock Schrödinger equation for Ψ, but they come nowhere near a point where they could attempt to solve it. Others think, with good reason, that the very nature of QM precludes such attempts at globalization. For one thing, although QM applies to single systems – one hydrogen atom, one pair of photons – it makes statistical predictions, so it must be tested on ensembles of such systems, that is, collections of noninteracting instances of them. Now, it does not seem likely that one will ever test a theory on an ensemble of worlds.[33] Moreover, the following consideration should make it clear that every scientifically significant application of QM inevitably refers to a more or less conventionally delimited fragment of the world. In QM one distinguishes between the deterministic evolution of chances and the random occurrence of chance events (see above, the quotation linked to note 28). But a chance distribution, deterministically evolving in isolation, does not translate into actual, particular, definite outcomes. These, it would seem, can only be effected as the deterministic system impinges on its boundaries with the rest of the world.[34]

[32] J. S. Bell (1990, p. 19) places "system" at the head of a "list of bad words", followed by "apparatus" and "environment". Such words, he says, "imply an artificial division of the world, and an intention to neglect, or take only schematic account of, the interaction across the split". Such artificial divisions, however, are inevitable in *experimental* physics and the key to its success, as opposed, say, to Presocratic or Aristotelian cosmology. Bell's curious stance contrasts with that of the mathematicians Birkhoff and von Neumann: "The concept of a physically observable 'physical system' is present in all branches of physics, and we shall assume it" (1936, p. 823).

[33] Peirce expressed it forcefully (CP 2.684): "The relative probability of this or that arrangement of Nature is something which we should have a right to talk about if universes were as plenty as blackberries, if we could put a quantity of them in a bag, shake them well up, draw out a sample, and examine them to see what proportion of them had one arrangement and what proportion another. But even in that case, a higher universe would contain us, in regard to whose arrangements the conception of probability could have no applicability."

[34] Think of the little balls dancing in a modern lottery urn. There is a digit on each ball. A prize is awarded to the number formed by the digits on, say, the first six balls to jump out of the urn. Evidently, if the urn encompassed the entire world, there could be no winners.

The duality of determinism and randomness underlies one of QM's most remarkable features. QM provides a separate, mathematically distinct representation for the state of a physical system and for the observable physical quantities by measuring which we get to know the system. By this expedient the theory can predict future states with certainty while assigning definite probabilities to future results of measurement. The probability distributions for the possible measured values of the different physical quantities are encoded in the mathematical representation of the state and can be readily retrieved from it.

In Hamilton's formulation of classical mechanics (see remark D at the end of §2.5.3), a system with n degrees of freedom is associated with a copy of $\mathbb{R}^{2n}$, the system's $2n$-dimensional *phase space*. The state of the system at any given time is represented by a point P in this space. The coordinates of P are the (generalized) position and momentum coordinates of the system, and every other physical quantity of interest to classical mechanics must be defined as a function of these coordinates. The past and future evolution of the state is represented by the unique solution through P of the Hamilton equations for the system. This solution encapsulates the past and future values of every physical quantity of interest.

A quantum-mechanical system is associated with a – generally infinitely dimensional – Hilbert space $\mathcal{H}$.[35] The state of the system is represented by a nonzero vector in $\mathcal{H}$. If $|\psi\rangle$ represents the state, so does $a|\psi\rangle$, where a is any complex number. From now on, unless otherwise indicated, it is understood that any vector $|\psi\rangle$ chosen as state representative is normalized, that is, such that $|\psi|^2 = 1$. It is assumed that every vector in $\mathcal{H}$ represents a possible state of the system. This implies that, if $|\psi_1\rangle$ and $|\psi_2\rangle$ represent two possible states, and a and b are any complex numbers, the linear combination $a|\psi_1\rangle + b|\psi_2\rangle$ represents a third state, the "superposition" of the former two (for which one can readily find a normalized representative). The physical quantities of interest – usually called 'observables' – are represented by self-adjoint operators in $\mathcal{H}$. The seemingly plausible assumption that every such operator represents a measurable quantity has turned out to be untenable (Wigner 1952; cf. Stein and Shimony 1971).

To see how these representations of states and observables work

---

[35] See Supplement I.7 and §6.2.3. If this will help you, you may think of $\mathcal{H}$ as the space $\mathfrak{F}_\Omega$ of complex-valued square-integrable functions on a suitable configuration space Ω.

together, consider for the moment a Hilbert space $\mathscr{H}$ with finitely many dimensions. Suppose that a certain observable Q is represented by the self-adjoint operator Q on $\mathscr{H}$. The possible values of Q are the proper values of Q. If $\mathscr{H}$ has n dimensions and there are n such values $q_1, \ldots, q_n$, there are corresponding proper vectors, $|\psi_1\rangle, \ldots, |\psi_n\rangle$, that form an orthonormal basis of $\mathscr{H}$. Any vector $|\psi\rangle$ representing a state of our system can be expressed as a linear combination of these vectors, viz., $|\psi\rangle = \Sigma_{i=1}^{n}|\psi_i\rangle\langle\psi_i|\psi\rangle$. If no two proper values correspond to the same proper vector, we say that Q is *nondegenerate*. Then, $|\langle\psi_i|\psi\rangle|^2$ is the probability that, if Q is measured on a system that happens to be in state $|\psi\rangle$, the measured value of Q is found to be q_i.[36] If Q is degenerate and the proper value q_k corresponds, say, to r linearly independent normalized proper vectors $|\psi_{k1}\rangle, \ldots, |\psi_{kr}\rangle$, the probability that the value of Q is q_k if Q is measured on a system in state $|\psi\rangle$ is equal to $\Sigma_{i=1}^{r}|\langle\psi_{ki}|\psi\rangle|^2$. If $\mathscr{H}$ is infinite-dimensional, there are other complications. In the first place, the concept of a basis must be widened (see Supplement I, after eqn. (S5)). Even so, not every self-adjoint operator on $\mathscr{H}$ will have a family of proper vectors constituting an orthonormal basis. It is usually assumed that only those that do are apt for representing observables (Messiah 1961, vol. I, p. 188; Cohen-Tannoudji et al. 1977, vol. I, p. 137). Finally, the basic notions of *proper vector* and *linear combination of vectors* must be adjusted – or replaced – to cope with the fact that some observables have a continuous spectrum of possible values (Supplement I.7).

The above gives an inkling of QM's probabilistic account of actual

---

[36] The condition "if Q is measured" will surely trouble some readers. They expect a physical theory to predict, at least with probability, the value that a physical quantity also *has* when it is *not* measured. QM generally cannot do this, essentially due to the fact that many pairs of self-adjoint operators p, q, representing physical quantities, satisfy the relations (6.14). Cf. Dirac 1958, pp. 46–47:

> The expression that an observable 'has a particular value' for a particular state is permissible in quantum mechanics in the special case when a measurement of the observable is certain to lead to the particular value, so that the state is an eigenstate of the observable.
>
> . . . .
>
> In the general case we cannot speak of an observable having a value for a particular state, but we can speak of its having an average value for the state. We can go further and speak of the probability of its having any specified value for the state, meaning the probability of this specified value being obtained when one makes a measurement of the observable.

measurement results. I turn now to the system's evolution in time. In the so-called Heisenberg picture, the state $|\psi\rangle$ of the system is fixed and self-adjoint operators representing the quantum-mechanical analogues of classical position and momentum coordinates evolve deterministically according to eqns. (6.15). This is equivalent – *via* a suitable mathematical transformation – to the more manageable Schrödinger picture in which the state $|\psi\rangle$ of the system evolves deterministically in $\mathcal{H}$ according to Schrödinger's time-dependent equation (note 22) and the observables are represented by fixed self-adjoint operators. Writing H for the Hamiltonian operator of the system and $|\psi(t)\rangle$ for the state of the system at time t, Schrödinger's equation can be written schematically:

$$\frac{d}{dt}|\psi(t)\rangle = -\frac{i}{\hbar}H|\psi(t)\rangle \qquad (6.20)$$

Therefore, if $|\psi(0)\rangle$ is the state of the system at some chosen time $t = 0$, its state $|\psi(t)\rangle$ at any arbitrary time t satisfies the relation:

$$|\psi(t)\rangle = \exp\left(-\frac{i}{\hbar}Ht\right)|\psi(0)\rangle \qquad (6.21)$$

where $\exp(-(i/\hbar)Ht)$ is a unitary operator on $\mathcal{H}$, commonly denoted by U_t. Since $\exp(0)$ is the identity and

$$\exp\left(-\frac{i}{\hbar}Ht_1\right)\exp\left(-\frac{i}{\hbar}Ht_2\right) = \exp\left(-\frac{i}{\hbar}H(t_1+t_2)\right)$$

– that is, $U_0 = I$ and $U_{t_1}U_{t_2} = U_{t_1+t_2}$ – it is clear that the operators U_t form a continuous group parametrized by t. The evolution of states in the Hilbert space $\mathcal{H}$ under eqn. (6.20) is appropriately described by the action of this group of unitary operators on $\mathcal{H}$, and it is therefore often referred to as 'unitary evolution'.

As even this bald sketch shows, the quantum-mechanical scheme of description and prediction is one of the most elegant creations of the human spirit. But it has implications that numerous philosophers and a few physicists find objectionable. I shall deal with their views in §§6.3–6.4. But let me show here, by means of a simple example, why the sort of situation that they have difficulty in assimilating is inevitable in the quantum-mechanical scheme. Consider a suitably prepared physical system S that, according to QM, will be in state $|\psi\rangle$ at time t. Arrangements are made to measure at t on S the observable (represented by the operator) Q. Assume that $|\psi\rangle$ is not a proper vector of Q and that Q is nondegenerate, with proper values a_i uniquely associated with the elements of an orthonormal basis of proper vectors $|\varphi_i\rangle$ ($i = 1$, $2, \ldots$). Under such circumstances, QM predicts that the measurement will record the particular value a_k with probability $|\langle\varphi_k|\psi\rangle|^2$. This is typically a real number greater than 0 and smaller than 1. However, as soon as the measurement is completed, its result turns out to be either a_k with certainty or some other value $a_j \neq a_k$, so that a_k is excluded. There is no contradiction here, for the prediction is compatible with both results. The prediction would be questionable only if a set of Q-measurements performed on a very large ensemble of systems in state $|\psi\rangle$ yielded a proportion of a_k-values that is significantly different from $|\langle\varphi_k|\psi\rangle|^2$. Still, a difficulty apparently arises if the matter is put as follows. Suppose that the measured value in effect is a_k. Then, unless the system S is destroyed during measurement (a common occurrence in microphysics), one ought to say in accordance with QM that, right after the measurement, S is in state $|\varphi_k\rangle$, the single proper state of Q corresponding to the proper value a_k. Thus S "jumps" at t from state $|\psi\rangle = \sum_{i=1}^{\infty}|\varphi_i\rangle\langle\varphi_i|\psi\rangle$ to state $|\varphi_k\rangle$. Such a "jump" cannot be derived from eqn. (6.20). So – it appears – the state of a quantum-mechanical system evolves, in different circumstances, in one or the other of two different ways, governed by different laws. One of these laws, the Schrödinger equation, is well known and similar to other familiar laws. The other law is unknown and apparently unfathomable. I dare say that this way of putting things involves either unwillingness to accept the reality of chance or inability to intellectually cope with it. The "jump" in question is the transition from an initial situation in which a chance event is being expected to a final situation in which the outcome of chance is already given. QM describes the former by means of a state vector from which different probability distributions can be extracted, one for each physical quantity that one might be willing to measure. Thus we have the following statement:

(A) 'S will attain state $|\psi\rangle$ at time t'.

The final situation, however, does not require such an abstruse description. The definite outcome of a definite measurement can be straightforwardly described like any ordinary matter of fact:

(B) 'Value a_k was obtained when Q was measured on S at t'.

If S was not destroyed by the measurement and Q is a nondegenerate operator with a discrete spectrum, (B) entails in QM a description of S at t by means of a state vector, viz.,

(C) 'The state of S at t, right after measurement, is $|\varphi_k\rangle$'.

Note that – within QM – (C) is strictly weaker than (B). (C) entails that 'the value of Q measured on system S at t is a_k with probability $|\langle\varphi_k|\varphi_k\rangle|^2 = 1$', but this is not equivalent to (B) (although philosophers tend to forget it). Thus, if we only wished to express what we know about S right after the measurement at t, (C) would be pointless. Its real utility is in predicting the outcome of subsequent measurements on S. (This is done by calculating the evolution of $|\varphi_k\rangle$ under Schrödinger's equation and expressing the resulting vector as linear combinations of proper vectors of the relevant operators.) In other words, (C) – like (A) – is appropriate for speaking about expected chance events. Now if (A) and (C) are read, in straight succession, omitting (B), it sounds like the story of a state evolution that is not subject to Schrödinger's equation. At first blush it might seem that physics should be committed to explain this story. However, (C) is just the predictive content of (B), under our special assumptions regarding S and Q. To say why and how (A) leads to (C) one must explain the "jump" from (A) to (B). To give such an explanation, however, would amount to denying that (B) describes the outcome of chance.

✦

In the remainder of this section I introduce some notions that we shall need in §§6.2.6, 6.3, and 6.4, namely, (*a*) the handling of compound systems, (*b*) projectors, (*c*) the expected value and the variance of an observable, (*d*) mixtures, and (*e*) the statistical operator. The explanations are dry, and I suggest that readers turn to them as and when they need them.

(*a*) If the states of systems A and B are represented by vectors in spaces $\mathcal{H}_A$ and $\mathcal{H}_B$, respectively, the states of the compound system $A + B$ are represented in the tensor product space $\mathcal{H}_A \otimes \mathcal{H}_B$. See Supplement I.8. Remember that $\mathcal{H}_A \otimes \mathcal{H}_B$ contains every conceivable linear combination of the tensor products $|\alpha\rangle \otimes |\beta\rangle$, for every vector $|\alpha\rangle$ in $\mathcal{H}_A$ and $|\beta\rangle$ in $\mathcal{H}_B$.

(*b*) Projectors are self-adjoint *operators* with peculiar properties that make them suitable for representing *states*. Let P be a linear operator on the Hilbert space $\mathcal{H}$, with adjoint $P^\dagger$. P is a *projector* if and only if $P^\dagger P = P$. This implies that $P = P^\dagger$ (P is self-adjoint) and that PP

= P (P is idempotent).[37] It can be shown that a projector P maps –
"projects" – the whole of $\mathcal{H}$ onto a subspace of it. This range of P is
not shared by any other projector and is therefore uniquely associated
with P. In particular, if $|\psi\rangle$ is a normalized vector, there is a unique
projector $P_{|\psi\rangle}$ that sends every vector in $\mathcal{H}$ to the subspace spanned by
$|\psi\rangle$. The projection of an arbitrary vector $|\varphi\rangle$ along the direction of $|\psi\rangle$
is equal to the result of multiplying $|\psi\rangle$ by the inner product of $|\psi\rangle$ and
$|\varphi\rangle$, viz.,

$$P_{|\psi\rangle}|\varphi\rangle = |\psi\rangle\langle\psi|\varphi\rangle \qquad (6.22)$$

We may therefore write

$$P_{|\psi\rangle} = |\psi\rangle\langle\psi| \qquad (6.23)$$

Equation (6.23) defines $P_{|\psi\rangle}$ uniquely for each normalized $|\psi\rangle$ in $\mathcal{H}$ and
may therefore be used for representing the same state as $|\psi\rangle$. Note
further that if the vectors ψ_i (for every index i in some set $\mathcal{I}$) form a
basis, the operator defined by adding their respective projectors sends
each vector in $\mathcal{H}$ to itself. In other words,

$$\sum_{i\in\mathcal{I}}|\psi_i\rangle\langle\psi_i| = I \qquad (6.24)$$

where I denotes the identity operator on $\mathcal{H}$.

Consider now an observable Q represented by an operator Q whose
proper value q_k corresponds to r linearly independent normalized
proper vectors $|\psi_{k1}\rangle, \ldots, |\psi_{kr}\rangle$. Let $P(q_k|\psi)$ be the probability that the
value of Q is q_k if Q is measured on a system in state $|\psi\rangle$. Then

$$\begin{aligned}
P(q_k|\psi) &= \sum_{i=1}^{r}|\langle\psi_{ki}|\psi\rangle|^2 \\
&= \sum_{i=1}^{r}\langle\psi|\psi_{ki}\rangle\langle\psi_{ki}|\psi\rangle \\
&= \sum_{i=1}^{r}\langle\psi|P_{|\psi_{ki}\rangle}|\psi\rangle \\
&= \langle\psi|P_k|\psi\rangle
\end{aligned} \qquad (6.25)$$

where P_k stands for the projector onto the subspace spanned by the
vectors $|\psi_{k1}\rangle, \ldots, |\psi_{kr}\rangle$ corresponding to the proper value q_k.

(c) The predictions of a probabilistic theory must be tested on very
large collections of like systems (ideally, on infinitely many of them).
One compares the predicted mean or expectation value of a quantity

---

[37] If $P = P^\dagger P$, then $P^\dagger = (P^\dagger P)^\dagger = P^\dagger P^{\dagger\dagger} = P^\dagger P = P$. So P is self-adjoint. Therefore $P^\dagger P =$
PP= P. So P is idempotent.

with the actual average of a long series of measurements. One also compares the predicted and the actual spread of the particular results about the mean. Suppose that we measure an observable Q, represented by the operator Q, on a large collection of identical systems – for example, silver atoms, or circularly polarized photons – prepared in a state represented by the normalized vector $|\psi\rangle$. The expectation value, denoted by $\langle Q \rangle_{|\psi\rangle}$, is then given by:

$$\langle Q \rangle_{|\psi\rangle} = \langle \psi | \mathsf{Q}\psi \rangle = \langle \mathsf{Q}\psi | \psi \rangle = \langle \psi | \mathsf{Q} | \psi \rangle \qquad (6.26)$$

(The second and third expressions are equal because Q is self-adjoint; the last expression serves as a conventional reminder of it.) Equation (6.26) can be easily proved if Q has a discrete, nondegenerate spectrum of proper values $\{q_i : i \in \mathcal{I}\}$ corresponding to the orthonormal basis $\{\psi_i : i \in \mathcal{I}\}$. Then the probability $\mathsf{P}_Q(q_i|\psi)$ of obtaining the value q_i in a measurement of Q on a system in state ψ is, as we know, equal to $|\langle \psi_i | \psi \rangle|^2$. Thus,

$$\begin{aligned} \langle Q \rangle_{|\psi\rangle} &= \sum_{i \in \mathcal{I}} q_i \mathsf{P}_Q(q_i|\psi) = \sum_{i \in \mathcal{I}} q_i |\langle \psi_i | \psi \rangle|^2 = \sum_{i \in \mathcal{I}} q_i \langle \psi_i | \psi \rangle^* \langle \psi_i | \psi \rangle \\ &= \sum_{i \in \mathcal{I}} \langle \psi | q_i \psi_i \rangle \langle \psi_i | \psi \rangle = \sum_{i \in \mathcal{I}} \langle \psi | \mathsf{Q} | \psi_i \rangle \langle \psi_i | \psi \rangle \\ &= \left\langle \psi \middle| \mathsf{Q}\left[\sum_{i \in \mathcal{I}} |\psi_i\rangle\langle\psi_i| \right] \middle| \psi \right\rangle = \langle \psi | \mathsf{Q} | \psi \rangle \end{aligned} \qquad (6.27)$$

where the last step in the derivation is based on eqn. (6.24).

A good measure of the dispersion of measured values about the mean is the *variance* (called "uncertainty" in some older books). The variance ΔQ_ψ of the quantity Q as measured on systems in state $|\psi\rangle$ is the square root of the expectation value of $(\mathsf{Q} - \langle Q \rangle_{|\psi\rangle})^2$. Hence, by eqn. (6.27),

$$\begin{aligned} (\Delta Q_\psi)^2 &= \left\langle \left(\mathsf{Q} - \langle Q \rangle_{|\psi\rangle} \right)^2 \right\rangle_{|\psi\rangle} = \langle \psi | (\mathsf{Q} - \langle \psi | \mathsf{Q} | \psi \rangle)^2 | \psi \rangle \\ &= \langle \psi | \mathsf{Q}^2 | \psi \rangle - 2\langle \psi | \mathsf{Q} | \psi \rangle \langle \psi | \mathsf{Q} | \psi \rangle + \langle \psi | \mathsf{Q} | \psi \rangle^2 \langle \psi | \psi \rangle \qquad (6.28) \\ &= \langle \psi | \mathsf{Q}^2 | \psi \rangle - \langle \psi | \mathsf{Q} | \psi \rangle^2 \end{aligned}$$

(*d*) It is not always possible to prepare a large collection of objects in a state that is representable by a particular vector. A method of preparation might yield a mixture of like objects in different states, represented by the vectors $|\psi_1\rangle, \dots, |\psi_n\rangle$, each with a known – or conjectured – relative frequency. If the collection is very large, the relative frequency of a given state approaches the probability that a randomly chosen object in the collection is in effect in that state. It would seem, at first sight, that the probability that an object picked at random from such a mixture is in the "pure" state represented, say, by $|\psi_k\rangle$ ($1 \le k \le n$) ought to be carefully distinguished from the probability $|\langle\varphi_i|\psi_k\rangle|^2$ that an object that happens to be in the "pure" state represented by $|\psi_k\rangle$ will sport, upon measurement of an observable Q with discrete non-degenerate spectrum $\{|\varphi_i\rangle: i \in \mathcal{I}\}$, the proper value corresponding to $|\varphi_i\rangle$. Yet these two kinds of probabilities must surely obey the same mathematical rules, for they are freely combined in QM calculations. The probability p_k that, say, the kth proper value of an observable will be measured on a mixture is computed by adding up the probabilities $p_{k1}, \ldots, p_{kr}$ that the said value will turn up in each of the r pure states in the mixture, weighted by the probabilities $w_1, \ldots, w_r$ that an object in the mixture is in one of these states: $p_k = \Sigma_{i=1}^{r} w_i p_{ki}$.

(e) Pure states and mixtures can be represented in a uniform way by means of the *statistical operator* (or *density operator*), which I shall now define. Consider first a system that at time t is in the pure state represented by the normalized vector $|\psi(t)\rangle$. The statistical operator $\rho(t)$ representing this state is then simply the projector $\mathsf{P}_{|\psi(t)\rangle}$:

$$\rho(t) = \mathsf{P}_{|\psi(t)\rangle} = |\psi(t)\rangle\langle\psi(t)| \tag{6.29}$$

Pick an orthonormal basis $\{\psi_i: i \in \mathcal{I}\}$, such that $|\psi(t)\rangle = \Sigma_{i\in\mathcal{I}}\, c_i(t)|\psi_i\rangle$, where $c_i(t)$ stands for $\langle\psi_i|\psi(t)\rangle$. Then the matrix of $\rho(t)$ in the basis $\{|\psi_i\rangle: i \in \mathcal{I}\}$ has the typical element:

$$\rho_{ij}(t) = \langle\psi_i|\rho(t)|\psi_j\rangle = c_i(t)c_j^*(t) \tag{6.30}$$

We have that

$$\mathrm{Tr}\rho(t) = \sum_{i\in\mathcal{I}}\rho_{ii}(t) = \sum_{i\in\mathcal{I}}|c_i(t)|^2 = 1 \tag{6.31}$$

since $|\psi(t)\rangle$ is normalized. The expectation value $\langle Q\rangle_{|\psi(t)\rangle}$ of an observable Q represented by the operator Q can now be expressed in terms of the statistical operator $\rho(t)$:

$$
\begin{aligned}
\langle Q\rangle_{|\psi(t)\rangle} &= \langle\psi(t)|\mathsf{Q}|\psi(t)\rangle \\
&= \sum_{i\in\mathcal{I}}\sum_{j\in\mathcal{I}}\langle\psi(t)|\psi_i\rangle\langle\psi_i|\mathsf{Q}|\psi_j\rangle\langle\psi_j|\psi(t)\rangle \\
&= \sum_{i\in\mathcal{I}}\sum_{j\in\mathcal{I}}\langle\psi_j|\psi(t)\rangle\langle\psi(t)|\psi_i\rangle\langle\psi_i|\mathsf{Q}|\psi_j\rangle \\
&= \sum_{i\in\mathcal{I}}\sum_{j\in\mathcal{I}}\langle\psi_j|\rho(t)|\psi_i\rangle\langle\psi_i|\mathsf{Q}|\psi_j\rangle \\
&= \sum_{j\in\mathcal{I}}\langle\psi_j|\rho(t)\mathsf{Q}|\psi_j\rangle = \mathrm{Tr}(\rho(t)\mathsf{Q})
\end{aligned}
\tag{6.32}
$$

By combining eqn. (6.32) with (6.25) we obtain an expression in terms of $\rho(t)$ for the probability $P_Q(q_k|\psi(t))$ that the observable Q takes the value q_k if measured on a system in state $|\psi\rangle$:

$$P_Q(q_k|\psi(t)) = \langle\psi(t)|P_k|\psi(t)\rangle = \text{Tr}(\rho(t)P_k) \qquad (6.33)$$

The evolution of the statistical operator $\rho(t)$ can be derived from the Schrödinger equation (6.20):

$$\begin{aligned}
\frac{d}{dt}\rho(t) &= \left(\frac{d}{dt}|\psi(t)\rangle\right)\langle\psi| + |\psi(t)\rangle\left(\frac{d}{dt}\langle\psi(t)|\right) \\
&= -\frac{i}{\hbar}H|\psi(t)\rangle\langle\psi(t)| + \frac{i}{\hbar}|\psi(t)\rangle\langle\psi(t)|H \\
&= -\frac{i}{\hbar}[H,\rho(t)]
\end{aligned} \qquad (6.34)$$

I now turn to an ensemble E in a mixture of states $|\psi_1\rangle, \ldots, |\psi_n\rangle$ and denote by p_r the probability that a randomly chosen object be in state $|\psi_r\rangle$ ($0 \le p_r \le 1$; $1 \le r \le n$; $\Sigma_{i=1}^n p_i = 1$). Let $P_Q(q_k|E)$ be the probability that the observable Q takes the value q_k if measured on an object of this ensemble. To calculate $P_Q(q_k|E)$ we multiply the probability p_r that an object is in state $|\psi_r\rangle$ by the probability $P_Q(q_k|\psi_r)$ that Q takes the value q_k if measured on an object in that state, and add up the products thus obtained. Writing ρ_r for the statistical operator representing the state $|\psi_r\rangle$ and ρ for the weighted sum $\Sigma_{i=1}^n p_i\rho_i$ of such statistical operators, we have that

$$\begin{aligned}
P_Q(q_k|E) &= \sum_{i=1}^n p_i P_Q(q_k|\psi_i) = \sum_{i=1}^n p_i\text{Tr}(\rho_i P_k) \\
&= \text{Tr}\left(\sum_{i=1}^n p_i\rho_i P_k\right) = \text{Tr}(\rho P_k)
\end{aligned} \qquad (6.35)$$

Evidently ρ, as a linear combination of projectors, is a self-adjoint operator (although generally not a projector). This is, *by definition*, the statistical operator that represents our mixture:

$$\rho = \sum_{i=1}^n p_i\rho_i \qquad (6.36)$$

Equations (6.31)–(6.34) continue to hold if $\rho(t)$ is the time-dependent statistical operator of a mixture (see Cohen-Tannoudji et al. 1977, vol. I, pp. 301f.).

The uniform representation of mixtures and pure states by statistical operators paved the way for a different conception of quantum-mechanical states. One regards "the state of a system, whether pure or not, as defined by its previous history, i.e., by the method of its preparation" (Fano 1957, p. 76). Thus, the general concept of a quantum-mechanical state is that of a mixture, while the so-called pure state is a limiting case. "A pure state is characterized by the existence of an experiment that gives a result predictable with certainty when performed on a system in that state and in that state only".[38] Such an experiment, providing a maximum of information about the state prepared, is called 'complete'. "A pure state can then be identified by specifying the complete experiment that characterizes it uniquely. Mathematically one can construct a variety of [self-adjoint] operators which have a given pure state as a [proper state]" (p. 75). But there are systems "for which no complete experiment gives a unique result predictable with certainty".[39] Still, the state of such systems is "fully identified by any data adequate to predict the (statistical) results of all conceivable observations of the system. [. . .] Indeed, 'state' means whatever information is required about a specific system, in addition to physical laws, in order to predict its behavior in future experiments" (p. 76). Nonpure states are called 'mixtures' because they can be represented by the incoherent superposition of pure states.[40] But one must bear in mind that such a representation is nonunique. "There is in general no reason, for example, why unpolarized light should be described as a mixture of two linear polarizations rather than of two circular ones". From this standpoint, one considers "only the broader set of all fluctuations among the experimental results obtained with an ensemble of systems prepared according to identical specifications", without analyzing those fluctuations into subsets "when this analysis is not unique and does not correspond to observable characteristics of the situation". In other words, one deals "with a *single statistical ensemble* of quantum mechanical systems *pre-*

---

[38] "For example, linear polarization of a light beam in a given plane is characterized by 100% transmission of each photon through a suitably oriented Nicol prism; no other state of polarization is fully transmitted by the same prism" (Fano 1957, p. 75).

[39] "For example, no polarization analyzer admits or rejects with certainty photons of partially polarized light" (Fano 1957, p. 76).

[40] "Incoherent superposition means, by definition, that to calculate the probability of finding a certain experimental result with a system in the mixed state one must first calculate the probability for each of the pure states and then take an average, attributing to each of the pure states an assigned 'weight'" (Fano 1957, p. 76). See the main text, under (d).

348 Quantum Mechanics

pared by identical procedures, not with a statistical ensemble of quantum mechanical ensembles" (Ibid.).

6.2.6 Heisenberg's Indeterminacy Relations

Heisenberg's paper, "The Intuitive Content of Quantum-Theoretical Kinematics and Mechanics" (1927), is regarded by some as the beginning of a new era in physics. He showed there that, if QM is right, it is impossible to establish with perfect or near-perfect accuracy the position and momentum of one or more particles at a given time. Heisenberg's language, both in this paper and in his Chicago lectures (1930), suggests that this impossibility reflects a limit to the precision attainable by experimental measurement, due to the existence of the finite quantum of action h. To the artless reader it might seem as if all that is being proved is that human observers cannot get to know accurately the position and momentum of a physical object because every successful attempt to measure the former will blur the latter, and vice versa, although the positions and momenta are still there, inscribed in real things, in their full classical sense. But the relations of indeterminacy (or "uncertainty") between certain pairs of observables in QM are a corollary of the theory's mathematics. Thus the simultaneous assignment of dispersion-free values to any such pair is incompatible with the theory's conceptual framework and therefore meaningless in its context.[41] Heisenberg's thought experiments were definitely not meant as *proofs* of the indeterminacy relations. As indicated by the title of his paper of 1927, they were proposed in order to make QM intuitively plausible despite its – for the early twentieth-century physicist – implausible implications.

Consider any two quantum-mechanical observables Q and P, represented by the self-adjoint operators Q and P. Write [Q,P] for QP – PQ. It can be easily proved that, for any ensemble of systems in an arbitrarily chosen state $|\psi\rangle$,

---

[41] Heisenberg (1927, pp. 179f.) puts this rather differently, but the upshot, I think, is the same: The definition of physical quantities rests on the experiments by which we measure them; "every experiment which we can use for defining these words ['electron position' and 'velocity'] necessarily contains the imprecision indicated by the equation $[\Delta Q_\psi \cdot \Delta P_\psi \sim h]$". Equation (6.14) could not be valid if this imprecision did not obtain. "If there were any experiments which allowed a 'sharper' simultaneous determination of P and Q ..., Quantum Mechanics would be impossible."

$$\Delta Q_\psi \cdot \Delta P_\psi \geq \frac{1}{2}\left|\langle i[Q,P]\rangle_{|\psi\rangle}\right| \tag{6.37}$$

Thus, if Q and P satisfy eqn. (6.14) – as they must if they happen to be the quantum-mechanical analogues of a pair of conjugate classical position and momentum coordinates – the product $\Delta Q_\psi \cdot \Delta P_\psi \geq \frac{1}{2}\hbar$. Therefore the variance of one of these observables is infinite if that of the other is zero.

To prove the inequality (6.37) recall eqn. (6.27) and the definition of variance, put $A = Q - \langle Q\rangle_{|\psi\rangle}$, $B = P - \langle P\rangle_{|\psi\rangle}$, note that A and B are self-adjoint operators, and bear in mind that $\langle Q\rangle_{|\psi\rangle}$ and $\langle P\rangle_{|\psi\rangle}$ are complex numbers that commute with every observable (so that $[A,B] = [Q,P]$). Let $|\varphi\rangle$ denote the vector $A|\psi\rangle + i\lambda B|\psi\rangle$, with λ an arbitrary real number:

$$\begin{aligned}
\langle\varphi|\varphi\rangle &= \langle\psi|(A - i\lambda B)(A + i\lambda B)|\psi\rangle \\
&= \langle\psi|A^2|\psi\rangle + \lambda\langle\psi|i[A,B]|\psi\rangle + \lambda^2\langle\psi|B^2|\psi\rangle \\
&= \langle A^2\rangle_{|\psi\rangle} + \lambda\langle i[A,B]\rangle_{|\psi\rangle} + \lambda^2\langle B^2\rangle_{|\psi\rangle} \\
&= (\Delta Q_\psi)^2 + \lambda\langle i[Q,P]\rangle_{|\psi\rangle} + \lambda^2(\Delta P_\psi)^2
\end{aligned} \tag{6.38}$$

The last expression is a second-degree polynomial in λ of the standard form $c + b\lambda + a\lambda^2$. Since $\langle\varphi|\varphi\rangle \geq 0$ (Supplement I.4), the polynomial has either no zeroes or equal zeroes and the discriminant $b^2 - 4ac \leq 0$. Thus, $\langle i[Q,P]\rangle_{|\psi\rangle}^2 \leq 4(\Delta Q_\psi)^2(\Delta P_\psi)^2$. The inequality (6.37) follows.

6.3 Philosophical Problems

6.3.1 The EPR Problem

It was Einstein who taught Bohr and his followers to conceive the transitions between atomic stationary states as chance events (see §6.1.2). But he could never bring himself to see this view as final. For him, one resorted to probability – in quantum as in classical physics – due to human ignorance, not because the physical world is ultimately a gigantic chance setup.[42] So he favored the notion that QM, despite its great successes, is an "incomplete" theory, in other words, that there must be some "elements of reality" that the theory ignores. In 1935 Einstein published, together with his younger collaborators Podolsky and Rosen, the

---

[42] Whence Einstein's insistence, in the very paper in which he gave chance a leading role in the quantum theory of radiation, that this was a "weakness of the theory" and a feature of its "present state" (1916n, p. 62).

famous EPR paper in which they purport to show this. Their allegation
was promptly impugned by Bohr (1935), with an argument that effectively persuaded the great majority of physicists (cf. §6.4.1). The present
fame of EPR is not due, however, to its immediate effect on physical
research – which was virtually nil – or to its logical excellence – which
is questionable – but to the proof, found 30 years later by J.S. Bell (1964)
that a theory that met the EPR demands would not just teach us *more*
things than QM, but would make predictions that flatly contradict QM.
It is not clear how such a theory could cope with the enormous mass of
empirical evidence that supports QM. Leaving this question aside,
several research teams designed and performed clever experiments to
test specific predictions in which QM and an EPR-style theory must run
foul of each other. The results were ambiguous at first, but in the end,
as one might well have expected, the superiority of QM was established
by the group led by Alain Aspect (references in note 45).

Einstein, Podolsky, and Rosen built their case on two seemingly
obvious premises:

1. If a physical theory gives a complete description of reality, every
   element of the physical reality must have a counterpart in the theory.
2. "If, without in any way disturbing a system, we can predict with
   certainty (i.e with probability equal to unity) the value of a physical quantity, then there exists an element of physical reality corresponding to this physical quantity" (1935, p. 777).

They proposed a thought experiment that, according to QM, enables
one to predict with certainty, without in any way disturbing a quantum
system, the value of certain physical quantities pertaining to that
system. Since the ψ-function representing such a system in QM does
not furnish us with a definite counterpart for all these quantities, Einstein, Podolsky, and Rosen concluded that the ψ-function "does not
provide a complete description of the physical reality" (p. 780). The
EPR thought experiment involves noncommuting operators with
continuous spectra. Bohm replaced it with a conceptually equivalent
thought experiment involving noncommuting operators with two-
valued nondegenerate spectra (viz., mutually perpendicular spin components). This was the subject of Bell's proof of 1964 and also the
ideal model for the experiments actually performed. It is also easier to
explain, so I shall refer to it.

This is how Bohm (1951, p. 614) describes his experimental setup:
A molecule consisting of two atoms with spin $\pm\hbar/2$ (see §6.1.4) is in a
state in which the total spin is 0. The molecule is disintegrated by a

process that does not change the total spin. The atoms move apart and cease to interact appreciably. Spin is conserved, so the spin of the two atoms continues to add up to 0. The spin component of either atom in a given direction takes one of the two possible values, $+\hbar/2$ or $-\hbar/2$. Thus, if we measure the spin component of the first atom in a particular direction and find it to be $+\hbar/2$, we can predict with certainty and without in any way interfering with the second atom that its spin component in the said direction is $-\hbar/2$. Therefore, the spin component of the second atom in the chosen direction is an element of reality according to premise 2. Since the direction in question was arbitrary, this is true of every direction. The trouble is that, for a particle with spin $\pm\hbar/2$, the spin components in mutually perpendicular directions are represented in QM by noncommuting linear operators, and the state of the particle at a given time can be represented by a proper vector of one or the other of two such operators, but not of both. Therefore, given a system of Cartesian coordinates x, y, and z, if we know with certainty that the second atom has spin $-\hbar/2$ in the direction parallel to the z-axis, QM will not allow us to assign to it spin $-\hbar/2$ or spin $+\hbar/2$ in the direction parallel to the x-axis. Yet these are the only two possible values of the spin component in this direction, and, according to the EPR argument, the spin component of the second atom in *every* direction is an element of reality. Therefore, by premise 1, QM does not give a complete description of reality.

This can be explained more formally as follows. If one temporarily forgets all physical properties except spin, the state of a spin $\pm\hbar/2$ particle α can be represented in a two-dimensional Hilbert space $\mathcal{H}_\alpha$. The spin components parallel to the Cartesian coordinate axes x, y, and z are represented by three linear operators on $\mathcal{H}_\alpha$ that I shall denote by S_x, S_y, and S_z, respectively.[43] Each one of these operators has two proper vectors, corresponding, respectively, to the proper values $+\hbar/2$ and $-\hbar/2$. To simplify formulas, I choose units such that $\hbar/2 = 1$ and denote the proper vectors of each operator by $|+\rangle$ and $|-\rangle$, followed by a subscript that indicates the direction. In this notation, $S_z|+\rangle_z = |+\rangle_z$ and $S_z|-\rangle_z = -|-\rangle_z$ and similarly for x and y. The two proper vectors of each operator span the space $\mathcal{H}_\alpha$. The proper vectors of S_x and S_y are therefore linear combinations of $|+\rangle_z$ and $|-\rangle_z$. In particular,

$$|+\rangle_x = \frac{1}{\sqrt{2}}(|+\rangle_z + |-\rangle_z) \qquad |-\rangle_x = \frac{1}{\sqrt{2}}(|+\rangle_z - |-\rangle_z) \qquad (6.39)$$

---

[43] These operators do not commute. In effect, $[S_x, S_y] = i\hbar S_z$, $[S_y, S_z] = i\hbar S_x$ and $[S_z, S_x] = i\hbar S_y$.

Now let α and β be two spin $\pm\hbar/2$ particles forming together – as in Bohm's setup – a 0-spin system that we denote by $\alpha + \beta$. Insofar as spin alone is concerned, the state of $\alpha + \beta$ is represented by a vector $|\psi\rangle$ in the product space $\mathcal{H}_\alpha \otimes \mathcal{H}_\beta$. This space is spanned by the four vectors $|+ +\rangle_z = |+\rangle_z \otimes |+\rangle_z$, $|- -\rangle_z = |-\rangle_z \otimes |-\rangle_z$, $|+ -\rangle_z = |+\rangle_z \otimes |-\rangle_z$, and $|- +\rangle_z = |-\rangle_z \otimes |+\rangle_z$, so ψ is a linear combination of these vectors.[44] $\alpha +$ β has an even chance of being found on measurement in state $|+ -\rangle_z$ or $|- +\rangle_z$ and no chance at all of being found in state $|+ +\rangle_z$ or $|- -\rangle_z$. We assume that system $\alpha + \beta$ is in the so-called singlet state, in which

$$|\psi\rangle = \frac{1}{\sqrt{2}}|+ -\rangle_z - \frac{1}{\sqrt{2}}|- +\rangle_z \qquad (6.40)$$

Suppose that α and β are now far apart and that we measure β's spin component parallel to the z-axis. There is the same probability $1/2$ that the measurement will show that β is in spin state $|+\rangle_z$ – so that α is in state $|-\rangle_z$ – or that β is in spin state $|-\rangle_z$ – so that α is in state $|+\rangle_z$. In either case it is evidently impossible, by eqn. (6.39), that α is in state $|+\rangle_x$ or in state $|-\rangle_x$. However, according to the EPR argument, one of these two states is no less an element of α's reality than one of the two states $|+\rangle_z$ and $|-\rangle_z$.

We are thus driven to the conclusion that, if the EPR argument is right, the description of reality provided by QM is not just incomplete – as Einstein and his associates claimed – but simply wrong. As I mentioned earlier, this is also the upshot of Bell's discovery. Bell (1964) proved by a fairly straightforward mathematical argument that statistical predictions based on the EPR conception of elements of reality are subject to an inequality that the statistical predictions of QM do not satisfy. Laboratory tests of Bell's inequality showed that real elementary particles in Bohm-type experiments do not satisfy them either.[45] More recently, Greenberger, Horne, and Zeilinger (1989) proved that by considering a system formed by three or more correlated particles of spin $\pm\hbar/2$ – instead of just two, as in Bohm's thought experiment and in Bell's original proof – a contradiction between the

---

[44] This sentence and the next remain true, of course, when the subscript z is consistently replaced with x (or with y).

[45] Aspect, Grangier, and Roger (1982) and Aspect, Dalibard, and Roger (1982). Their results are not immune to nitpicking, but – combined with all the other available evidence for QM – they would in the normal practice of science have laid the issue to rest if the dispute were purely about matters of fact and did not impinge on deeply seated metaphysical prejudices.

EPR assumptions and the predictions of QM can be derived not just in the case of statistical, but also of perfect, correlations (by which the authors mean arrangements in which the result of a measurement on one particle can be predicted with certainty given the outcomes of measurements on the other particles of the system). A satisfactory explanation of these matters would take us too long, and it would also be somehow redundant, for there are several good expositions in the literature.[46]

From a philosophical standpoint it is important to realize that the EPR reasoning – as formulated by its authors – is fallacious. According to premise 2, a quantitative attribute of a physical object is – or, in EPR jargon, "corresponds to" – an element of the reality of that object if it is *possible* to predict its value with certainty without in any way disturbing the object. In the Bohm-type situation described above, this criterion of reality was applied to the spin components of atom α in two mutually perpendicular directions. But, of course, according to QM, precisely when it is possible to predict with certainty the z-component of α – because one has measured the z-component of β –, it is *not* possible to predict with certainty the x-component of α, and vice versa. Thus, of two mutually perpendicular spin components of our atom α, no more than one at a time can satisfy the EPR criterion of reality under the principles of QM. Surely, while the z-component of α is being predicted with certainty without disturbing α, one can also predict with certainty the x-component of a similar atom α' – which is entangled with an atom β' like α is with β – again without in any way disturbing α'. But this fact about α' does not say anything about the present reality of the x-component of α; to think otherwise presupposes a different criterion of reality that the EPR authors do not make explicit.[47]

---

[46] My own favorite is still Wigner (1970) (but read '$\theta_{31} \geq \pi$' on p. 1,008, line 3 from below). See also Shimony (1990). A very readable analysis of the Aspect experiments will be found in Ruhla (1992), Ch. 8. Greenberger et al. (1990) give a perspicuous, self-contained exposition of the Greenberger–Horne–Zeilinger theorem and its background; see also Clifton, Redhead, and Butterfield (1991a, 1991b).

[47] It lurks perhaps in the following sentence: "Since at the time of measurement the two systems no longer interact, no real change can take place in the second system in consequence of anything that may be done to the first system" (Einstein, Podolsky, and Rosen 1935, p. 779). Yet this manner of change is common in social systems. For example, Judge Jones becomes Smith's father-in-law, and therefore incompetent – in some legal systems – to judge a case in which Smith is the plaintiff, as soon as, unbe-

There is one critical aspect of the EPR problem that I have side-stepped until now. I have unquestioningly accepted that in Bohm's experiment the measurement of a spin component on particle β does not "in any way" disturb particle α. This was the general attitude while the discussion was concerned only with thought experiments: One simply assumed that the condition prescribed in premise 2 was fulfilled. But since Bell's inequality was tested by real experiments, we often hear that its effective violation reveals the presence of a faster-than-light action-at-a-distance of some sort. Now the said experiments are designed so that no light signal can link the measurements on β with the correlative measurements on α (by which the QM predictions based on the former are confirmed). However, this precaution was not taken to test for faster-than-light signals, but to ensure that no influence went covertly from β to α; for if it did the data would be worthless. Indeed, if the experiments – as designed – did show the existence of such influence, this alone would lead to the conclusion that Einstein and his collaborators had in mind, albeit by a path that was different from theirs. Since QM itself does not contemplate any form of physical action through which the measurement performed on particle β might bring about a change in particle α, if such an action does exist, the quantum-mechanical description of reality is certainly incomplete. But what kind of action would this be? Shall we postulate a new kind of field through which the act of measuring on β the z-component of spin fixes the z-component of α? And devise some costly artifact that is capable of detecting the corresponding "particle"? And try to persuade the Congress of a wealthy and intellectually avid nation to finance its construction? In Bohm-type experiments the measurement of a particular spin component on particle β is deliberately arranged so that it does not *cause* any physical change in α. The act of measurement does however bring about a resetting of the *terms* in which – pursuant to QM – the spin of α must be described. Given that the total spin of α

---

knownst to Jones, Smith marries his daughter. However, before the advent of QM, it was unfamiliar to physicists. Right after the quoted sentence, Einstein et al. remark that it "is, of course, merely a statement of what is meant by the absence of interaction between the two systems". Of course, if one understands 'interaction' in this broad sense, instead of taking the word in its usual physical meaning of 'energy-and-momentum exchange', the sentence in question does indeed become trivially true, but it also becomes impossible to ensure by physical means – say, by separation, or shielding – that our two particles "no longer interact".

+ β is 0, as soon as, say, the z-component of β's spin is definitely known, so is the z-component of α. That the latter, if measured, obligingly complies with the quantum-mechanical description is one more reason for adopting QM.[48]

The violation of Bell-type inequalities in Bohm-type experiments according to QM has led philosophers to say that QM is a *nonlocal* theory. This is their way of expressing the fact – noted above for particle α – that the quantum-mechanical state description of an object sometimes must be changed even though one is not prompted to do so by what actually goes on in its immediate surroundings. I tend to think that this novel terminology is unnecessary, if not downright misleading. The correlation between the spin components of α and β discussed above was quite fitly described by Bohm himself as *noncausal* (1951, p. 430). And surely we ought not to expect that QM, having so famously made a mockery of causation, should submit to its traditional demands in this particular case.[49]

6.3.2 The Measurement Problem

QM began with Heisenberg's decision to pay attention exclusively to observable – that is, measurable – quantities and to refrain from trying to reconstruct in thought the unobservable "reality" that supposedly lies beneath them. It is therefore vexing – although perhaps only natural – that the problems regarding the microstructure of matter and radiation – quantum jumps, wave–particle duality, and so on – that beset the Old Quantum Theory and which Heisenberg's approach dis-

---

[48] Falkenburg (1995, pp. 295, 296) fittingly associates "the non-local EPR-correlations of particles detached from a coupled quantum system" with the breakdown of the classical concept of a particle and, more generally, of "*all the traditional representations of natural philosophy regarding the constitution of empirical reality in the small*".

[49] Niels Bohr, who understood that "the very existence of the quantum of action" entails "the necessity of a final renunciation of the classical ideal of causality" (1935, p. 697), had this to say about the EPR experiment: "It obviously can make no difference as regards observable effects obtainable by a definite experimental arrangement, whether our plans of constructing or handling the instruments are fixed beforehand or whether we prefer to postpone the completion of our planning until a later moment when the particle is already on its way from one instrument to another" (1949, p. 230). This trenchant remark can be profitably contrasted with recent quibbling over "quantum nonlocality" (and its purported conflict with Relativity).

pelled should be replaced in QM with a problem of measurement. QM is the source of several such problems – indeed the EPR problem might count as one of them – but '*the* measurement problem' designates in the literature the problem complex that I shall now discuss.[50] It arises at the interface between the deterministic evolution of the state of a quantum system and the measurement on it of a particular physical quantity with definite, probabilistically distributed values. As Wigner pithily remarks, "this is hardly surprising because the deterministic nature of the equations of motion prevents them from accounting for a probabilistic result" (1983, p. 285). The measurement problem is especially severe if one views the state vector and its unitary evolution as a representation of underlying reality. But it also demands consideration and solution if the observations results are the sole "reality" of interest for QM, and the state vector is only a store of information from which to calculate the probabilities of the several possible outcomes of different types of measurement. As we shall see, when QM is applied to the composite physical object formed by a system under study and the apparatus employed for studying it, the theory has – or so it seems – bewildering implications.

To make the following discussion easier to read – and to write – I shall no longer distinguish between physical states and observables, on the one hand, and the Hilbert space vectors and self-adjoint operators that represent them, on the other. Unless otherwise noted, I shall write as if every observable had a discrete, nondegenerate spectrum. This simplification – which is quite common in philosophical literature – is permissible because the fact that in real life some important observables have a degenerate spectrum, or a partly or completely continuous one, does not make the measurement problem any easier to deal with.[51]

---

[50] For a mathematically more advanced and thereby more satisfactory and much more complete up-to-date discussion of *the* measurement problem, see Busch, Lahti, and Mittelstaedt (1996). This is also the chief subject of Jeffrey Bub's prize-winning book (1997). For other problems of measurement in QM see Wigner (1983, pp. 275–76, 297–313).

[51] Indeed, an observable Q with a continuous spectrum cannot be measured exactly, for every measurement process yields rational numbers differing among themselves by multiples of some rational $\varepsilon > 0$, fixed by the instrument's power of resolution. So what one measures in fact is an observable $f(Q)$ with discrete spectrum, where f is a step function. For example, the typical clinical thermometer does not give us the values of the continuous quantity *temperature*, but of a discontinuous function of it, which jumps in steps of 0.1°C. I owe this important remark to Daneri, Loinger, and Prosperi (1962, p. 659).

Let S be a quantum-mechanical system, and let $\mathcal{H}_S$ denote its state space. If Q is an observable, with proper vectors $\{|\psi_i^Q\rangle: i \in \mathcal{I}\}$ forming an orthonormal basis of $\mathcal{H}_S$, the state of S is always equal to a linear combination $\Sigma_{i \in \mathcal{I}}\, c_i|\psi_i^Q\rangle$, such that, for each $i \in \mathcal{I}$, $|c_i|^2$ is the probability that, if Q is measured on S, it sports the proper value q_i corresponding to the proper vector $|\psi_i^Q\rangle$. Only in special circumstances will it happen that $c_k = 1$ for some $k \in \mathcal{I}$ and $c_i = 0$ for every other index i, in which case the state of S is a proper vector of Q, viz., $|\psi_k^Q\rangle$. In the general case, in which $c_i \neq 0$ for two or more indexes, I shall say that the state described by $\Sigma_{i \in \mathcal{I}}\, c_i|\psi_i^Q\rangle$ is a *nontrivial superposition* of the states $|\psi_i^Q\rangle$. To measure Q on S one must make S interact – perhaps very briefly – with an apparatus A. Let $m(Q,S,A)$ denote the interaction by which Q is measured on S with A. It makes good sense to think of A as a quantum-mechanical system with state space $\mathcal{H}_A$, which together with S forms a larger system $S + A$ whose state space is $\mathcal{H}_S \otimes \mathcal{H}_A$. To furnish measurements of Q the apparatus A should admit a zero state – represented by a normalized vector $|\alpha_0\rangle$ – in which it does not display any value of Q, and a set of states $\{|\alpha_i\rangle: i \in \mathcal{I}\}$ – orthogonal among themselves[52] and with $|\alpha_0\rangle$ – that are such that, for each $k \in \mathcal{I}$, $m(Q,S,A)$ brings A from state $|\alpha_0\rangle$ into state $|\alpha_k\rangle$ if and only if the value of Q measured by A on S is q_k.[53] We view the states $|\alpha_0\rangle$ and $\{|\alpha_i\rangle: i \in \mathcal{I}\}$ as the proper vectors of an operator R on $\mathcal{H}_A$; since A does its job of measuring Q only by adopting one of these states, I refer to them as the *characteristic states* – or *characteristic vectors* – of A.

Standard discussions of the measurement problem presuppose that, if S enters the measurement interaction in a proper state of the observable Q, then it retains this state right after the measurement.[54] This

---

[52] In order that A can do its measurement job, $\langle\alpha_j|\alpha_i\rangle$ must be 0 whenever $j \neq i$; otherwise, there would be a nonzero probability $|\langle\alpha_j|\alpha_i\rangle|^2$ that when A is in state $|\alpha_i\rangle$ it is also in a different state $|\alpha_j\rangle$.

[53] This corresponds to the ideal case in which every possible value of Q is reflected uniquely by a state of A. Usually, of course, a distinct state $|\alpha_k\rangle$ of A – for example, a distinct position of the arrow in A's dial – will correspond to a subset $\{|\psi_i^Q\rangle: i \in \mathcal{I}_k \subset \mathcal{I}\}$ of the proper states of Q. This is, indeed, inevitable if we drop the simplifying assumption that Q has only a discrete spectrum.

[54] See London and Bauer (1939, §11), van Fraassen (1972, p. 327), Wigner (1983, p. 281), Sudbery (1986, p. 186), Redhead (1987, p. 52), and Bub (1997, p. 52). Wigner stresses the "highly idealized" nature of the description provided by eqns. (6.41) and (6.43), and mentions some problems that this raises (1983, p. 284). On the other end of the spectrum, Shimony (1963, p. 5), simply *defines* a proper state of an observable F as "a state which is unchanged when a sufficiently careful measurement of F

excludes the common case in which the measurement procedure destroys the object on which a measurement is performed (e.g., a photon is converted into chemical energy upon arrival in the photographic plate that records its final position). The standard analysis does not apply to such measurements, or, in general, to measurements that significantly disturb the object, but only to *repeatable* measurements, which, when performed twice on the same system within a very short time, yield with certainty, on the second performance, the same result as on the first.[55]

Therefore, we assume that, if S and A go into $m(Q,S,A)$ in states $|\psi_k^Q\rangle$ and $|\alpha_0\rangle$, respectively, the state of $S + A$ evolves through $m(Q,S,A)$ from $|\psi_k^Q\rangle \otimes |\alpha_0\rangle$ to $|\psi_k^Q\rangle \otimes |\alpha_k\rangle$. So, if $m(Q,S,A)$ is a quantum-mechanical interaction governed by Schrödinger's equation and lasting a short time ε, the apparatus A must be so contrived that

$$U_\varepsilon(|\psi_k^Q\rangle \otimes |\alpha_0\rangle) = |\psi_k^Q\rangle \otimes |\alpha_k\rangle \qquad (6.41)$$

Now consider the general case, in which S goes into the interaction $m(Q,S,A)$ in a state $|\psi\rangle$ that is a nontrivial superposition of the $|\psi_i^Q\rangle$. In other words, $|\psi\rangle = \Sigma_{i \in 1} c_i |\psi_i^Q\rangle$, with $c_i \neq 0$ for more than one index i. The initial state of the compound system $S + A$ is then:

$$|\psi\rangle \otimes |\alpha_0\rangle = \sum_{i \in \mathcal{I}} c_i |\psi_i^Q\rangle \otimes |\alpha_0\rangle = \sum_{i \in \mathcal{I}} c_i(|\psi_i^Q\rangle \otimes |\alpha_0\rangle) \qquad (6.42)$$

Since the unitary operator U_ε is linear, the state of $S + A$ when the interaction ceases will be

$$U_\varepsilon\left(\sum_{i \in \mathcal{I}} c_i(|\psi_i^Q\rangle \otimes |\alpha_0\rangle)\right) = \sum_{i \in \mathcal{I}} c_i U_\varepsilon(|\psi_i^Q\rangle \otimes |\alpha_0\rangle)$$
$$= \sum_{i \in \mathcal{I}} c_i(|\psi_i^Q\rangle \otimes |\alpha_i\rangle) \qquad (6.43)$$

is performed". If we adopt this physical definition of proper state, the usual algebraic definition – 'a proper state of observable F represented by linear operator F is a state represented by a vector that F maps to a multiple of itself' – becomes a *prescription* for the due representation of proper states. (Compare Shimony's definition with the mathematical theorem stated in note 55.)

[55] Note the following important mathematical result: "*No continuous observable admits a repeatable measurement.*" This is Corollary 8.1.1 in Busch, Lahti, and Mittelstaedt (1996), who comment: "This result causes difficulties in our understanding of the operational definition of continuous observables, among them position, momentum and energy – observables which are most important for the concept of a particle in quantum physics" (p. 84).

To appreciate the implications of eqn. (6.43), let us put $\Sigma_{i \in \mathcal{I}}\, c_i(|\psi_i^Q\rangle \otimes |\alpha_i\rangle) = |\Psi\rangle$ and figure out the expectation value of the operator $Q' = Q \otimes R$ on $\mathcal{H}_S \otimes \mathcal{H}_A$ when $S + A$ is in state $|\Psi\rangle$. This is given by

$$\langle Q'\rangle_{|\Psi\rangle} = \langle \Psi | Q' | \Psi \rangle$$

$$= \sum_{i \in \mathcal{I}} |c_i|^2 \langle Q'\rangle_{|\psi_i^Q\rangle \otimes |\alpha_i\rangle} + \sum_{j \neq i} \sum_{i \in \mathcal{I}} c_j^* c_i \langle \psi_j^Q | \langle \alpha_j | Q' | \alpha_i \rangle | \psi_i^Q \rangle \qquad (6.44)$$

The last term on the right-hand side – viz., the double summation over $i, j \in \mathcal{I}$ $(j \neq i)$ – conveys the interference between the several proper states of Q'. Since the set $\{|\psi_i^Q\rangle \otimes |\alpha_i\rangle : i \in \mathcal{I}\}$ is orthonormal and consists of proper vectors of Q', every term in this summation vanishes. Thus, according to eqn. (6.44), the probability that system $S + A$ will be found right after $m(Q,S,A)$ in state $|\psi_k^Q\rangle \otimes |\alpha_k\rangle$ is $|c_k|^2$. $|\Psi_k^Q\rangle \otimes |\alpha_k\rangle$ is a proper state of Q' in which S is in state $|\psi_k^Q\rangle$ and A is in state $|\alpha_k\rangle$. The latter is the state that the apparatus reaches when it measures the value q_k of observable Q. So our result agrees with the QM prediction that if Q is measured on S when the latter is in state $|\psi\rangle = \Sigma_{i \in \mathcal{I}}\, c_i |\psi_i^Q\rangle$, the probability of obtaining the value q_k is precisely $|c_k|^2$.

Despite the pleasant consistency of these results, the transition described in eqns. (6.42) and (6.43) has often been judged objectionable. Equation (6.43) equates the final state of the macroscopic system $S + A$ with a nontrivial superposition of states; yet in real life – it is said – macroscopic objects are, at any given time, in one definite state or another, never in a superposition of them. To my mind, this objection reveals unreadiness to live with the quantum-mechanical concept of physical states. If physical states are properly represented by nonzero vectors in a Hilbert space of more than one dimension, every physical system is under every circumstance in nontrivial superpositions of different sets of states (some of which sets may well consist of proper states of some interesting observable). This follows at once from the elementary mathematical fact that every nonzero vector in such a space can be expressed in infinitely many different ways as a sum of two or more noncollinear vectors. Therefore, it is not through its entanglement with the – usually microscopic – observed system S that the macroscopic apparatus A becomes "infected with superposition". If superposition is a disease, it is one from which A suffers congenitally, if it is in effect a quantum system.

It is true, on the other hand, that according to QM, when one measures a discrete nondegenerate observable on an individual physical system – no matter what its size – one will always find the system in a definite proper state of that observable and never in a nontrivial superposition of such states. But the tension, if there is one, between this truth and the information contained in eqns. (6.42) and (6.43) merely reflects a certain laxity in the expression 'to find a system in a proper state of an observable'. What one actually *finds* through the measurement carried out on the system is a definite proper value of the measured observable. Based on this value, one *assigns* to the system, at the time of measurement, the corresponding proper state, which can then be used for calculating the probability of finding this or that proper value of the same or a different observable if it is measured on the system soon thereafter or subsequently. In the context of QM we must inevitably distinguish between the continuous evolution of states and the punctual detection of values. We care for the former only insofar as it provides the means of anticipating the probabilistic distribution of the latter; but ultimately what matters to us are the values. And, of course, no differential law can bridge the gap between a deterministic evolution and its chancy manifestations: this is precisely what real chance consists in (see pp. 340–42).

John von Neumann, who apparently was the first to notice a problem here, regally cut the Gordian knot. He postulated that the interactions between quantum-mechanical systems and measurement apparatuses are not governed by Schrödinger's equation, but instead they constitute a different sort of quantum-mechanical change. One must therefore distinguish between "two fundamentally different types of interventions which can occur in a system S or in an ensemble $\{S_1, \ldots, S_n\}$. First, the arbitrary changes by measurements [. . .]. Second, the automatic changes which occur with passage of time" (1955, p. 351). The latter is the unitary evolution that we have considered to this point, while the former results from "discontinuous, nondeterministic[56] and instantaneously acting experiments or measurements" (1955, p. 349). By virtue of it, when a system in the state $|\psi\rangle$ interacts with an

---

[56] Instead of "nondeterministic" the English translation has "non-causal". Given the connotations of 'kausal' in German philosophical and scientific literature *c*. 1930, I believe that my rendering is more accurate (see note 28 and Chapter Three, note 33; cf. von Neumann's own discussion of "causality" in his (1955, pp. 302f.)). Two lines later, the adverbs "continuously and causally" are applied in one breath to unitary evolution. In the light of what I said in §3.4.3, it should be clear that 'causally' here means 'deterministically', for causal change, in the ordinary acceptance of this expression, cannot obey a differential equation. Indeed, it is the other type of change, by "discontinuously and instantaneously acting experiments", that better fits the commonsense idea of causation.

instrument designed to measure the observable Q, the vector $|\psi\rangle$ at once becomes – "the wave function collapses into" – one of the proper vectors of Q. Collapse into the proper vector $|\psi_i\rangle$ occurs with probability $|\langle\psi_i|\psi\rangle|^2$ if the spectrum of Q is discrete and nondegenerate. If the spectrum of Q is discrete but degenerate, and there are r normalized linearly independent vectors $|\psi_{k1}\rangle, \ldots, |\psi_{kr}\rangle$ corresponding to the proper value q_k, $|\psi\rangle$ collapses with probability $\Sigma_{i=1}^{r}|\langle\psi_{ki}|\psi\rangle|^2$ into *some* normalized vector in the subspace spanned by these r vectors.[57] Things get more complicated if Q has a partially or totally continuous spectrum. But besides such purely technical – and manageable – difficulties, von Neumann's solution faces serious obstacles of another sort.

To begin with, the very suggestion that QM countenances physical interactions that are not governed by Schrödinger's equation is disconcerting. If QM is any good, it should apply – within a satisfactory margin of error – to *every* physical situation in which h differs significantly from 0 and the speed of light can be regarded as practically infinite. If the system S accidentally interacts with the apparatus A, say, in a laboratory junkyard, then, according to QM, $S + A$ should undergo a unitary evolution. It would seem therefore that von Neumann's alternative form of evolution, involving the "collapse" of S's latest state to a proper state of the measured observable, can only occur if the compound system somehow knows that the process it is going through is meant to serve as a measurement. Evidently this can only happen if the compound includes a knowing mind. Of course, measurements are contrived by human beings and are completed only when one of them takes note of the results. Only they can *find* the value of a certain observable to be this or that when it is measured on a physical system. So perhaps one ought to include the human observer O, besides the system S and the apparatus A, among the interacting parts of a proper measurement process.

Regardless of what metaphysical materialists may think, the truth is that there is no evidence that interactions involving O are governed by Schrödinger's equation, nor have we the slightest inkling of what a differential equation for the system $S + A + O$ would be like. On the other hand, we do know that a measurement interaction involving an observer O should normally lead him or her to a state of awareness

---

[57] According to the much quoted – and disputed – Lüders' Rule (Lüders 1951), in the degenerate case $|\psi\rangle$ collapses with the stated probability precisely into its projection on the subspace spanned by the proper vectors corresponding to the proper value in question.

that reflects the finding of a particular proper value of the observable that the interaction was supposed to measure on S. In the light of these facts, it is not altogether unreasonable to distinguish between the unitary evolution of $S + A$ and the "collapse of the wave function" brought about by the interaction between $S + A$ and O. The unanalyzed concept of "collapse" nicely fits both our knowledge and our ignorance. Indeed, if the information content of the state vector $|\psi\rangle$ consists of probability distributions – one for each family of commuting observables – it is not extravagant to say that, when the observer perceives a definite proper value of a particular observable Q, $|\psi\rangle$ collapses forthwith to a matching proper vector of Q – just as a gambler's hopes collapse when he sees the roulette ball stop, say, at double-zero.

Still, these considerations show only that von Neumann's solution is not so wanton as some philosophers have made it out to be, not that it solves the measurement problem. Indeed, the transition described in eqns. (6.42) and (6.43) has an unwanted consequence that von Neumann's solution does nothing to explain away. It turns out that, following the entanglement of apparatus A with system S, the predicted state of $S + A$ precludes the assignment of definite probabilities to the proper values of some observables. Let T be an observable of A that does not commute with the observable R that we considered above. Let I denote the identity operator on $\mathscr{H}_S$. By replacing Q' with $I \otimes T$ in eqn. (6.44) we obtain the expectation value of this operator on $\mathscr{H}_S \otimes \mathscr{H}_A$ when $S + A$ is in state $|\Psi\rangle$:

$$\langle I \otimes T \rangle_{|\Psi\rangle} = \langle \Psi | I \otimes T | \Psi \rangle$$
$$= \sum_{i \in \mathscr{I}} |c_i|^2 \langle I \otimes T \rangle_{|\psi_i^Q\rangle \otimes |\alpha_i\rangle} + \sum_{j \neq i} \sum_{i \in \mathscr{I}} c_j^* c_i \langle \psi_j^Q | \langle \alpha_j | I \otimes T | \alpha_i \rangle | \psi_i^Q \rangle$$

$$(6.44^*)$$

Due to the condition we imposed on T, the interference terms – the summands led by the factors $c_j^* c_i$ ($i \neq j$) – do not all vanish, so there is no definite probability that $S + A$, in state Ψ, sports a particular proper value of $I \otimes T$. Such a predicament is not easy to reconcile with our experience of macroscopic objects.[58]

---

[58] In my view, Schrödinger's renowned cat paradox constitutes a serious problem only if it is made to turn around the difficulty just stated. He imagined the following "diabolical device": A cat is penned up in a steel box together with a Geiger counter containing "a tiny bit of radioactive substance, *so* small, that *perhaps* in the course of

It is often said that Bohr managed to sidestep this difficulty by asserting that QM applies solely to microscopic objects and that the behavior of macroscopic systems must be described in classical terms. Bohr's actual views were subtler (see §6.4.1), but let me just say here why the said assertion cannot solve the measurement problem. For one thing, the border between "small" and "big" objects is fuzzy. What physical laws apply in this no-man's land? Moreover, physicists have discovered and incorporated in their research equipment many macroscopic effects that can only be understood in quantum-physical terms (e.g., the laser). Indeed many laboratory procedures are currently used that depend explicitly on quantum interactions (see Braginsky and Khalili 1992).

One must therefore search for another way out of this conundrum. Von Neumann himself was well aware of this need and pointed out the direction to follow:

> We must show that [the unitary evolution of the compound system] gives the same result for S as the direct application of [wave function collapse] on S. If this is successful, then we have achieved a unified way of looking at the physical world on a quantum mechanical basis.
>
> (Von Neumann 1955, p. 352; my notation)

Right after these lines he refers us to the last section of the book (VI.3), but I do not see that the discussion there really proceeds in the said direction. It rests entirely on the conception of measurement as an interaction of the observed system with an observing system that includes a human consciousness, and von Neumann obviously knows no way of incorporating the latter in a unitary evolution. However, we do not expect physics to account for the occurrence of states of awareness – such as the perception by someone of a pointer over a dial mark or of a row of digits in a computer printout – but only for the behavior of the physical objects involved in measurement. To perform a measurement on a quantum system we must put it in interaction with another system that undergoes, through the interaction, macroscopic modifica-

---

an hour one of the atoms decays, but also, with equal probability, perhaps none; if it happens, the counter tube discharges and through a relay releases a hammer which shatters a small flask of hydrocyanic acid". Schrödinger comments: "If one has left this entire system to itself for an hour, one would say that the cat still lives *if* meanwhile no atom has decayed. The first atomic decay would have poisoned it. The ψ-function of the entire system would express this by having in it the living and the dead cat (pardon the expression) mixed or smeared out in equal parts" (1935; quoted from Wheeler and Zurek 1983, p. 157).

tions depending on the state of the first system. Thus, the aim of QM is "essentially to make predictions on the trace that our microscopic body will leave in the macroscopic world, when the trace left at a certain time is known" (Daneri, Loinger, and Prosperi 1962, p. 298). Clearly, then, "in order to build up a satisfactory quantum theory of measurement, it is necessary to have a theory, at least schematic, of the large bodies, i.e. a theory which gives the connection between the macro-properties of these bodies and their microscopic structure described by quantum mechanics" (Ibid.). This theory ought to prove, from the laws of quantum mechanics and the structure of macroscopic bodies, that states which are not compatible with the actual macroscopic observations are, in effect, impossible. In other words, it ought to imply that the interference terms in eqn. (6.44*) must promptly become zero, at least for all practical purposes.

Günther Ludwig did valuable work along these lines since the 1950s.[59] The paper by Daneri et al. (1962), from which I have just quoted, was an important step forward. Recent efforts by Gell-Mann and Hartle, Omnès, and others turn around *decoherence*, "a dynamical effect taking place in the bulk of matter, [which] happens so quickly that one cannot catch it while it is acting and one can only accede most of the time to a situation where it has already occurred" (Omnès 1994, p. 269). The name 'decoherence' is linked to 'coherences' – which is what some authors call the off-diagonal elements of the matrix (6.30) (cf. Cohen-Tannoudji et al. 1977, I, 303) – and signifies the vanishing of the coherences of some suitably chosen statistical operators as a consequence of the said effect.[60] There is still no general theory of decoherence, but detailed studies of several models have produced promising results. I cannot dwell on them here. The following indications may stimulate some readers to pursue this fascinating matter on their own.

In this approach, the solution of the measurement problem is acces-

[59] Ludwig (1953, 1954). A revised and enlarged second edition of Ludwig (1954) was published in English in 1984/85. Together with Ludwig (1985/87) it offers a grand synthesis of Ludwig's lifework on the foundations of QM, in four imposing volumes, which regrettably, I must confess, I have not been able to penetrate. A less "gründlich" but much more accessible presentation of his ideas on the subject is given in vols. 3 and 4 of his *Introduction to the Foundations of Theoretical Physics* (1984, 1979; in German).

[60] See Gell-Mann and Hartle (1993, p. 3347n.), on another related but significantly different sense of 'decoherence'.

sory to that of another, more general problem, which can be stated thus: Since the objects that surround us are supposed to be aggregates of mutually interacting quantum systems, and yet they obey the laws of classical mechanics to an excellent approximation, QM should account for this "quasiclassical" behavior of everyday objects. In particular, it should explain why things of our size such as a chair or a ship are always found in states that, quantum-mechanically, can only be conceived as proper vectors of pairwise commuting observables, and why no interference of state vectors is ever noted in such objects (except in some specially devised superconducting systems and, of course, in the familiar case of radiation). By solving this problem one would obtain, as a supplementary bonus, a quantum-mechanical explanation of the peculiar behavior of apparatuses. Gell-Mann and Hartle deal with the problem in their paper "Classical Equations for Quantum Systems" (1993); and Omnès deals with it in Chapter 6, "Recovering Classical Physics", of his book (1994). Omnès begins by noting that only a few properties are normally sufficient for describing a macroscopic object as such and its motion. "They are easily identified when one actually sees a real object: they are the coordinates of the center of mass, some orientation angles, some distances between outstanding reference points (as, for instance, the distance between the ends of a spring), the orientation of a wheel in a clockwork, the electric charge on a capacity, and so on" (1994, p. 205). He calls them "the *collective observables*". He denotes by H_c the part of the Hamiltonian operator that depends only on the collective observables. If there is no thermal dissipation, the total Hamiltonian H of the object – regarded as a closed system S – is the sum of the "collective Hamiltonian" H_c and a "microscopic Hamiltonian" H_e that depends only on the noncollective or "microscopic" observables. H_c represents the mechanical energy and H_e the internal energy (in the sense of thermodynamics). However, if dissipation occurs, $H = H_c + H_e + H_{int}$, where the coupling Hamiltonian H_{int} depends on both kinds of observables and expresses the exchange between the two kinds of energy. Obviously, the coupling "must be taken into account in order to describe the existence of friction effects in classical physics. Far less trivial is the fact that it has also a dramatic effect upon a superposition of two macroscopically different states, which is called *decoherence*. It destroys quantum interferences at a macroscopic level . . ." (Omnès 1994, p. 269). Decoherence is formally derived for several special cases by constructing a complete statistical operator ρ for the system under consideration and then extracting from it a "reduced" or "collective" statistical operator ρ_c (by performing a partial trace of ρ). Under unitary evolution the off-diagonal terms of the matrix of ρ_c vanish very rapidly. Omnès calculates the decoherence time for a pendulum with a mass of 1 gram, a period of 1 second, and damping time of 1 hour. If the pendulum is let go at zero temperature in a superposition of two states with initial positions differing only by 1 micron, 1 nanosecond later the off-diagonal matrix terms of the reduced statistical operator will be smaller than $\exp(-10^3)$. Indeed the decoherence time is less the shorter the damping time and the greater the mass, the initial temperature, the squared frequency, and the squared separation between the initial positions of the superposed states. "No effect so spontaneous, so efficient, and of so frequent occurrence is known in the whole field of physics", says Omnès (1994, p. 291).

The mathematical procedure that I have adumbrated only shows that the systems to which it is applicable cannot display any interference of proper vectors of a collective observable. Still, someone could object that

> there always exist more subtle observables that would be able to show it. By measuring such an observable, as always was assumed to be possible in principle, one will exhibit the existence of a surviving quantum superposition of macroscopically different states. The answer provided by decoherence is therefore valid "for all practical purposes", because one can only measure in practice a collective observable of a macroscopic object, but it is not a valid answer as far as basic principles are concerned.
>
> (Omnès 1994, p. 305)

This objection, which was actually made by Bell (1975),[61] is firmly rejected by Omnès. He estimates that a typical measurement apparatus A has 10^{27} continuous degrees of freedom (including the noncollective ones). He reckons roughly that a second apparatus B capable of measuring an arbitrary, noncollective observable on A would require

---

[61] Commenting on "a very elegant and rigorous paper" by K. Hepp (1972), Bell conceded that certain formal results show "that any *fixed* observable Q will eventually give a very poor (zero, in [the case considered]) measure of the persisting coherence". However, "nothing forbids the use of different observables as time goes on". Thus, "while for any given observable one can find a time for which the unwanted interference is as small as you like, for any given time one can find an observable for which it is as big as you do *not* like" (1975, in Bell 1987, pp. 48–49).

no less than $10^{10^{18}}$ degrees of freedom. B would therefore be too large to fit within our horizon.[62] According to Omnès this implies that such an observable cannot be measured *in principle*. Philosophers will prefer to use the words 'in principle' more strictly, but surely impossibility for *all* practical purposes should be good enough for them. As Cini aptly noted in a paper that anticipated many of the above ideas, "this situation implies a close analogy with the second law of thermodynamics, which is valid to a very high degree of approximation in spite of its being incompatible with the time reversibility of the equations of motion" (1983, p. 30). We may rest assured that we shall never see a cup of white tea spontaneously split into separate layers of milk and tea, although this is possible in principle. Analogously, even though every macroscopic object should teem with interferences between the proper states of some observable or other, this state of affairs cannot come to light if the conditions for decoherence are satisfied.

6.4 Meta-Physical Ventures

The conceptual difficulties of QM have prompted a wide variety of purported solutions. Those suggested in §6.3 are perhaps the most economical, namely, to suppress the EPR problem by accepting that the idea of causation – despite its invaluable contribution to the intelligent handling of our everyday life – is quite insufficient for the description of natural phenomena, and to tame the measurement problem by strictly physical means, developed as far as possible from within QM. In this section I shall comment on a sample of other conceptions, chosen mainly for their influence and their diversity. I collect them under the epithet 'meta-physical' (*meta* = 'beyond') for they view the meaning and scope of QM from standpoints outside empirical science. Such is clearly the case of Bohr's epistemology (§6.4.1) and of quantum logic (§6.4.3), if indeed this is more than a misnomer; but it is also the case of Bohm's theory of hidden variables (§6.4.2), which is, no doubt, a proper physical theory, but was proposed as a substitute for QM not

---

[62] Omnès does not explain whether he refers to our so-called *particle* horizon or to our *event* horizon. If the center of gravity of B stood within the Solar System, its outer rim would, in the former case, be further away than the most distant galaxies from which a light signal can – in principle – reach us now; in the latter case, the outer rim of B would be beyond the most distant galaxies from which a light signal will ever reach the earth.

because of any experimental results unaccountable by the latter, but on unabashedly metaphysical grounds. As to Everett's "theory of the universal wave function" (§6.4.4), we shall see that – at any rate in Bryce DeWitt's "many-universes" version – it makes QM into a metaphysical theory of the less reputable sort.

6.4.1 Complementarity

The most influential philosophy of quantum mechanics is the "Copenhagen interpretation" put forward by Niels Bohr in Como at the Volta centennial conference of 1927, and further elucidated and elaborated by him in numerous lectures and papers.[63] The very difficulty that philosophical readers usually find in ascertaining Bohr's exact meaning probably made it easier for other outstanding physicists – Heisenberg, Born, von Weizsäcker – to equate their views with his. Bohr extended his philosophy from microphysics to biology, psychology, and cultural anthropology. In its most general version it turns around the idea that no single coherent system of human concepts can cope with the complexity of things, so that in each field of intellectual endeavor we must resort to *pairs* of concepts that afford mutually inconsistent but complementary perspectives (e.g., "thoughts" and "feelings", "instinct" and "reason" – Bohr 1958, p. 27). On the face of it, this looks like the perfect recipe for woolly thinking, and it is fondly cherished by some as such. But in its specific application to quantum physics Bohr's conception of complementarity achieved a quite definite and, for most practicing physicists, quite convincing formulation. Thus, in his book on Bohr, the distinguished particle physicist Abraham Pais says that "Bohr's exegesis of the quantum theory is the best we have to date" (1991, p. 435). According to Pais, Bohr became in effect, through his idea of complementarity, the successor to Kant in philosophy (Ibid., p. 23). Although somewhat exaggerated, this statement contains a valuable suggestion, viz., that we use Bohr's tacit link with Kant to throw light on complementarity.

Kant maintained that natural phenomena can be perceived and described as such only if the sense impressions that disclose them are combined and ordered under concepts, the most general of which are contributed by human reason. Kant distinguished two kinds of ratio-

---

[63] Many of them were collected in Bohr (1934, 1958, 1963), recently reprinted as *The Philosophical Writings of Niels Bohr* (Woodbridge, CN: Ox Bow Press, 1987; 3 vols.).

nal principles controlling the constitution of human experience, viz., the "mathematical" principles of the "axioms of intuition" and the "anticipations of perception" (§3.3), and the "dynamical" principles of the "analogies of experience" (§3.4). The former preside over the determination of distances, areas, and volumes in space and of intervals in time, and prescribe the continuity of all intensive magnitudes. The latter regulate the world wide web of causal relations, subject to major conservation principles and resting on the universal interaction of all things. This conception of experience beautifully fits Newtonian science, as it was shaped in the eighteenth and early nineteenth centuries (§2.5.2 and 2.5.3). But Kant and his followers were apparently convinced that our everyday experience is also organized – less precisely, perhaps, but no less consistently – after the same patterns. The "mathematical" and "dynamical" principles of reason are supposed to work together, and Kant does not for a moment imagine that there could arise a conflict or incompatibility among them. Moreover, although he emphatically asserted the role of sensation in providing the "matter" that is to fill the structure projected by reason – so that, for example, the gravitational constant or the boiling point of alcohol certainly could not be ascertained otherwise than by experiment –, he does not seem to have seriously contemplated that the "manifold of sense" might turn out, at some point, not to fit that structure. Yet this is precisely what, according to Bohr, has happened with the discovery of the quantum of action.

The progress of experience has compelled physicists to acknowledge that physical action never occurs in quantities less than Planck's constant h. This implies that the interference of laboratory equipment with the physical objects under study cannot be indefinitely reduced but shall always have a lower bound. According to Bohr, this is the source of Heisenberg's indeterminacy relations, by virtue of which there is a lower bound – viz., h (more precisely, $h/4\pi$) – to the precision with which positions and momenta can be simultaneously assigned to particles and bodies. As an illustrative example, Bohr cites the "acute difficulties" that stood in the way of a "consistent description" of the Compton effect (see note 9 and §6.1.3):

Any arrangement suited to study the exchange of energy $[E]$ and momentum $[P]$ between the electron and the photon must involve a latitude in *the space–time description* of the interaction sufficient for the definition of wave-number $[\sigma]$ and frequency $[\nu]$ which enter into the relation

$[E = h\nu, P = h\sigma]$. Conversely, any attempt of locating the collision
between the photon and the electron more accurately would, on account
of the unavoidable interaction with the fixed scales and clocks defining
the space–time reference frame, exclude all closer account as regards *the
balance of momentum and energy*.

(Bohr 1949, p. 210; my italics)

The situation thus revealed has momentous consequences:

> On the one hand, the definition of the state of a physical system, as ordi-
> narily understood, claims the elimination of all external disturbances.
> But in that case, according to the quantum postulate, any observation
> will be impossible, and, above all, *the concepts of space and time lose
> their immediate sense*. On the other hand, if in order to make observa-
> tion possible we permit certain interactions with suitable agencies of
> measurement, not belonging to the system, an unambiguous definition
> of the state of the system is naturally no longer possible, and *there can
> be no question of causality* in the ordinary sense of the word.

(Bohr 1928, in Bohr 1934, p. 54; my italics)

Thus, it turns out that "our usual causal space–time description" – that
is, the joint application of Kant's mathematical and dynamical principles – has hitherto been "appropriate" only because h is negligible "as
compared to the actions involved in ordinary sense perceptions" (Ibid.,
p. 55). Still, Bohr does not think that we must now scuttle the Kantian
categories and create a new framework for the description of phenomena. Although we must be prepared, as our knowledge grows, "to
expect alterations in the points of view best suited for the ordering of
our experience" (1934, p. 1), and although "the modern development
of physics" is forcing on us a "thoroughgoing revision of our conceptual means of comparing observations" (1949, p. 239), it would be,
according to Bohr,

> a misconception to believe that the difficulties of the atomic theory may
> be evaded by eventually replacing the concepts of classical physics with
> new conceptual forms. [...] The recognition of the limitation of our
> forms of perception by no means implies that we can dispense with our
> customary ideas or their direct verbal expressions when reducing
> our sense impressions to order. No more is it likely that the fundamen-
> tal concepts of the classical theories will ever become superfluous for the
> description of physical experience.

(Bohr 1934, p. 16)

Not only does the very recognition of the quantum of action and the determination of its magnitude "depend on an analysis of measurements based on classical concepts, but it continues to be the application of these concepts alone that makes it possible to relate the symbolism of the quantum theory to the data of experience" (Ibid.). On this point, Bohr is emphatic: "*However far the phenomena transcend the scope of classical physical explanation, the account of all evidence must be expressed in classical terms*". The reason for this is simple:

> By the word "experiment" we refer to a situation where we can tell others what we have done and what we have learned and [. . .], therefore, the account of the experimental arrangement and of the results of the observations must be expressed in unambiguous language with suitable application of the terminology of classical physics.
>
> (Bohr 1949, p. 209)

For "only with the help of classical ideas is it possible to ascribe an unambiguous meaning to the results of observation" (1934, p. 17). Therefore, "the space–time coordination and the claim of causality, the union of which characterizes the classical theories" must now be regarded "as complementary but exclusive features of the description" (1928, in 1934, p. 54). Because it is impossible "in the field of quantum theory" to accurately control "the reaction of the object on the measuring instruments, i.e., the transfer of momentum in case of position measurements, and the displacement in case of momentum measurements," one has to renounce in each experimental arrangement to "one or the other of two aspects of the description of physical phenomena, the combination of which characterizes the method of classical physics, and which therefore in this sense may be considered as *complementary* of each other" (1935, p. 699).

> In fact, it is only the mutual exclusion of any two experimental procedures, permitting the unambiguous definition of complementary physical quantities, which provides room for new physical laws, the coexistence of which might at first sight appear irreconcilable with the basic principles of science. It is just this entirely new situation as regards the description of physical phenomena, that the notion of *complementarity* aims at characterizing.
>
> (Bohr 1935, p. 700)

According to Bohr, "the quantum-mechanical formalism" precisely offers "an adequate tool" for this kind of description, inasmuch as it is "a purely symbolic scheme permitting only predictions, on lines of the correspondence principle, as to results obtainable under conditions specified by means of classical concepts" (1949, pp. 210f.). But "there can be no question of any unambiguous interpretation of the symbols of quantum mechanics other than that embodied in the well-known rules which allow to predict the results to be obtained by a given experimental arrangement described in a totally classical way" (1935, p. 701). "The appropriate physical interpretation of the symbolic quantum-mechanical formalism amounts only to predictions, of determinate or statistical character, pertaining to individual phenomena[64] appearing under conditions defined by classical physical concepts" (1949, p. 238).[65]

Bohr's approach entails, of course, that the unitary evolution of state vectors in Hilbert space does not represent a process in the real world, but is merely a computational link between two sets of data described in classical terms, viz., those concerning the preparation of an experiment and those concerning its results. This view has offended the metaphysical conscience of most philosophers and some physicists. In particular, the utilization of QM in astrophysics and cosmology seems questionable or at any rate funny if the talk about vectors and linear operators in Hilbert space is meaningful only as it relates to experimental arrangements describable by classical concepts that do not fit the conditions in the early universe and in the interior of stars. This has led some to "interpret" unitary evolution as more than just an algorithmic bridge between data input and data output (§6.4.4), and others to seek for a proper physico-mathematical description of the real processes that supposedly underlie the statistical correlations correctly predicted by QM (§6.4.2). Their qualms are not altogether unjustified for, although every physical theory obtains its meaning and truth

---

[64] Note that Bohr advocates "the application of the word *phenomenon* exclusively to refer to the observations obtained under specified circumstances, including an account of the whole experimental arrangement" (1949, pp. 237f.).

[65] In other words, "quantum mechanics speaks neither of particles the positions and velocities of which exist but cannot be accurately observed, nor of particles with indefinite positions and velocities. Rather, it speaks of experimental arrangements in the description of which the expressions 'position of a particle' and 'velocity of a particle' can never be employed simultaneously" (Frank 1935; quoted in Jammer 1974, p. 200).

through those few points where it comes in contact with human experience, the theories of Newton and Einstein provided a conceptual framework encompassing both this experience and the natural world beyond it, while QM, as seen from Copenhagen, depends on the admittedly insufficient classical framework for describing macroscopic events in the laboratory and has nothing to say about anything else.

How do the philosophical problems of §6.3 fare under Bohr's approach? He had little trouble dispelling the EPR paradox (Bohr 1935), but that is no great feat if, as Bohr himself implied, the EPR argument is fallacious. At first blush it might seem that Bohr also dissolves the measurement problem. As David Bohm pointedly notes, "according to Bohr's interpretation nothing is measured in the quantum domain [. . .] because all 'unambiguous' concepts that could be used to describe, define, and think about the meaning of the results of such a measurement belong to the classical domain only" (1980, p. 75). On a closer look, however, I find that the measurement problem remains very much alive. The interference terms in eqn. (6.44*) are predicted by the quantum-mechanical formalism, so they ought to show up somehow in laboratory results. Bohr owes us an explanation of why this does not happen.

6.4.2   Hidden Variables

In the "final remarks" of the paper in which he established the probabilistic interpretation of Schrödinger's ψ-function (§6.2.4), Born says that anyone who will not rest content with indeterminism is naturally free "to assume that there are additional parameters, not yet introduced in the theory, which determine the individual event" (1926b, p. 825). In a lecture delivered at Oxford 20 days after submitting that paper, Born compared the use of probability in QM and in the classical theory of gases (§4.3.3). The latter conceives a gas as a collection of molecules whose positions and momenta are determined at every instant by their initial values and their evolution according to the differential equations of classical mechanics (§2.5.3). The theory, however, works with probabilistic assumptions leading to statistical predictions, because it is impossible to know the exact positions and momenta of the molecules at any given time. Thus "the classical theory introduces the microscopic coordinates which determine the individual process, only to eliminate them because of ignorance by averaging over their values; whereas the new theory gets the same results without introducing them at all. Of course, it is not forbidden to believe in the existence of these coordinates; but they will only be of physical significance when methods have been devised for their experimental observation" (1927b, in Born 1969, p. 10). Some physicists, however, have exercised the freedom allowed by Born, despite the total absence of such experimental methods. The "hidden variables" – or "hidden parameters" – approach to QM consists in postulating the existence of hitherto unobserved and presumably unobservable physical quantities whose evolution under suitably designed laws exactly determines the outcome of the individual quantum processes.[66] In 70 years, theories countenancing such hidden variables have not led to the discovery of a single new physical effect. Indeed, their partisans are content if the consequences of their hypotheses agree with the predictions of QM. I shall refer to hidden variables theories that satisfy this requirement as 'HV-extensions of QM'.

In his book of 1932 (IV, §§1–2), von Neumann argued that no HV-extension of QM is possible. He discussed the following situation, which is characteristic of QM: The measurement of a particular quantity on an ensemble of systems is found to yield different values, although every system is in the same quantum-mechanical state. There are two explanations for this: Either (a) the systems are in different states, which QM is incapable of distinguishing, so it represents them all by the same ψ-function; or (b) the systems are really in the same state, and the dispersion of measured values is due not to our ignorance and to the grossness of the quantum-mechanical representation, but to "nature itself, which has disregarded the 'principle of sufficient cause'" (von Neumann 1955, p. 302). Alternative (a) would imply that our ensemble comprises as many subensembles as there are different results of the said measurement, and that every system in any one subensemble is in a particular dispersion-free state, characterized by a distinct value of a list of "hidden variables" unknown to QM. Von Neumann proved from assumptions he considered plausible that, under QM, no dispersion-free states are possible. The main assumption is that, if the quantities $A, B, \ldots, K$ are represented by operators A, B, . . . , K, and $a, b, \ldots, k$ are any real numbers, then (i) the quantity $aA + bB + \ldots + kK$ is represented by the operator $a\mathsf{A} + b\mathsf{B} + \ldots$

---

[66] See the massive survey by Belinfante (1973). It is perhaps worth noting that, with the exception of Louis de Broglie (1956), Nobel prize winners have stood aloof of such efforts.

+ kK, and (ii) its expectation value $\langle aA + bB + \ldots + kK \rangle_{|\psi\rangle}$ in an ensemble prepared in a given state $|\psi\rangle$ is equal to $a\langle A \rangle_{|\psi\rangle} + b\langle B \rangle_{|\psi\rangle} + \ldots + k\langle K \rangle_{|\psi\rangle}$. Now, this assumption holds, without question, for quantummechanical states, but there is no reason why it should also apply to the dispersion-free states foreign to QM that von Neumann is arguing about, and to the "hidden variables" that would specify them. Therefore, von Neumann's formal proof does not justify his informal conclusion that "the present system of quantum mechanics would have to be objectively false, in order that another description of the elementary processes than the statistical one be possible" (von Neumann 1955, p. 325; see Bell's discussion in his 1987, pp. 4–5).

Although von Neumann's argument is unable to prove that QM precludes the existence of well-defined, deterministically evolving "hidden variables", other mathematical theorems discovered by Gleason (1957), Bell (1964), and Kochen and Specker (1967) impose severe conditions on HV-extensions of QM.[67] However, the theory put forward by David Bohm (1952) meets those conditions and makes the same predictions as QM. To get a feeling of how it works, let us take a look at Bohm's treatment of a system consisting of a single particle. Schrödinger's equation for one particle of mass m moving in a classical potential V is eqn. (ES1*) of note 23, which Bohm writes thus:

$$ i\hbar \frac{d\psi}{dt} = -\frac{\hbar^2}{2m}\nabla^2\psi + V(\mathbf{x})\psi \tag{6.45} $$

ψ is a complex-valued function on three-dimensional Euclidean space. It can be expressed in terms of two real-valued functions, R and S, on the same space:

$$ \psi = R\exp(iS/\hbar) \tag{6.46} $$

By substituting eqn. (6.46) in (6.45) and separating real and imaginary parts we obtain the equations:

---

[67] To agree with QM, a hidden variables theory must be "nonlocal" and "contextual", in the senses indicated in notes 68 and 69. The theorems of Gleason, and of Kochen and Specker are often mentioned and discussed in philosophical literature, and they deserve more than a bare mention here. However, to state them intelligibly I would have to fill a few pages with mathematical definitions. I refer the curious reader to Redhead (1987) (who devotes pp. 119–52 to "the Kochen–Specker paradox") and Hughes (1989a) (who reprints a very readable proof of Gleason's theorem on pp. 321–46). As to Bell's theorem, it is none other than the famous inequality that was mentioned in §6.3.1; for references, see note 46.

$$\frac{dR}{dt} = -\frac{1}{2m}(R\nabla^2 S + 2\nabla R \cdot \nabla S)$$

$$\frac{dS}{dt} = -\left(\frac{(\nabla S)^2}{2m} + V(\mathbf{x}) - \frac{\hbar}{2m}\frac{\nabla^2 R}{R}\right) \tag{6.47}$$

Writing P for R^2, eqns. (6.47) become

$$\frac{dP}{dt} + \nabla\left(P\frac{\nabla S}{m}\right) = 0 \tag{6.48}$$

$$\frac{dS}{dt} + \frac{(\nabla S)^2}{2m} + V(\mathbf{x}) - \frac{\hbar^2}{4m}\left(\frac{\nabla^2 P}{P} - \frac{1}{2}\frac{(\nabla P)^2}{P^2}\right) = 0 \tag{6.49}$$

Now, if $\hbar = 0$, eqn. (6.49) reduces to eqn. (2.50), the classical Hamilton–Jacobi equation for our system. Then, as Bohm recalls, given an ensemble of particles whose trajectories are solutions of the classical equations of motion, every one of which is normal to a given surface S = const., every surface of constant S is normal to every trajectory and $\nabla S(\mathbf{x})/m$ equals the velocity vector $\mathbf{v}(\mathbf{x})$ of any particle passing the point $\mathbf{x}$. Equation (6.48) can therefore be rewritten thus:

$$\frac{dP}{dt} + \nabla(P\mathbf{v}) = 0 \tag{6.48a}$$

$P\mathbf{v}$ may then be regarded as the mean current of particles in the ensemble and $P(\mathbf{x})$ as the probability density, so eqn. (6.48a) expresses the conservation of probability. Bohm's decisive move was to extend this interpretation to the case $\hbar > 0$. He did it by assuming that, besides the classical potential V, there is a quantum potential U acting on the particles that is given by the last term on the left-hand side of eqn. (6.49):

$$U(\mathbf{x}) = -\frac{\hbar^2}{4m}\left(\frac{\nabla^2 P}{P} - \frac{1}{2}\frac{(\nabla P)^2}{P^2}\right) = -\frac{\hbar^2}{2m}\frac{\nabla^2 R}{R} \tag{6.50}$$

"The Eq. [6.49] can still be regarded as the Hamilton–Jacobi equation for our ensemble of particles, $\nabla S(\mathbf{x})/m$ can still be regarded as the particle velocity, and Eq. [6.48] can still be regarded as describing conservation of probability in our ensemble. Thus, it would seem that we have here the nucleus of an alternative interpretation for Schroedinger's equation" (Bohm 1952, p. 170). In this interpretation, particles are endowed with "precisely definable and continuously varying values of position and momentum"; they move "under the action of a force which is not entirely derivable from the classical potential, $V(\mathbf{x})$, but which also obtains a contribution from the 'quantum-mechanical' potential, $U(x)$"; since the latter is a function of $R(x)$, which is the modulus of the wave function $\psi(x)$ – see eqn. (6.46) – "we have effectively been led to regard the wave function of an individual [particle] as a mathematical representation of an objectively real field" that acts on the particle in a way that is similar to, although not identical with, the way the electromagnetic field acts on a charge (p. 170).

Bohm lists three assumptions under which his theory yields the same predictions as QM, viz., (1) the ψ-field satisfies Schrödinger's equation; (2) the particle's momentum equals $\nabla S(x)$; and (3) we do not predict or control the precise location of the particle but have, in practice, a statistical ensemble with probability density $P(x) = |\psi(x)|^2$ (1952, p. 171). He emphasizes, however, that these assumptions do not belong to the conceptual structure of his theory and may therefore be relaxed. He takes pride in the fact that "there are an infinite number of ways of modifying the mathematical form of the theory that are consistent with our interpretation and not with the usual interpretation" (p. 179). In particular, by modifying the differential equation for ψ – say by making it inhomogeneous or by including nonlinear terms that are large only for processes involving small distances – one could perhaps account for "phenomena associated with distances of the order of 10^{-13} cm or less, which are not now adequately understood in terms of the existing theory" (p. 179). Such modifications would generate discrepancies between the predictions of QM and of Bohm's theory, which, if testable, would facilitate a choice between them. Still, the enormous flexibility of the latter can hardly count as an asset in the actual practice of physics.

Bohm (1952) extends his approach to the many-body case and presents in Part II a theory of measurement. According to it, "the at present 'hidden' precisely definable particle positions and momenta determine the results of each individual measurement process, but in a way whose precise details are so complicated and uncontrollable, and so little known, that one must for all practical purposes restrict oneself to a statistical description of the connection between the values of these variables and the directly observable results of measurements. Thus, *we are unable at present to obtain direct experimental evidence for the existence of precisely definable particle positions and momenta*" (p. 183; my italics). This reminds one of the Cartesian adversaries of Newton, who, to avoid gravity's "magical" action-at-a-distance, tried to explain planetary motions by means of fantastic hypotheses. On the

other hand, Bohm's new force is in a way more "magical" than
Newton's. The quantum potential U is not linked to any known source
and, in the n-body case, it is defined, of course, on $3n$-dimensional
space. Although logically consistent, these notions "are difficult to
understand from a physical point of view" and should be regarded "as
schematic or preliminary representations of certain features of some
more plausible physical ideas to be obtained later" (Bohm 1980, pp.
80f.). Moreover, "the 'quantum-mechanical' forces may be said to
transmit uncontrollable disturbances instantaneously from one particle to another through the medium of the ψ-field", as Bohm notes in
connection with the EPR experiment (1952, p. 186).[68] As a consequence of the mutual entanglement of all interacting particles through
the ψ-field, the results of measurement depend on context: Measurement of the same observable A on two systems prepared in the same
way may lead to different results if A is measured on one of them
together with an observable B and on the other together with an
observable C, and B and C are noncommuting.[69] Therefore, "the measurement of an 'observable' is not really a measurement of any physical property belonging to the observed system alone. Instead, the value
of an 'observable' measures only an incompletely predictable and con-
trollable potentiality belonging just as much to the measuring apparatus as to the observed system itself" (1952, p. 183).[70]

6.4.3   Quantum Logic

Since Lobachevsky and Bolyai achieved posthumous fame by breaking
the Euclidian monopoly of truth in geometry, some philosophers have
dreamed of going one step further and shattering the uniqueness of
logic. The first nonstandard logic was formulated – according to
Jammer (1974, p. 342) – in 1910 by N.A. Vasil'ev, a professor of philosophy at Kazan University. Vasil'ev's "imaginary logic" – as he called
it to draw attention to the analogy with Lobachevsky's "imaginary
geometry" – was based on the denial of the so-called law of bivalence,
by virtue of which every statement must have one and one only of the

---

[68] Bohm's theory is "nonlocal"; see note 67.
[69] Bohm's theory is "contextual"; see note 67.
[70] Bohm's last thoughts were published, after his death, in Bohm and Hiley (1993). I
have not been able to find in this book any important idea that was not already
present in Bohm's brilliant paper of 1952.

two admissible truth values *true* and *false*. Vasil'ev added a third value, *indifferent*, and, denying the law of contradiction, concluded that a sentence of the form '*S* is *P* and *S* is not *P*' is neither true nor false, but indifferent. A more plausible three-valued logic was proposed by Jan Łukasiewicz (1920), who saw in it the proper way of formally dealing with contingent statements about the future, such as 'A sea-battle will be fought tomorrow' (cf. Aristotle, *De Interpretatione*, Chapter 9). The following are his truth tables for negation ($\neg$) and material implication ($\to$), with the three truth values by 1, 0 and $\frac{1}{2}$:[71]

p	1	$\frac{1}{2}$	0
$\neg p$	0	$\frac{1}{2}$	1

p	1	1	1	$\frac{1}{2}$	$\frac{1}{2}$	$\frac{1}{2}$	0	0	0
q	1	$\frac{1}{2}$	0	1	$\frac{1}{2}$	0	1	$\frac{1}{2}$	0
$p \to q$	1	$\frac{1}{2}$	0	1	$\frac{1}{2}$	$\frac{1}{2}$	1	1	1

In 1931, Zygmunt Zawirski recommended using a three-valued logic to cope with the baffling features of QM, such as the apparent impossibility of simultaneously assigning a precise position and a precise momentum to a particle. This suggestion was later taken up by Paulette Février (1937a, 1937b, 1951) and by Hans Reichenbach (1944), who developed it in different ways. Their systems found a cool reception and I shall not comment on them.[72]

Quantum logic, as it is usually understood today, stems from a joint paper by Garrett Birkhoff and John von Neumann on "The Logic of Quantum Mechanics" (1936). They saw that the subspaces of a Hilbert space $\mathcal{H}$ form a lattice that I shall call $L(\mathcal{H})$.[73] This is similar to but sig-

---

[71] Let me recall that in standard logic, with two truth-values, T ('true') and F ('false'), the truth table of material implication is

p	T	T	F	F
q	T	F	T	F
$p \to q$	T	F	T	T

The importance of material implication in standard logic lies in its close connection with entailment: Given ($p \to q$), p entails q; that p entails q entails ($p \to q$).

[72] Jammer (1974, pp. 361–79) discusses the systems of multivalued quantum logic of Février, Reichenbach, and von Weizsäcker and the chief objections leveled against them. See also Putnam (1957), Feyerabend (1958), and Haack (1974, Ch. 8).

[73] Supplement II contains all the information about lattices that is required to understand the remainder of this section.

nificantly different from the Boolean lattice of subsets of the phase space of a classical physical system, which in turn is isomorphic to the lattice of propositions asserting or denying that the state of the system lies in this or that part of the phase space. Birkhoff and von Neumann noted that the propositions concerning the state of a quantum system with Hilbert space $\mathcal{H}$ formed a lattice isomorphic to $L(\mathcal{H})$. Now, the Boolean structure of the lattice of propositions concerning the state of a classical system reflects the meaning of the classical logical operators 'and', 'or', and 'not'. So, by analogy, the non-Boolean structure of the newly built lattice of propositions about quantum states in $\mathcal{H}$ was said to reflect – or to constitute – a nonclassical logic appropriate for thinking about micro-objects. The novelty of the logic could then explain away some of the conceptual perplexities associated with QM.

To see what this means, let us consider the lattices involved, beginning with the Boolean lattice of propositions in classical logic. (For symbolism and terminology, see Supplement II.) By proposition I mean a class of logically equivalent sentences. I say that the proposition p is true if any – and therefore every – sentence in p is true, and that p is false if any sentence in p is false. The denial of p, symbolized by $\neg p$, is the proposition that is true if and only if p is false, and is false if and only if p is true (law of bivalence). The lattice of propositions is defined by: (i) For any propositions p and q, $p \le q$ if and only p entails q, that is, if no assignment of truth values is possible in which p is true and q is false; (ii) the maximal element **1** is the class of tautologies, that is, the sentences that are true in every possible assignment of truth values; (iii) the maximal element **0** is the class of contradictions, that is, the sentences that are false in every possible assignment of truth values; and (iv) the orthocomplement $p' = \neg p$. It follows at once from condition (i) that the meet of propositions p and q, denoted in lattice theory by '$p \wedge q$', is the proposition that is true if and only if both p and q are true and otherwise is false; and that the join of p and q, denoted by '$p \vee q$', is the proposition that is false if and only if both p and q are false and otherwise is true. (Due to this, the same symbols $\wedge$ and $\vee$ employed in lattice theory to signify, respectively, meet and join, are used for conjunction and disjunction in standard logic.) Let $(p \leftrightarrow q)$ stand for the proposition that is true if and only if p and q are either both true or both false. It is easily proved by means of truth tables that both $(p \vee (q \wedge r)) \leftrightarrow ((p \vee q) \wedge (p \vee r))$ and $(p \wedge (q \vee r)) \leftrightarrow ((p \wedge q) \vee (p \wedge r))$ are tautologies, so the distributive laws $D_{\vee/\wedge}$ and $D_{\wedge/\vee}$ hold in the classical lattice of propositions, which thus is a Boolean lattice.

The Boolean lattice $B(\Sigma)$ of subsets of a set Σ is defined in Supplement II.4. $B(\Sigma)$ is defined on the power set $\mathscr{P}(\Sigma)$ and is partially ordered by the relation of inclusion $\subseteq$. Note in particular that the meet of two subsets P and Q is their intersection $P \cap Q$ and the join of P and Q is their union $P \cup Q$. We are interested in the case in which Σ is the phase space of a classical system S, which, if S has n degrees of freedom, is $\mathbb{R}^{2n}$. I say that a proposition is a state description of S if it contains a sentence of the form 'the state of S is in P', for some $P \subseteq \mathbb{R}^{2n}$. Let $B(S)$ be the classical Boolean lattice of such propositions. Let p be the equivalence class of 'the state of S is in P' (likewise for q, etc.). Since 'the state of S is in P' is not logically equivalent to 'the state of S is in Q' unless $P = Q$, the mapping $f\colon p \mapsto P$ is a one–one mapping of $B(S)$ onto $B(\mathbb{R}^{2n})$. This mapping is an isomorphism of lattices, as the following considerations will show. $P \subseteq Q$ if and only if every conceivable state of S that lies in P also lies in Q; this is necessary and sufficient for p to entail q; thus, $p \leq q$ in $B(S)$ if and only if $f(p) \leq f(q)$ in $B(\mathbb{R}^{2n})$. Moreover, 'the state of S is in $\mathbb{R}^{2n}$' is true under every circumstance and therefore belongs to the maximal element of $B(S)$, and 'the state of S is in $\varnothing$' in not true under any circumstance and therefore belongs to the minimal element of $B(S)$. Finally, 'the state of S is not in P' is logically equivalent to 'the state of S is in P'', where P' denotes the complement of P in $B(\mathbb{R}^{2n})$, so $f(p') = f(\neg p) = P' = (f(p))'$.[74]

I now turn to the lattice $L(\mathscr{H})$ formed by the subspaces of the Hilbert space $\mathscr{H}$. Its maximal element is $\mathscr{H}$, its minimal element is the zero vector, the meet $A \wedge B$ of two subspaces A and B is their intersection $A \cap B$, and their join $A \vee B$ is the subspace $A \oplus B$ spanned by them ($|\psi\rangle$ is in $A \oplus B$ if $|\psi\rangle$ is a linear combination of vectors in A and vectors in B). If A is a subspace of $\mathscr{H}$, there is a subspace A' formed by all the vectors that are orthogonal to A (that is, orthogonal to each vector in

---

[74] To simplify my exposition I have adopted above a rather unrealistic form of state description for classical systems. In real life, neither the data obtained by observation nor the predictions achieved by solving the classical equations of motion for a system of more than two bodies enable us to locate the state of a system at a point of phase space. In the actual practice of physics, especially in classical statistical mechanics, two measurable subsets of phase space are equated when their symmetric difference – i.e., the set of points belonging to one or the other but not to both – has measure zero. Anyway, the collection of (classes of purportedly equal) subsets obtained in this way is again a Boolean lattice, which is isomorphic to the classical lattice of state descriptions referred to them. Thus, for our purposes, a less simplistic presentation would not make any difference.

A). A' satisfies conditions OC1–OC3 in Supplement II.3 and is therefore the orthocomplement of A. So $L(\mathcal{H})$ is an orthocomplemented lattice, but it is not Boolean, as the following example shows. Let $|\varphi\rangle$ and $|\psi\rangle$ be orthogonal vectors in a Hilbert space $\mathcal{H}$ and put $|\theta\rangle = a|\varphi\rangle + b|\psi\rangle$, where the scalars a and b are both different from 0. Let A_φ, A_ψ, and A_θ be the subspaces of $\mathcal{H}$ spanned, respectively, by $|\varphi\rangle$, $|\psi\rangle$, and $|\theta\rangle$. Then, clearly, $A_\theta \wedge (A_\varphi \vee A_\psi) = A_\theta \neq 0$, but $(A_\theta \wedge A_\varphi) \vee (A_\theta \wedge A_\psi) = 0 \vee 0 = 0$. Thus, $L(\mathcal{H})$ does not obey the distributive law $D_{\wedge\vee}$. The structure of $L(\mathcal{H})$ can be reproduced in an obvious way in the set of projectors of $\mathcal{H}$, that is, the linear operators that map the whole of $\mathcal{H}$ onto each of its subspaces (see §6.2.5). Let P_A denote the projector onto subspace A. Put $P_A \vee P_B = P_{A \vee B}$, $P_A \wedge P_B = P_{A \wedge B}$, $(P_A)' = P_{A'}$. These conditions define a lattice isomorphic to $L(\mathcal{H})$ with minimal element P_0 and maximal element $P_{\mathcal{H}}$.

The very idea of a quantum logic depends on the construction of a lattice that

(L1) contains all the propositions suitable for conveying information – data, predictions – about a quantum system S,

(L2) is partially ordered by entailment – or at least by something that can pass for entailment if one makes allowance for a change in logic – and

(L3) is isomorphic with the lattice $L(\mathcal{H}_S)$ of subspaces of the system's Hilbert space $\mathcal{H}_S$.

By virtue of (L3) the lattice in question is non-Boolean and therefore essentially different from the classical lattice of propositions. Without (L2) there could hardly be any reason for regarding the quantum lattice of propositions as a logical system. Birkhoff and von Neumann's proposal for a lattice meeting these three requirements is somewhat less perspicuous than one might wish. G. W. Mackey (1963), who brought great clarity and rigor into this subject, proceeded in the opposite direction: He constructed a partially ordered structure that satisfies (L1) and (L2) and then *postulated* that it is isomorphic with the lattice of subspaces of a Hilbert space. From this standpoint, the lattice of propositions concerning a quantum system appears as a structured expression of experience that is more basic and hence, presumably, more lasting than the Hilbert space of standard QM. Unfortunately, it is not possible to deal here with Mackey's work. Anyway, his followers are not in the business of revolutionizing logic; they explore an alternative mathematical structure for the formulation and eventual advancement of QM, which they call "quantum logic" mainly in deference to Birk-hoff and von Neumann (cf. Jauch 1968, p. 77; Beltrametti and Casinelli 1981, p. xxiii).

A simpler scheme was adopted by Putnam (1969, 1974), with the explicit aim of revising logic proper – comprising, I presume, the general theory of deductive inference – in the light of physical experience.

> Suppose we are willing to adopt the heroic course of changing our logic. What then? It turns out that there is a very natural way of doing this. Namely, *just read the logic off from the Hilbert space* $\mathcal{H}_S$. Two proposi-tions are to be considered equivalent just in case they are mapped onto the same subspace of $\mathcal{H}_S$, and a proposition p is to be considered as 'implying' a proposition q just in case A_p is a subspace of A_q.
>
> (Putnam 1979, p. 179; my notation)

Putnam's intent can, I think, be spelled out as follows: Let $A, B, \ldots$ denote the subspaces of the Hilbert space $\mathcal{H}_S$, and let $a, b, \ldots$ be the propositions whose informative content is that the state vector of system S lies in $A, B, \ldots$, respectively. Such propositions form a lattice $L(S)$ if meet, join, and orthocomplement are defined thus:

$a \wedge b$ is the proposition that the state of S is in $A \cap B$;

$a \vee b$ is the proposition that the state of S is in $A \oplus B$;

a' is the proposition that the state of S is orthogonal to A.

The mapping $a \mapsto A$ is an isomorphism of $L(S)$ onto $L(\mathcal{H}_S)$, so $L(S)$ fulfills requirement (L3). It also satisfies (L2): Partial order in $L(S)$ con-notes a form of entailment, for $a \leq b$ is tantamount to 'the state of S is in A only if it is in B and the state of S is orthogonal to B only if it is orthogonal to A'. Does it also meet (L1)? Do the elements of $L(S)$ represent, as Putnam says, "all possible physical propositions about S" (1979, p. 179)? In standard QM, information about a system S has to do chiefly with the probability that this or that physical quantity rep-resented by a self-adjoint operator in $\mathcal{H}_S$ sports a certain value – or a value in a certain range – if it is measured on S when S is prepared in such-and-such a state. Surely, such information is not conveyed by propositions that do no more than report with certainty to what sub-spaces of $\mathcal{H}_S$ the state of S belongs or is orthogonal. But in Putnam's view, *"probability enters in quantum mechanics just as it entered in classical physics, via considering large populations* [and] *whatever problems may remain in the analysis of probability, they have nothing* special to do with quantum mechanics" (1979, p. 186; see the explanation preceding this conclusion). So the primary – and proper – quantum-mechanical information about S concerns only matters that can be asserted with certainty.

An example will show how this works. Let P and Q be two nondegenerate observables on $\mathcal{H}_S$ that do not commute with each other. Let each one have only two linearly independent proper vectors. Forgetting all other observables, we view $\mathcal{H}_S$ as two-dimensional. Put $P|\psi_i\rangle = p_i|\psi_i\rangle$ and $Q|\phi_i\rangle = q_i|\phi_i\rangle$ $(i = 1,2)$. I denote by ψ the subspace spanned by the vector $|\psi\rangle$ and by $\psi(t)$ the proposition that S is in ψ at time t. Clearly, $\mathcal{H}_S = \psi_1 \oplus \psi_2 = \phi_1 \oplus \phi_2$; therefore, we may assert with certainty that $(\psi_1(t) \vee \psi_2(t)) \wedge (\phi_1(t) \vee \phi_2(t))$ at all times. If, at time t_m, Q is measured on S with result q_1, this certifies that $((\psi_1(t_m) \vee \psi_2(t_m)) \wedge \phi_1(t_m)$. But, of course, this does not entail that $(\psi_1(t_m) \wedge \phi_1(t_m)) \vee (\psi_2(t_m) \wedge \phi_1(t_m))$, for the lattice of quantum propositions is not Boolean. Thus, from the fact disclosed by measurement, that the state of S at t_m was in subspace ϕ_1 and, of course, trivially, also in $\psi_1 \oplus \psi_2$, it does not follow that the said state was then *either* in $\psi_1 \cap \phi_1$ *or* in $\psi_2 \cap \phi_1$. This would indeed be absurd, for $\psi_i \cap \phi_1 = \{0\}$.

Putnam insisted that the logical operators symbolized by $\vee$, $\wedge$, and $\neg$ (or $'$) have essentially the same meaning in the new and the old logic. This is surprising, for the classical operators are truth-functional – the truth value of any proposition formed by using one of these operators follows, by definition, from the truth value of the proposition or propositions to which the said operator is applied – but the quantum operators are not.[75] But Putnam surely had in mind that the meaning of the logical operators can be gathered – after Gentzen (1934) and Jaśkowski (1934) – from the rules that govern their use in inference. For, if entailment is represented in each logic by partial order in the respective lattice, then clearly the following laws of entailment hold in both: (i) p entails $p \vee q$, and q entails $p \vee q$; (ii) if p entails r and q entails r, then $p \vee q$ entails r; (iii) p and q jointly entail $p \wedge q$; (iv) $p \wedge q$ entails both p and q; and (v) $(p')'$ entails and is entailed by p. In classical logic, (i)–(v) translate into sound rules of inference – for example, 'given p, infer $p \vee q$' – sufficient to establish every logical truth. However, if the same rules of inference were applicable in quantum logic, the distrib-

[75] No truth-functional characterization of meet, join, and orthocomplement can be given in a lattice isomorphic to $L(\mathcal{H})$ if the Hilbert space $\mathcal{H}$ has more than two dimensions (Kochen and Specker 1967; for a related result, see Jauch and Piron 1963).

utive laws $D_{\vee/\wedge}$ and $D_{\wedge/\vee}$ could without more ado be established by means of them; therefore, some important change must have affected the very meaning of entailment. Moreover, despite (v), there must be some significant difference in the meaning of negation represented by orthocomplementation in both $B(S)$ and $L(S)$, for in a Boolean lattice orthocomplementation is fixed by the partial order, whereas in a non-Boolean lattice it must be introduced on its own (Bell and Hallett 1982; cf. Supplement II.4).

For Putnam the main attraction of quantum logic was the hope it offered of preserving the purportedly realist commonsense view of physical properties. For common sense, each thing is characterized by a series of properties – shape, color, weight, and so on – that change with the circumstances but are supposed to be there, fully determined at all times while the thing exists. Philosophy and physics have undermined this view by conceiving most – if not all – properties of a thing C as the manifestation of relations that it has with other things, or with surrounding fields, and which must vanish with the latter, even if C remains. Thus, in Newtonian physics, a particle can be meaningfully said to possess weight only in the presence of gravitational sources. Mach and Einstein formed a relational conception of inertia, a quantity that Newton still conceived as intrinsic to each thing. Even so, since some relations are never lacking, classical physics could apparently retain the core of commonsense "realism", by conceiving things through the properties that correspond to such universal relations and treating the others as derivative or even as subjective. But according to QM it is not possible to assign at the same time definite values to certain pairs of physical quantities, including the two – admittedly relational but universal – properties of position and momentum, which in classical mechanics jointly characterize the state of each thing. What shall we make of a particle that has a definite velocity but is not located at any particular place? Or one that occupies a definite position but neither rests there nor moves with a definite velocity? Should we believe that physical properties are *had* only insofar as they are *observed*? Putnam (1969) thought that quantum logic would rescue us from this undesirable conclusion.

To facilitate his exposition, he invited the reader to "pretend that all physical magnitudes have finitely many values, instead of continuously many" (1979, p. 178). Suppose, then, that for our system S the admissible values of position and momentum are, respectively, $p_1, \ldots, p_m$ and $q_1, \ldots, q_n$. Let $P_i (Q_i)$ denote the subspace of $\mathcal{H}_S$ spanned by the proper vectors of position (momentum) corresponding to p_i (q_i). Write $p_i(t)$ and $q_i(t)$ for 'S has position p_i at time t' and 'S has momentum q_i at time t', respectively. Then, if it is verified that S has position p_k at time t, we may confidently assert that $p_k(t) \wedge (q_1(t) \vee \ldots \vee q_n(t))$ but certainly not that $(p_k(t) \wedge q_1(t)) \vee \ldots \vee (p_k(t) \wedge q_n(t))$, for the distributive laws are not valid in quantum logic. Putnam (1974) informally introduces the existential quantifier (by using which, indeed, one could drop the pretense that physical quantities have only finitely many values). With it, $(q_1(t) \vee \ldots \vee q_n(t))$ is shortened to $\exists x q_x(t)$, which may be read as "there is a definite value of momentum that S has at t." Interestingly, in quantum logic, $\exists x q_x(t)$, $\exists x p_x(t)$, and hence $\exists x q_x(t) \wedge \exists x p_x(t)$ are *logically true*, for $P_1 \vee \ldots \vee P_m = Q_1 \vee \ldots \vee Q_n = \mathcal{H}_S$. However, from $\exists x q_x(t) \wedge \exists x p_x(t)$ one may not infer $\exists x \exists z(q_x(t) \wedge p_z(t))$. Moreover, the last expression is *logically false*, since, for every pair of indices j and k, $Q_j \wedge P_k = 0$. Thus, thanks to quantum logic we can rest assured that, at any time t, there exists a definite momentum possessed by S at t *and* there exists a definite position possessed by S at t, and yet assent to the quantum-mechanical theorem that S does not possess a definite position and a definite momentum at the same time t. I wonder how many common-sense realists will feel that their craving for determinate properties is relieved by this result. Like other Putnam efforts to save this or that face of realism, his quantum logic is a tantalizing *jeu d'esprit*.[76]

[76] Hilary Putnam acknowledges his debt to David Finkelstein (1962/63, 1969; see also Finkelstein 1973) for some of the above ideas on quantum logic. For extensive, enlightening criticism of both authors, see Stachel (1986). In the sixties and seventies, Peter Mittelstaedt sought to vindicate quantum logic in the context of the "dialogical logic" of Lorenzen and Lorenz (cf. their book of 1978). In this approach to the foundations of logic, an implication is proved logically true through a schematic dialogue in which someone proposes it, an opponent proves the antecedent yet questions the consequent, and the proponent succeeds in showing that the consequent has already been granted in the course of the opponent's argument. (For example, someone proposes $p \rightarrow (q \rightarrow p)$; the opponent asserts p, whereby the proponent is constrained to assert $(q \rightarrow p)$; the opponent asserts q and questions p, only to be told that p has been already asserted by him.) In Mittelstaedt's quantum-logical dialogues the proponent is often unable to turn the opponent's own words against him, because they are no longer "available" when the proponent requires them. For example, if the opponent has invoked at some stage of the dialogue a definite position value, the proponent cannot quote him later if a definite value of momentum has in the meantime been asserted of the same system. Because of this, implications that have been dialogically proved under more familiar conditions – such as the two-way implications in the distributive laws – cannot be established in a quantum-logical setting. See Mittelstaedt (1978, 1979).

6.4.4   "Many Worlds"

My last example of meta-quantum-physics is Everett's solution of the measurement problem. It was the subject of his doctoral thesis (1957), written under the supervision of J. A. Wheeler. He developed it at greater length in a manuscript that was not published until 1973. I shall follow both. According to Everett, his approach provides "a reformulation of quantum theory in a form believed suitable for application to general relativity" (1957, p. 454). He believes that it will enable one to "consider the state function of the whole universe", from which "all of physics is presumed to follow" (1973, p. 9). (Pity he does not deign to offer us any quantum-mechanical predictions regarding "the whole universe", a system on which indeed no quantum-mechanical observable can be measured.)

Everett recalls the two fundamentally different ways in which quantum states can change according to von Neumann (see §6.3.2), viz., "*Process* 1: The discontinuous change brought about by the observation of a quantity with eigenstates $|\phi_1\rangle$, $|\phi_2\rangle$, . . . , in which the state $|\psi\rangle$ will be changed to the state $|\phi_i\rangle$ with probability $|\langle\psi|\phi_i\rangle|^2$"; and "*Process* 2: The continuous, deterministic change of state of the isolated system with time according to a wave equation $\partial|\psi\rangle/\partial t = U|\psi\rangle$, where U is a linear operator" (1957, p. 454; cf. 1973, p. 3; I supply Dirac notation). After vigorously criticizing the very idea of Process 1, Everett undertakes to describe observation processes "completely by the state function of the composite system which includes the observer and his object-system, and which at all times obeys the wave equation" (1973, p. 8). From his standpoint, there is no fundamental distinction between "measurement apparatuses" and other physical systems. "A measurement is simply a special case of interaction between physical systems – an interaction which has the property of *correlating* a quantity in one subsystem with a quantity in another" (1973, p. 53). Since almost every interaction between systems produces some correlation, one could indeed take the view "that the two interacting systems are continually 'measuring' one another" (Ibid.). However, this view does not correspond closely to our intuitive idea of measurement as a process yielding measured values. So Everett specifies several conditions that characterize measurements. Without going into the technical details, let me just say that his description of the interaction between an observed system S and a measuring apparatus A is quite similar to that given above, in the text surrounding eqns. (6.41)–(6.44). From this he concludes "that for any possible measurement, for which the initial system state is not an eigenstate, the resulting state of the composite system leads to *no* definite system state nor any definite apparatus state. The system will not be put into one or another of its eigenstates with the apparatus indicating the corresponding value, and nothing resembling Process 1 can take place" (1973, p. 60). Thus, "it seems as though nothing can ever be settled by such a measurement" (p. 61). Moreover, this conclusion has nothing to do with the size of the apparatus. And yet, "macroscopic objects always appear to us to have definite positions" (Ibid.).

However, before dismissing his objective reading of QM "because the actual states of systems as given by [it] seem to contradict our observations" one ought to investigate – says Everett – "what the theory itself says about the *appearance* of phenomena to observers" (p. 63). So he summons us to "the task of making deductions about the appearance of phenomena on a *subjective* level, to observers which are considered as *purely physical* systems and are treated within the theory" (Ibid.; my italics). Reading these words one thinks uncharitably: "Ain't this a bid to *square* the *circle?*" But we must be patient. Everett himself concedes that "in order to accomplish this it is necessary to identify some objective properties of such an observer (states) with subjective knowledge (i.e., perceptions)."[77] He is content, however, to let his observers have "memories" (in the clearly nonsubjective way in which computer hard disks have them). "When the state ψ^O describes an observer whose memory contains representations of the events A, B, ..., C we shall denote this fact by appending the memory sequence in brackets as a subscript, writing: $\psi^O_{[A,B,\,...\,,C]}$" (pp. 64f.). I shall combine this notation with the Dirac brackets that we have been using. Let an observer O in initial state $|\psi^O_{[\,...\,]}\rangle$ measure on system S the observable Q with proper vectors $\{|\varphi_i\rangle\}_{i \in \mathscr{I}}$. If S is initially in state $|\varphi_k\rangle$ ($k \in \mathscr{I}$), then, through the measurement interaction, the compound system $S + O$ undergoes the unitary evolution described in eqn. (6.41). In our new notation:

---

[77] "Thus, in order to say that an observer O has observed the event *a*, it is necessary that the state of O has become changed from its former state to a new state which is dependent upon *a*" (Everett 1973, pp. 63f.). Necessary indeed; but is it sufficient? Certainly not. Why should such a change of state generate a *perception* of *a*? In fact "the theory itself" has not a single word to say about the subjective appearances of phenomena. To make his deductions Everett must act the ventriloquist.

$$U_\varepsilon\left(|\varphi_k\rangle \otimes |\psi^O_{[\ldots]}\rangle\right) = |\varphi_k\rangle \otimes |\psi^O_{k[\ldots,\alpha_k]}\rangle \qquad (6.51)$$

where α_k denotes the memory mark associated with state $|\varphi_k\rangle$ (say, a recording of the proper value corresponding to $|\varphi_k\rangle$). In the general case, in which S is not initially in a proper state of the observable being measured, the compound system undergoes the evolution described in eqn. (6.43). The state reached by $S + O$ is suitably described in the new notation by:

$$|\Psi^{S+O}\rangle = \sum_{i \in \mathcal{I}} c_i\left(|\varphi_i\rangle \otimes |\psi^O_{i[\ldots,\alpha_j]}\rangle\right) \qquad (6.52)$$

Everett comments:

> There is no longer any independent system state or observer state, although the two have become correlated in a one–one manner. However, in each *element* $[|\varphi_i\rangle \otimes |\psi^O_{i[\ldots,\alpha_j]}\rangle]$ of the superposition [6.52], the object–system state is a particular eigenstate of the observation [*sic*], and *furthermore the observer–system state describes the observer as definitely perceiving that particular system state*. This correlation is what allows one to maintain the interpretation that a measurement has been performed.

> (Everett 1957, p. 459; cf. 1973, p. 68)

The words italicized by Everett deserve careful study, for his solution of the measurement problem turns on them. We must bear in mind that the right-hand side of eqn. (6.52) expresses the state vector $|\Psi^{S+O}\rangle$ as a linear combination of vectors belonging to a particular basis of $\mathcal{H}_{S+O}$, the Hilbert space of the (object + observer)-system. The elements of that basis are all the tensor products formed by a proper vector of the measured observable $\mathbf{Q}$, on the one hand, and a characteristic vector of the observer O, on the other (cf. §6.3.2). Interestingly, in the said linear combination every basis vector of the form $|\varphi_k\rangle \otimes |\psi^O_h\rangle$, with $h \neq k$, is multiplied by the scalar 0. This condition is part of the standard description of QM measurement (cf. London and Bauer 1939, §11). $|\Psi^{S+O}\rangle$ can of course be expressed as a superposition in terms of each and every basis of $\mathcal{H}_{S+O}$, infinitely many of which are formed from some other, arbitrarily chosen basis of $\mathcal{H}_S$ and the $\mathcal{H}_O$-basis of characteristic states of O. But few if any of these alternative expressions will share with eqn. (6.52) this remarkable feature: Each nonzero term combines a basis vector of $\mathcal{H}_S$ with a single basis vector of $\mathcal{H}_O$, which occurs in the superposition only in that term. By virtue of it, the superposition on the right-hand side of eqn. (6.52) is certainly privileged among the countless superpositions equal to $|\Psi^{S+O}\rangle$. Yet it might not be the only one to be so privileged. For all we know, there may be other bases of $\mathcal{H}_S$ and $\mathcal{H}_O$ that combine together in a similar way to yield linear combinations equal to $|\Psi^{S+O}\rangle$. This is of no consequence in the standard formulation of QM, in which a group of persons, the physicists, express $|\Psi^{S+O}\rangle$ in terms of the basis $\{|\varphi_i\rangle \otimes |\psi_j^O\rangle\}_{i,j \in \mathscr{J}}$ because the $|\varphi_i\rangle$'s are the proper vectors of the observable that they propose to measure and the $|\psi_j^O\rangle$'s represent the characteristic states marked on the dial of their equipment. But in Everett's purportedly cosmic, anti-anthropocentric formulation of QM, every expression of $|\Psi^{S+O}\rangle$ as a superposition with the said feature stands on a par with the one in eqn. (6.52), and if the latter is not unique, his "deductions about the appearance of phenomena on a subjective level" become highly questionable.

I now turn to Everett's key statement: *The observer-system state $|\psi_{i[\ldots,\alpha_i]}^O\rangle$ describes the observer as definitely perceiving the particular object–system state $|\varphi_i\rangle$.* Never mind the idiosyncratic use of 'perceiving'. It stands here for the fact that the physical object O bears the mark, denoted by α_i, of its interaction with another object S when S was in state $|\varphi_i\rangle$. That bearing such a mark should involve some awareness of it surely does not follow from the theory. More important is this: The theory does not predict that O will bear that mark after interacting with S when the latter is in a state different from $|\varphi_i\rangle$. In the notation the bracketed suffix α_i in $|\psi_{i[\ldots,\alpha_i]}^O\rangle$ denotes the most recent mark that O bears right after interacting with S while the latter is in state $|\varphi_i\rangle$. This is not a mark that O should sport as the latest when $S + O$ reaches the state $|\Psi^{S+O}\rangle \neq |\varphi_i\rangle \otimes |\psi_{i[\ldots,\alpha_i]}^O\rangle$. (If it did so, O could not pass for a good observer.) The suffix α_i occurs in eqn. (6.52) simply because, following Everett, I have affixed it once and for all to the symbol for the ket $|\psi_i^O\rangle$, which is a factor in one of the tensor products in terms of which eqn. (6.52) spells out $|\Psi^{S+O}\rangle$. But in this use the suffix α_i says nothing about the marks present in O. It is just not true that eqn. (6.52) displays "a superposition of states . . . for *each of which* the apparatus has recorded a definite value" α_i (1957, p. 457); all we can say is that eqn. (6.52) displays a superposition of states for each of which the apparatus *would record* the definite value α_i *if* the compound system *were in that state* – which, of course, cannot be the case when it is in the state described in eqn. (6.52), unless we assume that all the scalars $c_j = 0$, except one.

I apologize for explaining at such length these fairly obvious minutiae. However, I feel that they are neglected by Everett and by the numerous writers who either endorse his views or treat them with deference. Everett refers to expressions like the right-hand side of eqn. (6.52) as *"the* superposition" that is *"the* final result" of a measurement interaction. This is not inaccurate, but it glosses over the fact that there are in every such case many equally valid expressions of the said result as a superposition of other vectors. He also speaks of the vectors $\{|\psi^O_{i[\ldots,\alpha_i]}\rangle\}_{i \in \mathcal{J}}$ that occur as factors of the summands on the right-hand side of eqn. (6.52) as if they represented states in which the observer O actually finds him- or herself after the interaction with S.[78] After discussing the case of repeated measurements – after which "every element of the resulting final superposition will describe the observer with a memory configuration [. . .] in which the earlier memory coincides with the later" (1957, p. 459) – Everett writes:

> We thus arrive at the following picture: Throughout all of a sequence of observation processes there is only one physical system representing the observer, yet there is no single unique *state* of the observer (which follows from the representations of interacting systems). Nevertheless, there is a representation in terms of a *superposition*, each element of which contains a definite observer state and a corresponding system state. Thus with each succeeding observation (or interaction), the observer state "branches" into a number of different states. Each branch represents a different outcome of the measurement and the *corresponding* eigenstate for the object–system state. All branches exist simultaneously in the superposition after any given sequence of observations.
>
> (Everett 1957, p. 459)

In the footnote appended to this paragraph Everett remarks:

> From the viewpoint of the theory *all* elements of a superposition (all "branches") are "actual", none any more "real" than the rest. It is unnecessary to suppose that all but one are somehow destroyed, since all the separate elements of a superposition individually obey the wave equation with complete indifference to the presence or absence ("actuality" or not) of any other elements. This total lack of effect of one branch on another also implies that no observer will ever be aware of any "splitting" process.
>
> (Everett 1957, pp. 459f., note ‡)

These quotations give the gist of Everett's solution to the measurement problem. To establish quantitative results he skillfully develops a

---

[78] Besides the two passages I quote next, see the footnote in Everett (1973, p. 68), which was mercifully omitted in the published dissertation.

measure applicable to the elements of a superposition of orthogonal vectors and applies it to ensembles of observation sequences. He concludes that the statistical assertions of the standard formulation of QM "will appear to be valid to *almost all*[79] observers described by separate elements of the superposition [. . .], in the limit as the number of observations goes to infinity" (1973, p. 74).

Everett's reformulation of QM was taken up by Bryce DeWitt, who contributed the metaphor of the splitting universe, under which it is mostly known. He wrote:

> Our universe must be viewed as constantly splitting into a stupendous number of branches, all resulting from the measurement like interactions between its myriads of components. Because there exists neither a mechanism within the framework of the formalism nor, by definition, an entity outside the universe that can designate which branch of the grand superposition is the 'real' world, all branches must be regarded as equally real.
>
> To see what this multiworld concept implies one need merely note that because every cause, however microscopic, may ultimately propagate its effects throughout the universe, it follows that every quantum transition taking place on every star, in every galaxy, in every remote corner of the universe is splitting our local world on earth into myriads of copies of itself.
>
> (DeWitt 1971, p. 222)

DeWitt contrasts "the mixture of metaphysics with physics" (1970, in DeWitt and Graham 1973, p. 35) that he attributes to the Copenhagen interpretation with the "remarkable metatheorem" proved by Everett, viz., that "*the mathematical formalism of the quantum theory is capable of yielding its own interpretation*" (1971, p. 212). I confess that I cannot understand how the formalism of vectors and linear operators in complex Hilbert space can yield the following key elements of the interpretation proposed by Everett:

(i) An *observer* – in Everett's sense, that is, a *purely physical automaton* – *perceives* – in the ordinary sense, that is, *becomes conscious of* – each "memory mark" associated to the tensor products of correlated object–system and observer–system states, a linear combination of which is the state of the compound object–observer system right after the measurement.

---

[79] "Except for a set of memory sequences of measure zero" (Everett 1957, p. 461).

(ii) If there is more than one tensor product with a nonzero scalar factor in the said linear combination, the observer develops a separate stream of consciousness for each, which is absolutely impervious to the others.

Everett's bizarre reading of vector sums may have been prompted by the classical Fourier analysis of waves in 3-space. Telephone companies literally superpose the electromagnetic renderings of many simultaneous long-distance messages in a single wave train that is echoed by a satellite and then automatically analyzed at the destination exchange into its several components, each one of which is transmitted over a separate private telephone line. No doubt we may speak in this case of genuine splitting of what was initially added up. Still, the signal could also be split into other, meaningless components if the analysis were not guided by human interests and aims.

6.5   A Note on Relativistic Quantum Theories

The relativistic corrections ably introduced by Sommerfeld in Bohr's hydrogen atom model (§6.1.1) contributed not only to the early success of the Old Quantum Theory but also, decisively, to the general acceptance of SR by physicists. Since QM does not comply with SR requirements, its creators knew from the outset that QM was a provisional first approximation that would not be tenable in the presence of high energies. Indeed Schrödinger sought for a Lorentz invariant wave equation before settling for the Lorentz noninvariant one that bears his name (§6.2.2). And it was clear to all that only a relativistic quantum theory could suitably deal with radiation.

Efforts toward such a theory began as early as 1926, while nonrelativistic QM had not yet settled down. A useful starting point was supplied by Jordan in the final section of the three-men paper (Born et al. 1926, Section 4.3). Jordan complained later that nobody read that section, nobody took notice of it, and nobody wanted to believe it.[80] However, it probably exerted some influence on Dirac's paper, "The Quantum Theory of the Emission and Absorption of Radiation" (1927), which in turn was utilized by Jordan in his trailblazing work

---

[80] 1963 interview with T. S. Kuhn, quoted in Schweber (1994, p. 11).

on matter fields.[81] Dirac's Lorentz invariant equation for the electron
was a major breakthrough (Dirac 1928). The fine structure of the
hydrogen spectrum – the subject of Sommerfeld's relativistic corrections – flowed from it at once. More surprisingly, so did electron spin
(cf. §6.1.4). Yet the Dirac equation involved such difficulties that
Heisenberg – in a letter to Pauli of 31 July 1928 – referred to it as "the
saddest chapter in modern physics" (quoted in Miller 1994, p. 30). The
following comment, made by Heisenberg in 1963, explains his initial
gloom:

> Until that time I had the impression that in quantum theory we had come
> back into the harbor, into the port. Dirac's paper threw us out into the
> open sea again. Everything got loose again and we got into new diffi-
> culties. Of course at the same time, I saw that we had to go that way.
> *There was no escape from it because relativity was true.*
>
> (Interview with T. S. Kuhn, quoted in Schweber 1994,
> p. 5; my italics)

The theory centered in Dirac's equation is known as Quantum Electrodynamics (QED). Its most conspicuous difficulty led to its greatest
triumph. The equation has solutions that represent particles with
negative energy. Dirac conjectured that every negative energy state is
normally occupied by an electron, so that transitions to such states are
forbidden by the Exclusion Principle and therefore are never recorded
in spectrographs. However, transitions from negative to positive energy
states are allowed. If, perchance, such a transition occurs, creating a
hole, so to speak, in the sea of negative energy electrons, this would
show up as "a new kind of particle, unknown to experimental physics,
having the same mass and opposite charge to an electron. We may call
such a particle an anti-electron" (Dirac 1931, p. 61). The theory
implied that such positively charged, low-mass particles would be difficult but not impossible to observe with the then available tools. Blackett and Occhialini, who worked like Dirac in Cambridge, had in fact
found some evidence of their existence and told Dirac about it, but
they did not dare to print it without further confirmation, so the first
photograph of a positron track to appear in print was taken at Caltech
by Anderson (1932). Blackett's diffidence was true to the spirit of
his working place, as expressed by Rutherford: "They [the theorists]

---

[81] Papers by Jordan and Klein, Jordan and Wigner, and Jordan and Pauli; for references,
see Miller (1994, p. 115).

play games with their symbols but we in the Cavendish turn out the
real facts of nature" (Schweber 1994, p. 67). Yet it was Dirac who,
playing with symbols, brought to light ordinary matter's elusive
Doppelgänger.[82]

The other major difficulty besetting QED is, if I may say so, intrinsic to the Dirac equation and cannot be solved by experimental
research. The differential equations of physics usually cannot be solved
exactly, but one can approximate a solution as closely as one wishes
by evaluating the successive terms of an infinite series. However, in the
case of the Dirac equation, the second term of the relevant series is
already apt to be a divergent integral. A term of this sort is not only
impossible to calculate but, it would seem, also physically meaningless.
Before the Second World War this difficulty caused most theorists to
be pessimistic about QED. In 1930 Bohr wrote to Dirac that he
expected the theory to fail "for energies of order $137\,mc^2$", and in 1935
Heisenberg wrote to Pauli that, "with respect to QED we are still at
the stage in which we were in 1922 with regard to quantum mechanics: We know that everything is wrong" (letters quoted in Schweber
1994, p. 84). However, right after the war, Tomonaga, Schwinger,
Feynman, and Dyson developed a method for handling the divergences
that tranquilized most physicists and generated predictions of unrivaled
precision.[83] With the flair for euphemism that is so typical of our time,
the method was dubbed 'renormalization'. The gist of it is well
explained – for philosophical use – by Teller (1995, Chapter 7; 1988).

Thus refurbished, QED was celebrated as the most accurate theory
in physics. It set the pattern after which two other successful relativistic quantum theories were built in the 1960s: the unified theory of electromagnetic and nuclear weak interactions and the "chromodynamics"
of nuclear strong interactions. Celebration, however, was not unani-
mous. The following remarks were made by Dirac himself, in one of
his last papers, as he spoke of the "infinite factors" that appear in QED
"when we try to solve the equations":

---

[82] The story has been variously told. See, for instance, de Maria and Russo, "The discovery of the positron" (1985) and Roqué, "The manufacture of the positron"
(1997). For a fine philosophical analysis, see Falkenburg 1995, pp. 119–27.

[83] I take the following information from Weinberg (1995/96, I, 490). Let m_μ and μ_μ
denote the muon's mass and magnetic moment, respectively, while e stands for the
charge of the electron. Then, the predicted value of μ_μ, to fourth order, is 1.00116546
$e/2m_\mu$, and its current experimental value is 1.001165923 $e/2m_\mu$.

These infinite factors are swept into renormalization procedures. The
result is a theory which is not based on strict mathematics, but is rather
a set of working rules. Many people are happy with this situation
because it has a limited amount of success. But this is not good enough.
Physics must be based on strict mathematics. One can conclude that the
fundamental ideas of the existing theory are wrong. A new mathematical basis is needed.

                              (Dirac 1984; quoted in Schweber 1994, p. 71)

According to Steven Weinberg, "we have learned in recent years to
think of our successful quantum field theories, including quantum electrodynamics, as 'effective field theories', low-energy approximations to
a deeper theory that may not even be a field theory, but something different"; therefore, "we think differently now about some of the problems of quantum field theories [. . .] that used to bother us when we
thought of these theories as truly fundamental" (1995/96, I, xxi).

I cannot go further into this subject. To discuss it meaningfully, even
at a superficial level, requires a much greater proficiency in mathematical analysis than I expect from the readers of this book and can
myself offer. Partly because of it, but partly, no doubt, because they are
sickened by untidy math, most philosophers of physics tend to neglect
QED.[84] Yet there are at least three reasons why they ought to pay heed
to it, and to the kindred quantum field theories of chromodynamics
and the electroweak field. First, quantum field theories have been the
working theories at the frontline of physics for over 30 years. Second,
these theories appear to do away with the familiar conception of
physical systems as aggregates of substantive individual particles. This
conception was already undermined by Bose–Einstein and Fermi–Dirac
statistics (§6.1.4), and greatly impaired by QM, according to which
the so-called particles cannot be assigned a definite trajectory in ordinary space. But quantum field theories go a long step further and – or
so it would seem – conceive "particles" as excitation modes of the field.
This, I presume, motivated Howard Stein's saying that "the quantum
theory of fields is the contemporary locus of metaphysical research"

---

[84] There have been exceptions, of course; see Redhead (1988) and Teller (1995). Auyang
(1995) is a philosophical book by a physicist who is exceptionally well versed in good
philosophy. *Added in proof*: After writing the above, I received a copy of Tian Yu
Cao's *Conceptual Developments of 20th Century Field Theories* (1997), a remarkable historico-critical study, two-thirds of which are devoted to the quantum field
theories.

(1970, p. 285). Finally, the very fact that physicists consciously and fruitfully resort to unperspicuous theories can teach us something about the aim and reach of their science. Here is how physicists work, dirty-handed, in their everyday practice, a far cry from what is taught at the Sunday school of the "scientific worldview".

CHAPTER SEVEN

✦

Perspectives and Reflections

7.1 Physics and Common Sense

Modern mathematical physics began in open defiance of common sense. Galileo declared – through his spokesman Salviati – that he could not "sufficiently admire the outstanding acumen" of the heliocentrist astronomers, who, "through sheer force of intellect," had "done such violence to their own senses as to prefer what reason told them over that which sensible experience plainly showed them to the contrary" (EN VII, 355; Drake translation). Furthermore, he judged color and sound, heat and cold to be mere affections of the human senses, like the tickling one feels when a feather is introduced into one's nose, which, of course, lies not on the feather but on the nerves stimulated by it (1623, §48). The most conspicuous features by which we perceive and classify in everyday life the objects that surround us were thus pronounced mind-dependent or "subjective" and banished from the stock of notions that the new physics would employ to describe and understand the real, "objective" nature of things.[1] Physical discourse retained many terms from ordinary language – arithmetic terms familiar in

---

[1] The quotation marks surrounding *objective* and *subjective* are meant to make the reader wary of these terms, not to indicate that Galileo used them in this way. In his time, those words meant the opposite of what they mean today. "The Scholastic Philosophy made the distinction between what belongs to things subjectively (Lat. *subjective*), or as they are 'in themselves', and what belongs to them objectively (Lat. *objective*), as they are presented to consciousness" (*OED, s.v.* 'objective', 2). However, in the first quarter of the eighteenth century the modern meaning was already established. The *OED* cites Watts 1725: "Objective certainty, is when the proposition is certainly true in itself; and subjective, when we are certain of the truth of it. The one is in things, the other is in our minds" (ii. ii. § 8).

housekeeping and trade; geometric terms developed in carpentry, architecture, and land-surveying; chronometric terms used in lithurgy, seafaring, and, increasingly since the Renaissance, also in daily business; terms applicable to machines found in harbors and building sites –, which physicists sought only to define better and to apply with greater precision. However, with hindsight we are bound to view the wholesale dismissal of core ingredients of ordinary language right at the start of modern physics as being only a first step, a preparation for and anticipation of what would come later. As we saw in Chapters Five and Six, since 1900 the notions of "classical" geometry, chronometry, and dynamics have been found wanting and have been replaced in physics by concepts drawn from novel mathematical theories that are entirely foreign to commonsense "intuitions" and the ordinary conduct of human life.

More significantly perhaps, modern physics did away with the view of nature as forming – much like our social environment – a network of references and meanings, means and ends, values and countervalues. This perception of nature, shared until 1600 by all nations, is now regarded as "primitive" and superstitious by people educated in our civilization. One of the first to say so clearly and forcefully was Spinoza, in the Appendix to Book I of his *Ethics*, in which he sought to discredit the common supposition that "all natural things, just like men themselves, work to some end" and "indeed . . . that God Himself directs everything to some sure end", and to show "why so many assent to this prejudice, and why all are naturally inclined to embrace it".

> After men have persuaded themselves that everything which happens, happens for their sake, they must judge that which is most useful to them to be what matters most in each thing, and they must esteem that to be most excellent by which they are most beneficially affected. In this way they must form those notions by which they explain nature, namely *good*, *evil*, *order*, *confusion*, *heat*, *cold*, *beauty* and *ugliness*.

. . . .

> For example, if the motion which the nerves receive from objects represented through the eyes conduces to good health, the objects by which it is caused are called *beautiful*; while those exciting a contrary motion are called *ugly*. Those things, too, which move the senses through the nostrils are called odorous or fetid; those which do so through the taste are called sweet or bitter, savory or insipid; those which act through the touch, hard or soft, heavy or light; those, lastly, which stimulate hearing

are said to make noise, sound, or harmony, the latter having deranged
men to such a point that they have come to believe that even God is
delighted with it.

(Spinoza, *Opera* I, 81–82)

Spinoza's radical stance was not endorsed by seventeenth-century
physicists, who, however, professed that physical inquiry must ignore
natural goals and values. These were held to be accessible to God alone,
and out of bounds for human science, although the pervasive beauty
and occasional ugliness of nature surely was no less evident then than
it is now. Indeed, as the quick progress of physics showed, it was wise
to *define* its subject matter by abstracting from good and evil, order
and confusion, the beauty and the ugliness of things. A difficulty arises,
of course, if one is committed to give a physical explanation of the
presence of mind and values in nature. If what we call 'physical' has
been deliberately defined so as to exclude them, it stands to reason that
only magic can put them back into it. Fortunately, no such explanation is needed for the fruitful pursuit of physics. (This, I dare say, is
the redeeming insight in Descartes's otherwise perverse mind–body
dualism.)

The human mind certainly takes pride of place in the physical laboratory, through the purposeful design and performance of experiments and the value-laden appraisal of results. But physics managed to
ignore this bewildering presence at the core of its practice for 300 years,
and the philosophy of physics generally turned a blind eye on it. Only
in the last few decades have some cosmologists argued that the actual
existence of physicists must make a difference to physics – and they
have been fiercely resisted by philosophers. The argument involves the
so-called Anthropic Principle, in several, more and less sober versions
(for encyclopedic treatment see Barrow and Tipler 1986). They all turn
around the following fact: According to the currently accepted theories of physics, living organisms can exist in the universe in some places
and for some time only if the values of certain fundamental constants
of nature lie within certain fairly narrow intervals. Their measured
values do, of course, fulfill this requirement. The Anthropic Principle
is meant to explain this arguably unlikely coincidence. The "strong"
version demands the intervention of an intelligent demiurge to fix those
constants at the beginning of time and thus prepare the terrain for life
on earth. The "weak" version plays the Darwinian spoilsport: the coin-

cidence requires no explanation at all, for obviously, if it did not happen, we would not be here to wonder at it.

In their great treatise, *The Large Scale Structure of Space-Time* (1973), Hawking and Ellis invoke the prerequisites of laboratory life to settle a specific physical question. The Einstein field equations (eqns. (5.29) and (5.31)) admit solutions containing closed timelike world-lines (Gödel 1949):

> However the existence of such curves would seem to lead to the possi-bility of logical paradoxes: for, one could imagine that with a suitable rocketship one could travel round such a curve and, arriving back before one's departure, one could prevent oneself from setting out in the first place. Of course there is a contradiction only if one assumes a simple notion of free will; but this is not something which can be dropped lightly since the whole of our philosophy of science is based on the assumption that one is free to perform any experiment. It might be possible to form a theory in which there were closed timelike curves and in which the concept of free will was modified [. . .] but one would be much more ready to believe that space-time satisfies what we shall call the *chronol-ogy condition*: namely, that there are no closed timelike curves.
>
> (Hawking and Ellis 1973, p. 189)

Hawking and Ellis's remark is apt and trenchant, yet few have been stirred by it, probably because not many working physicists pay atten-tion to closed timelike worldlines except when they read science fiction on holidays. On the other hand, in the context of QM a debate has been raging for two-thirds of a century over the need and the right to include the decisions and perceptions of experimentalists in nature's bookkeeping. There is a difficulty in combining the deterministic evo-lution of chances that is so brilliantly described by QM with the occur-rence of definite chance events. Predicted probabilities must indeed eventually give way to actual outcomes, but physicists and philoso-phers have a hard time conceiving the transition. The incongruence of expectation and fulfillment is one of the commonest facts of life, yet not one that modern physics was well prepared to handle. As we saw in §6.3.2 and §6.4.4, some philosophies of quantum mechanics try to deal with it in terms of the subjective–objective dichotomy (and ulti-mately as a mode of mind–body dualism). However, the classical oppo-sition of subjective prediction versus objective realization is now turned upside down: Expectations are epitomized in the evolving ψ-function, which stands for objective becoming, while fulfillment, the chance realization of one or the other definite outcome, comes in the guise of subjective awareness.

The QM debate, however, lay still far in the future when Descartes and his followers conceived the subject matter of physics as a mindless and aimless realm. Initially they had hoped to pierce through the veil of "obscure and confused" sense appearances by "sheer force of intellect". However, it was soon realized that the "clear and distinct ideas" of men would not carry physics very far. Except for at most a few general truths, physics must learn from experience. The problem is that in human experience, as Berkeley forcefully noted, the purportedly objective attributes of bodies are inextricably interwoven with the allegedly subjective features of perception (see §3.1). The Cartesian quest for certainty then took a different direction. The positivist tradition stemming from Berkeley sought to build the foundations of science not on abstract ideas of space, time, and matter, but on the rock-solid ground of elementary sense impressions, unpolluted by intellect and free, of course, from the "idols" of common sense. Every scientific truth ought to be derivable by deductive or inductive reasoning from elementary statements about irreducible sense data. This was easier said than done. As we saw in §4.4.3(iii), Mach's valuable critical work did precious little to further his sensationist ideology. Carnap's *Aufbau* (1928) and its sequels (Goodman 1951; Moulines 1973) dauntlessly undertook to carry out the foundationist program of positivism. When completed it would have provided a procedure for inferring say, eqns. (5.29) and (6.20) from premises in sense-datum language. Needless to say, all three books, for all their rebarbative display of misplaced precision, do not come anywhere near this utopia.

In the early 1930s, under pressure from Otto Neurath, Carnap and his friends in the Vienna Circle converted from "logical positivism" to "logical empiricism". The results of observation and experiment that form the evidence on which scientific generalizations rest must be shared by the community of researchers. Therefore, they should be stated in the public – "intersubjective" – language in which we talk to each other about things and events in the world, not in terms of private sense impressions. Neurath insisted that the "physicalistic" language that he advocated is only a fragment of ordinary language, for, in the interests of communication, it ought to be understandable by those born blind and deaf. But this requirement is surely exaggerated and laboratory protocols rarely comply with it. Laboratory assistants will probably avoid color words when reporting to a color-blind professor, but such situations are exceptional. Although its name suggests innovative artifice, physicalism in effect signified the philosophical reinstatement of commonsense language in scientific discourse (from which, indeed, it had never been excluded in real life). Neurath saw that this spelled the end of foundationism. His seminal paper, "Protocol Sentences", includes the famous ship metaphor:

> *There is no way of taking conclusively established pure protocol sentences as the starting point of the sciences.* No *tabula rasa* exists. We are like sailors who must rebuild their ship on the open sea, never able to dismantle it in dry-dock and to reconstruct it there out of the best materials. [. . .] Vague linguistic conglomerations always remain in one way or another as components of the ship.
>
> (Neurath 1932/33; in Ayer 1959, p. 201)

But most Vienna Circle philosophers still hoped to satisfy their craving for certainty. Carnap demanded that scientific discourse be reduced to *thing-language* – that is, "that language which we use in everyday life in speaking about the perceptible things surrounding us" (1936/37, p. 466) – and, more specifically, to the "observable predicates of the thing-language" (p. 467). The key notion of an *observable predicate* was explicated thus:

> A predicate 'P' of a language L is called *observable* for an organism (e.g. a person) N, if, for suitable arguments, e.g. 'b', N is able under suitable circumstances to come to a decision with the help of a few observations about a full sentence, say 'P(b)', i.e. to a confirmation of either 'P(b)' or '~P(b)' of such a high degree that he will either accept or reject 'P(b)'.
>
> (Carnap 1936/37, pp. 454ff.)

Presumably such predicates ought to be value-free, although there are suitable circumstances in which anyone acquainted with this or that physical instrument is able to decide after a few observations that *it is not working well* (so its data output should be thrown out). Carnap granted that this notion of observable predicate was imprecise, for different persons are more or less able to decide a given sentence quickly. He would, however, conventionally draw a sharp distinction between observable and nonobservable predicates (p. 455). Many important terms of physics are nonobservable; they were dubbed 'theoretical'. For the new foundationism to work, the meaning and use of theoretical terms should fully rest on that of observable terms. Carnap's attempts to "reduce" the former to the latter involved ever less stringent criteria and resulted in ever more tenuous links (see in particular Carnap 1956). Pliable and open-minded though they were, the logical empiricists remained adamant on one point: Cognitive meaning could not accrue to observational terms through their joint employment with theoretical terms. This rigidity may have been due in part to their general disregard of the history of science and the details of scientific practice, but it was quite essential to the foundationist orthodoxy. As a consequence of it, when Feyerabend, Hanson, and Kuhn made clear c. 1960 that the meaning and use of observational terms in physics depend on the theories in which they are embedded, logical empiricism fell down like a house of cards.

The quick transition from Carnapian foundationism to Kuhnian historicism generated a pseudoproblem that made philosophy of science the laughing stock of practicing physicists. Carnap conceived a physical theory as a seamless linguistic structure anchored to experience through its observational predicates. But if these – as the saying goes – are "theory-laden", they secure no outside links. The theories of physics rise then as close, self-contained edifices that cannot communicate or be compared with one another on shared terms. On this view, a physicist cannot teach Hamiltonian mechanics from 9 to 10 and quantum mechanics from 11 to 12 unless he is afflicted with a double personality. Such a ridiculous conclusion could, of course, have been avoided by taking Neurath's metaphor more seriously. In Neurath's ship, the neat steel turrets of theory are built on and bridged by the wooden planks of common sense, which may be worn and musty yet are indispensable to keep afloat the enterprise of knowledge. Physicists who advocate different theories do not "practice their trades in different worlds" (Kuhn 1962, p. 149), for there is but one world for them to wake up to, namely, the world they are in together with the persons they love and the goods they yearn for, some aspects and fragments of which are represented by physical theories in their abstractive and simplifying fashion while its concrete reality is talked about – often unperspicuously yet mostly to the point – in everyday language. If our ordinary understanding were a systematic affair – the "theory of common sense" that some philosophers have tried in vain to make explicit – it would stand aloof, impermeable to intellectual innovation and incapable of furthering it. However, thanks to the vices of ambiguity, vagueness, and ductility – which the Vienna Circle sought to correct through the use of formal language – it affords the nourishing ground on which physical theories and other such intellectual systems can grow and across which they can communicate.

7.2   Laws and Patterns

We have all heard that the aim of physics is the discovery and formulation of the *laws of nature*. The phrase 'law of nature' is first attested in Plato's dialogue *Gorgias* (483e), where it is pointedly used as an oxymoron by Callicles, an otherwise unknown young Athenian right-winger. Fifth-century sophists regularly opposed 'nature' (φύσις) and 'law' (νόμος), the latter being the product of human consensus and the defining mark of civilized society.[2] Callicles complains that laws are made by the majority who are weak, and therefore decree that the ambitions of the strong are unjust, and remain content with equality; whereas "nature itself proclaims that it is just that the better man have more than the meaner one and the abler more than the less able". After asking rhetorically on what principle of justice Xerxes campaigned against Greece and Xerxes's father against the Scythians, Callicles offers this answer: "They acted according to the nature of what is just (κατὰ φύσιν τὴν τοῦ δικαίου) and indeed, by Zeus, according to the very law of nature (κατὰ νόμον γε τὸν τῆς φύσεως)." Despite this inauspicious beginning, the Greek expression νόμος φύσεως ('law of nature') – and its translations into Latin and the modern European languages – made a splendid career. It fitted admirably with the Stoic idea of a rational world order of which legitimate human laws are a manifestation, "the right reason of nature . . . being a divine law by which whatever belongs to each thing or concerns it was assigned to it" (Chrysippus, FM 337; cf. FM 323: "The laws of the several *poleis* are but appendages to the right reason of nature"). Independently of this, it was also used to convey the medical concept of normality, of the workings of the animal body in good health.[3] The expression νόμος φύσεως ('law of nature') is often found in the Greek Church fathers,

---

[2] Cf. Plato (*Gorg.* 482e): "Nature and law are contrary to one another in many respects"; (*Protag.* 337d): "Law, the tyrant of mankind, enforces many things contrary to nature". Antiphon, fr. 44 (DK), is the locus classicus on "necessary nature" versus "adventitious law".

[3] This use, repeatedly attested in Galenus (second century A.D.), is found already in Plato (*Timaeus* 83e). Note that the "laws of nature" in this sense can be actually violated, just like the laws of man.

who readily associated it with their idea of God as creator and sovereign of the world;[4] but they also fondly employed it to denote the familiar necessity of things to which God Himself was not bound, although He voluntarily submitted to it in the person of Jesus Christ.[5]

The founders of modern physics took Plato's metaphor quite literally, and set out to find the articles of nature's legal code. In the writings of these Christian authors the word 'law' does not signify the universal scope of the prescribed regularities, but rather the legislative authority of their divine source; thus, 'Kepler's Laws' became the standard name for what at most might qualify as local traffic regulations. However, all the founders shared the quaint belief that God, although infinitely powerful, is a paradigm of good husbandry and therefore has used the thriftiest means to achieve the richest variety and abundance of effects. They understood this to imply that all the lowly local laws of nature must follow from a few all-embracing ones. Descartes proclaimed three universal laws of nature, two of which he derived from the very fact that they were willed by God (1644, II, arts. 37, 39, quoted in §1.3). In a similar vein, Newton put forward his three laws of motion, and showed that Kepler's Laws (duly corrected) could be derived from a specification of the second. Thus, the particular *facts* of nature came to be conceived as instances of universal statements or *laws* organized into a deductive system or *theory*. From the writings of Galileo, Descartes, Huygens, and Newton we gather that there was one exemplary theory in everyone's mind, viz., the *Elements* of Euclid, in which a wealth of geometrical laws, exactly obeyed – or so it seemed

---

[4] Cf. St. Basil, Epistle 302, a letter of condolence. The saint reminds a recent widow of "the legislation (νομοθεσία) of our God, in force from the beginning," by which "whoever comes to birth must necessarily depart from life at the proper time" – and advises her not to be "vexed with the common laws of nature (μὴ ἀγανακτῶμεν ἐπὶ τοῖς κοινοῖς τῆς φύσεως νόμοις)".

[5] St. John Chrysostomus, commenting on *Genesis* 1:1 – "In the beginning God made heaven and earth" – bids us to "observe the order and sequence: first the roof, and then the ground. As I said, He was not slave to the law of nature, or the order of art, but [exercised] the authority of power" (Migne PG, 59, 507). The Council of Ephesus (431 A.D.) emphatically asserted – against Nestorianism – that Christ was "under the law of nature", and was therefore really and fully born of a woman and experienced "the taste of death" (Schwartz ACO, vol. 1.1.6, pp. 38 [line 43], 19 [line 35], 103 [line 39]). See also St. Athanasius, *Exp. in Psalmos* (Migne PG, 27, 460 [lines 26–32]); St. Gregory Nazianzenus, *In sanctum pascha*, orat. 45 (Migne PG, 36, 660 [lines 31–33]): "as a man," Christ "toiled and hungered and thirsted and agonized, by law of nature".

– by all bodies, was inferred – or so it was thought – from a small number of fairly simple principles.

The scheme of ascent *from facts to laws, from laws to theories* was still canonic in the philosophy of physics long after its theological motivation had been discarded. For John Herschel (1830) a law of nature is either "a general proposition, announcing, in abstract terms, a whole group of particular facts relating to the behavior of natural agents in proposed circumstances," or "a proposition announcing that a whole class of individuals agreeing in one character agree also in another" (p. 100). Like his French contemporary, Comte, Herschel enjoins us to concentrate our attention on such laws, "dismissing then, as beyond our reach, the enquiry into causes" (p. 91). To find the laws of nature we must rely entirely on experience, on "the observation of facts and the collection of instances" (p. 118). "As particular inductions and laws of the first degree of generality are obtained from the consideration of individual facts, so Theories result from a consideration of these laws" (p. 190). Ascent to theory is now justified on purely methodological grounds:

> The analysis of phenomena, philosophically speaking, is principally useful, as it enables us to recognize, and mark for special investigation, those which appear to us simple; to set methodically about determining their laws, and thus to facilitate the work of raising up general axioms, or forms of words, which shall include the whole of them; which shall, as it were, transplant them out of the external into the intellectual world, render them creatures of pure thought, and enable us to reason them out a priori.
>
> (Herschel 1830, p. 97)

Herschel's description of the direct objects of a physical theory as "the creatures of reason rather than of sense" (p. 190) betrays his professional acquaintance with them. Still, it is clear that – as was customary in his century and during most of the next – Herschel viewed physical theories as collections of statements intended to be true of the actual course of the world. This is the distinctive feature of what Putnam (1962) dubbed "the received view of theories". On this view, physics ultimately aims to fuse all such collections into a single consistent one, a "theory of everything". For, as John Stuart Mill explained, "the question, What are the laws of nature? may be stated thus: What are the fewest and simplest assumptions, which being granted, the whole existing order of nature would result? [or] thus:

What are the fewest general propositions from which all the unifor-
mities which exist in the universe might be deductively inferred?"
(*System of Logic* III.iv.1; 1874, p. 230).

I do not plan to discuss the "received view" in detail.[6] I only wish
to stress one aspect of its twentieth-century history that, I think, has
not been sufficiently emphasized. As noted above, on the "received
view" physical theories should follow the example of Euclid's
Elements. Indeed, until the end of the nineteenth century, almost everyone took for granted that the final theory of everything would include
Euclidian geometry as a subtheory, that is, that the latter's axioms
would either follow deductively from the "fewest general propositions"
put forward by the former, or would be counted among them.
However, the evolution of geometry during that same century (§4.1)
led to a completely different reading of mathematical axiom systems
in general and of Euclid's in particular.

The new understanding of axioms and their consequences is implicit
in Hilbert's masterly *Foundations of Geometry* (1899) and caused his
clash with Frege.[7] Hilbert invites us to conceive "three different systems
of things". He calls them "points", "straights", and "planes", but only
as lip service to tradition. He does not explain what these things are,
but introduces five groups of axioms that impose conditions on their
mutual relations. (Examples: Any two *points* determine a unique
straight on which they are said to lie; any three *points*, not all of which
lie on the same *straight*, determine a unique *plane* on which they are
said to lie.) Frege thought that these unexplained terms referred to
things that Hilbert assumed to be well-known and that the axioms
were intended as true statements about such things. In a letter of 29
December 1899 Hilbert disabused him: "I do not wish to presuppose
anything as known; I see in my declaration in §1 the definition of the
concepts 'points', 'straights', 'planes', provided that one adds all the
axioms in axiom groups I–V as expressing the defining characters"
(Frege KS, p. 411). Frege had complained that Hilbert's concepts were

---

[6] I refer the reader to the excellent critical surveys by Frederick Suppe (1977, pp.
   3–118, 619–32; 1989, pp. 38–77).
[7] The Frege–Hilbert controversy has been repeatedly discussed by philosophers who,
   however, are loath to acknowledge that on this occasion their sharp-witted hero Frege
   displayed unusual obtuseness. A commendable exception is Shapiro (1997, pp.
   157–70). Cf. Torretti (1978, pp. 249–52).

equivocal, the predicate 'between' being applied to genuine geometrical points in §1 and to real number pairs in §9. Hilbert replied:

> Of course every theory is only a scaffolding or schema of concepts together with their necessary mutual relations, and the basic elements can be conceived in any way you wish. If I take for my points any system of things, for example, the system love, law, chimney-sweep, ... and I just assume all my axioms as relations between these things, my theorems – for example, Pythagoras's – also hold of these things. In other words: every theory can always be applied to infinitely many systems of basic elements. One needs only to apply an invertible one–one transformation and to stipulate that the axioms for the transformed things are respectively the same. [. . .] This feature of theories can never be a shortcoming[8] and is in any case inevitable.
>
> (Hilbert to Frege, 29.12.1899; in Frege KS, pp. 412–13)

Besides the undefined terms 'point', 'straight', and 'plane', Hilbert employs five undefined relational predicates, viz., 'between' (a ternary relation among points), two kinds of incidence (binary relations between a point and a straight, and between a point and a plane), and two kinds of congruence (between segments and between angles, respectively). Axiom groups I–V jointly characterize a structure built from the said three systems by means of these five relations. By using the idioms of a later age this characterization can be formulated in the traditional shape of a definition, as follows: *A Euclidian 3-space is an ordered octuple* $\langle \Sigma, \Lambda, \Pi, B, I_1, I_2, C_1, C_2 \rangle$ *satifying the Hilbert axioms I–V*, where Σ, Λ, and Π are three nonempty, nonintersecting sets, B is a triadic relation among elements of Σ, I_1 is a dyadic relation between an element of Σ and an element of Λ, I_2 is a dyadic relation between an element of Σ and an element of Π, C_1 is a dyadic relation between two unordered pairs ("segments") of elements of Σ, and C_2 is a dyadic relation between certain equivalence classes ("angles") formed by ordered triples of elements of Σ.[9] There are indeed many conceiv-

---

[8] "But a rather powerful advantage" (Hilbert's footnote ad loc.).

[9] Hilbert defines an angle as an (unordered) pair of rays issuing from the same point. The clumsier definition that I have in mind when I describe angles as equivalence classes of ordered point triples may be stated as follows: Let $Bxyz$ stand for 'point y is between point x and point z'. Assume that B satisfies Hilbert's axioms for betweenness (group II, which adequately characterizes the intuitive relations of three collinear points, one of which lies between the other two). I say that two ordered triples of

able octuples that satisfy this definition, and yet, surprisingly, the Hilbert axioms are so contrived that all these octuples embody one and the same structure, in the following precise sense: If S and S' are any two such octuples, there are one–one mappings f_1, f_2, and f_3 that map the first, second, and third terms of S onto, respectively, the first, second, and third terms of S', sending congruent segments to congruent segments and congruent angles to congruent angles.[10]

On this novel approach, as Paul Bernays neatly expressed it, "an axiom system is regarded not as *a system of statements* about a subject matter but as *a system of conditions* for what might be called a relational structure" (1967, p. 97; my italics), that is, as the specification of a concept of a certain type. Thus, if philosophers had promptly assimilated Hilbert, the "received view" of theories would not have made it to the twentieth century. But its modern advocates, notably the logical empiricists, although thoroughly acquainted with Hilbert's book and the developments in mathematics that motivated it, did not understand it in the manner explained above and so were able to defend the "received view" in good conscience. For them, an axiomatic mathematical system is not the *specification of a concept*, which may or may not have one or more real instances, but an *uninterpreted calculus*, that is, a system of meaningless ink or chalk marks forming strings – "sentences" – and strings of strings – "proofs" – according to strict rules of syntax. Never mind that the stuff printed in mathematical books and journals does not satisfy this description. Anyone who doubted that all of mathematics was ideally reducible to uninterpreted calculi was invited to read the three volumes of *Principia Mathematica* by Whitehead and Russell.[11] From this standpoint, the

---

points $\langle x,y,z \rangle$ and $\langle u,v,w \rangle$ are A-*equivalent* if and only if $x = u$ and either $Bxyv$ and $Bxzw$, or $Bxyv$ and $Bxwz$, or $Bxvy$ and $Bxzw$, or $Bxvy$ and $Bxwz$, or $Bxyw$ and $Bxzv$, or $Bxyw$ and $Bxvz$, or $Bxwy$ and $Bxzv$, or $Bxwy$ and $Bxvz$. An *angle* is a class of A-equivalent point triples. (The reader should verify that $\angle yxz$ is identical with $\angle vuw$ if and only if vertex x = vertex u and one of the eight stated alternatives is fulfilled, and that A-equivalence is indeed an equivalence, i.e., a reflexive, symmetric, and transitive dyadic relation.)

[10] It is not hard to see that the mappings in question will not preserve the congruences unless they also preserve the relations of incidence and betweenness (for betweenness this condition follows at once from my definition of 'angle' in note 9).

[11] Of course, Whitehead and Russell did not understand their formal system as an uninterpreted calculus. Quite the contrary. Following Frege, they construed the numerals as *proper names* (of certain classes). The good thing about this was that one could

"received view" can be straightforwardly restated thus: A physical theory – and, indeed, any properly formulated scientific theory – is an axiomatic mathematical system suitably interpreted so that its "sentences" are meaningful sentences and its "proofs" are valid logical proofs concerning a definite domain of reality. Now, the rules of syntax of the standard calculi – stemming from Frege (1879) and Peano (1895–1908) through Whitehead and Russell (1910–13) – were devised to ensure that the "proofs" would work in just this way. Thus, to make any such calculus into a physical theory in the stated sense, it is sufficient to bestow a physical meaning on the so-called nonlogical symbols of the calculus.[12] According to logical empiricism, this can only be done by associating such symbols with terms and predicates of the physicalistic language (§7.1). However, even the informal versions of accepted physical theories teem with notions that cannot be rendered accurately in physicalistic terms;[13] and this situation cannot be remedied by translating those theories, say, to the formal language of *Principia Mathematica*. (Imagine, say, that eqn. (6.44) has been rewritten as a sentence in this formal language; try to imagine a physicalistic rendering of this sentence.) So the logical empiricists settled in the end for a diminished version of the "received view": They equated physical theories with *partially interpreted calculi* (Carnap 1956). Briefly, it comes down to this. While the logical symbols of an uninterpreted calculus receive their intended logical force from the rules of syntax, the nonlogical symbols are divided into two classes: Those in the first class, dubbed "observational", are associated with terms and predicates of the physicalistic language, but those in the second class,

now be quite certain that 'two' does not denote Julius Caesar (cf. Frege 1884, p. 68). The bad thing was that in Whitehead and Russell's construal, 'two' – at least in unregimented ordinary English – is – like 'John' – the proper name of countless entities, a different one for each level in the logical hierarchy of types introduced by Russell to avoid the contradiction that he had discovered in Frege's system. (In a duly regimented language one might distinguish the different twos by means of indices, usually numerals; one ought to refrain, however, under pain of circularity, from giving them the said meaning in this use.)

[12] Evidently, in an uninterpreted calculus the distinction between logical and nonlogical symbols is wholly out of place. On the other hand, Frege (1879) made the semantic intent of his "conceptual script" clear to all by using quaint geometrical figures as logical symbols and the letters of various alphabets as nonlogical symbols.

[13] Consider acceleration (as in classical mechanics). How would you explicate it in physicalistic language? By equating it with average increases of average speed over a short sequence of short time intervals?

dubbed "theoretical", remain uninterpreted. So, strictly speaking, the "theoretical" symbols mean nothing, although they do acquire a shadow of meaning through their logical connections with the "observational" symbols. Typically, the "theoretical" symbols of such a partially interpreted calculus would include all the undefined primitives that occur in the axioms, the "observational" symbols – which turn up in remote logical consequences of the axioms – being defined in the calculus in terms of those primitives, through usually long chains of "theoretical" intermediaries. This version of the "received view" is a far cry from Herschel's. A logical empiricist can surely reach for a "theory of everything" in which every observable regularity of nature can be deductively inferred. However, the axioms of such a theory could not pass for true statements about the universal order of things. They would just be master recipes for the computation of empirical predictions (and retrodictions) from empirical data. To call them "laws of nature" would be a colossal abuse of language.

In the 1960s and 1970s a different view of physical theories came to the fore in America. As is usual in philosophy, there is more than one version of it. However, for simplicity's sake, I shall bring them all together under the single portmanteau term *structuralism*.[14] This term does justice to what I think is the gist of this view, which may be summarily stated thus: At the core of a physical theory there is always a coherent piece of mathematics that is intended to throw light on the processes and states of affairs in the theory's chosen field of study. Any coherent piece of mathematics can be articulated – as Hilbert did for Euclidian geometry – as the conception of a relational system or structure. Such a conception throws light on a physical domain when this is grasped as an instance or family of instances of the structure.

Structuralism originated with Patrick Suppes (1960, 1962, 1967, 1969) and gained strength through the early writings of Joseph Sneed (1971) and Bas van Fraassen (1972).[15] Suppes took his cue from

---

[14] Frederick Suppe (1967, 1977, 1989) calls it 'the semantic conception of theories'. I think that this appellation can only make sense at a time when there are people who are willing to countenance a conception of theories that is *not* semantic. In saner times, the term will not pick out a specific conception of theories.

[15] A structuralist approach to physics can be discerned at a much earlier date in the physical writings of a few great mathematicians who taught in Hilbert's Göttingen. Thus, Minkowski (1909) went straight for the structure at the core of Special Relativity, Weyl's *Space–Time–Matter* (1918) can be readily given a structuralist reading,

Bourbaki, whose reconstruction of mathematics as the science of abstract structures achieved the peak of its influence and began to spill out into undergraduate textbooks in the 1950s.[16] Bourbaki endeavored to bring all the seemingly disparate branches of contemporary mathematics under a single unifying perspective by treating each as the study of a particular "espèce de structure". Bourbaki gave a precise set-theoretic definition of "species of structure". It would be out of place to repeat it here.[17] I think the reader can get a clearer and firmer idea of Bourbaki's meaning from my presentation of several such species of structure in the Supplements (*group* and *field*s in I.2, *vector space* in I.3, *inner product space* in I.4, *Hilbert space* in I.7, *poset* in II.1, *lattice* in II.2, and *topological space* in III.1). Another example is Euclidian 3-space as defined above, viz., as an ordered octuple whose first three components are abstract sets and whose remaining five components are drawn, in fulfillment of Hilbert's axioms, from some of their Cartesian products and their power sets. As explicated by Suppes and his followers, a physical theory consists of a species of structure plus the "empirical claim" that certain physical systems – the theory's "applications" – are instances of that species.

Some qualifications must be added to this streamlined statement to appreciate its actual import and its clarifying power. First, this is what a physical theory is *as explicated* by the structuralists, not as it exists

and von Neumann (1932) established the equivalence between matrix mechanics and wave mechanics by proving that the underlying structures are isomorphic (see §6.2.3).

[16] Nicholas Bourbaki is the pseudonym of a group of French mathematicians – including Henri Cartan, Claude Chevalley, Jean Dieudonné, Laurent Schwartz and André Weil, among others – who undertook a systematic exposition of the whole of mathematics, which has been appearing since 1939 as *Élements de mathématiques* in an open-ended series of fascicles. I must say that I first saw a Bourbakian approach applied to all the major theories of physics in Bunge's *Foundations of Physics* (1967). I am therefore inclined to regard this book as a classic of structuralism, although the author would probably disavow me.

[17] I tried my hand at it in Torretti (1990, pp. 84–86). For the original, see Bourbaki (1970, ch. IV, § 1). The main idea is, roughly speaking, that any instance of a given species of structure can be described as an ordered n-tuple $\langle A_1, \ldots, A_k, A_{k+1}, \ldots, A_n \rangle$, where k and n are positive integers ($k < n$), $A_1, \ldots, A_k$ are arbitrary nonempty sets, and $A_{k+1}, \ldots, A_n$ are distinguished elements meeting definite conditions and picked out from $A_1, \ldots, A_k$ or from other sets obtained from the former by way of the set-theoretic operations of forming Cartesian products and ascending to power sets. (These two operations are defined in Supplement I.1.)

in real life, in the discussions and writings of the physicists. Not every physical theory in circulation has been given a structuralist explication. Moreover, the treatment is only available for theories that have attained sufficient clarity and precision. Physical theories do not fall like ripe fruits from a "place beyond heaven" (Plato, *Phaedrus* 247c), but are gropingly fashioned by physicists on earth. Their concepts are not all found ready-made on the smorgasbord of extant mathematics, but have often been created by the physicists themselves, or, at any rate, adapted to their needs. This is done with a view to the applications that they have in mind, which, in turn, are delineated with increasing distinctness in nature's flux by their being grasped with those concepts. The structuralist explication pairs a structure specified by axioms with a family of physical complexes conceived as particular instances of it. Live physics is murkier. Still, structuralism offers a picture of physical theories in an idealized state of maturity that throws much light on their actual development, their mutual relations, and their semantic links with their referents in the real world.

There is one more qualification that I should add before proceeding any further. Suppes and his followers present the mathematical core of a physical theory as a Bourbakian species of structure. This is a fairly straightforward method of presentation that was readily available to them and facilitated a uniform approach to the mathematical concepts employed in physics. But they do not hold to Bourbakism as a dogma. Any viable way of understanding mathematics as the study of abstract patterns can yield good structuralist explications of the mathematical ingredient in physical theories (cf. Balzer, Moulines, and Sneed 1983; Mormann 1996). Structuralism does not depend on the set-theoretic reformulation of the mathematical theories of physics. Its mainspring is the vision of physics as an endeavor to grasp the patterns of nature as concrete embodiments of the abstract ones conceived by such theories.[18]

Having made this clear, I shall assume, for definiteness, that struc-

---

[18] Two further qualifications express only my own judgment: (i) The structuralist analysis of physical theories by Joseph Sneed (1971) was constrained by the desire to solve his "problem of theoretical terms"; I do not think that this is a real problem (see Torretti 1990, pp. 109–30). (ii) Sneed's followers claim that in sciences that are very different from physics, such as biology, sociology, and even the history of science, every proper scientific theory can be explicated as a Bourbakian species of structure coupled with an empirical claim; I do not share this view.

turalist explications are made in the usual set-theoretic way. Imagine for a moment that a demon of uncommon intelligence undertook to do physics in this style. The set-theoretic hierarchy is so rich that he or she could well build from it a species of structure of which nature in all its complexity is an instance. Human physicists, however, are bound to be more modest. Indeed, modern mathematical physics was born of the realization that earlier attempts to grasp the entire fabric of the world in one fell swoop had failed altogether. Since the seventeenth century, physicists have been focusing on particular patterns that they isolate and grasp by means of suitably contrived mathematical concepts. Surely, they have thrived not only by their modesty, but also and chiefly by their uncanny eye for patterns. Thanks to it they ably discerned those features in actual phenomena that make together a distinctive pattern from the rest, which only mess it up, and they recognized structural affinities and identities where formerly one had seen only abysmal differences (e.g., between the fall of apples and the circulation of the moon). In this they were well served by their decision to geometrize physics, to conceive the patterns in nature as instances of mathematical structures.[19] However, to do it they had to replace in their minds the concrete physical processes that they intended to study – which were unmanageably complex and entangled with the rest of the world – with simplified models that truly instantiated such abstract mathematical structures as they could handle.[20] The gap between representation and reality was then filled by tactfully introducing correc-

---

[19] Thus, as the reader will recall, Maxwell was able to bring, as if with a magic wand, all the phenomena of optics into the fold of electrodynamics, because his mathematical theory of the latter predicted waves propagating *in vacuo* with the speed of light.

[20] The word 'model' is used in current philosophical literature in two somehow contrary senses. In the sense used above – and in many other passages of this book – a *model* is a representation of an individual or generic object by a real or ideal object of a different sort (e.g., a plastic model of an airport terminal; the familiar model of a pendulum, consisting of a weightless inextensible string with a massive dimensionless particle at one end, affixed by the other end to a frictionless nail). In the sense that is common in logic, models are models of structures, and a *model* of a structure of a given species is any set endowed with structural features satisfying the requirements of that species. Both senses can be nicely combined in one breath in a modified version of the above text, to wit: "Physicists had to replace concrete physical processes with simplified models (first sense) which are in effect models (second sense) of such abstract structures as they could handle." In this book I usually say 'realization' or 'instance' for 'model' in the second sense.

tions and agreeing on admissible margins of error. This devious approach to truth, which Aristotle would have spurned, enabled modern physics to rise far above Aristotle's level of ignorance. Thanks to it we can cope with two inescapable human limitations, namely, our incapacity to make error-free measurements and to work with mathematics that is too complex. But it also entails that the "empirical claim" of a physical theory mentioned above, viz., that the theory's applications are actual examples of its characteristic mathematical structure, must be taken with a sizable pinch of salt, for the structure is instantiated by the models, not by the physical realities that the models represent only to within some specified or unspecified but anyway finite degree of approximation. Thus, no astronomer in his right mind would claim that the solar system is exactly pictured by the Newtonian model that he uses to calculate the future motions of the planets. (The model certainly does not include representatives of every existing asteroid and comet.) What he claims instead – with abundant empirical support – is that the planetary positions and velocities recorded in fact and those computed from the model will not differ, within some agreed future time, by more than some agreed margin. Until the end of the nineteenth century one could maintain in good faith that the real solar system can be approximated with increasing accuracy by a series of ever more complex models instantiating the Newtonian theory of gravity. However, we cannot say this any longer. It is generally agreed that for predictions surpassing a certain level of accuracy the astronomer must resort to models that are instances of General Relativity. Indeed, in the case of Mercury, due to its smallness and its short distance from the Sun, Newtonian models fail already at a level of accuracy that astronomical observation had attained by 1860 (§5.4, note 64).

To account for such facts philosophers of an older persuasion tried to develop a concept of approximate truth. If a physical theory is a system of statements that are not exactly true and yet are not committed to the trash can of science, it must be because those statements are true approximately, to a degree that scientific methodology ought be capable of measuring. However, all attempts to calibrate approximate truth have ended in failure, essentially, I dare say, because, unless you twist the meaning of English words beyond recognition, the truth of a statement cannot be a matter of degree. Structuralism need not distort the grammar of 'truth'. From its standpoint, a theory is – or is not – exactly true of its purported models, which, in turn, are good or bad representatives of the physical phenomena for which they are supposed to stand. And, of course, a particular model of a particular process may be good enough for one purpose and not for another one.

A major strength of the structuralist view is its ability to elucidate, from this standpoint, the relations between the different and often incompatible physical theories that vie but also cooperate with one another. On the "received view" two theories can vie for our acceptance only if they contradict each other on one or more issues; but then they cannot cooperate, for how could two mutually inconsistent sets of statements jointly contribute to the solution of a scientific problem? Structuralism nimbly avoids this dilemma. We shall see in §7.3 how it dissolves the difficulty that the "received view" had to face after Kuhn's analysis of the succession of rival theories. Here I shall describe in structuralist terms two well-known examples of collaboration between theories.[21]

My first example is Einstein's work on Mercury's perihelion· advance, mentioned in §5.4 and alluded to above. In November 1915, Einstein (1915h) showed that an approximate solution of eqns. (5.28) – the second set of gravitational field equations published by him in that month – accounted for the difference of about 43″ per century between the observed precession of Mercury's perihelion and the value predicted by astronomers using the Newtonian theory of gravity. The difference is also accounted for by Schwarzschild's exact solution of eqns. (5.29) – Einstein's third and final set of November 1915 –, which agree with eqns. (5.28) in empty space. Using more recent figures (Weinberg 1972, p. 199) the said difference arises as follows: The secular precession actually measured is

$$\Delta\alpha = 5{,}600.73 \pm 0.41'' \tag{7.1}$$

The observed perihelion of Mercury must advance some 5,025″ per century due to the fact that the astronomical coordinate system is affixed to the moving Earth. Of the remaining 575.73″, some 532″ can

---

[21] Another example that deserves being studied from this point of view is the remarkable assortment of theories involved in the analysis of experimental results in contemporary particle physics. For a lucid description of this conceptual pot-pourri, see Falkenburg (1995, Ch. 5). As she aptly notes, the "theory of measurement" standardly employed in ascertaining the measurable attributes of the particle tracks, scattering events and resonances recorded in high-energy experiments, is "anything but a strict deductive system of special cases of one-and-the-same theory" (p. 183).

be attributed, in accordance with the Newtonian theory, to gravitational perturbations caused by the other planets, chiefly Venus, Earth, and Jupiter. So, according to pre-relativistic astronomy, the secular precession of Mercury's perihelion should not be larger than

$$\Delta\theta = 5,557.62 \pm 0.20''$$ (7.2)

The difference $\Delta\alpha - \Delta\theta = 43.11 \pm 0.45''$ agrees beautifully with the secular precession of $43.03''$ predicted by General Relativity for a spinless test particle of negligible mass simulating Mercury's motion in an otherwise empty spherically symmetric spacetime such as would correspond to the presence of a solar mass at the axis of symmetry.[22] Now, neither Einstein nor Schwarzschild figured out the precession of Mercury's perihelion in a relativistic model of the solar system, complete with all the planets. That was far beyond their mathematical ability. So they worked in fact with three models, namely, (i) a vacuum solution of Einstein's eqns. (5.28) and (5.29); (ii) the standard Newtonian model of the solar system, perfected by Laplace and Poincaré; and (iii) a model that we may call Ptolemaic or, more exactly, Eudoxian, consisting of an imaginary sphere affixed to the Earth and homocentric with it, on which the observed positions of Mercury and the Sun are recorded. Since the question at issue does not depend on the accurate measurement of very short time intervals, all three models share the standard (Newtonian?) time of astronomy. It is in model (iii) that Mercury's perihelion – that is, the point on the sphere where it comes closest to the Sun – advances by $\Delta\alpha$ in 100 years. This model is purely kinematic and can be converted to either (i) or (ii) by a suitable coordinate transformation. In both cases, the transformation wipes out approximately $5,025''$ from the perihelion advance. The rest, amounting to some $575.5''$, must be explained dynamically. As we saw, model (i) takes care of some $43''$ and model (ii) of the remaining $532''$. Seeing these figures, the reader may wonder why this was regarded as a triumph of Einstein's theory, instantiated by model (i), over Newton's, instantiated by model (ii). Now Einstein and his followers were certainly unable to build a relativistic model of the solar

---

[22] Apart from the given numerical values, this prediction is significant because in the Newtonian theory a particle with Mercury's mass and simulating its motion around an otherwise lonely Sun would describe an ellipse with the center of gravity of the system at one focus and its perihelion *would not precess at all*. Thus, Einstein's discrepancy with Kepler's picture of the Solar System is steeper than Newton's.

system that accounted for the full dynamically conditioned precession of 575.5″. But still, it was clear to everybody that in such a model – given the formal relation between eqns. (5.29) and the Poisson equation of Newtonian theory (cf. point (D) in §5.4) – the precession increase over 43″ due to the presence of massive planets would not differ significantly from the 532″ attributed to their influence in the Newtonian model.

My second example is more speculative. I mentioned earlier the singularity theorems proved in the 1960s by Penrose, Hawking, and Geroch (§5.5, note 73). They imply that in a typical GR spacetime there are black holes, that is, spacetime regions that absorb all the energy they receive without ever releasing any. Although the evidence for the actual existence of such regions is not entirely beyond question even now, the theoretical study of the properties of black holes soon became a flourishing academic industry. Bekenstein suggested that certain properties of black holes might be connected with entropy, and other important links to thermodynamics were proposed. However, it did not seem possible to assign a finite entropy to a black hole, since that would imply that it had a finite temperature and therefore, if surrounded by thermal radiation at the same temperature, would remain in equilibrium with it. This seemed absurd, for the black hole would inevitably absorb some of the radiation but could not emit anything in return. To overcome this difficulty, Hawking (1974, 1975) constructed a QM model of a black hole. According to QM, energy trapped inside a potential well has a finite chance of tunneling through the barrier surrounding it. This is, for example, what happens in radioactivity, which QM conceives as the sudden, uncaused ejection of certain quanta of matter and radiation from the atomic nucleus to which they are normally bound. Hawking applied the same general idea to black holes and calculated the time that it would take for a black hole of given mass m to evaporate completely. (If m is the mass of the Sun, the time in years is of the order of 10^{63}.) On the "received view" of theories, Hawking's procedure plainly defies logic. If GR and QM are regarded as two sets of statements, their union is an inconsistent set that therefore cannot be satisfied. Worse still, according to GR there are no black holes in the flat Minkowski spacetime underlying Quantum Field Theory, let alone in the Newtonian spacetime underlying nonrelativistic QM. However, from the structuralist standpoint, both theories can very well cooperate on this issue, as on many others. If there is a real black hole in the world, it can be represented both in a GR model and in a QM model. A difficulty would arise only if the degrees of approximation claimed for the models cannot be jointly attained because the incompatibility between the theories forbids it.

7.3 Rupture and Continuity

In the Preface to the second edition of the *Critique of Pure Reason*, Kant speaks of the moment when, in the full light of history, physics entered at last "upon the secure path of science" after centuries of "random groping" (1787, p. xiv). That was in the first half of the seventeenth century, when Galileo and Torricelli carried out their experiments. They understood "that reason has insight only into that which she produces after a plan of her own, and that she must not allow herself to be kept, as it were, in nature's leading-strings, but must herself show the way with principles of judgment based upon fixed laws, constraining nature to give answer to questions of reason's own determining. [. . .] Reason, holding in one hand her principles [. . .] and in the other hand the experiment which she has devised in conformity with them, must approach nature in order to learn from it, but not as a pupil who must listen to everything the teacher chooses to say, but as an appointed judge who compels the witnesses to answer the questions which he puts to them" (p. xiii). Kant was persuaded that this "revolution in the manner of thinking" had infallibly secured progress "in endless expansion throughout all time" (p. xi). There is no indication that he ever considered the following question, which his judicial metaphor immediately suggests: What would happen to the secure progress of science if reason, the judge, on her own initiative or at her sovereign's behest, would change the terms of her questions or the rules for evaluating the answers? Although reason began to wield her principles at a given time and place, Kant regarded them as being above history. The interrogation of nature by reason between Arno and Clyde, Elbe and Ebro, in the seventeenth and eighteenth centuries of the Christian era, was, in his eyes, only an expression of reason's timeless nature.

The unchangeable principles by which, according to Kant, the human understanding must "spell out appearances in order to read them as experience" (1781, p. 314) include Euclidian geometry and Newtonian chronometry, mass conservation, strict determinism, and instantaneous interaction at a distance (§§3.3 and 3.4). The advent of

Relativity and quantum physics brought these principles to nought. Quite reasonably, many philosophers saw here a practical refutation of Kant's philosophy. Former Kantians, like Carnap and Reichenbach, and other advocates of logical empiricism sought to separate the informative contents of science, provided by the senses alone, from its conceptual scaffolding. Neglecting Kant's warning that, without concepts, the senses are blind, they viewed scientific concepts as a mere instrument, of arbitrary design, for codifying sense data in a way that is economical and easy to handle. Having failed in his attempt to reduce scientific discourse to the language of sense impressions, Carnap adopted Neurath's physicalistic language (§7.1) and tried to show that its "observational" vocabulary sufficed for conveying the full cognitive contents of physical theories. The permanence of that vocabulary, despite the drastic changes of theories, secured a connection between the latter and facilitated their comparison. When Carnap and his friends regarded everyday laboratory talk as a factor of continuity in physics they certainly were on the right track. But their distinction between *observational* terms – that are applicable to real life things and processes – and *theoretical* terms – to be used only as chips in calculation – drove them into a blind alley, for the observations of greatest interest for today's physics can only be precisely described and profitably interpreted in a language that is loaded with theoretical terms (and assumptions).

As I mentioned in §7.1, the "theory-ladenness" of observation was rediscovered by Hanson (1958) and Feyerabend (1958, 1962) and played a key role in the devastating criticism of logical empiricism that they and other authors undertook *c.* 1960. By that same time, Thomas Kuhn developed a vision of the history of every science as a succession of ruptures or "revolutions" separated by periods of continuity or "normal science". According to Kuhn (1962), scientific revolutions break the continuity of "normal science" because they bring about a change in the manner of thinking by virtue of which any postrevolutionary theory is "incommensurable" with the pre-revolutionary theory that it displaces. This radical conclusion is inevitable if the relevant facts of observation can be articulated as such only in the context and within the perspective of one or the other theory. In this case, evidently, there is no independent data base with which the theories could be compared. On the other hand, as Shapere aptly remarked (1966; cf. 1964), if successive theories are incommensurable for the stated reason, one cannot even claim that they concern the same subject matter, for the breakdown of the earlier conceptual system entails the loss of all references.

To counter this shocking result, Hilary Putnam (1970, 1973, 1975b) put forward his doctrine of reference without sense. In contrast with classical Fregean semantics, in which the concept one thinks fixes the kind of things one means, Putnam's semantics anchors the reference of each term to an original act of name-giving – in which, for instance, Adam points to a river in Paradise and says: "*that* is *water*" – linked to the word's present use by a causal chain of communications. If this were so, scientific revolutions could not alter the meaning of the main terms of science, and successive theories could easily be compared with one another and with experience. But the doctrine of senseless reference, although perhaps admissible in the case of the proper names of individuals, is inapplicable to general terms, because the recognition of different individuals as instances of the same class depends on the concept by which one grasps them. Thus, we could not regard the fall of heavy bodies toward the center of the earth, the circulation of the planets about the sun, and the universal recession of distant galaxies as phenomena of the same kind if our concept of gravity were the same as Aristotle's.

I do not intend here to fight a theory of meaning that is no longer upheld by its author.[23] I refer to it because it shows how seriously this distinguished philosopher took the alleged incommensurability of scientific theories and to what extremes he was ready to go to avoid it. In my view, however, the thesis of incommensurability rests on a misunderstanding. Certainly, there are ruptures in the history of physics, as indeed there must be if there is genuine intellectual novelty; but the ruptures heal because the same factors that promote them and make them possible contribute to restore the continuity. I shall clarify this view with examples, but I must warn first against a common mistake. In the literature prompted by Kuhn's book (1962), the eponymous scientific revolution of the seventeenth century and the great innovations in twentieth-century physics are often treated on par with the minirevolutions that, it is said, are happening all the time in every puny scientific specialty. This leveling approach threw light on the similarities between major and minor sciences, but it also generated confusion concerning the significance of their respective revolutions.

---

[23] I did so, with apologies to Putnam, in Torretti (1990, §2.6).

Surely it is an exaggeration to say that "after a revolution the scientists respond to a different world" (Kuhn 1962, p. 110) if the revolution in question concerns, for example, the origin of heart disease or the displacement of continental plates. Indeed, when the geologists, discarding the very idea of terra firma, see continents gliding over magma like skating boards they radically alter the interpretation of countless phenomena in their area of study, but they can go on describing them with the same terms of physics, chemistry, and common sense used by their predecessors; the continental plates move in the same world in which America, Eurasia, and Africa were previously thought to rest on the earth.

A true "revolution in the manner of thinking" in Kant's sense makes no allowance for a general – scientific or nonscientific – form of discourse in which the rival theories are embedded and through which they can communicate. If there is a single coherent global way of organizing our experience and this suffers mutation, the theories that precede and follow such change will not even acknowledge one another as theories. A system of reason like the one proposed by Kant is absolute and self-contained and leaves no room for a "manner of thinking" that is different from its own (cf. Davidson 1974). The statement that scientific theories separated by a scientific revolution are incommensurable rests on the following misunderstanding: Its supporters continue to see human – or, at any rate, scientific – thought in the guise of all-encompassing unitary reason à la Kant, although they regard it as a product of history and liable to mutations à la Darwin. The fact is that, if – as Kant taught – there is an architecture of human reason, it is not all of one piece, like that of Escorial or Versailles, but more like the cathedral of Santiago de Compostela, combining many styles that do not spoil but magnify the beauty of the whole. A more illuminating architectural metaphor was proposed by Wittgenstein, who compared "language" to a city with a labyrinth of short streets with little squares and houses of diverse ages at the center – which he assimilated to everyday speech – , surrounded by modern suburbs with straight and regular avenues and monotonous housing – resembling the idiolects of science and technology (1958, I, §18). Note that this metaphor does not imply that scientific languages must be translatable to ordinary language nor that they draw their meanings from it; but rather that prescientific language is older and more stable and remains afoot when one of the special jargons is remodeled or removed, and so can maintain a connexion that would otherwise be interrupted. In

Perspectives and Reflections

particular, physics, while its *theories* can be adequately expressed only in a language teeming with mathematical terms and formulas, must continually resort to ordinary language in its *practice*, whenever instructions are given to set up or to handle an experiment.

Take an example. In §5.1 I mentioned Michelson's interferometer (see Fig. 12), an instrument designed to measure the velocity of the earth relative to aether, the massless carrier of light waves postulated by nineteenth-century physics. The instrument was built, in agreement with the aether theory, so that the pattern of fringes generated by mixing light traveling from a common source along two different paths would manifest the path-dependence of the speed of light. However, no change in the fringe pattern was detected when the paths were interchanged. To explain this null result FitzGerald (1889) and Lorentz (1892) assumed that solid bodies contract, due to some action of the aether, in the direction in which they move through it. On the other hand, Einstein's Special Relativity (1905r) implied a completely different reading of the interferometer results. According to it, Michelson's instrument cannot measure the speed of the earth in the aether, for the new theory denies the very existence of an aether. So what the interferometer experiment compares is the speed of light along a path at rest on a laboratory that, at least for this purpose, may be regarded as an inertial system, and its speed along another path, perpendicular to the former, but equally at rest on the same laboratory. It is no wonder that the experiment fails to disclose any differences. Einstein's revolution turned Michelson and Morley's perplexing experiment into an innocuous banality. This was paid for with a drastic change in all the fundamental concepts of physics: time, distance, mass, force. But if the sense of these terms has changed, how can they refer to the same phenomena? If physicists talked only of what they grasp with their theories, the jump between two conceptually irreconcilable theories would sever their references. Fortunately it is not so. Physical theories grasp only an abstract and idealized aspect of life, but physicists often think of other more concrete aspects of it, for example, when they buy cheese and wine at the supermarket. When they ask the janitor to dust the interferometer they will refer to the apparatus in ways that do not in the least depend on their understanding of what goes on when it is put to work. Such modes of reference are shared, of course, by relativists and physicists of the Lorentz–FitzGerald persuasion.

This state of affairs fits well into the structuralist scheme, according to which a physical theory consists of a mathematical structure and a collection of intended applications, that is, of aspects or fragments of the world that are modeled by instances of that structure. If the modeling is successful – within the admissible margin of imprecision – the fragments or aspects in question are grasped and understood by the theory, and its terms refer in effect to them.[24] But there must be a way of referring to each one of those fragments and aspects without the theory, for otherwise one could not designate them as candidates for modeling. Such forms of reference are probably confused and do not properly delineate their referents or contribute in any way to understand them; but they provide a semantic handle by which to keep a hold on the referents even if the modeling fails and the theory's concepts merely glide over them. A physicist must be able to speak of that which he does not understand; otherwise, a baffling observation or an experimental result contrary to prediction would fall outside the reach of discourse and could not act as a catalyzer of conceptual novelty. The insufficiencies of accepted theory, the anomalies that will lead to its breakdown, can only be signaled in a paratheoretic language that acts as a factor of continuity.[25]

---

[24] For instance, NASA engineers who speak of a spaceship "entering the gravitational field of Mars" grasp both ship and planet as massive bodies that are subject to mutual gravitational attraction in accordance with Newtonian theory.

[25] D. G. Mayo's philosophy of experiment provides further evidence that physical theories must communicate among themselves to do their jobs and should therefore be embedded in a common discourse. Taking her cue from Suppes's 1962 criticism of "philosophers of science [who] overly simplify the structure of science", Mayo argues convincingly that every experiment involves the interplay of a host of models (1996, Ch. 5 and 7). Although the breakdown and classification of the models used in an experimental inquiry are not "a cut and dried affair" (p. 129), she groups them for elucidation in three levels: (a) the *primary model* of the physical theory being tested; (b) the *experimental model* that, on the one hand, specifies the key features of the experiment and states the primary question or questions with respect to it and, on the other hand, specifies analytical techniques for linking the data to that question or questions; (c) the *data models* that articulate and "massage" the raw data of observation in the manner required for the application of the experimental model's analytical tools. Take for example the confrontation of General Relativity and other modern theories of gravity with experimental data drawn from the Solar System. Solar system experiments are modeled after a theory known as the parametrized post-Newtonian (PPN) formalism. "The PPN framework takes the slow motion, weak field, or post-Newtonian limit of metric theories of gravity, and characterizes that limit by a set of 10 real-valued parameters. Each metric theory of gravity has particular values for the PPN parameters" (Will 1981, p. 10). Suitably massaged results of astronomical observations, organized into appropriate data models, supply the

There is another factor of continuity in the history of physics that I believe is responsible, more than anything else, for the unity displayed by modern mathematical physics since Galileo, despite all the changes that it has gone through. It lies at the core of every physical theory, which, as we have seen, can be explicated as a mathematical species of structure. As such, it has its definite place in the ideal universe of structures and is clearly and distinctly related to other structures, including those at the core of other physical theories. For example, the cosmological models of General Relativity are smooth four-dimensional manifolds with a semi-Riemannian metric satisfying the Einstein field equations (5.31) and possibly other conditions – such as stable causation – that are often prescribed to exclude monsters. Thus, they form a subspecies of the much broader species of n-dimensional Riemannian manifolds, which is a part of the smooth manifolds, the full class of which may be regarded in turn as a distinctive kind of topological spaces. This example is particularly clear, because the said species of structure, of which the relativistic worlds are very peculiar instances, are by themselves the subject of important branches of mathematics. In most cases, however, the conceptual core of a physical theory must be explicated by a highly idiosyncratic species of structure, combining subspecies of several familiar species in a tangled web.[26] In all cases, however, the rich available stock of structures that pure

---

measured values of those parameters. The measured value of this or that parameter can then be compared with the different values assigned to it by the diverse theories of gravity. In this way, in each particular Solar System experiment the same PPN model of experiment mediates between the data and several alternative primary models, based on General Relativity and its rivals. Mayo's approach undercuts Kuhnian incommensurability. The theory instantiated by any given model is a creature of human thought and therefore, at some point, has come out of the blue. However, the theories that experimental physics teems with do not stand in isolation, but must be so conceived that they can be linked with other such theories via the several model levels. Nor are these theories closed worldviews. They range from the humble rules for drawing histograms and for calculating averages to the lofty conceptual frameworks of classical dynamics and electromagnetism, relativity and quantum mechanics. But even the latter, for all their *weltanschaulich* posturing, are made to work only in a piecemeal fashion, through the testing of detachable hypotheses in separate experiments. Two theories corresponding to diverse primary models may differ deeply in their basic concepts; but, if any one of them is meant to take the place of the other, they must both be linked to the same experimental models, or, at any rate, they must have application to the same data models. Thus, through such links, they are brought under a common standard.

[26] The rational reconstruction of such systems is a favorite subject of research in Sneed's school. See Moulines (1975) and Bartelborth (1987).

mathematics studies with great generality not only furnishes physicists with a practically inexhaustible source of revolutionary concepts, but it also facilitates the comparison between new theories and their predecessors. Thus the very factor that contributes the means for innovative rupture makes it possible to determine the conceptual link between the new and the old theories and to perceive the continuity that binds them.

I shall illustrate this idea with two examples. The first one is quite modest and therefore can be explained in detail. I shall compare the central notion of Lorentz's prerelativistic electrodynamics with the homologous notion in Minkowski's relativistic electrodynamics. Lorentz's theory is a modified version of Maxwell's (§4.2). In it, the whole universe is filled with a massless aether, none of whose parts can suffer displacement relative to its other parts. This substance penetrates the innermost recesses of massive matter. The aether is the seat of forces, subject to Newton's Second Law, that act on the electrically charged bodies that move or rest in it. The forces are represented by two time-dependent vector fields, the electric field $\mathbf{E}$ and the magnetic field $\mathbf{B}$. The total force $\mathbf{F}$ exercised at a particular time on a particle with charge q moving through the aether with velocity $\mathbf{v}$ is given by

$$\mathbf{F} = q(\mathbf{E} + \mathbf{v} \times \mathbf{B}) \qquad (7.3)$$

the values of $\mathbf{E}$ and $\mathbf{B}$ being taken at the point where the particle is located in the aether at that time. The said local, momentary values of the fields depend – with delay – on the positions and velocities of all the electric charges in the world. The relations between the electric and magnetic fields and the distribution of electric charges and currents are governed by the Maxwell equations (eqns. I–IV in Chapter Four, note 42). Let x, y, z be Cartesian coordinates for a reference frame at rest in the aether. The vectors $\mathbf{E}$ and $\mathbf{B}$ can be analyzed, as usual, into their components relative to the said coordinates (cf. eqn. 4.11):

$$\mathbf{E} = (E_x, E_y, E_z) \qquad \mathbf{B} = (B_x, B_y, B_z) \qquad (7.4)$$

The numerical value of the components can be obtained by measuring various electromagnetic effects in the laboratory.

In relativistic electrodynamics there is no aether. The Maxwell equations and the Lorentz force (7.3) take the same form relative to any inertial frame of reference. However, a charged particle of mass m evidently cannot experience the same acceleration $\mathbf{F}/m$ with respect to an inertial frame relative to which it moves with constant velocity $\mathbf{v}$ and with respect to the inertial frame in which it is at rest. Neither can the field vectors $\mathbf{E}$ and $\mathbf{B}$ be the same in two inertial systems in motion relative to each other, for the distribution of electric currents cannot be the same in both. Therefore, the electric and magnetic fields depend on the arbitrary election of a reference frame and do not represent self-subsisting physical realities. Nevertheless, as Minkowski showed, the components of $\mathbf{E}$ and $\mathbf{B}$ relative to a Cartesian coordinate system (x,y,z) anchored to an inertial frame happen to be the components of a geometric object on Minkowski spacetime relative to the spacetime chart (t,x,y,z) adapted to the same inertial frame and consisting of Einstein time t and the said Cartesian coordinates x, y, z. A geometric object is, as such, independent of the chosen reference frame, but its analysis into components relative to a particular coordinate system varies, with the latter, according to definite rules characteristic of each type of object. The object in question here can therefore be regarded as the appropriate mathematical representation of a single physical reality, the electromagnetic field, which from a human standpoint, bound to this or that inertial frame, is decomposed in a natural way into an electric field and a magnetic field. Minkowski conceived the electromagnetic field as a field of so-called polar vectors, which, on a 4-manifold, has precisely 6 components relative to any particular chart. Today we conceive it, in a mathematically equivalent way, as a field of antisymmetric tensors of type (0,2). On a 4-manifold, a geometric object of this type has 16 components relative to a given chart, but, due to antisymmetry, no more than 6 of them will be significantly different. Specifically, if we denote the electromagnetic field by $\mathscr{F}$ and its 16 components relative to (t,x,y,z) by $\mathscr{F}_{tt}$, $\mathscr{F}_{tx}$, and so on, we have that – with the sign conventions used by Feynman et al. (1964) –

$$\mathscr{F}_{tx} = -\mathscr{F}_{xt} = -E_x \qquad \mathscr{F}_{ty} = -\mathscr{F}_{yt} = -E_y \qquad \mathscr{F}_{tz} = -\mathscr{F}_{xt} = -E_z$$

$$\mathscr{F}_{xy} = -\mathscr{F}_{yx} = -B_z \qquad \mathscr{F}_{yz} = -\mathscr{F}_{zy} = -B_x \qquad \mathscr{F}_{zx} = -\mathscr{F}_{xz} = -B_y$$

$$\mathscr{F}_{tt} = \mathscr{F}_{xx} = \mathscr{F}_{yy} = \mathscr{F}_{zz} = 0$$

$$(7.5)$$

Therefore, the 16 components of $\mathscr{F}$ relative to the Lorentz chart (x,y,z,t) are given by the following matrix:

$$\begin{vmatrix} 0 & -E_x & -E_y & -E_z \\ E_x & 0 & -B_z & B_y \\ E_y & B_z & 0 & -B_x \\ E_z & -B_y & B_x & 0 \end{vmatrix} \qquad (7.6)$$

Thus, although the relativist and prerelativist conceptions of the physical world in general and of electrodynamics in particular are radically different, the numbers that determine electrodynamic phenomena in accordance with the Maxwell equations have a definite role to play in each theory. Measured by the same methods in the same laboratories, they will be the same numbers even though they stand for physical quantities conceived in a completely different way. This example suggests that, among the several adjectives available for describing the relation between a revolutionary physical theory and the theory displaced by it, 'incommensurable' was not a remarkably thoughtful choice.

My second example is more ambitious, and I shall barely sketch it.[27] I have just recalled that models of General Relativity are Riemannian manifolds. This concept was introduced by Riemann in order to furnish physicists with a generalization of Euclidian space, to which they could resort if, as Riemann anticipated, physical data obtained with methods of ever greater precision would not fit into the Procrustean framework of traditional geometry. Riemann was convinced that metrical relations in a physical continuum depend on the natural forces that act in it. He therefore thought it unlikely that the geometry that has served us so well at the human scale could also be adequate at the cosmic and the molecular scales. As a first step beyond Euclid, Riemann proposed a family of metrics that agree optimally with Euclidian metrics on a neighborhood of each point of space (see §4.1.3). The semi-Riemannian metrics of General Relativity are designed to agree optimally on a neighborhood of each point of *spacetime* with the Minkowski metric (see §§5.2 and 5.4). General Relativity is a geometrodynamic theory of gravity. A particle moving under the sole influence of gravity describes a spacetime geodesic: Its worldline depends on geometry alone. But geometry, in turn, depends on the forces of nature: The metric is determined by the distribution of matter. This theory of gravity has precious little to do with Newton's and is apparently incomparable with it. Nevertheless, a few years after Einstein published General Relativity (1915i, 1916e), Élie Cartan (1923) put forward a geometrodynamic formulation of Newton's theory that is equivalent to its classical formulation (by Laplace and Poisson; see §4.2). Cartan's formulation was further elaborated by Friedrichs (1927) and Havas (1964). In it, as in Einstein's theory, a material particle influenced solely by gravity describes a space-

---

[27] For details, see Earman and Friedman (1973).

time geodesic, and the metric that fixes what spacetime curves are geodesic depends on the distribution of matter. The differential equations governing this dependence lead to precisely the same predictions as the classical Newtonian theory. In this way, the versatile mathematical structure that Riemann invented on purpose to facilitate a finer adjustment of physical theories to experience has made it possible to reformulate an old theory in terms that allow a comparison with its revolutionary successor. The new formulation is of course anachronistic. (It is also needlessly clumsy, and nobody would dream of using it to calculate the orbit of a communications satellite.) However, – in stark contrast, say, with renderings of Greek poetry in modern languages – it exactly matches the original, because only neat mathematical concepts are at play.[28]

Thus, the tradition of physics has continued across all ruptures because the same factors that contribute to innovation also facilitate comparison between its successive stages. However, the historical continuity of physics does not imply that it advances on a "secure path" toward a preestablished goal, nor that it is impossible to develop a different physics that is mathematical and experimental like ours, but is incompatible and even incomparable with it. To prove that such an alternative physics is possible one usually invokes the so-called Duhem–Quine thesis, viz., that a given stock of empirical information can be approximated by an infinite number of different theories. This thesis is questionable, for a stock of data can hardly *exist* apart from the "manner of thinking" that articulates it. However, "manners of thinking" are not born, armed to the teeth, out of Zeus's brain, but grow contingently in the course of history. An alternative physics could arise elsewhere in the universe or issue from an unorthodox deviation of ours (see Cushing 1994). After some generations and a few turns of the road two intellectual traditions can become mutually incomprehensible even though they share the same roots.

---

[28] A historically more significant example is mentioned by Falkenburg: In a very influential paper, Eugen Wigner (1939) classed all the solutions of conceivable relativistic field equations – including, among others, the Maxwell equations and the Dirac equation – according to the irreducible representations of the Poincaré group. "This classification is *not* linked to a *particular* formulation of quantum theory [. . .], but embraces all relativistic theories of free fields, that is, it extends to a whole *class of physical theories* which are characterized by a certain mathematical structure of the theoretical entity 'field' " (Falkenburg 1995, pp. 229–30).

7.4   Grasping the Facts

Particular facts cannot be grasped as such unless general concepts are brought to bear on them. Neglect of this truism has been the source of endless blunders in philosophy. On the other hand, it borders paradox, for how can you pinpoint a fact to bring a concept to bear on it unless you have somehow grasped it? The classical philosophical solution of this difficulty, due to Aristotle, is that the individual things themselves, present to the senses, convey to any intelligent observer the concepts under which they, their properties, and relations properly belong. From the commonsense standpoint of someone steeped in the routines of civilized life this seems obvious. Don't tables signal us to eat or write on them, running water to quench our thirst, red traffic lights to stop? It is true that we often make mistakes, but those who consciously or unconsciously endorse the Aristotelian solution are not disturbed by them (although mistakes are facts too, which naturally do not announce themselves as such). The average adult members of a stable community, in firm possession of their mental faculties, discount errors in the classification of familiar objects as transient inconveniences, which are easily overcome by increased attention and a few simple tests. After all, how could such errors ever be corrected except by giving a closer look to the state of affairs and letting it speak for itself? As for the unfamiliar, there is a common human tendency – abundantly documented in the annals of exploration – to assimilate it to the familiar, and then, little by little, to make allowance for the differences. This suggests that novel facts do not sport the concepts by which we grasp them, but that we must draw these concepts from our stock-in-trade, refurbishing them as needs be for their new jobs. It is therefore unlikely that Aristotelianism could have risen as an explicit philosophical doctrine in a fast-changing social environment, such as that associated with modernity. Indeed, the unconscious Aristotelians we still meet everywhere tend to be the same people who methodically avoid the unfamiliar or refuse to acknowledge it as such.

As we saw in Chapter One, modern physics began in a thoroughly anti-Aristotelian mood, motivated by the perceived inadequacy of traditional concepts for dealing with some common phenomena, notably with the motion of projectiles. While Aristotelian science favored loving attention to detail, through which alone one could succeed in conceiving the real in its full concreteness, Galileo and his followers conducted their research with scissors and blinkers: Scissors to cut off not just the subjective (§7.1) but also the irrelevant; blinkers to fix the sight on what mattered and prevent distraction by what was being – provisionally? – left aside. The natural processes and states of affairs under study were represented by simplified models, manageable instances of definite mathematical structures. The inevitable discrepancies between the predicted behavior of such models and the observed behavior of the objects they stood for were ascribed to "perturbations" and observation errors.

Galileo proclaimed that the book of nature is written in mathematical language (1623, §6; quoted in §1.3). However, natural phenomena do not wear the script on their face. Kepler did not read off the sky the concept and the properties of the ellipse to which he fitted the observed positions of Mars; he found them in Apollonius's *Conica*. It was all right for Kepler to repeat that "God is always doing geometry".[29] But how do we mortals get to know the divine geometry of things? Philosophers of mathematics are still bedeviled by this question. However, in the seventeenth century the following reply, given by Kepler, was not immediately ruled out of court:

> Geometry, being part of the divine mind from time immemorial, from before the origin of things, being God Himself (for what is in God that was not God Himself), has supplied God with the models for the creation of the world and has been transferred to man together with the image of God.
>
> (Kepler GW, VI, 223; quoted in Caspar 1993, p. 271)

This is not too far from Descartes's thesis that we know bodies through our inborn, "clear and distinct" idea of extension; after all, innate ideas are supposedly implanted in us in the act of creation directly by God Himself. And Malebranche's contention that we "see" all corporeal things "in God" and are thus directly acquainted with their mathematical patterning may well pass for a more precise – and more daring – restatement of Kepler's.

Today few would countenance basing physics on our knowledge of God. However, one implication of seventeenth-century theological commitments has lingered on as a source of confusion. Like the smile

---

[29] τὸν θεὸν ἀεὶ γεωμετρεῖν. The Greek phrase, fondly quoted by Kepler (GW, VI, 298), was attributed in Antiquity to Plato; see Plutarch (*Qu. Conv.* 8.2).

of the Cheshire cat, the idea of a ready-made world continues to haunt a philosophical tradition from which the idea of its Maker has long vanished. Although everyone is aware that physicists work on selected aspects of reality using idealized models that cannot claim perfect adequacy, the dream persists of a final theory of everything representing the true mathematical structure of the universe.

Kepler believed that he had discovered that planets really describe ellipses with the Sun at one focus. He ascribed to observational inaccuracy the discrepancies between the elliptical path that he traced for Mars and the positions recorded by Tycho Brahe. With the introduction of the telescope – in Kepler's own lifetime – the quality of astronomic observation improved substantially and the said excuse was no longer viable. However, the theory of universal gravitation put forward in Newton's *Principia* (1687) satisfactorily explained both the successes and the failings of Kepler's "laws" (§2.3). In the course of the two following centuries Newtonian celestial mechanics was able to account, within the very narrow margins of error admissible in telescopic astronomy, for every motion tracked in the Solar System (except a small fraction of Mercury's perihelion advance). So physicists and astronomers got into the habit of thinking that Newton's law of gravity was exactly obeyed by every bit of matter, although all its applications to particular celestial bodies involved simplified models of the objects in question.

This understanding of physical theory weighed heavily on Kant's decision "to deny knowledge so as to make room for faith" (1787, p. xxx). He accepted the exact validity of Newtonian mechanics for *phenomena*, and even argued that some form of instantaneous action at a distance was a necessary prerequisite of human experience. But he denied that any branch of human science could apply to the *things-in-themselves* on which – he assumed – the phenomena are grounded. In this way, our rational will could have free play in the realm of the really real, even if it was excluded from everything we can see and touch by the universal determinism that Kant and his contemporaries believed was entailed by the exact validity of mechanics. What solace a serious moral thinker can get from this strange doctrine is not a subject I can discuss here. What matters to us is Kant's philosophy of phenomena. Although I already dealt with it at some length in Chapter Three, there are a few points that I wish to emphasize now.

First, Kant detheologized physics: No matter what its accuracy, its sole concern are the objects of human experience, subject to the specific conditions of human sensibility (space and time), and therefore it cannot in any way approach a God's eye view of reality.

Second, there is some tension between Kant's doctrine of the antinomies (§3.5) and his understanding of "nature" as "the connection of phenomena determining one another with necessity according to universal laws" (1787, p. 479). For him, the "thoroughgoing connection" of events according to unchangeable laws is a principle "from which no deviation is allowed and no phenomenon can be exempted" (1787, pp. 564, 570). And yet the antinomies entail that *the whole of nature* is only an Idea (in Kant's sense) that is incapable of realization although it is indispensable as a guiding principle of science (cf. 1783, §§40 and 56). So the connection of mutually determined phenomena cannot be thoroughgoing, but only on the way of becoming so. Indeed, Kant's solution to the First and Second Antinomies rests on the premise that phenomena are not fully determinate in every respect (*omnimode determinata*) like things-in-themselves, but instead they gradually approach determinacy in the course of the progressive construction of experience by the human understanding.

Finally, and most significantly for us, Kant offered a novel, powerful account of the use of mathematical concepts in grasping physical facts that does not depend on God's geometrizing nor on our being made privy to God's plan of creation. Concepts are needed to "spell out appearances in order to read them as experience" (1781, p. 314), in other words, to articulate the fugitive flickerings of sensory awareness into a steady display of objects, connected in space and lasting through time. As I noted in §3.4.1, this Kantian doctrine of the role of concepts in the constitution of objectivity has two sides. On the one hand, all concepts carry necessity with them and therefore, on Kant's account, bring it into the facts that are constituted and grasped by means of them. On the other hand, according to Kant, the very general concepts listed in his table of categories are themselves specifically necessary for there to be any facts. This second side of Kant's doctrine mattered most to him but is hard to defend now that physics has discarded the concepts and principles that Kant incorporated into his "metaphysics of experience". Philosophical attempts to salvage Kant's conception of a changeless human reason by watering it down and moving to weaker principles and more abstract categories than his do not look promising. One cannot easily forget the case of Bertrand Russell (1897) who, renouncing Kant's fixation with Euclidian geometry, equated the "form of externality" with the nonmetric projective space that, according to Felix Klein, supported equally well the parabolic (i.e., Euclidian), hyperbolic (Lobachevskian), and elliptic metric geometries (see §4.1.2). It then turned out that when physical geometry was actually revolutionized by Einstein the new metric adopted was none of the above and would not fit into Russell's scheme (or Klein's). A more likely way of preserving the gist of Kant's teaching was tried by Strawson. His "descriptive metaphysics" concerns the "massive central core of human thinking which has no history – or none recorded in histories of thought". This core comprises "categories and concepts which, in their most fundamental character, change not at all. Obviously these are not the specialties of the most refined thinking. They are the commonplaces of the least refined thinking; and are yet the indispensable core of the conceptual equipment of the most sophisticated human beings" (Strawson 1959, p. 10). Strawson's "massive core" may perhaps be equated with the commonsense background from which physical theories and other such intellectual systems grow and across which they communicate. But then such a background core is not so frightfully stable as Strawson would have it, and it does not enjoy any privilege over physical theories when it comes to setting the terms in which natural phenomena are to be described and understood.

Kant's changeless scheme of forms, categories, and Ideas suggests that he still believed that our species was created at one stroke by God (although he would not indulge in asserting this as a scientific or philosophical truth). After Darwin, we conceive human reason as a gradually evolving complex of various, possibly colliding, currents, that must be investigated by the hermeneutic methods of cultural anthropology and intellectual history, not by phenomenological introspection and transcendental deduction. The idea that reason is open-ended encourages the exercise of scientific freedom. The loss of certainty is more than compensated for by improved efficacy, as anyone can readily verify by comparing the last 400 years of physics with the preceding 2,000. Scientists remain bound by tradition but are free to radically reshape it. In this sense, they face the surprises of experience with unfettered minds. No longer constrained by the purportedly unconditional validity of their principles, they may turn to their genius or muse for the concepts wanted to grasp the facts.

Einstein was probably the first physicist to become fully aware of this freedom. As he put it, the "concepts and fundamental laws" of physics are "free inventions of the human mind (*freie Erfindungen des menschlichen Geistes*), which cannot be justified a priori by the nature of the human mind nor in any other way" (1933, p. 180). Physics "is in a state of evolution", and its "basis cannot be obtained through distillation by any inductive method from the experiences lived through, but [. . .] can only be attained by free invention" (1936, p. 96). Einstein noted that "Newton, the first creator of a comprehensive, operative system of theoretical physics, still believed that the fundamental concepts and fundamental laws of his system could be derived from experience" (1933, p. 181). Newton surely felt uncomfortable about the concept of absolute space, for he must have seen that nothing in experience appeared to correspond to it. He also felt uneasy about forces acting at a distance. "But the enormous practical success of his theory may well have prevented him and the physicists of the eighteenth and nineteenth centuries from perceiving the fictional character of the foundations of his system" (Ibid.). They were persuaded that "the fundamental concepts and fundamental laws of physics are not in a logical sense free inventions of the human mind, but can be derived from experiments by 'abstraction', i.e., by a logical procedure" (p. 182).

> Indeed, clear knowledge of the incorrectness of this view was first procured by the general theory of relativity. It showed that one could do justice to all the relevant empirical facts on a footing (*Fundament*) vastly different from Newton's, in an even more satisfactory and complete way than on his. But quite apart from the question of superiority, the fictional character of foundations (*Grundlagen*) was made perfectly evident by the fact that two essentially different foundations can be exhibited, both of which amply agree with experience; this proves at any rate that every attempt at a logical derivation of the fundamental concepts and fundamental laws of mechanics from elementary experiences is doomed to failure.
>
> (Einstein 1933, p. 182)

Like genetic mutations, the concepts of science come out of the blue. While mutations have been traced – "in principle" – to chemical changes attributable to random quantum interactions, we have no inkling of the origin of concepts. To suggest that the advent of Special Relativity – that is, the switch from Galilei invariance to Lorentz invariance – is related to modern aesthetic and moral relativism, or that Heisenberg's "uncertainty" relations reflect the monetary and political instability of the Weimar Republic, makes for entertaining science journalism but throws no light on the creation of scientific ideas. But, of course, what matters most is not the source of our thinking, but its aptness. So philosophers of science traditionally pay great attention to this question: How can we tell that our concepts fit the facts that we intend to grasp by means of them?

For Kepler and his contemporaries this amounted to: How can we know that the concepts of science are the same that God had in mind when he created the world? Using the metaphor that modern science unwittingly inherited from Callicles, via the Church fathers, the same question is often stated thus: How can we establish that a general statement of fact expresses a true law of nature? Having realized that we plainly cannot do it, philosophers set out to find clear rules and firm criteria by which to evaluate the *probability* that such general statements might express true laws, given the available evidence.

Research in this area – known as *inductive logic*, but also, more modestly, as *confirmation theory* – is one of the blindest alleys in twentieth-century philosophy. Until the 1950s it mainly turned around the logical notion of probability. Keynes (1921) conceived probability as a quantitative relation between statements based on their meanings. A similar idea was adumbrated by Wittgenstein (1922, 5.15–5.156). From this standpoint, given any two statements h and e, the probability that h is true if e is true is a real number $p(h,e)$ in the closed interval [0,1], which depends solely on what h and e say. Thus, a statement such as '$p(h,e) = 0.75$' is a logical truth. If h is a scientific hypothesis and e is the conjunction of all the available evidence, $p(h,e)$ can be regarded as a measure of confirmation of h by e. Clearly, a relation of this sort can only be made precise for statements couched in a formal language, with fixed, explicit syntactic and semantic rules. Carnap (1945a, 1945b, 1950, 1952) struggled steadfastly to meet this requirement. He was able to define logical probability only for languages too simple to be of much use in science, under fairly arbitrary conventions regarding the probability of basic alternative states of affairs. Most frustratingly, every function p in Carnap's "continuum of inductive methods" takes the value $p(h,e) = 0$ if h states a hypothetical law of nature and e is a finite conjunction of particular statements of fact.

In the second half of the twentieth century a different approach to the confirmation of scientific hypotheses by empirical evidence has won increasing support among philosophers. Known as *Bayesianism* – after Thomas Bayes, an eighteenth-century English cleric and statistician –, it actually turns on the conception of subjective probability independently introduced by Frank Ramsey (1926) and Bruno de Finetti (1930, 1931, 1937). This can be roughly summarized as follows. A real number $b_r(s)$ is assigned to the "degree of belief" of a person r in a statement s. This quantitative assignment reflects the odds that r is willing to give or take in a bet concerning s. Specifically, r will be ready to pay, if s comes true, $b_r(\text{not-}s)/b_r(s)$ times the amount she is to receive if s comes false. Ramsey and de Finetti proved that such betting behavior would lead to a loss, come what may, unless the function b_r complies with the following conditions:

B1 For any statement s, $0 \leq b_r(s) \leq 1$.

B2 For any two mutually incompatible statements s and t, $b_r(s \vee t) = b_r(s) + b_r(s)$.[30]

B3 In particular, $b_r(s \vee \text{not-}s) = 1$.

Therefore, our authors concluded, a rational person must adjust its betting rates or "degrees of belief" so that they agree with conditions B1–B3.[31] Suppose now that b_r is restricted to statements that assert the occurrence of some event or are truth-functions or first-order generalizations formed from such statements. Let S denote the set of events that occur if and only if statement s is true. If b_r complies with B1–B3, the function p, defined by $p(S) = b_r(s)$ for any such statement s, is a probability function on events, namely, r's personal or *subjective probability function*. Bruno de Finetti argued that this is the only concept of probability that makes sense and is not sheer fantasy (1974, I, 3–4). He was careful to insist that probability thus conceived applies to events only (Ibid., I, 139), and indeed only to *uncertain* – that is, future or unknown – events. Run-of-the-mill Bayesians are much less fastidious, and they are ready to speak about the subjective probability of major physical theories, such as General Relativity, and to evaluate how it changes in the light of incoming data. Since my aim here is not to defend Bayesianism, but just to cursorily inform about it, I take for granted that this manner of speaking is not plain rubbish, and I shall

---

[30] $s \vee t$ is the statement that is true unless both s and t are false. To express in English the exact force of the connective $\vee$, bureaucrats have created the conjunction 'and/or'.

[31] Let me note, in passing, that if this linkage between rationality and betting behavior holds good, a rational person must know every logical and mathematical truth. For if s is such a truth and r is uncertain about s, so that $b_r(s) < 1$, then, under the usual assumptions of the Ramsey–de Finetti argument, r can be cornered into making bets she must lose come what may.

spend no effort in trying to justify it. Therefore, like the Bayesians, I write $p(s)$ for the subjective probability of statement s even if s is not restricted in the way I indicated above. Following the usual practice in the theory of probability, if $p(s) > 0$, one defines the subjective probability $p(t|s)$ of statement t conditional on (the truth of) s by:

$$p(t|s) = \frac{p(t \wedge s)}{p(s)} \qquad (7.7)^{32}$$

If $p(t) > 0$,

$$p(s|t) = \frac{p(t \wedge s)}{p(t)} \qquad (7.8)$$

Therefore,

$$p(s|t) = \frac{p(t|s)p(s)}{p(t)} \qquad (7.9)$$

This result is the simplest form of Bayes's Theorem, an elementary proposition of the mathematical theory of probability on which Bayesians build.[33] Although Bayesianism has grown ever more complex, as its advocates tease their brains to overcome the difficulties that beset it, its basic ideas are easy – not to say facile – and can be explained briefly. I shall focus on three. Let p stand for the subjective probability function of a particular person at a particular time; if it is necessary to distinguish successive times, I use numerical subscripts, thus: p_1, p_2, and so on.

(i) Let h be a scientific hypothesis and e a statement of a particular fact such that h entails e. Let $p(h)$ and $p(e)$ be both greater than 0 and less than 1. Clearly, $p(e|h) = 1$. Therefore, $p(h|e) = p(h)/p(e)$. Thus, the probability of h is necessarily less than the probability of h given e, if e is a logical consequence of h and $0 < p(e) < 1$. If these conditions are

---

[32] $t \wedge s$ is the statement that is false unless both s and t are true. The connective $\wedge$ is well represented by the English conjunction 'and'. To avoid the restriction of conditional probabilities to cases in which the condition has non-zero probability, some authors favor taking *conditional probability* as a primitive concept, to be characterized by axioms. These axioms then replace the standard axiomatic characterization of absolute probability.

[33] Bayes's Theorem takes its name from Thomas Bayes not because he discovered it, but because of the way he wielded it in his famous paper of 1763. Earman (1992, Ch. 1) gives a sharp and most instructive assessment of the relevance of Bayes's work to contemporary Bayesianism.

met, we say that e is confirmatory of h. This result, and others like it, are touted by Bayesians as providing a satisfactory philosophical justification of the commonplace view that confirmation by empirical evidence increases the credibility of hypotheses in the eyes of rational agents. However, the preceding argument shows that $p(h|e) > p(h)$ only if e is still uncertain, that is, if the evidence in question is not yet available; for, if e were known to the subject of the probability function p, $p(e)$ would equal 1, in which case $p(h|e) = p(h)$.

(ii) Thus, the Bayesian proof that confirmatory evidence improves the credibility of scientific hypotheses cannot simply rest on a result like (i), which is solidly grounded on the mathematical theory of probability, but depends also on the following *principle of conditionalization*: If p_1 and p_2 are the probability functions of a rational agent right before and right after the acquisition of new knowledge e, then $p_2(h) = p_1(h|e)$ for every hypothesis h such that $0 < p_1(h) < 1$.[34] There have been clever attempts to prove that agents who do not comply with the principle of conditionalization can be constrained to make bets that they will lose come what may (Teller 1973; Skyrms 1987). If this is so, conditionalization is, arguably, a necessary ingredient of rationality. However, if the new knowledge e lies outside the domain of p_1 or, worse still, if it provokes a change in the very concepts from which statements are built, the principle of conditionalization cannot be applied.

(iii) The principle of conditionalization implies that rational agents will bet with increasing confidence for hypotheses that are persistently and consistently supported by incoming evidence, but not that they will achieve certainty about such hypotheses, nor that they will eventually agree on the betting rates. So, if (i) and (ii) were all there is to it, Bayesianism would cut a poor figure as a philosophical theory of the growth of knowledge, even if one turns a blind eye on the essential restriction to conditionalization mentioned at the end of (ii). Therefore, Bayesians take special pride in the mathematical theorems which establish that, under certain conditions, $p_n(h)$ converges to 1 as n grows beyond all bounds if, for each n, the transition from p_n to p_{n+1} satisfies the principle of conditionalization as applied to new evidence

---

[34] Of course, in real life a global revision of one's subjective probability function may be brought along not by the acquisition of some new certainty, but by a local change in the said function. This possibility is taken care of by Jeffrey conditionalization (so called after Richard Jeffrey). See Earman (1992, p. 34).

confirmatory of *h*. (See in particular Gaifman and Snir 1982, and the comments in Earman 1992, Chapter 6.) This implies, in turn, that all rational agents will ultimately agree. Still, the rate of convergence need not be the same for different agents, and the theorems provide no hint that might enable one to estimate this rate. They assure us only that the subjective beliefs of rational agents converge in the long run to practical certainty and unanimity. But then, as Keynes dryly remarked, in the long run we shall all be dead. Moreover, as I noted at the end of (ii), the Bayesian school, for all its mathematical sophistication, remains committed to the feckless assumption that concepts and meanings are fixed and that a rational agent will not be moved by empirical evidence to *see* things in a fundamentally different way.

Fortunately, the advancement of physics does not depend on the convergent evolution of the subjective probability functions entertained by rational human beings. It is enough that physics continues to produce ever more comprehensive theories for successfully conceiving the great families of phenomena we are wont to discern in nature. We cannot judge a physical theory by the way it apes God's view of its subject matter, for we have no idea how well it does it, or if it does it at all. But this does not mean that we have to measure a theory's success by how close to 1 is its average subjective probability, or by how fast this average is increasing via conditionalization on new evidence.[35] From the standpoint reached in §7.2, it makes no sense to ask about the probability of, say, Poisson's equation. For a Newtonian gravitational field, the equation holds by definition. A meaningful question is whether this body here – say, a comet, or a Saturnian moon – moves through such a field, that is, whether it is part of a Newtonian gravitational system or, more precisely, whether it, together with the whole relevant fragment of reality, can be suitably modeled by a Newtonian system. (Here 'relevant' means 'that must be taken into account in order to understand and predict those aspects of the body's behavior that interest us at present'; and 'suitably' means 'to within the admissible margin of inaccuracy'.) Analogous questions are being asked all the time in everyday life. Kant postulated a special faculty for answer-

---

[35] With these remarks I do not intend to belittle the importance of statistical inference in physics. For a powerful and mind-expanding statement of how it works in real science I refer the reader to Mayo (1996). Mayo's "error statistics" continues the tradition of C. S. Peirce and Egon Pearson, and keeps clear of the "Bayesian Way".

ing them, which he called 'judgment' (*Urteilskraft*).[36] In the enterprise of grasping the facts, it relies on whatever concept or combination of concepts looks most promising, yet it stands ready to replace them by other, more befitting ones if and when they become available. Everyday concepts, however, are flexible in a way that physical theories are not. According to the conventional wisdom, "if it walks like a duck, it swims like a duck, it quacks like a duck, then it *is* a duck". And a duck's deviant behavior will not easily lead one to reclassify it as something else. (More probably, one will pronounce it "mentally ill", or "acting under stress".) On the other hand, the unexplained perihelion advance of Mercury by a mere 43″ per century – less than eight-thousandths of the amount observed – is enough to conclude that neither the Sun–Mercury system, nor the entire Solar System, nor, for that matter, any gravitational system in the world can be represented to our full satisfaction by a Newtonian model.[37] Due to their great rigidity physical theories cannot supply us with resilient articles of belief that could vie in endurance with myths and superstitions. But this feature of mathematical physics is also a great source of power for those who wield it with circumspection to find their way about the world.

[36] "Lack of judgment is properly that, which is called stupidity (*Dummheit*), and for such infirmity there is no remedy" (Kant 1781, p. 133n). Certainly no algorithm to "wash it out".

[37] To be fair, I should recall that (i) the 43″ secular perihelion advance was not felt to be critical to the Newtonian theory of gravity until a viable replacement was available, viz., GR; and (ii) although Einstein had the Mercury anomaly in mind since he began working on a new theory of gravity (see note 64 in Chapter Five), his main motive for replacing Newton's theory was that it is not Lorentz invariant.

Supplements

I. On Vectors

1   **Set-theoretic concepts.** Let S be a set. We write $x \in S$ to indicate that x is an element of S. A set R is said to be a *subset* of S (symbolically: $R \subseteq S$) if every element of R is an element of S. The set of all subsets of a set S is called the *power set* of S and is denoted by $\mathscr{P}S$. Set-theoreticians assume that, given any set S, the set $\mathscr{P}S$ is also given. If A and B are sets, the *Cartesian product* $A \times B$ is the set of all ordered pairs $\langle a,b \rangle$, such that a belongs to A and b belongs to B. We write A^2 for $A \times A$, A^3 for $(A \times A) \times A$, and A^n for $A^{n-1} \times A$. If A and B are sets, a *mapping* f of A into B is a correspondence that assigns to each element $x \in A$ a unique element $f(x)$ of B. I usually refer to the correspondence thus described as 'the mapping $f: A \to B$ by $x \mapsto f(x)$'. f is said to be a *one–one* or *injective* mapping if, for any x and y, $f(x) = f(y)$ entails that $x = y$. f is said to be *surjective* if it maps A *onto* B, that is, if every element of B is assigned by f to some element of A. f is *bijective* if it is both injective and surjective (i.e., if it is one–one and onto).[1]

2   **Groups and fields.** A *group* is a set G that contains a distinguished *neutral* element e and on which two operations are defined, viz., (I) the *group product*, which is a mapping of $G \times G$ into G by $\langle a,b \rangle \mapsto ab$, such

---

[1] For those who have difficulty understanding the intuitive notions of 'assignment' and 'correspondence', mathematicians have contrived a definition of 'mapping' that is based solely on the concept of 'set-membership'. Here it goes. We begin with the purely set-theoretical definition of an ordered pair, proposed by Kuratowski (1921): The ordered pair $\langle x,y \rangle$ is the set $\{\{x\},\{x,y\}\}$. With this in hand, we define the ordered n-tuple $\langle x_1, \ldots, x_n \rangle$, for any integer $n > 2$, to be the ordered pair $\langle \langle x_1, \ldots, x_{n-1} \rangle, x_n \rangle$. Let A and B be sets. The mapping $f: A \to B$ by $x \mapsto f(x)$ is the ordered triple $\langle A,B,C \rangle$, where C is a set of ordered pairs $\langle x,f(x) \rangle$, such that $x \in A$, $f(x) \in B$, and each element of A occurs in one and only one ordered pair in C.

that (I.i) for any a in G, $ae = ea = a$, and (I.ii) for any a, b, c in G, $a(bc)$ $= (ab)c$; and (II) the *inverse operation*, a mapping of G into G that assigns to each a an element a^{-1} (the *inverse* of a), such that $aa^{-1} = a^{-1}a = e$. A group is said to be *abelian* if, for any group elements a and b, $ab = ba$. For further clarifications and examples, see §4.1.2.

A *field* can be readily described as a set F that incorporates two abelian groups. (I) The *additive group* consists of F itself, with a group product that is called *addition* and is symbolized by $+$; the neutral element is called 'zero' and denoted by 0; the inverse of an element a is denoted by $-a$. (II) The *multiplicative group* consists of $F\backslash\{0\}$ (i.e., what remains of F when one removes 0); the group product is called *multiplication* and symbolized by $\times$, or by $\cdot$, or by mere juxtaposition ($a \times b$ $= a \cdot b = ab$); the neutral element is called 'one' and is denoted by 1; the inverse of an element a is denoted by a^{-1} or by $1/a$. Multiplication is extended to the whole of F by the rule: $a \times 0 = 0 \times a = 0$ for any a in F. Addition and multiplication are meshed together by the following law of distribution: For any a, b, c in F, $a(b + c) = ab + ac$. The most frequently encountered fields are the field $\mathbb{Q}$ of rationals, the real number field $\mathbb{R}$, and the complex number field $\mathbb{C}$.

3   **Vector spaces.** The concept of a *vector space* is one of the most important in modern mathematical physics. Let $\mathcal{V}$ be an abelian group, and let $\mathbb{F}$ be a field (in physics usually either $\mathbb{R}$ or $\mathbb{C}$). The elements of $\mathcal{V}$ are called *vectors*; in this supplement, I represent them with boldfaced characters (in particular, **0** denotes the neutral element). The group product of $\mathcal{V}$ is called 'addition' and is symbolized by $+$. The inverse of $\mathbf{v} \in \mathcal{V}$ is denoted by $-\mathbf{v}$. The elements of $\mathbb{F}$ are called *scalars*. Arbitrary scalars will be represented with lowercase italics. We define a mapping of $\mathbb{F} \times \mathcal{V}$ into $\mathcal{V}$, called *scalar multiplication*, that assigns to each scalar a and vector $\mathbf{v}$ a vector $a\mathbf{v}$. $\mathcal{V}$ is a *vector space* over $\mathbb{F}$ – a *real vector space* if $\mathbb{F} = \mathbb{R}$, and a *complex vector space* if $\mathbb{F} = \mathbb{C}$ – if the addition of vectors meshes with the multiplication by scalars according to the following rules: (i) $a(\mathbf{v} + \mathbf{w}) = a\mathbf{v} + a\mathbf{w}$; (ii) $(a + b)\mathbf{v} = a\mathbf{v} + b\mathbf{v}$; (iii) $a(b\mathbf{v}) = (ab)\mathbf{v}$; (iv) $1\mathbf{v} = \mathbf{v}$ (where 1 denotes the one of $\mathbb{F}$). Let $\mathcal{W}$ be a nonempty subset of $\mathcal{V}$, closed under vector addition and scalar multiplication (i.e., such that, for any $\mathbf{v}$ and $\mathbf{w}$ in $\mathcal{W}$ and any a in $\mathbb{F}$, $\mathbf{v} + \mathbf{w}$ and $a\mathbf{v}$ are in $\mathcal{W}$). Then $\mathcal{W}$ – with the operations of $\mathcal{V}$ restricted to $\mathcal{W}$ – is a vector space over $\mathbb{F}$ on its own right and is said to be a *subspace* of $\mathcal{V}$. (*Exercise*: Prove that **0** belongs to $\mathcal{W}$.)

Given n distinct vectors $\mathbf{v}_1$, $\mathbf{v}_2$, . . . , $\mathbf{v}_n$ and any scalars a_1, a_2, . . . , a_n, the vector $\mathbf{v} = a_1\mathbf{v}_1 + a_2\mathbf{v}_2 + . . . + a_n\mathbf{v}_n$ is said to be a *linear combination* of the $\mathbf{v}_1$, $\mathbf{v}_2$, . . . , $\mathbf{v}_n$. The vectors $\mathbf{v}_1$, $\mathbf{v}_2$, . . . , $\mathbf{v}_n$ *span* the subset of $\mathcal{V}$ formed by all their linear combinations. $\mathbf{v}_1$, $\mathbf{v}_2$, . . . , $\mathbf{v}_n$ are *linearly inde-*

pendent if $a_1v_1 + a_2v_2 + \ldots + a_nv_n = 0$ only if $a_1 = a_2 = \ldots = a_n = 0$. The phrase 'linearly independent' conveys the fact that no vector belonging to such a set is a linear combination of the others. (*Exercise:* Infer this fact from the foregoing definition.) A finite set B of linearly independent vectors is a *basis* of $\mathcal{V}$ if $\mathcal{V}$ is spanned by B. If $B = \{v_1, v_2, \ldots, v_n\}$ is a basis and $v = \Sigma_{i=1}^{n} a_i v_i$, the scalars a_i are the *components* of v *relative to* B. Suppose that $\mathcal{V}$ has a basis formed by n vectors; this implies that no set of linearly independent vectors contains more than n vectors. Thus, every basis of $\mathcal{V}$ has exactly n vectors. We say then that $\mathcal{V}$ is an n-dimensional vector space. Given any integer n, all n-dimensional vector spaces over the same field $\mathbb{F}$ are isomorphic, that is, there are bijective mappings of each onto the others that preserve vector addition and multiplication by scalars. If no finite collection of vectors is a basis, $\mathcal{V}$ is said to be infinite-dimensional. In I.9 we shall see how the concept of basis can be extended to infinite-dimensional spaces.

4   **Inner product spaces.** A real or complex vector space $\mathcal{V}$ is an *inner product space* if it is endowed with an inner product, that is, a mapping that assigns to each pair of vectors v and w a scalar subject to the conditions IP listed below, and which I shall denote by $\langle v|w \rangle$. Let the field of scalars be $\mathbb{C}$, and let x^* stand for the complex conjugate of x (in other words, if $x = a + ib$, where a and b are real numbers, then $x^* = a - ib$). The inner product on $\mathcal{V}$ obeys the following rules, for any vectors u, v and w and any scalars a and b:

IP1 $\langle v|w \rangle = \langle w|v \rangle^*$;
IP2 $\langle v|aw + bu \rangle = a\langle v|w \rangle + b\langle v|u \rangle$;
IP3 $\langle v|v \rangle \geq 0$, equality holding if and only if $v = 0$.

IP1 and IP2 jointly entail that $\langle av|w \rangle = a^*\langle v|w \rangle$. If the field of scalars is $\mathbb{R}$, rule (i) naturally becomes $\langle v|w \rangle = \langle w|v \rangle$. Other notations often used for the inner product, instead of $\langle v|w \rangle$, are $v \cdot w$ and (v,w).

The *length* of a vector v in $\mathcal{V}$ is defined by $\|v\| = \sqrt{\langle v|v \rangle}$. Note that $\|v\|$ is necessarily a real number (by IP1). The *distance* between two vectors v and w is then given by $\|v - w\|$ (I write $v - w$ for $v + -w$). The mapping $v \mapsto \|v\|$ evidently satisfies conditions N1–N4 in §4.1.3 and is therefore a norm in $\mathcal{V}$. It is assumed that $\mathcal{V}$ has the topology induced by this norm, so that, for each v in $\mathcal{V}$ and ρ in $\mathbb{R}$ the set $\{u: u \in \mathcal{V}$ and $\|v - u\| < \rho\}$ is an open neighborhood of u. The inner product satisfies *Schwarz's inequality*: $|\langle v|w \rangle| \leq \|v\| \cdot \|w\|$.

By using the inner product, one defines the *angle* between two vectors a and b as

$$\sphericalangle(a,b) = \arccos\left(\frac{\langle a|b \rangle}{\|a\| \cdot \|b\|}\right)$$

This definition has a natural motivation in the space of directed segments or "arrows" in Euclidian space, which is of course the prototype of all vector spaces: Let **a** and **b** be two such arrows issuing from a point P; assume that $\|\mathbf{a}\|$ measures Euclidian length. Then, by the trigonometric "law of cosines", $\|\mathbf{a} - \mathbf{b}\|^2 = \|\mathbf{a}\|^2 + \|\mathbf{b}\|^2 - 2\|\mathbf{a}\| \cdot \|\mathbf{b}\| \cos\sphericalangle(\mathbf{a},\mathbf{b})$; so, to obtain $\|\mathbf{a} - \mathbf{b}\|^2 = \|\mathbf{a}\|^2 - 2\langle \mathbf{a}|\mathbf{b}\rangle + \|\mathbf{b}\|^2$ we must put

$$\cos\sphericalangle(\mathbf{a},\mathbf{b}) = \frac{\langle \mathbf{a}|\mathbf{b}\rangle}{\|\mathbf{a}\| \cdot \|\mathbf{b}\|}$$

The definition of angle entails that, for any scalar m and vector **v**, $\cos\sphericalangle(\mathbf{a},m\mathbf{a}) = 1$; therefore, $\sphericalangle(\mathbf{a},m\mathbf{a}) = 0$. Thus, it is fair to say that multiplication by a scalar preserves the "direction" of vectors. If $\cos\sphericalangle(\mathbf{a},\mathbf{b}) = 0$, $\sphericalangle(\mathbf{a},\mathbf{b}) = \pi/2$ and vectors **a** and **b** are said to be mutually perpendicular or *orthogonal*. This happens if and only if $\langle \mathbf{a}|\mathbf{b}\rangle = 0$. A set of vectors $(\mathbf{v}_i)$, indexed by a set of indices Ω, is said to be *orthonormal* if, for all indices i, j in Ω, $\|\mathbf{v}_i\| = 1$ and $\langle \mathbf{v}_i|\mathbf{v}_j\rangle = 0$ whenever $i \neq j$. An orthonormal set is said to be *complete* if it is not contained in any other such set. (*Exercise*: Prove that every orthonormal set is a linearly independent set.)

5   **Linear and multilinear functions.** Let $\mathcal{V}$ be a vector space over a field $\mathbb{F}$. A linear function on $\mathcal{V}$ is a mapping $\varphi: \mathcal{V} \to \mathbb{F}$ such that, for any **v** and **w** in $\mathcal{V}$ and any a and b in $\mathbb{F}$, $\varphi(a\mathbf{v} + b\mathbf{w}) = a\varphi(\mathbf{v}) + b\varphi(\mathbf{w})$. If φ and ψ are two linear functions on $\mathcal{V}$, their sum $\varphi + \psi$ is defined by the condition: $(\varphi + \psi)(\mathbf{v}) = \varphi(\mathbf{v}) + \psi(\mathbf{v})$ for every **v** in $\mathcal{V}$. Clearly, $\varphi + \psi$ is a linear function on $\mathcal{V}$, for $(\varphi + \psi)(a\mathbf{v} + b\mathbf{w}) = \varphi(a\mathbf{v} + b\mathbf{w}) + \psi(a\mathbf{v} + b\mathbf{w}) = a\varphi(\mathbf{v}) + b\varphi(\mathbf{w}) + a\psi(\mathbf{v}) + b\psi(\mathbf{w}) = a(\varphi(\mathbf{v}) + \psi(\mathbf{v})) + b(\varphi(\mathbf{w}) + \psi(\mathbf{w})) = a(\varphi + \psi)(\mathbf{v}) + b(\varphi + \psi)(\mathbf{w})$. Let the product $a\varphi$ of the linear function φ by the scalar a be defined by $(a\varphi)(\mathbf{v}) = a\varphi(\mathbf{v})$ for every **v** in $\mathcal{V}$. $a\varphi$ is also a linear function. It is an easy exercise to show that the set of linear functions on $\mathcal{V}$, with these operations, is a vector space over $\mathbb{F}$. We denote it by $\mathcal{V}^*$ and call it the *dual* of $\mathcal{V}$. (Show, in particular, that the function which assigns the 0 of $\mathbb{F}$ to every vector in $\mathcal{V}$ is linear and performs as the 0 of $\mathcal{V}^*$.)

Let $\mathcal{V}$ be an inner product space. If **v** is any vector in $\mathcal{V}$, the mapping of $\mathcal{V}$ into $\mathbb{F}$ by $\mathbf{u} \mapsto \langle \mathbf{v}|\mathbf{u}\rangle$ is a linear function (by IP2) and so belongs to $\mathcal{V}^*$. On the other hand, it can be proved that, for each φ in $\mathcal{V}^*$, there is a unique **v** in $\mathcal{V}$ such that, for any **u** in $\mathcal{V}$, $\varphi(\mathbf{u}) = \langle \mathbf{v}|\mathbf{u}\rangle$. Thus, there is a one–one correspondence between the vectors in $\mathcal{V}$ and the linear functions in $\mathcal{V}^*$. This is the origin of the notation that I use for the inner product. It was introduced by the physicist Dirac, who denoted an arbitrary vector of a vector space $\mathcal{V}$ by $|\mathbf{u}\rangle$ and called it a *ket*, an arbitrary vector of the dual space $\mathcal{V}^*$ by $\langle \mathbf{v}|$ and called it a *bra*, and let the *bra[c]ket*

$\langle v|u \rangle$ stand for the value of $\langle v|$ at $|u \rangle$. If $\langle v|$ is precisely the bra that the said one–one correspondence assigns to the ket $|v \rangle$, it turns out that $\langle v|u \rangle$ is identical with the inner product of $|v \rangle$ and $|u \rangle$.

An n-linear function on $\mathcal{V}$ is a mapping of $\mathcal{V}^n$ into $\mathbb{F}$ that is linear in each argument. For example, a *bi*linear function φ on $\mathcal{V}$ is a real-valued mapping defined on $\mathcal{V}^2$ that satisfies the following twofold condition of linearity: For every a and b in $\mathbb{F}$ and every u, v, and w in $\mathcal{V}$, $\varphi(au + bv, w) = a\varphi(u,w) + b\varphi(v,w)$ and $\varphi(u, av + bw) = a\varphi(u,v) + b\varphi(u,w)$. This definition can be readily extended to scalar-valued mappings defined on $\mathcal{V}^* \times \mathcal{V}^*$, on $\mathcal{V} \times \mathcal{V}^*$, and on $\mathcal{V}^* \times \mathcal{V}$. The last two would illustrate the notion of a mixed bilinear function on $\mathcal{V}$ *and* $\mathcal{V}^*$.

6   **Linear operators and matrices.** Let $\mathcal{V}$ and $\mathcal{W}$ be vector spaces over a field $\mathbb{F}$. A *linear operator* $\mathcal{V} \rightarrow \mathcal{W}$ is a mapping A of $\mathcal{V}$ into $\mathcal{W}$ such that, for any v and w in $\mathcal{V}$ and any a and b in $\mathbb{F}$, $A(av + bw) = A(av) + A(bw)$. This implies, in particular, that $A(0) = 0$. Linear operators can be added together and multiplied by scalars, much like linear functions. Thus, if A and B are linear operators $\mathcal{V} \rightarrow \mathcal{W}$ and a is any scalar, aA and $A + B$ are the mappings of $\mathcal{V}$ into $\mathcal{V}$ such that, for any v in $\mathcal{V}$, $(aA)(v) = A(av)$ and $(A + B)(v) = A(v) + B(v)$, respectively. Both aA and $A + B$ are linear operators $\mathcal{V} \rightarrow \mathcal{W}$, as the reader will readily show. Hereafter I write Av for $A(v)$.

In the remainder of this section, A, B, $C \ldots$ denote linear operators $\mathcal{V} \rightarrow \mathcal{V}$. In particular, I denotes the identity operator, defined by $Iv = v$ for each v in $\mathcal{V}$. If there is a $v \neq 0$ in $\mathcal{V}$ such that $Av = av$, that is, if A merely multiplies the vector v by the scalar a, we say that a is a *proper value* (or *eigenvalue*) of A and that v is the corresponding *proper vector* (or *eigenvector*). Note that if v is a proper vector of A and b is any scalar, bv is also a proper vector corresponding to the same proper value.[2] If two or more linearly independent vectors are proper vectors of A corresponding to the same proper value λ, A is said to be *degenerate*. Any nonzero linear combination of the said proper vectors is again a proper vector corresponding to λ. (Prove it!) The number of linearly independent proper vectors corresponding to λ is called the *multiplicity* or the *degree of degeneracy* of λ.

Assume now that $\mathcal{V}$ is finite-dimensional and let $V = \{v_1, v_2, \ldots, v_n\}$ be a basis of $\mathcal{V}$. Obviously, if A is a linear operator and we know Av_i for each v_i in V, we can determine Av for any vector v in $\mathcal{V}$, for we can calculate the components of Av relative to V from the components of v and of $Av_1, \ldots Av_n$, relative to V. Let $v = \Sigma_{i=1}^n a_i v_i$ and $Av_i = \Sigma_{j=1}^n A_{ji} v_j$ ($i = 1, \ldots, n$). Clearly,

---

[2] Let $Av = av$. Then $A(bv) = bAv = bav = a(bv)$.

$$\mathbf{A}\mathbf{v} = \sum_{i=1}^{n} a_i \mathbf{A}\mathbf{v}_i = \sum_{j=1}^{n} \sum_{i=1}^{n} A_{ji} a_i \mathbf{v}_j \qquad (\text{S1})$$

so the components of $\mathbf{A}(\mathbf{v})$ relative to V are $a_1' = \Sigma_{i=1}^{n} A_{1i} a_i$, $a_2' = \Sigma_{i=1}^{n} A_{2i} a_i$ ..., $a_n' = \Sigma_{i=1}^{n} A_{ni} a_i$. (Verify this for $n = 2$.) Thus, just like the vector $\mathbf{v}$ is represented, relative to the basis V, by the scalar n-tuple $\langle a_1, \ldots, a_n \rangle$ of its components, so the linear operator $\mathbf{A}$ can represented by the $n \times n$ scalar array

$$\begin{pmatrix} A_{11} & A_{12} & \ldots & A_{1n} \\ A_{21} & A_{22} & \ldots & A_{2n} \\ \ldots & \ldots & \ldots & \ldots \\ A_{n1} & A_{n2} & \ldots & A_{nn} \end{pmatrix}$$

This is the *matrix* of $\mathbf{A}$ in the basis V.

Generally speaking, an $n \times m$ matrix is an array of nm scalars arranged in n rows and m columns. Each scalar element of a matrix is labeled with two integers, the first of which indicates the element's row and the second its column. I designate a matrix by its typical element enclosed in parentheses, say, (A_{ij}). If $A_{ij} = 0$ whenever $i \neq j$, the matrix (A_{ij}) is said to be *diagonal*. The sum of two matrices (A_{ij}) and (B_{ij}) is the matrix $(A_{ij} + B_{ij})$ that is formed by adding each pair of elements with the same indices. Evidently, two matrices can be added only if both have the same number of rows and the same number of columns. We multiply a matrix by a number r (of the same kind as its elements) by multiplying each element by r: $r(A_{ij}) = (rA_{ij})$. The product of (A_{ij}) and (B_{ij}) – in that order – is the matrix

$$(A_{ij}) \times (B_{ij}) = (\Sigma_k A_{ik} B_{kj}) \qquad (\text{S2})$$

whose ij-th element is formed by multiplying the ith row of (A_{ij}), element-by-element, by the jth column of (B_{ij}) and adding together the products. Evidently, two matrices can be multiplied only if the first has as many columns as the second has rows. The $n \times n$ square matrix (δ_{ij}) – where $\delta_{ij} = 1$ if $i = j$ and 0 otherwise – acts as the "one" of matrix multiplication. We call it the *unit matrix* and denote it by $\mathbf{1}_n$, or simply by $\mathbf{1}$ (omitting the subscript does not usually create confusion). Note that an $n \times m$ matrix (A_{ij}) can be multiplied by $\mathbf{1}_m$ *on the left* and by $\mathbf{1}_n$ *on the right*; in either case, the product is (A_{ij}). Note further that $\mathbf{1}$ is the matrix of the identity operator $\mathbf{I}$ *in every basis*.

The rules of matrix addition and multiplication by a scalar sound natural enough. The peculiar rule of matrix multiplication can be readily justified in the light of what was said above about linear operators. Regard the components of $\mathbf{v}$ relative to V as an $n \times 1$ matrix (a_{j1}). Then, the components of $\mathbf{A}\mathbf{v}$ relative to V form the $n \times 1$ matrix $(\Sigma_{i=1}^{n} A_{ji} a_i)$, which is obtained by multiplying (a_{i1}) on the left by (A_{ji}), the matrix of A relative to V. Moreover, if B is another linear operator on $\mathcal{V}$ whose matrix relative to the basis V is (B_{ji}), the composite linear operator C = AB, which sends v to ABv, is represented, relative to V, by the matrix $(C_{ji}) = (\Sigma_{k=1}^n A_{jk}B_{ki}) = (A_{ji}) \times (B_{ji})$. (Again the reader should work this out for the case $n = 2$.)

I must mention two important scalar-valued functions that are defined on square matrices, viz., the *trace* and the *determinant*. The *trace* of the $n \times n$ matrix (A_{ij}) is the sum of its diagonal elements:

$$\text{Tr}(A_{ij}) = \Sigma_i A_{ii} \tag{S3}$$

It is worth noting that, if the linear operator A is represented, relative to different bases of $\mathcal{V}$, by the matrices (A_{ij}) and (A_{ij}'), $\text{Tr}(A_{ij}) = \text{Tr}(A_{ij}')$. We are therefore entitled to speak of the trace $\text{Tr}(A)$ of the operator A.

The reader has probably met 3×3 and even 4×4 determinants in high-school algebra. The determinant $\det(A_{ij})$ of our $n \times n$ matrix (A_{ij}) is constructed as follows: Let σ be a permutation of $(1,2,\ldots,n)$; form the product $A_{1\sigma(1)}A_{2\sigma(2)} \ldots A_{n\sigma(n)}$; prefix to it the sign + or −, depending on whether σ is odd or even; do this with each of the $n!$ permutations of $(1,2,\ldots,n)$; $\det(A_{ij})$ is the sum of the $n!$ terms thus obtained. Again, if A is represented, relative to different bases of $\mathcal{V}$, by the matrices (A_{ij}) and (A_{ij}'), we have that $\det(A_{ij}) = \det(A_{ij}')$.

Now look again at the system of n linear equations $\Sigma_{k=1}^n A_{ik}a_k = a_i'$ printed right after eqn. (S1). Clearly, the linear operator A is an injective mapping if and only if this system has unique solutions a_k ($1 \le i \le n$). The reader may recall that this is so if and only if $\det(A_{ij}) \ne 0$. This fact provides a useful way of characterizing the set of proper values of A. Let λ be any scalar. Consider the linear operator $(A - \lambda I)$. λ is a proper value of A if and only if there is a v in $\mathcal{V}$ such that $(A - \lambda I)v = 0$. If $(A - \lambda I)$ is not injective, there are vectors u_1 and u_2 such that $(A - \lambda I)u_1 = (A - \lambda I)u_2$ and $u_1 - u_2 = v \ne 0$. Thus v is a proper vector of A, corresponding to the proper value λ. As we know, if there is one such proper vector v, there are many more (viz., av, for any scalar a). Thus λ is a proper value of A if and only if $(A - \lambda I)$ is not injective, that is, if and only if $\det(A_{ij} - \lambda\delta_{ij}) = 0$. Since $\det(A_{ij} - \lambda\delta_{ij})$ is an nth degree polynomial in λ, there can be at most n different scalars λ satisfying the equation $\det(A_{ij} - \lambda\delta_{ij}) = 0$. They are the proper values of A. The set of proper vectors that correspond to them is, of course, infinite; it includes a set of n linearly independent ones. If $\mathcal{V}$ is an inner product space, this set can be so chosen that it constitutes an orthonormal set, in effect an orthonormal basis of $\mathcal{V}$.

If $\mathcal{V}$ is an arbitrary vector space and A is a linear operator $\mathcal{V} \to \mathcal{V}$, the set of scalars λ such that $(A - \lambda I)$ is not injective is known as the

spectrum of **A**. (This name is motivated by physical applications.) We
have just shown that, for finite-dimensional $\mathcal{V}$, the spectrum of **A** is precisely the set of its proper values.

7   **Hilbert spaces.** I shall now define a class of vector spaces of central
importance in Quantum Mechanics (see §6.2.5). A *Hilbert space* is a
complete inner product space. Inner product spaces were defined in I.4.
An inner product space $\mathcal{H}$ is said to be *complete* if every Cauchy sequence
of vectors in $\mathcal{H}$ converges to a vector in $\mathcal{H}$. An infinite sequence v_1, v_2,
... of vectors in $\mathcal{H}$ is a *Cauchy sequence* if for every positive real number
δ there is an integer N such that, for every m, $n > N$, $\|v_m - v_n\| < \delta$. The
sequence v_1, v_2, ... is said to *converge* to the vector v if for every positive real number δ there is an integer N such that, for every $n > N$,
$\|v - v_n\| < \delta$.[3] In the rest of this section, $\mathcal{H}$, $\mathcal{H}'$, and so on, denote complex
Hilbert spaces.

Given our definitions, any *finite-dimensional* inner product space is a
Hilbert space. But, of course, it is the infinite-dimensional ones that
demand a special study. If $\mathcal{H}$ is infinite-dimensional and we are given any
list v_1, v_2, ... , v_n, no matter how long, of linearly independent vectors,
there is always in $\mathcal{H}$ some other vector v_{n+1} that is linearly independent
of them. Still, due to the completeness of $\mathcal{H}$, it makes sense to speak of
linear combinations of infinitely many vectors in $\mathcal{H}$. Consider a sequence
v_1, v_2, ... of vectors, none of which is a linear combination of those preceding it. Suppose that a_1v_1, $a_1v_1 + a_2v_2$, ... , $\Sigma_{k=1}^{n}a_kv_k$, ... is a Cauchy
sequence converging to a vector v. We then regard the infinite series
$\Sigma_{k=1}^{\infty}a_kv_k = v$ as a linear combination – in an expanded sense – of the
vectors v_1, v_2, v_3, ... By using the Axiom of Choice one can readily prove
that any Hilbert space $\mathcal{H}$ contains a complete orthonormal set of vectors
$\{v_i|\ i \in \Omega\}$ such that, for any v in $\mathcal{H}$, $v = \Sigma_{k\in\Omega}\ a_kv_k$, with $a_i = \langle v_i|v \rangle$ for
each $i \in \Omega$. Or, in Dirac notation,

$$|v\rangle = \sum_{k\in\Omega}\langle v_k|v\rangle|v_k\rangle = \sum_{k\in\Omega}|v_k\rangle\langle v_k|v\rangle \qquad (S4)$$

Applying the bra $\langle v|$ on the left- and right-hand sides of eqn. (S4)
we obtain at once the Theorem of Pythagoras in arbitrarily many
dimensions:

$$\|v\|^2 = \langle v|v\rangle = \sum_{k\in\Omega}\langle v|v_k\rangle\langle v_k|v\rangle = \sum_{k\in\Omega}\langle v_k|v\rangle^*\langle v_k|v\rangle = \sum_{k\in\Omega}|\langle v_k|v\rangle|^2 \qquad (S5)$$

---

[3] Convergence as defined here is sometimes qualified as *strong*. In contrast with it, *weak*
convergence is defined as follows: The sequence v_1, v_2, ... is said to *converge* to the
vector v if, for every positive real number δ, there is an integer N such that, for every
$n > N$, $|\langle v|v\rangle - \langle v_n|\ v_n\rangle| < \delta$. Strong convergence implies but is not implied by weak
convergence.

If the orthonormal set $\{v_i|\ i \in \Omega\}$ spans $\mathcal{H}$ in this way, we are surely justified to call it a basis of $\mathcal{H}$. The cardinality of Ω provides a "dimension-number" for $\mathcal{H}$. In particular, eqns. (S4) and (S5) can be proved for countable Ω, without resorting to the Axiom of Choice, if $\mathcal{H}$ is *separable*, that is, if $\mathcal{H}$ contains a countable subset $\mathcal{S}$ that is dense in $\mathcal{H}$. (The subset $\mathcal{S}$ is *dense* in $\mathcal{H}$ if, for any v in $\mathcal{H}$, there is an element of $\mathcal{S}$ in every neighborhood of v.) Indeed, $\mathcal{H}$ is separable if and only if it contains a countable complete orthonormal set of vectors.

Let $\mathcal{K}$ be a nonempty subset of a Hilbert space $\mathcal{H}$. If $\mathcal{K}$ contains every linear combination of vectors in $\mathcal{K}$, $\mathcal{K}$ is said to be a *linear submanifold* of $\mathcal{H}$. $\mathcal{K}$ is then a vector space (with the operations of $\mathcal{H}$ restricted to $\mathcal{K}$). However, $\mathcal{K}$ need not be a Hilbert space (with the said operations), for there might be a Cauchy sequence of vectors in $\mathcal{K}$ that does not converge to a vector in $\mathcal{K}$. If $\mathcal{K}$ is a Hilbert space in its own right when the operations that characterize $\mathcal{H}$ are restricted to $\mathcal{K}$, we say that $\mathcal{K}$ is a *subspace* of $\mathcal{H}$.

Linear operators $\mathcal{V} \to \mathcal{H}$ (where $\mathcal{V}$ is any vector space) are defined as in I.6. If $\mathcal{V}$ is a linear submanifold of $\mathcal{H}$, I speak of linear operators *in* $\mathcal{H}$ (*on* $\mathcal{H}$, if $\mathcal{V} = \mathcal{H}$). Let A be a linear operator in $\mathcal{H}$. Suppose that there exists a linear operator $A^\dagger$ such that, for all v and w in $\mathcal{H}'$, $\langle v|Aw \rangle = \langle A^\dagger v|w \rangle$; $A^\dagger$ is called the *adjoint* of A. A is a *bounded* operator if there is a real number $C \geq 0$ such that, for every u at which A is defined, $\|Au\| \leq C\|u\|$. The least such number is called the *norm* of A and is denoted by $\|A\|$. If A is bounded, its *adjoint* $A^\dagger$ exists and is also a bounded operator, with the same norm as A. If A and $A^\dagger$ exist and are defined on all of $\mathcal{H}$, they are both bounded. Continuous linear operators on $\mathcal{H}$ are bounded. If the linear operator $A: \mathcal{V} \to \mathcal{H}$ is bounded and $\mathcal{V}$ is dense in $\mathcal{H}$, A can be extended uniquely to a bounded operator on $\mathcal{H}$.

A is said to be *unitary* if its adjoint $A^\dagger$ is the inverse operator A^{-1}, in other words, if $A^\dagger A = AA^\dagger = I$. Unitary operators preserve the inner product; for, if A is unitary, then, for any vectors v and w in $\mathcal{H}$, $\langle Av|Aw \rangle = \langle A^\dagger Av|w \rangle = \langle v|w \rangle$.

A is said to be *self-adjoint* if $A = A^\dagger$. If A is self-adjoint, all its proper values are real. For suppose that for some v in $\mathcal{V}$, $Av = av$; then $a\langle v|v \rangle = \langle v|av \rangle = \langle v|Av \rangle = \langle Av|v \rangle = \langle av|v \rangle = a^*\langle v|v \rangle$; so $a = a^*$; therefore, a is real. If A is self-adjoint, any two proper vectors of A that correspond to different proper values are mutually orthogonal. Suppose that $Av = av$, $Aw = bw$, and $a \neq b$; then $a\langle w|v \rangle = \langle w|Av \rangle = \langle Aw|v \rangle = b^*\langle w|v \rangle = b\langle w|v \rangle$; thus $(a - b)\langle w|v \rangle = 0$; since $(a - b) \neq 0$, we have that $\langle w|v \rangle = 0$; therefore, w and v are orthogonal. These properties of self-adjoint operators motivate their use in Quantum Mechanics.

The study of linear operators in an infinite-dimensional Hilbert $\mathcal{H}$ must face two difficulties that do not arise in the finite-dimensional case:

(i) The spectrum of such an operator does not consist simply of proper values corresponding to its proper vectors, and (ii) the operator can be unbounded. "It is a fact of life that many of the most important operators which occur in mathematical physics are not bounded" (Reed and Simon 1980, p. 249). Because of this one must pay attention to the operator's precise domain of definition (whereas, if the operator is bounded, one can always deal simply with its unique extension defined on the whole $\mathcal{H}$). With regard to (i), I shall consider only the case of a *self-adjoint* operator A in $\mathcal{H}$. Its spectrum $s(A)$ typically comprises a discrete collection of real numbers that are proper values of A in the usual sense, and its complement, the continuous spectrum, which cannot be associated simply with proper vectors. (Indeed, some such operators – like the position operator in Quantum Mechanics – only have a continuous spectrum.) The continuous spectrum was dealt with differently by Dirac and von Neumann. Dirac's approach amounts to a bold generalization of the concepts of proper vector and proper value that makes the latter applicable to the continuous spectrum. For this purpose, it uses the notorious Dirac "delta function", which was spurned as nonsense until Schwartz's theory of distributions (1957/1959) made it respectable. Von Neumann's approach attains rigor within the bounds of classical analysis. Mathematicians and philosophers prefer it to this day. Neither approach can be properly explained within the bounds of this book.[4]

8   **Tensor product of two Hilbert spaces.** Let $\mathcal{H}_1$ and $\mathcal{H}_2$ be complex Hilbert spaces. It can be shown that there is a Hilbert space $\mathcal{H}$ and a mapping $\otimes$: $\langle v,w \rangle \mapsto v \otimes w$ of $\mathcal{H}_1 \times \mathcal{H}_2$ into $\mathcal{H}$ such that, for every scalar a, b and suitable vectors **u**, **v**, **w**, **t**,

(i) $(a\mathbf{u} + b\mathbf{v}) \otimes \mathbf{w} = a(\mathbf{u} \otimes \mathbf{w}) + b(\mathbf{v} \otimes \mathbf{w})$;
(ii) $\mathbf{u} \otimes (a\mathbf{v} + b\mathbf{w}) = a(\mathbf{u} \otimes \mathbf{v}) + b(\mathbf{u} \otimes \mathbf{w})$;
(iii) $\langle \mathbf{u} \otimes \mathbf{t} | \mathbf{v} \otimes \mathbf{w} \rangle = \langle \mathbf{u}|\mathbf{v}\rangle\langle \mathbf{t} |\mathbf{w}\rangle$;
(iv)   every neighborhood of every vector in $\mathcal{H}$ contains a linear combination of vectors in the range of $\otimes$ (i.e., a linear combination of vectors of the form $\mathbf{v}_1 \otimes \mathbf{v}_2$, with $\mathbf{v}_i$ in $\mathcal{H}_i$, $i = 1,2$).

Moreover, if the pairs $\langle \mathcal{H},\otimes \rangle$ and $\langle \mathcal{H}',\otimes' \rangle$ both satisfy the above requirements, there is a unique isomorphism f: $\mathcal{H} \to \mathcal{H}'$ such that $\mathbf{v} \otimes' \mathbf{w} = f(\mathbf{v} \otimes \mathbf{w})$ for every pair $\langle v,w \rangle$ in $\mathcal{H}_1 \times \mathcal{H}_2$. By *the tensor product* $\mathcal{H}_1 \otimes \mathcal{H}_2$ of spaces $\mathcal{H}_1$ and $\mathcal{H}_2$ I mean the structure that is realized by any such pair $\langle \mathcal{H}, \otimes \rangle$ by dint of conditions (i)–(iv). We call $\mathbf{v} \otimes \mathbf{w}$ the tensor product

---

[4] Brief descriptions are given by Redhead (1987, pp. 6–16).

of vectors v and w. If A_1 and A_2 are linear operators on $\mathcal{H}_1$ and $\mathcal{H}_2$, respectively, the linear operator $A_1 \otimes A_2$ on $\mathcal{H}_1 \otimes \mathcal{H}_2$ is defined by: $A_1 \otimes A_2(v \otimes w) = A_1 v \otimes A_2 w$, for every suitable v and w.

II. On Lattices

1 **Partial ordering.** A relation R is said to be *binary* (or *dyadic*) if it is apt to hold between the elements of an ordered pair. I write aRb for 'object a has relation R to object b' (or, synonymously: 'R holds between the first and the second element of the ordered pair $\langle a,b \rangle$'). A set S is *partially ordered* by the binary relation $\leq$ if the following conditions are satisfied for any elements a, b, and c of S:

PO1. $a \leq a$ (the relation $\leq$ is reflexive).

PO2. If $a \leq b$ and $a \neq b$, it is not the case that $b \leq a$ ($\leq$ is antisymmetric).

PO3. If $a \leq b$ and $b \leq c$, then $a \leq c$ ($\leq$ is transitive).

If S is partially ordered by $\leq$, we say that $\langle S, \leq \rangle$ is a *poset* (for 'partially ordered set'). Depending on the nature of the objects in S, we read '$a \leq b$' as 'a is less than or equal to b' or as 'a precedes or is equal to b'; alternatively, we read '$a \leq b$' as 'b is greater than or equal to a' or as 'b follows or is equal to a' Consider any subset $\{a_i \in S: i \in \mathcal{I}\}$. Suppose that there is an element b in S such that (i) $b \leq a_i$ for every $i \in \mathcal{I}$, and (ii) $c \leq b$ for every c in S such that, for every $i \in \mathcal{I}$, $c \leq a_i$. We say then that b is the *infimum* of $\{a_1, \ldots, a_n\}$. Suppose that there is an element b in S such that (i) $a_i \leq b$ for every $i \in \mathcal{I}$, and (ii) $b \leq c$ for every c in S such that, for every $i \in \mathcal{I}$, $a_i \leq c$. We say then that b is the *supremum* of $\{a_i \in S: i \in \mathcal{I}\}$.

2 **Lattices.** The poset $\langle L, \leq \rangle$ is a *lattice* if the following conditions are fulfilled:

L1. L contains a minimal element **0** that is less than or equal to every element of L.

L2. L contains a maximal element **1** that is greater than or equal to every element of L.

L3. If a and b are any two elements of L, L contains an element that is the infimum of $\{a, b\}$ and an element that is the supremum of $\{a, b\}$.

When speaking of lattices, the supremum of a and b is called their *join* and is denoted by $a \vee b$, and the infimum of a and b is called their *meet* and is denoted by $a \wedge b$. Our definitions imply that, for any $a, b \in L$, $a \leq b$ if and only if $a = a \wedge b$ and $b = a \vee b$. From PO1–PO3 and L1–L3 the following statements can be readily inferred, for every $a, b, c \in L$:

L4$_\wedge$. $a \wedge b = b \wedge a$.   L4$_\vee$. $a \vee b = b \vee a$.
L5$_\wedge$. $a \wedge (b \wedge c) = (a \wedge b) \wedge c$.   L5$_\vee$. $a \vee (b \vee c) = (a \vee b) \vee c$.
L6. $a = a \vee (b \wedge a) = a \wedge (b \vee a)$.

An *atom* in a lattice $\langle L,\leq,\mathbf{0},\mathbf{1}\rangle$ is an element $a \in L$ such that $a \neq \mathbf{0}$ and, for any $x \in L$, if $\mathbf{0} \leq x \leq a$, then either $\mathbf{0} = x$ or $x = a$. In other words, a is an atom if there are no elements in the lattice between a and $\mathbf{0}$. The lattice $\langle L,\leq,\mathbf{0},\mathbf{1}\rangle$ is said to be *atomic* if every element of L, except $\mathbf{0}$, is either greater than or equal to an atom.

3   **Orthocomplementation.** Let a and b belong to the lattice $\langle L,\leq,\mathbf{0},\mathbf{1}\rangle$. If $a \vee b = \mathbf{1}$ and $a \wedge b = \mathbf{0}$, b is said to be a *complement* of a. $\langle L,\leq,\mathbf{0},\mathbf{1}\rangle$ is a *complemented* lattice if every element of L has a complement. Consider now an injective mapping of L into L that assigns to each $a \in L$ an element $a' \in L$ such that

OC1. a' is a complement of a,
OC2. $(a')' = a$, and
OC3. $b' \leq a'$ for any $b \in L$ such that $a \leq b$.

The mapping meeting these requirements can be defined on a lattice in only one way (or not at all). The object a' is called the *orthocomplement* of a. A lattice $\langle L,\leq,\mathbf{0},\mathbf{1},'\rangle$ on which such a mapping is defined is said to be *orthocomplemented*. Note that an element a of an orthocomplemented lattice may have several complements, although only one of them will jointly satisfy OC2 and OC3.

Orthocomplementation can obviously be defined in any poset that is not a lattice, provided that it contains a maximal and a minimal element (i.e., provided that it satisfies conditions L1 and L2, even if it does not satisfy L3). We speak then of an *orthocomplemented poset*.

4   **Boolean lattices.** A *Boolean* lattice is a lattice $\langle L,\leq,\mathbf{0},\mathbf{1}\rangle$ in which the two following *distributive laws* hold for any three a, b, $c \in L$:

D$_{\vee\wedge}$. $a \vee (b \wedge c) = (a \vee b) \wedge (a \vee c)$;
D$_{\wedge\vee}$. $a \wedge (b \vee c) = (a \wedge b) \vee (a \wedge c)$.

In a Boolean lattice every element a has one and only one complement, which I denote by a', because it satisfies the rules of orthocomplementation OC1–OC3. a' is the maximum of the lattice elements whose meet with a is $\mathbf{0}$. In other words, $x = a'$ if and only if $y \wedge a = \mathbf{0}$ for every $y \leq x$. On the other hand, any orthocomplemented lattice in which the orthocomplement of an element is its only complement is necessarily a Boolean lattice.[5]

---

[5] There is no essential difference between a Boolean lattice and what is known as Boolean algebra. Take any set B, endowed with two binary algebraic operations $\vee$ and

Given any set Σ, its power set $\mathscr{P}\Sigma$ is partially ordered by the relation of inclusion $\subseteq$. $\langle \mathscr{P}\Sigma,\subseteq \rangle$ is a lattice with maximal element Σ and minimal element $\varnothing$. Take the union of two sets for their join and the intersection of two sets for their meet. Define the complement of any $A \subseteq \Sigma$ by $A' = \Sigma\backslash A$ (the set of all elements of S that do not belong to A). The lattice $\langle \mathscr{P}\Sigma,\subseteq,\Sigma,\varnothing,' \rangle$ is Boolean; we call it the *Boolean lattice of subsets* of Σ. Every Boolean lattice $\langle L,\leq,0,1,' \rangle$ can be represented by the Boolean lattice of subsets of some fundamental set Σ_L by virtue of a mapping $f: L \to \mathscr{P}\Sigma_L$ which is such that (i) $f(0) = \varnothing$, (ii) $f(1) = \Sigma$, and, for any $a, b \in L$, (iii) $f(a') = \Sigma\backslash f(a)$ and (iv) $a \leq b$ if and only if $f(a) \subseteq f(b)$. These conditions imply, of course, that for any $a, b \in L$, $f(a \wedge b) = f(a) \cap f(b)$, and $f(a \vee b) = f(a) \cup f(b)$.

5   **Modularity and orthomodularity.** An orthocomplemented lattice $\langle L,\leq,0,1,' \rangle$ is *modular* if it satisfies the following condition:

M. For any $a, b, c \in L$, such that $a \leq b$, $a \vee (c \wedge b) = (a \vee c) \wedge b$.

In a modular lattice, any permutation of a, b, and c satisfies the distributive laws $D_{\vee/\wedge}$ and $D_{\wedge/\vee}$, provided that $a \leq b$.

An orthocomplemented lattice $\langle L,\leq,0,1,' \rangle$ is *orthomodular* if it satisfies the following condition:

OM. For any $a, b \in L$, such that $a \leq b$, $a \vee (b \wedge a') = b$.

In an orthomodular lattice, any permutation of a, a', and b satisfies the distributive laws $D_{\vee/\wedge}$ and $D_{\wedge/\vee}$, provided that $a \leq b$.

Conditions M and OM can hold also for an orthocomplemented poset $\langle L,\leq,0,1,' \rangle$ that is not a lattice.

III. Terms from Topology

1   **Topological spaces.** Let X be any set of objects, which we call *points*, and T a collection of subsets of X, which we call *open sets*. T is a *topology* on X and $\langle X,T \rangle$ a *topological space* if and only if the following three conditions are met:

$\wedge$ (i.e., two mappings of $B \times B$ into B) that satisfy conditions L4–L6 and D. If both operations have "neutral elements" **0** and **1** such that for any $a \in B$, $a \vee \mathbf{0} = a = a \wedge \mathbf{1}$, and there is a one-one mapping of B into B by $x \mapsto x'$ such that for any $a \in B$, $a \vee a' = \mathbf{1}$ and $a \wedge a' = \mathbf{0}$, $\langle B,\vee,\wedge,\mathbf{0},\mathbf{1},' \rangle$ is a *Boolean algebra*. B is partially ordered by the relation $\leq$ defined by: $a \leq b$ if an only if $a \wedge b = a$ and $a \vee b = b$. The reader should verify that, pursuant to the definitions in the main text, $\langle B,\leq,\mathbf{0},\mathbf{1},' \rangle$ is indeed a Boolean lattice.

T1. The entire set X and the empty set $\emptyset$ are open sets.
T2. The union of any collection of open sets is an open set
T3. The intersection of any two open sets is an open set.

If A is an open set in $\langle X,T \rangle$, its complement $X \backslash A$ (the set of all points in X but not in A) is said to be *closed*. If A is any subset of X, the intersection of all closed sets that include A is clearly a closed set,[6] called the *closure* of A. If p is a point of X, a *neighborhood* of p is any subset of X that includes an open set that contains p. (This idea of neighborhood is central to topology, which is naturally regarded as the theory of neighborhood structures; a definition of topology that is equivalent to the one above can be given purely in terms of neighborhood systems attached to the points of X.) Let $\langle X,T \rangle$ be a topological space, let Y be a subset of X, and consider the collection U of all sets that are the intersection of a member of T with Y. $\langle Y,U \rangle$ is a topological space. (*Exercise*: Prove this.) U is the *subset topology* induced in Y by T. Topologies are often defined by fixing a *base*, that is, a collection of easily characterized sets. The topology *generated* from a given base is the smallest topology that includes the base plus all other sets required by conditions T1, T2, and T3. The standard topology of Euclidian space is generated from the open balls (a set of points in Euclidian space is an open ball if it contains all the points that lie at less than a certain distance from a certain point, and only them). The concept of *compactness* provides an exact characterization of finiteness in purely topological terms, without resorting to metric concepts (such as *distance* or *volume*). An *open covering* of the topological space $\langle X,T \rangle$ is a collection C of sets in T such that each point in X belongs to one member of C (at least). An open covering C of $\langle X,T \rangle$ is said to have a *subcovering* C' if C' is also an open covering of $\langle X,T \rangle$ and every member of C' is a member of C. A topological space $\langle X,T \rangle$ is said to be *compact* if and only if every open covering of $\langle X,T \rangle$ has a finite subcovering (i.e., a subcovering consisting of finitely many sets). Consider an ordinary circle S and a straight line R; let T_S and T_R be the subset topologies induced, respectively, in S and R by the standard Euclidian topology. Then, $\langle S,T_S \rangle$ is compact and $\langle R,T_R \rangle$ is not. (*Exercise*: Prove it.) By using the concept of topological space we can achieve a precise general definition of continuity. Suppose that $\langle X,T \rangle$ and $\langle Y,U \rangle$ are two topological spaces. The mapping $f:X \to Y$ is said to be *continuous* if and only if every open set of $\langle Y,U \rangle$ is the image under f of an open set of $\langle X,T \rangle$. If f is a continuous bijective mapping with continu-

---

[6] Each closed set that includes A has an open complement that excludes A. The union of these open sets is open by T2. The intersection of the closed sets that include A is precisely the complement of this union, so it is a closed set.

ous inverse f^{-1}, X and Y obviously share the same topological structure that is faithfully portraited by f (and by f^{-1}). We then say that f is an *homeomorphism* and that X and Y are *homeomorphic*.

2   **Topological manifolds.** Let a and b be real numbers such that $a < b$. The sets $\{x \in \mathbb{R}: a < x < b\}$, $\{x \in \mathbb{R}: a < x\}$, and $\{x \in \mathbb{R}: x < b\}$ are *open intervals*; and every open interval falls under one of these three descriptions for some pair of real numbers such as a and b. The standard topology of the real line $\mathbb{R}$ is generated from the base formed by the open intervals.[7] The standard topology of $\mathbb{R}^n$ can be generated from the base formed by the Cartesian products of n open intervals. A topological space $\langle X, T \rangle$ is said to be a *topological n-dimensional manifold* if every point of X has an open neighborhood that is homeomorphic to $\mathbb{R}^n$ (with the standard topology). Comparing this definition with the notion of an *n*-manifold (or real *n*-dimensional smooth manifold) explained in §4.1.3, we see that every *n*-manifold is also a topological *n*-dimensional manifold.

3   **Borel sets of the real line.** Consider any set S. Consider a collection $\mathcal{B}(S)$ of subsets of S such that (i) $S \in \mathcal{B}(S)$; (ii) if $A \in \mathcal{B}(S)$, its complement $S\backslash A \in \mathcal{B}(S)$; and (iii) if $A_1, A_2, \ldots$ is any countable list of subsets in $\mathcal{B}(S)$, their union $\cup_{k=1}^{\infty} A_k \in \mathcal{B}(S)$. Then $\mathcal{B}(S)$ is a *Borel field* of S. If M is *any* nonempty collection of subsets of S, the smallest Borel field of S that contains M is said to be the Borel field *generated* by M. Let $\mathcal{B}(\mathbb{R})$ denote the Borel field of $\mathbb{R}$ generated by the open intervals. The sets in $\mathcal{B}(\mathbb{R})$ are called the *Borel sets* of $\mathbb{R}$. They are precisely the Lebesgue measurable subsets of the real line. $\mathcal{B}(\mathbb{R})$ is important in the theory of probability because, by dint of the Axiom of Choice – which most mathematicians accept, as one told me, "in order to survive" – a probability function satisfying the standard Kolmogorov axioms cannot be consistently defined on the entire power set $\mathcal{P}\mathbb{R}$; so, normally, a function that expresses the probability that, say, the value of a physical quantity lies in a particular subset of $\mathbb{R}$ is not defined for *all* the subsets of $\mathbb{R}$ but only for the Borel sets.

[7] *Exercise*: Prove that if $\mathbb{R}$ is endowed with the said standard topology, a function $f:\mathbb{R} \to \mathbb{R}$ is continuous according to the familiar definition of a continuous real-valued function of a real variable if and only if it is also continuous in accordance with the topological definition of continuity given at the end of III.1.

References

Alexander, H. G., ed. (1956). *The Leibniz-Clarke Correspondence*. Manchester: Manchester University Press, 1956.

Alpher, R. A. and R. C. Herman (1948). "Evolution of the universe". *Nature*. 162:774–75.

Alpher, R. A., H. Bethe, and G. Gamow (1948). "The origin of chemical elements". *Physical Review*. 73:803–804.

Ampère, A. M. (1827). "Mémoire sur la théorie des phénomènes électrodynamiques uniquement déduite de l'expérience, dans lequel se trouvent réunis les Mémoires que M. Ampère a communiqués à l'Académie royale des Sciences dans les séances des 4 et 26 décembre 1820, 10 juin 1822, 22 décembre 1823, 12 septembre et 21 novembre 1825". *Mémoires de l'Académie Royale des Sciences*. 6:175–388 (1823). (Volume 6 of the *Mémoires de l'Académie des Sciences* corresponds to 1823, but was actually issued in 1827; both dates are given on the title page.).

Anderson, C. D. (1932). "The apparent existence of easily deflectable positives". *Science*. 76:238–39.

Anscombe, G. E. M. (1971). *Causality and Determination: An Inaugural Lecture*. Cambridge: Cambridge University Press.

Appell, P. (1893). *Traité de mécanique rationelle*. Tome I. Paris: Gauthier-Villars.

Arabatzis, T. (1996). "Rethinking the 'discovery' of the electron". *Studies in History and Philosophy of Modern Physics*. 27B:405–35.

Arana, J. (1990). *Apariencia y verdad: Estudio sobre la filosofía de P. L. M. de Maupertuis (1698–1759)*. Buenos Aires: Charcas.

Aristotle. *Aristotelis Opera*. Ex recognitione I. Bekkeris edidit Academia Regia Borussica. Berlin: G. Reimer, 1831. 2 vols. (My quotations refer, as is customary, to the pages, columns, and lines of this edition; however, they are taken from various more recent editions and may not always agree with Bekker's text.)

Aristotle. *De Anima*. Books II and III. Translated with Introduction and Notes by D. W. Hamlyn. Oxford: Clarendon Press.

Aspect, A., J. Dalibard, and G. Roger (1982). "Experimental tests of Bell's inequalities using time-varying analyzers". *Physical Review Letters*. 49:1804–7.

Aspect, A., P. Grangier, and G. Roger (1982). "Experimental realization of Einstein-Podolsky-Rosen Gedankenexperiment: A new violation of Bell's inequalities". *Physical Review Letters*. 48:91–94.

Auyang, S. Y. (1995). *How Is Quantum Field Theory Possible?* New York: Oxford University Press.

Ayer, A. J., ed. (1959). *Logical Positivism*. New York: Free Press.

Baade, W. (1956). "The period-luminosity relation of the cepheids". *Publications of the Astronomical Society of the Pacific*. 28:5–16.

Baldwin, F. J. (1901/1902). *Dictionary of Philosophy and Psychology*. New York: Macmillan. 2 vols.

Balzer, W., C. U. Moulines, and J. D. Sneed (1983). "The structure of empirical science: Local and global". In R. Barcan Marcus et al., eds., *Proceedings of the 7th International Congress of Logic, Methodology and Philosophy of Science, 1983*. Amsterdam: North-Holland, 1986, pp. 291–306.

Barbour, J. B. (1989). *Absolute or Relative Motion? A Study from a Machian Point of View of the Discovery and the Structure of Dynamical Theories. Part 1: The Discovery of Dynamics*. Cambridge: Cambridge University Press.

Barrow, J. D. and F. J. Tipler (1986). *The Anthropic Cosmological Principle*. Oxford: Clarendon Press.

Bartelborth, T. (1988). *Eine logische Rekonstruktion der klassischen Elektrodynamik*. Frankfurt a. M.: Peter Lang.

Bartels, A. (1994). *Bedeutung und Begriffsgeschichte: Die Erzeugung wissenschaftlichen Verstehens*. Paderborn: Ferdinand Schöningh.

Basilius (St. Basil). *Epistulae*. Edited by Y. Courtonne. Paris: Belles Lettres, 1957–1966.

Bayes, T. (1763). "An essay towards solving a problem in the doctrine of chances". *Royal Society of London Philosophical Transactions*. 53:370–418.

Bechler, Z. (1991). *Newton's Physics and the Conceptual Structure of the Scientific Revolution*. Dordrecht: Kluwer.

Belinfante, F. J. (1973). *A Survey of Hidden-Variables Theories*. Oxford: Pergamon.

Bell, J. and M. Hallett (1982). "Logic, quantum logic and empiricism". *Philosophy of Science*. 49:375–79.

Bell, J. S. (1964). "On the Einstein–Podolsky–Rosen paradox". *Physics*. 1:195–200 (reprinted in Bell 1987, pp. 14–21).

———(1966). "On the problem of hidden variables in Quantum Mechanics". *Review of Modern Physics*. 38:447–52 (reprinted in Bell 1987, pp. 1–13).

———(1975). "On wave packet reduction in the Colemann–Hepp model". *Helvetica Physica Acta*. 48:93–98 (reprinted in Bell 1987, pp. 44–51).

———(1987). *Speakable and Unspeakable in Quantum Mechanics*. Cambridge: Cambridge University Press.

———(1990). "Against 'measurement'". In A. I. Miller, 1990, pp. 17–31.

Beltrametti, E. G. and G. Cassinelli (1981). *The Logic of Quantum Mechanics*. Reading, MA: Addison-Wesley.

Berkeley, G. (1710). *A Treatise Concerning the Principles of Human Knowlege* [sic]. Dublin: Printed by A. Rhames, for J. Pepyat.

————(1744). *Siris: A Chain of Philosophical Reflexions and Inquiries Concerning the Virtues of Tar-water, and Divers other Subjects Connected Together and Arising One from Another.* Dublin: Printed by M. Rhames, for R. Gunne.

————(1752). *Alciphron: or, the Minute Philosopher. In Seven Dialogues containing an Apology for the Christian Religion, against those who are called Free-thinkers.* 3rd Ed. London: J. and R. Tonson and S. Draper. (First edition, 1732).

Bernays, P. (1967). "Hilbert, David". In Edwards 1967, vol. 3, pp. 496–504.

Bernoulli, D. (1738). *Hydrodynamica.* Argentorati [Strasbourg]: Dulsecker. (English translation by T. Carmodus and H. Cobus, *Hydrodynamics*, New York: Dover, 1968.)

Bernoulli, J. (1713). *Ars conjectandi: Opus posthumum.* Basilea [Basle]: Impensis Thurnisiorum, fratrum.

Birkhoff, G. and J. von Neumann (1936). "The Logic of Quantum Mechanics". *Annals of Mathematics.* 37:823–43 (reprinted in C. A. Hooker 1975).

Birkhoff, G. D. (1923). *Relativity and Modern Physics.* Cambridge, MA: Harvard University Press.

Blau, S. K. and A. H. Guth (1987). "Inflationary cosmology". In S. Hawking and W. Israel 1987, pp. 524–603.

Bohm, D. (1951). *Quantum Theory.* Englewood Cliffs, NJ: Prentice-Hall.

————(1952). "A suggested interpretation of the Quantum Theory in terms of 'hidden variables'". *Physical Review.* 85:166–79, 180–93.

————(1980). *Wholeness and the Implicate Order.* London: Routledge.

Bohm, D. and B. J. Hiley (1993). *The Undivided Universe: An ontological interpretation of quantum theory.* London: Routledge.

Bohr, N. (CW). *Collected Works.* Amsterdam: North-Holland, 1972– .

————(1913). "On the constitution of atoms and molecules". *Philosophical Magazine.* 26:1–25, 476–502, 857–75.

————(1918). "On the quantum theory of line-spectra. Part I. On the general theory". *Det Kongelige Danske Videnskabernes Selskab, Matematisk-fysiske Meddelser.* (8) 4 1:1–36 (reprinted in B. L. Van der Waerden 1967, pp. 95–137).

————(1922). *The Theory of Spectra and Atomic Constitution.* Cambridge: Cambridge University Press.

————(1924). *On the Application of the Quantum Theory to Atomic Structure, Part I: The Fundamental Postulates.* Supplement to the *Cambridge Philosophical Society Proceedings.*

————(1928). "The quantum postulate and the recent development of atomic theory". *Nature.* 121:580–90 (printed version of Bohr's Como lecture of 1927. Reprinted in Bohr 1934, pp. 51–91).

————(1934). *Atomic Theory and the Description of Nature.* Cambridge: Cambridge University Press.

————(1935). "Can quantum-mechanical description of physical reality be considered complete?". *Physical Review.* 48:696–702 (reprinted in J. A. Wheeler and W. H. Zurek 1983, pp. 145–51; the text printed on p. 148 should go on p. 149, and vice versa).

———(1949). "Discussion with Einstein on epistemological problems in atomic physics". In P. A. Schilpp 1949, pp. 199–241.

———(1958). *Atomic Physics and Human Knowledge*. New York: Wiley.

———(1963). *Essays 1958–1962 on Atomic Physics and Human Knowledge*. New York: Wiley.

Bohr, N., H. Kramers, and J. C. Slater (1924). "The quantum theory of radiation". *Philosophical Magazine*. 47:785–822 (reprinted in B. L. Van der Waerden 1967, pp. 159–76).

Boltzmann, L. (WA). *Wissenschaftliche Abhandlungen*. Herausgegeben von F. Hasenöhrl. Leipzig: Barth, 1901. 3 vols. (reprint: New York, Chelsea, 1968).

———(1871a). "Einige allgemeine Sätze über Wärmegleichgewicht". *Sitzungsberichte der Akademie der Wissenschaften zu Wien*. 63:679–711 (reprinted in Boltzmann WA, I, 260–87).

———(1871b). "Analytischer Beweis des zweiten Hauptsatzes der mechanischen Wärmetheorie aus den Sätzen über das Gleichgewicht der lebendigen Kraft". *Sitzungsberichte der Akademie der Wissenschaften zu Wien*. 63:712–32 (reprinted in Boltzmann WA, I, 288–308).

———(1872). "Weitere Studien über das Wärmegleichgewicht unter Gasmolekülen". *Sitzungsberichte der Akademie der Wissenschaften zu Wien*. 66:275–370 (reprinted in Boltzmann WA, I, 316–402).

———(1896a). *Vorlesungen über Gastheorie*. Teil I. Leipzig: Barth.

———(1896b). "Entgegnung auf die wärmetheoretischen Betrachtungen des Hrn. E. Zermelo". *Annalen der Physik*. (3) 57:773–84 (reprinted in Boltzmann WA, III, 567–78).

———(1897). "Zu Hrn. Zermelos Abhandlung 'Über die mechanische Erklärung irreversibler Vorgänge'". *Annalen der Physik*. (3) 60:392–98 (reprinted in Boltzmann WA, III, 579–86).

———(1898). *Vorlesungen über Gastheorie*. Teil II. Leipzig: Barth.

———(1905). *Populäre Schriften*. Leipzig: Barth.

———(LGT). *Lectures on Gas Theory*. Translated by S. G. Brush. Berkeley: University of California Press, 1964 (unaltered reprint: New York, Dover, 1995).

Bondi, H. and T. Gold (1948). "The steady-state theory of the expanding universe". *Monthly Notices of the Royal Astronomical Society*. 108:252–70.

Born, M. (1924). "Über Quantenmechanik". *Zeitschrift für Physik*. 26:379–95.

———(1926a). "Zur Quantenmechanik der Stossvorgänge. (Vorläufige Mitteilung)". *Zeitschrift für Physik*. 37:863–67.

———(1926b). "Quantenmechanik der Stossvorgänge". *Zeitschrift für Physik*. 38:803–27.

———(1927a). "Das Adiabatenprinzip in der Quantenmechanik". *Zeitschrift für Physik*. 40:167–92.

———(1927b). "Physical aspects of quantum mechanics". In M. Born 1969, pp. 6–12. (First published in *Nature* 119:354–57 (1927).)

———(1962). *Atomic Physics*. 7th Ed. From the original translation of J. Dougall revised by the author in collaboration with R. J. Blin-Stoyle. London: Blackie & Son.

———(1968). *My life and my views*. New York: Scribner's.

462   References

———(1969). *Physics in my generation.* 2nd rev. ed. New York: Springer.
Born, M. and P. Jordan (1925a). "Zur Quantentheorie aperiodischer Vorgänge". *Zeitschrift für Physik.* 33:479–505.
———(1925b). "Zur Quantenmechanik". *Zeitschrift für Physik.* 34:858–78. (English translation of §§1–3 in B. L. Van der Waerden 1967, pp. 277–306.)
Born, M., W. Heisenberg, and P. Jordan (1926). "Zur Quantenmechanik II". *Zeitschrift für Physik.* 35:557–615. (English translation in B. L. Van der Waerden 1967, pp. 321–85.)
Boskovic, R. J. (1758). *Philosophiae naturalis theoria redacta ad unicam legem virium in natura existentium.* Vindobonae [Vienna].
———(1763) *Theoria philosophiae naturalis redacta ad unicam legem virium in natura existentium.* Venetiis [Venice]: ex Typographia Remonidniana.
Bose, S. N. (1924). "Plancks Gesetz und Lichtquantenhypothese". *Zeitschrift für Physik.* 26:178–81.
Bourbaki, N. (1970). *Théorie des ensembles.* Paris: Hermann.
Boyle, R. (1666). *The origine of formes and qualities (according to the corpuscular philosophy) illustrated by considerations and experiments.* Oxford: Printed by H. Hall printer to the University, for Ric. Davis.
———(1674). *De hypothesis mechanicæ excellentia et fundamentis considerationes quædam amico propositæ.* London: Typis T.N., impensis Henrici Herringman.
———. (SPP). *Selected Philosophical Papers.* Edited with an introduction by M. A. Stewart. Indianapolis: Hackett, 1991.
Bradley, J. (1728). "A Letter from the Reverend Mr. James Bradley Savilian Professor of Astronomy at Oxford and F.R.S. to Dr. Edmund Halley Astronom. Roy. etc. giving an Account of a new discovered Motion of the Fix'd Stars". *Royal Society of London Philosophical Transactions.* 35:637–61.
Braginsky, V. B. and F. Y. Khalili (1992). *Quantum Measurement.* Edited by K. S. Thorne. Cambridge: Cambridge University Press.
Braginsky, V. B. and V. I. Panov (1972). "Verification of the equivalence of inertial and gravitational mass". *Soviet Physics JETP.* 44:463–66 (English translation; the Russian original was published in *Zh. Eksp. Teor. Fiz.*, 71:873–79 (1971).)
Braithwaite, R. B. (1953). *Scientific Explanation: A Study of the Function of Theory, Probability and Law in Science. Based Upon the Tarner Lectures, 1946.* Cambridge: Cambridge University Press.
Brown, H. R. and R. Harré, eds. (1988). *Philosophical Foundations of Quantum Field Theory.* Oxford: Clarendon Press.
Brown, R. (1828). "A brief account of observations made in the months of June, July, and August 1827, on the particles contained in the pollen of plants; and on the general existence of active molecules in organic and inorganic bodies". *Edinburgh New Philosophical Journal.* 5:358–71 (reprinted in *Philosophical Magazine* 4:161–73 (1828)).
Bub, J. (1997). *Interpreting the Quantum World.* Cambridge: Cambridge University Press.
Buchwald, J. Z. (1985). *From Maxwell to Microphysics: Aspects of Electro-*

magnetic Theory in the Last Quarter of the Nineteenth Century. Chicago: University of Chicago Press.

———(1994). *The Creation of Scientific Effects: Heinrich Hertz and Electric Waves.* Chicago: University of Chicago Press.

———, ed. (1995). *Scientific Practice: Theories and Stories of Doing Physics.* Chicago: University of Chicago Press.

Buijs-Ballot, C. H. D. (1858). "Über die Art der Bewegung, welche wir Wärme und Elektrizität nennen". *Annalen der Physik.* (2) 103:240–59.

Bunge, M. (1967). *Foundations of Physics.* Berlin: Springer.

Burbury, S. H. (1894). "Boltzmann's minimum function". *Nature.* 51:78.

Busch, P., P. J. Lahti, and P. Mittelstaedt (1996). *The Quantum Theory of Measurement.* 2nd rev. ed. Heidelberg: Springer.

Butts, R. E. (1968). *William Whewell's Theory of Scientific Method.* Pittsburgh, PA: University of Pittsburgh Press (reprint: Indianapolis, Hackett, 1989).

Campbell, L. and W. Garnett (1884). *The Life of James Clerk Maxwell.* 2nd Ed. London: Macmillan.

Campbell, N. R. (1920). *Physics: The Elements.* Cambridge: Cambridge University Press (reprinted as *Foundations of Science: The Philosophy of Theory and Experiment,* New York, Dover, 1957).

Cao, Tian Yun (1997). *Conceptual Developments of 20th Century Field Theories.* Cambridge: Cambridge University Press.

Carnap, R. (1928). *Der logische Aufbau der Welt.* Berlin: Weltkreis-Verlag.

———. (1936/37). "Testability and meaning". *Philosophy of Science.* 3:419–71; 4:1–40.

———(1945a). "On inductive logic". *Philosophy of Science.* 12:72–97.

———(1945b). "The two concepts of probability". *Philosophy and Phenomenological Research.* 5:513–32.

———(1950). *Logical Foundations of Probability.* Chicago: University of Chicago Press.

———(1952). *The Continuum of Inductive Methods.* Chicago: University of Chicago Press.

———(1956). "The methodological character of theoretical concepts". *Minnesota Studies in the Philosophy of Science.* 1:38–76.

Carnot, S. (1824). *Réflexions sur la puissance motrice du feu et sur les machines propres a développer cette puissance.* Paris: Chez Bachelier (facsimile reprint: Paris, Jacques Gabay, 1990).

Cartan, É. (1923). "Sur les variétés à connexion affine et la théorie de la relativité généralisée (première partie)". *Annales de l'École Normale Supérieure.* 40:325–412.

Caspar, M. (1993). *Kepler.* Translated and edited by C. D. Hellman. New York: Dover.

Chandrasekhar, S. (1995). *Newton's Principia for the Common Reader.* Oxford: Clarendon Press.

Charlton, W. (1970). *Aristotle's Physics, Books I and II.* Translated with introduction and notes. Oxford: Clarendon Press.

Chrysippus (FM). "Fragmenta Moralia". In von Arnim, SVF. Vol. 3, pp. 3–191.

Cini, M. (1983). "Quantum theory of measurement without wave packet collapse". *Il Nuovo Cimento*. 73B:27–55.

Ciufolini, I. and J. A. Wheeler (1995). *Gravitation and Inertia*. Princeton, NJ: Princeton University Press.

Claggett, M. (1959a). *The Science of Mechanics in the Middle Ages*. Madison, WI: University of Wisconsin Press.

———, ed. (1959b). *Critical Problems in the History of Science*. Madison, WI: University of Wisconsin Press.

Clausius, R. (1857). "Über die Art der Bewegung, welche wir Wärme nennen". *Annalen der Physik*. (2) 100:353–80.

———(1858). "Über die mittlere Länge der Wege, welche bei der Molecularbewegung gasförmigen Körper von den einzelnen Molecülen zurückgelegt werden; nebst einigen anderen Bemerkungen über die mechanische Wärmetheorie". *Annalen der Physik*. (2) 105:239–58.

Clifton, R. K., M. L. G. Redhead, and J. N. Butterfield (1991a). "Generalization of the Greenberger–Horne–Zeilinger algebraic proof of nonlocality". *Foundations of Physics*. 21:149–84.

———(1991b). "A second look at a recent algebraic proof of nonlocality". *Foundations of Physics Letters*. 4:395–403.

Cohen, I. B. (1980). *The Newtonian Revolution: With illustrations of the transformation of Scientific Ideas*. Cambridge: Cambridge University Press.

Cohen-Tannoudji, C., B. Diu, and F. Laloë (1977). *Quantum Mechanics*. Translated from the French by S. R. Hemley, N. Ostrowsky, and D. Ostrowsky. Paris: Hermann. 2 vols.

Colodny, R. G., ed. (1965). *Beyond the Edge of Certainty*. Englewood Cliffs, NJ: Prentice-Hall.

———, ed. (1966). *Mind and Cosmos*. Pittsburgh, PA: University of Pittsburgh Press.

———, ed. (1972). *Paradigms and Paradoxes: The Philosophical Challenge of the Quantum Domain*. Pittsburgh, PA: University of Pittsburgh Press.

———, ed. (1986). *From Quarks to Quasars: Philosophical Problems of Modern Physics*. Pittsburgh, PA: University of Pittsburgh Press.

Cornford, F. M. (1936). "The invention of space". In *Essays in Honour of Gilbert Murray*. London: Allen & Unwin, pp. 215–235 (reprinted in M. Čapek, ed., *The Concepts of Space and Time*, Dordrecht: Reidel, 1976, pp. 3–16).

Costabel, P. (1960). *Leibniz et la dynamique: Les textes de 1692*. Paris: Hermann. (Leibniz's *Essay de dynamique* is printed on pp. 97–106.)

Courant, R. (1936). *Differential and Integral Calculus*. Volume II. Translated by E. J. McShane. London: Blackie & Son.

Cushing, J. T. (1994). *Quantum Mechanics: Historical Contingency and the Copenhagen Hegemony*. Chicago: University of Chicago Press.

d'Alembert, J. (1758). *Traité de dynamique dans lequel les loix de l'équilibre et du mouvement des corps sont réduites au plus petit nombre possible, et démontrées d'une maniere nouvelle, et où l'on donne un principe général pour trouver le mouvement de plusieurs corps qui agissent les uns sur les autres d'une*

References465

manière quelconque. Nouvelle Edition, revûe et fort augmentée par l'Auteur. Paris: Chez David (facsimile reprint: Paris, Jacques Gabay, 1990).

Damerow, P., G. Freudenthal, P. McLaughlin, and J. Renn (1992). *Exploring the Limits of Preclassical Mechanics. A Study of Conceptual Development in Early Modern Science: Free Fall and Compounded Motion in the Work of Descartes, Galileo, and Beeckman*. New York: Springer.

Daneri, A., A. Loinger, and G. M. Prosperi (1962). "Quantum theory of measurement and ergodicity conditions". *Nuclear Physics*. 33:297–319.

Darrigol, O. (1992). *From c-Numbers to q-Numbers: The Classical Analogy in the History of Quantum Theory*. Berkeley, CA: University of California Press.

Davidson, D. (1974). "On the very idea of a conceptual scheme". *Proceedings and Addresses of the American Philosophical Society*. 47:5–20.

de Broglie, L. (1923a). "Ondes et quanta". *Comptes rendus de l'Académie des Sciences*. 177:507–10.

———(1923b). "Quanta de lumière, diffraction et interférences". *Comptes rendus de l'Académie des Sciences*. 177:548–50.

———(1925). "Recherches sur la théorie des quanta". *Annales de physique*. 3:22–128.

Debs, T. A. and M. L. G. Redhead (1996). "The twin 'paradox' and the conventionality of simultaneity". *American Journal of Physics*. 64:384–92.

de Finetti, B. (1930). "Funzione caratteristica di un fenomeno aleatorio". *Memorie della Reale Accademia dei Lincei*. IV 5:86–133.

———(1931). "Sul significato soggettivo della probabilità". *Fundamenta Mathematicae*. 17:298–329.

———(1937). "La prevision: ses lois logiques, ses sources subjectives". *Annales de l'Institut Henri Poincaré*. 7:1–68 (translated as "Foresight: Its logical laws, its subjective sources" in Kyburg and Smokler 1980, pp. 53–118).

———(1970). *Teoria della probabilità*. Torino: Einaudi.

de Gandt, F. (1995). *Force and Geometry in Newton's Principia*. Translated by C. Wilson. Princeton, NJ: Princeton University Press.

de Maria, M. and A. Russo (1985). "The discovery of the positron". *Rivista di Storia delle Scienze*. 2:237–86.

Descartes, R. (AT). *Œuvres*. Edited by C. Adam and P. Tannery. Paris: Cerf, 1897–1912. 12 vols.

———(1637). *Discours de la méthode pour bien conduire sa raison, et chercher la verité dans les sciences. Plus La dioptrique, Les méteores, et La géometrie, qui sont des essais de cette méthode*. Leyde: de l'imprimerie de I. Maire.

———(1641). *Meditationes de prima philosophia, in quibus Dei existentia & animae à corpore distinctio demonstratur*. Amstelodami: apud Ludovicum Elzevirium.

———(1644). *Principia philosophiae*. Amstelodami: apud Ludovicum Elzevirium.

de Sitter, W. (1917a). "On the relativity of inertia: Remarks concerning Einstein's latest hypothesis". *K. Nederlandse Akademie van Wetenschappen. Proceedings*. 19:1217–25.

———(1917b). "On the curvature of space". *K. Nederlandse Akademie van Wetenschappen. Proceedings*. 20:229–43.

————(1917c). "On Einstein's theory of gravitation, and its astronomical consequences. III". *Monthly Notices of the Royal Astronomical Society.* 78:3–28.
————(1933). "On the expanding universe and the time-scale". *Monthly Notices of the Royal Astronomical Society.* 93:628–34.
d'Espagnat, B., ed. (1971). *Foundations of Quantum Mechanics.* Proceedings of the International School of Physics "Enrico Fermi", Course IL. New York: Academic Press.
Destouches-Février, P. (1951). *La structure des théories physiques.* Paris: Presses Universitaires de France.
DeWitt, B. S. (1970). "Quantum mechanics and reality". In B. S. DeWitt and N. Graham 1973, pp. 155–65. (Originally published in *Physics Today,* 23 9:30–35 (September 1970).)
————(1971). "The many-universes interpretation of quantum mechanics". In B. d'Espagnat 1971, pp. 211–62 (reprinted in B. S. DeWitt and N. Graham 1973, pp. 167–218).
DeWitt, B. S. and N. Graham, eds. (1973). *The Many-Worlds Interpretation of Quantum Mechanics.* A Fundamental Exposition by Hugh Everett, III, with Papers by J. A. Wheeler, B. S. DeWitt, L. N. Cooper, D. van Vechten, and N. Graham. Princeton, NJ: Princeton University Press.
Dicke, R. H., P. J. E. Peebles, P. G. Roll, and D. T. Wilkinson (1965). "Cosmic black-body radiation". *Astrophysical Journal.* 142:414–19.
Diels, H. and W. Kranz (DK). *Die Fragmente der Vorsokratiker.* Siebente Auflage. Berlin: Weidmannsche Verlagsbuchhandlung. 3 vols. (items are identified by a number designating an author or school, followed by a letter – A for testimonies, B for direct quotations, C for spuria –, followed by a number).
Dirac, P. A. M. (1926a). "On the theory of quantum mechanics". *Royal Society of London Proceedings.* A112:661–77.
————(1926b). "The physical interpretation of the quantum dynamics". *Royal Society of London Proceedings.* A113:621–41.
————(1927). "The quantum theory of the emission and absorption of radiation". *Royal Society of London Proceedings.* A114:243–65 (reprinted in J. Schwinger 1958, pp. 1–23).
————(1928). "The quantum theory of the electron". *Royal Society of London Proceedings.* A117:610–24.
————(1930). *The Principles of Quantum Mechanics.* Oxford: Clarendon Press.
————(1931). "Quantized singularities in the electromagnetic field". *Royal Society of London Proceedings.* A133:60–72.
————(1958). *The Principles of Quantum Mechanics.* 4th Ed. (revised). Oxford: Clarendon Press (first edition 1930, second edition 1935, third edition 1947).
————(1984). "The requirements of a fundamental physical theory". *European Journal of Physics.* 5:65–67.
Drake, S. (1978). *Galileo at Work: His Scientific Biography.* Chicago: University of Chicago Press.
Dugas, R. (1955). *A History of Mechanics.* Translated by J. R. Maddox. Neuchâtel: Éditions du Griffon (the original French was published in 1950, also by Éditions du Griffon).

Duhem, P. (1905–1906). *Les origines de la statique*. Paris: Hermann. 2 vols.

———(1906). *La théorie physique: son objet, sa structure*. Paris: Chevalier et Rivière.

———(1906–1913). *Études sur Léonard de Vinci, ceux qu'il a lus, ceux qui l'ont lu*. Paris: Hermann. 3 vols.

———(1908). "La valeur de la théorie physique, à propos d'un livre récent". *Révue génerale des Sciences pures et appliquées*. 19:7–19 (reprinted in Duhem 1914, pp. 473–509).

———(1913). "Examen logique de la théorie physique". *Revue scientifique*. 51:713–40. (English translation in Duhem HPS, pp. 232–38.)

———(1913–58). *Le système du monde: Histoire des doctrines cosmologiques de Platon à Copernic*. Paris: Hermann. 10 vols. (Vols. 1–5 appeared in 1913–1916; vols. 6–10 were posthumously published between 1954 and 1958).

———(1914). *La théorie physique: son objet, sa structure*. Deuxième édition, revue et augmentée. Paris: Marcel Rivière (reprint of first edition of 1906, supplemented with two appendices).

———(HPS). *Essays in the History and Philosophy of Science*. Translated and edited, with introduction, by R. Arew and P. Barker. Indianapolis: Hackett, 1996.

Earman, J. (1992). *Bayes or Bust? A Critical Examination of Bayesian Confirmation Theory*. Cambridge, MA: The MIT Press.

———(1995). *Bangs, Crunches, Whimpers, and Shrieks: Singularities and Acausalities in Relativistic Spacetimes*. New York: Oxford University Press.

Earman, J. and M. Friedman (1973). "The meaning and status of Newton's Law of Inertia and the nature of gravitational forces". *Philosophy of Science*. 40:329–59.

Earman, J. and J. Mosterín (1999). "A critical analysis of inflationary cosmology". *Philosophy of Science*. 66 (forthcoming).

Earman, J. and J. Norton (1987). "What price spacetime substantivalism? The hole story". *British Journal for the Philosophy of Science*. 38:515–25.

Earman, J., A. I. Janis, G. J. Massey, and N. Rescher, eds. (1993). *Philosophical Problems of Internal and External Worlds: Essays on the Philosophy of Adolf Grünbaum*. Pittsburgh, PA: University of Pittsburgh Press.

Eddington, A. S. (1924). *The Mathematical Theory of Relativity*. 2nd Ed. Cambridge: Cambridge University Press.

Edwards, P., ed. (1967). *The Encyclopedia of Philosophy*. New York: Macmillan. 8 vols.

Ehlers, J. F., A. E. Pirani, and A. Schild (1972). "The geometry of free fall and light propagation". In L. O'Raifeartaigh 1972, pp. 63–84.

Ehrenfest, P. (CSP). *Collected Scientific Papers*. Edited by M. J. Klein. Amsterdam: North-Holland, 1959.

Ehrenfest, P. and T. Ehrenfest (1906). "Über eine Aufgabe aus der Wahrscheinlichkeitsrechnung, die mit der kinetischen Deutung der Entropievermehrung zusammenhängt". In P. Ehrenfest CSP, pp. 128–30 (originally published in *Mathematisch-Naturwissenschaftliche Blätter*, Nr. 11–12, 1906).

Einstein, A. (CP). *The Collected Papers of Albert Einstein*. Princeton, NJ: Princeton University Press, 1987– .

————(1902b). "Kinetische Theorie der Wärmegleichgewichtes und des zweiten Hauptsatzes der Thermodynamik". *Annalen der Physik.* (4) 9:417–33 (reprinted with editorial notes in Einstein CP 2, 56–75).

————(1903). "Eine Theorie der Grundlagen der Thermodynamik". *Annalen der Physik.* (4) 11:170–87 (reprinted with editorial notes in Einstein CP 2, 76–97).

————(1904). "Zur allgemeinen molekulären Theorie der Wärme". *Annalen der Physik.* (4) 14:354–62 (reprinted with editorial notes in Einstein CP 2, 98–108).

————(1905i). "Über einen die Erzeugung und Verwandlung des Lichtes betreffenden heuristischen Gesichtspunkt". *Annalen der Physik.* (4) 17:132–48 (reprinted with editorial notes in Einstein CP 2, 149–69).

————(1905k). "Über die von der molekularkinetischen Theorie der Wärme geforderte Bewegung von in ruhenden Flüssigkeiten suspendierten Teilchen". *Annalen der Physik.* (4) 17:549–60 (reprinted with editorial notes in Einstein CP 2, 223–36).

————(1905r). "Zur Elektrodynamik bewegter Körper". *Annalen der Physik.* (4) 17:891–921 (reprinted with editorial notes in Einstein CP 2, 275–310).

————(1905s). "Ist die Trägheit eines Körpers von seinem Energiegehalt abhängig". *Annalen der Physik.* (4) 18:639–41 (reprinted with editorial notes in Einstein CP 2, 311–15).

————(1906b). "Zur Theorie der Brownschen Bewegung". *Annalen der Physik.* (4) 19:371–81 (reprinted with editorial notes in Einstein CP 2, 333–45).

————(1906e). "Das Prinzip von der Erhaltung der Schwerpunktsbewegung und die Trägheit der Energie". *Annalen der Physik.* (4) 20:627–33 (reprinted with editorial notes in Einstein CP 2, 359–66).

————(1907j). "Über das Relativitätsprinzip und die aus demselben gezogenen Folgerungen". *Jahrbuch der Radioaktivität und Elektronik.* 4:411–62 (reprinted with editorial notes in Einstein CP 2, 433–88).

————(1915g). "Zur allgemeinen Relativitätstheorie (Nachtrag)". *K. Preussische Akademie der Wissenschaften Sitzungsberichte,* pp. 799–801 (reprinted with editorial notes in Einstein CP 6, 225–29).

————(1915h). "Erklärung der Perihelbewegung des Merkurs aus der allgemeinen Relativitätstheorie". *K. Preussische Akademie der Wissenschaften Sitzungsberichte,* pp. 831–39 (reprinted with editorial notes in Einstein CP 6, 233–43).

————(1915i). "Die Feldgleichungen der Gravitation". *K. Preussische Akademie der Wissenschaften Sitzungsberichte,* pp. 844–47 (reprinted with editorial notes in Einstein CP 6, 244–49).

————(1916e). "Die Grundlagen der allgemeinen Relativitätstheorie". *Annalen der Physik.* (4) 49:769–822 (reprinted with editorial notes in Einstein CP 6, 283–339).

————(1916j). "Strahlungs-Emission und -Absorption nach der Quantentheorie". *Verhandlungen der Deutschen Physikalischen Gesellschaft.* 18:318–23 (reprinted with editorial notes in Einstein CP 6, 363–70).

————(1916n). "Zur Quantentheorie der Strahlung". *Mitteilungen der Physikalische Gesellschaft Zürich.* 18:47–62 (reprinted with editorial notes in

Einstein CP 6, 381–98; the same paper appeared, with a minor correction, in *Physikalische Zeitschrift* 18:121–28 (1917)).

———(1917a). *Über die spezielle und die allgemeine Relativitätstheorie. (Gemeinverständlich).* Braunschweig: Vieweg (reprinted with editorial notes in Einstein, CP 6, 521–39).

———(1917b). "Kosmologische Betrachtungen zur allgemeinen Relativitätstheorie". *K. Preussische Akademie der Wissenschaften Sitzungsberichte*, pp. 142–52 (reprinted with editorial notes in Einstein CP 6, 540–52).

———(1922). "Bemerkung zu der Arbeit von A. Friedmann, 'Über die Krümmung des Raumes'". *Zeitschrift für Physik.* 11:326.

———(1923). "Notiz zu der Arbeit von A. Friedmann, 'Über die Krümmung des Raumes'". *Zeitschrift für Physik.* 16:228.

———(1924). "Quantentheorie des einatomigen idealen Gases". *Preussische Akademie der Wissenschaften. Physisch-mathematische Klasse Sitzungsberichte*, pp. 261–67.

———(1925a). "Quantentheorie des einatomigen idealen Gases. 2. Abhandlung". *Preussische Akademie der Wissenschaften. Physisch-mathematische Klasse Sitzungsberichte*, pp. 3–14.

———(1925b). "Quantentheorie des idealen Gases". *Preussische Akademie der Wissenschaften. Physisch-mathematische Klasse Sitzungsberichte*, pp. 18–25.

———(1933). "Zur Methode der theoretischen Physik". In Einstein 1934, pp. 176–187. (The Herbert Spencer Lecture, delivered at Oxford, 10 June 1933.)

———(1934). *Mein Weltbild.* Zweite Auflage. Amsterdam: Querido Verlag.

———(1935). "Elementary derivation of the equivalence of mass and energy". *Bulletin of the American Mathematical Society.* 41:223–30.

———(1936). "Physics and Reality". In Einstein, *Out of my later years*. Revised reprint edition. Secaucus, NJ: Citadel Press, 1956, pp. 59–97. (Originally published in *The Journal of the Franklin Institute*, vol. 221, no. 3, March 1936.)

Einstein, A. and M. Besso (1972). *Correspondance 1903–1955.* Traduction, notes et introduction de P. Speziali. Paris: Hermann.

Einstein, A., H. Born, and M. Born (1969). *Briefwechsel 1916–1955.* München: Nymphenburger Verlagshandlung.

Einstein, A. and M. Grossmann (1913). "Entwurf einer verallgemeinerten Relativitätstheorie und einer Theorie der Gravitation". *Zeitschrift für Mathematik und Physik.* 62:225–61.

Einstein, A. and W. de Sitter (1932). "On the relation between the expansion and the mean density of the universe". *Proceedings of the National Academy of Sciences.* 18:213–14.

Einstein, A. and L. Infeld (1949). "On the motion of particles in general relativity theory". *Canadian Journal of Mathematics.* 1:209–41.

Einstein, A., B. Podolsky, and N. Rosen (1935). "Can quantum-mechanical description of physical reality be considered complete?". *Physical Review.* 47:777–80 (reprinted in J. A. Wheeler and W. H. Zurek 1983, pp. 138–41).

Einstein, A., L. Infeld, and B. Hoffmann (1938). "The gravitational equations and the problem of motion". *Annals of Mathematics.* 39:65–100.

Ellis, G. F. R. and R. M. Williams (1988). *Flat and Curved Space-Times*. Oxford: Clarendon Press.

Ellis, R. L. (1849). "On the foundations of the theory of probabilities". *Transactions of the Cambridge Philosophical Society*. 8:1–6.

Euler, L. (1736). *Mechanica, sive motus scientia analytica exposita*. Petropoli [St. Petersburg]: Academia scientiarum.

——— (1748). "Reflexions sur l'espace et le tems". *Histore de l'Académie Royale des Sciences et Belles Lettres (Berlin)*. 4:324–33.

——— (1765). *Theoria motus corporum solidorum seu rigidorum ex primis nostrae cognitionis principiis stabilita et ad omnes motus qui in huiusmodi corpora cadere possunt, accommodata*. Rostock: A. F. Röse.

——— (1768–74). *Lettres à une princesse d'Allemagne sur divers sujets de physique et de philosophie*. St. Petersbourg: Académie Impériale des Sciences. 3 vols.

Everett (III), H. (1957). "'Relative state' formulation of quantum mechanics". *Reviews of Modern Physics*. 29:454–62. (Everett's Princeton doctoral dissertation; reprinted in B. S. DeWitt and N. Graham 1973, pp. 141–49.)

——— (1973). "The theory of the universal wave function". In B. S. DeWitt and N. Graham 1973, pp. 3–140 (this detailed presentation of Everett's ideas was written about the same time as his brief Ph.D. thesis of 1957, but it remained unpublished until 1973).

Falkenburg, B. (1995). *Teilchenmetaphysik: Zur Realitätsauffassung in Wissenschaftsphilosophie und Microphysik*. 2. Auflage. Heidelberg: Spektrum.

Fano, U. (1957). "Description of states in Quantum Mechanics by density matrix and operator techniques". *Reviews of Modern Physics*. 29:74–93.

Faraday, M. *Diary*. London: G. Bell & Sons, 1932–1936. 7 vols.

——— (EREM). *Experimental Researches in Electricity and Magnetism*. London: Richard & John E. Taylor, 1839–1855. 3 vols.

Fermi, E. (1926). "Zur Quantelung des idealen einatomigen Gases". *Zeitschrift für Physik*. 36:902–12.

——— (1937). *Thermodynamics*. Englewood Cliffs, NJ: Prentice-Hall.

Février, P. (1937a). "Les relations d'incertitude de Heisenberg et la logique". *Comptes rendus de l'Académie des Sciences*. 204:481–83.

——— (1937b). "Sur une forme générale de la définition d'une logique". *Comptes rendus de l'Académie des Sciences*. 204:958–59.

——— (1951). See Destouches-Février, P.

Feyerabend, P. K. (1958). "An attempt at a realistic interpretation of experience". *Proceedings of the Aristotelean Society*. 58:143ff (reprinted in Feyerabend 1981, I, 17–36).

——— (1962). "Explanation, reduction and empiricism". *Minnesota Studies in the Philosophy of Science*. 3:28–97 (reprinted in Feyerabend 1981, I, 44–96).

——— (1970). "Against Method: Outline of an Anarchistic Theory of Knowledge". *Minnesota Studies in the Philosophy of Science*. 4:17–130 (enlarged version in book form: London, New Left Books, 1975; revised edition: London, Verso, 1988).

————(1981). *Philosophical Papers*. Cambridge: Cambridge University Press. 2 vols.

Feynman, R. P., R. B. Leighton, and M. Sands (1964). *The Feynman Lectures on Physics. Volume II: Mainly Electromagnetism and Matter.* Reading, MA: Addison-Wesley.

Finkelstein, D. (1962/63). "The logic of quantum physics". *Transactions of the New York Academy of Sciences.* 25:621–37.

————(1969). "Matter, space and logic". In R. S. Cohen and M. Wartofsky, eds., *Boston Studies in the Philosophy of Science.* Dordrecht: Reidel, Vol. 5, pp. 199–215.

————(1972). "The physics of logic". In R. G. Colodny 1972, pp. 47–66.

FitzGerald, G. F. (1885). "On the structure of mechanical models illustrating some properties of the aether". In G. F. FitzGerald 1902, pp. 157–62 (originally published in *Philosophical Magazine*, 19:438–43 (1885)).

————(1889). "The ether and the earth's atmosphere". *Science.* 12:390.

————(1902). *Scientific Writings of the Late George Francis FitzGerald.* Edited by J. Larmor. Dublin: Hodges and Figgis.

Fourier, J. (1822). *Théorie analytique de la chaleur.* Paris: Firmin Didot (facsimile reprint: Paris, Jacques Gabay, 1988).

Fox, R. (1971). *The Caloric Theory of Gases from Lavoisier to Regnault.* Oxford: Clarendon Press.

Frank, P. (1932). *Das Kausalgesetz und seine Grenzen.* Wien: Julius Springer.

————(1935). "Philosophische Deutungen und Missdeutungen der Quantentheorie". *Erkenntnis.* 6:303–17.

Frege, G. (KS). *Kleine Schriften.* Herausgegeben von I. Angelelli. Darmstadt: Wissenschaftliche Buchgesellschaft, 1967.

————(1879). *Begriffsschrift, eine der arithmetischen nachgebildete Formelsprache.* Halle a.S.: Louis Nebert.

————(1884). *Die Grundlagen der Arithmetik: Eine logisch mathematische Untersuchung über den Begriff der Zahl.* Breslau: Wilhelm Koebner.

Friedman, M. (1983). *Foundations of Space-Time Theories: Relativistic Physics and the Philosophy of Science.* Princeton, NJ: Princeton University Press.

————(1992). *Kant and the Exact Sciences.* Cambridge, MA: Harvard University Press.

Friedmann, A. (1922). "Über die Krümmung des Raumes". *Zeitschrift für Physik.* 10:377–86.

————(1924). "Über die Möglichkeit einer Welt mit konstanter negativer Krümmung des Raumes". *Zeitschrift für Physik.* 21:326–32.

Friedrichs, K. (1927). "Eine invariante Formulierung des Newtonschen Gravitationsgesetzes und des Grenzüberganges vom Einsteinschen zum Newtonschen Gesetz". *Mathematische Annalen.* 98:566–75.

Funkenstein, A. (1986). *Theology and the Scientific Imagination from the Middle Ages to the Seventeenth Century.* Princeton, NJ: Princeton University Press.

Gaifman, H. and M. Snir (1982). "Probabilities over rich languages". *Journal of Symbolic Logic.* 47:495–548.

Galileo Galilei (EN). *Le Opere.* Nuova ristampa della Edizione Nazionale. Firenze: G. Barbera, 1964–1966. 20 vols.

————(1623). *Il saggiatore nel quale con bilancia esquisita e giusta si ponderano le cose contenute nella Libra astronomica e filosofica di Lotario Sarsi Sigensano, scritto in forma di lettera all' ill.^{mo} et rever.^{mo} mons.^{re} d. Virginio Cesarini*. Roma: Appresso Giacomo Mascardi.

————(1638). *Discorsi e dimostrazioni matematiche intorno à due nuove scienze attenenti alla Mecanica & i Movimenti Locali*. Leida: Appresso gli Elsevirii.

————(1974). *Two New Sciences, including Centers of Gravity and Force of Percussion*. Translated by S. Drake. Madison, WI: University of Wisconsin Press.

Gamow, G. (1948). "The evolution of the universe". *Nature*. 162:680–82.

Gauss, C. F. (WW). *Werke*. Achter Band. Herausgegeben von der Königlichen Gesellschaft der Wissenschaften zu Göttingen. Leipzig: Teubner, 1900.

Geiger, H. and E. Marsden (1909). "On a diffuse reflection of the α-particles". *Royal Society of London Proceedings*. 82:495–99.

Gell-Mann, M. and J. B. Hartle (1993). "Classical equations for quantum systems". *Physical Review* D. 47:3345–82.

Gentzen, G. (1934). "Untersuchungen über das logische Schliessen". *Mathematische Zeitschrift*. 39:176–210, 405–31.

Gibbs, J. W. (1902). *Elementary Principles in Statistical Mechanics developed with especial reference to the Rational Foundations of Thermodynamics*. New York: Scribner's (reprint: Woodbridge, CN, Ox Bow Press, 1981).

Gilson, É. (1930). *Étude sur le rôle de la pensée médiévale dans la formation du système cartesien*. Paris: Vrin.

Gleason, A. M. (1957). "Measures on the closed subspaces of a Hilbert space". *Journal of Mathematics and Mechanics*. 6:885–93 (reprinted in C. A. Hooker 1975, pp. 123–33).

Gödel, K. (1949). "An example of a new type of cosmological solutions of Einstein's field equations of gravitation". *Reviews of Modern Physics*. 21:447–50.

Goldstein, H. (1950). *Classical Mechanics*. Reading, MA: Addison-Wesley.

Gooding, D. (1981). "Final steps to the field theory: Faraday's study of magnetic phenomena, 1845–1850". *Historical Studies in the Physical Sciences*. 11:231–75.

Gouy, L. G. (1888). "Note sur le mouvement brownien". *Journal de physique théorique et appliquée*. 7:561–64.

Greenberger, D. M., M. A. Horne, and A. Zeilinger (1989). "Going beyond Bell's Theorem". In M. Kafatos, ed., *Bell's Theorem, Quantum Theory, and Conceptions of the Universe*. Dordrecht: Kluwer; pp. 73–76.

Greenberger, D. M., M. A. Horne, A. Shimony, and A. Zeilinger (1990). "Bell's Theorem without inequalities". *American Journal of Physics*. 58:1131–43.

Guth, A. H. (1982). "Inflationary universe: A possible solution of the horizon and flatness problems". *Physical Review* D. 23:347–56.

————(1997). *The Inflationary Universe: The Quest for a New Theory of Cosmic Origins*. Reading, MA: Addison-Wesley.

Hacking, I. (1965). *Logic of Statistical Inference*. Cambridge: Cambridge University Press.

Hafele, J. C. and R. E. Keating (1972). "Around-the-world atomic clocks". *Science*. 177:166–70.

Hall, A. R. and M. B. Hall, eds. (1962). *Unpublished Scientific Papers of Isaac Newton. A Selection from the Portsmouth Collection in the University Library, Cambridge*. Cambridge: Cambridge University Press.

Hamilton, W. R. (1835). "Second Essay on a General Method in Dynamics". *Royal Society of London Philosophical Transactions*. 125:95ff. (reprinted together with the – first – essay "On a General Method in Dynamics" in Hamilton, *Collected Papers*, Vol. II. Cambridge: Cambridge University Press, 1940, pp. 103–211).

Hannabuss, K. (1997). *An Introduction to Quantum Theory*. Oxford: Clarendon Press.

Hanson, N. R. (1958). *Patterns of Discovery*. Cambridge: Cambridge University Press.

Harman, P. M. (1982a). *Metaphysics and Natural Philosophy: The Problem of Substance in Classical Physics*. Sussex: Harvester.

——(1982b). *Energy, Force, and Matter: The Conceptual Development of Nineteenth-Century Physics*. Cambridge: Cambridge University Press.

——(1987). "Mathematics and reality in Maxwell's dynamical physics". In R. Kargon and P. Achinstein 1987, pp. 267–97.

Havas, P. (1964). "Four-dimensional formulations of Newtonian mechanics and their relation to the special and the general theory of relativity". *Reviews of Modern Physics*. 36:938–65.

Hawking, S. W. (1974). "Black hole explosions?". *Nature*. 248:30–31.

——(1975). "Particle creation by black holes". *Communications of Mathematical Physics*. 43:199–220.

Hawking, S. W. and G. F. R. Ellis (1973). *The Large Scale Structure of Space-Time*. Cambridge: Cambridge University Press.

Hawking, S. and W. Israel, eds. (1987). *Three Hundred Years of Gravitation*. Cambridge: Cambridge University Press.

Hawking, S. W. and R. Penrose (1970). "The singularities of gravitational collapse and cosmology". *Royal Society of London Proceedings*. A 314:529–48.

Heisenberg, W. (1925). "Über quantentheoretische Umdeutung kinematischer und mechanischer Beziehungen". *Zeitschrift für Physik*. 33:879–93 (English translation in B. L. Van der Waerden 1967, pp. 261–76).

Heisenberg, W. (1927). "Über den anschaulichen Inhalt der quantentheoretischen Kinematik und Mechanik". *Zeitschrift für Physik*. 43:172–98.

——(1930). *The Physical Principles of the Quantum Theory*. Translated into English by C. Eckart and F. C. Hoyt. Chicago: University of Chicago Press.

——(1969). *Der Teil und das Ganze*. München: Piper.

Helmholtz, H. (WA). *Wissenschaftliche Abhandlungen*. Leipzig: Barth, 1882–1895.

——(1847). *Über die Erhaltung der Kraft*. Berlin: Reimer (reprinted, with addenda, in Helmholtz WA, I, 12–75).

——(1866). "Über die tatsächlichen Grundlagen der Geometrie". *Verhandlungen des naturhistorisch-medicinischen Vereins zu Heidelberg*. 4:197–202.

——(1887). "Zählen und Messen, erkenntnistheoretisch betrachtet". In *Philosophische Aufsätze Eduard Zeller gewidmet*. Leipzig: Verlag Fues (reprinted in H. von Helmholtz, *Die Tatsachen in der Wahrnehmung/Zählen und Messen.* Darmstadt: Wissenschaftliche Buchgesellschaft, 1959; pp. 77–112).

Hempel, C. G. and P. Oppenheim (1948). "Studies in the logic of explanation". *Philosophy of Science.* 15:135–75.

Hepp, K. (1972). "Quantum theory of measurement and macroscopic observables". *Helvetica Physica Acta.* 45:237–48.

Herapath, J. (1821). "A mathematical inquiry into the causes, laws and principal phenomenae of heat, gases, gravitation, etc.". *Annals of Philosophy.* (2) 1:273–93, 340–51, 401–06.

Herivel, J. (1965). *The Background to Newton's Principia: A Study of Newton's Dynamical Researches in the Years 1664–1684.* Oxford: Clarendon Press.

Herschel, J. F. W. (1830). *A Preliminary Discourse on the Study of Natural Philosophy.* London: Longman, Rees, Orme, Brown, & Green, and John Taylor (facsimile reprint: Chicago, University of Chicago Press, 1987).

Hertz, H. (1894). *Die Prinzipien der Mechanik in neuem Zusammenhang dargestellt.* Leipzig: Barth (facsimile reprint: Darmstadt, Wissenschaftliche Buchgesellschaft, 1963).

Hilbert, D. (1899). "Grundlagen der Geometrie". In *Festschrift zur Feier der Enthüllung des Gauss-Weber-Denkmals in Göttingen.* Leipzig: Teubner, pp. 3–92.

Hintikka, J. (1967). "Kant on the mathematical method". *The Monist.* 51:352–75 (reprinted in L. W. Beck, ed., *Kant Studies Today.* LaSalle, IL: Open Court, 1959, pp. 117–40).

Hooke, R. (PW). *Posthumous Works.* London: Frank Cass & Co., 1974.

Hooker, C. A., ed. (1975). *The Logico-Algebraic Approach to Quantum Mechanics.* Volume I. Dordrecht: Reidel.

Hopf, L. (1948). *Introduction to the Differential Equations of Physics.* Translated by W. Nef. New York: Dover.

Hoyle, F. (1948). "A new model for the expanding universe". *Monthly Notices of the Royal Astronomical Society.* 108:372–82.

Hubble, E. P. (1929). "A relation between distance and radial velocity among extragalactic nebulae". *Proceedings of the National Academy of Sciences.* 15:168–73.

Hughes, R. I. G. (1989). *The Structure and Interpretation of Quantum Mechanics.* Cambridge, MA: Harvard University Press.

Hume, D. (THN). *A Treatise of Human Nature.* Edited by L. A. Selby-Bigge. Oxford: Clarendon Press, 1888.

Hunt, B. J. (1987). "'How my model was right': G. F. FitzGerald and the reform of Maxwell's theory". In R. Kargon and P. Achinstein 1987, pp. 299–321.

Huygens, C. (OC). *Œuvres complètes.* Publiées par la Société Hollandaise des Sciences. La Haye: Martinus Nijhoff, 1888–1950. 22 vols.

———(1690). *Traité de la lumiere. Avec un discours de la cause de la pésanteur.* Leide.

Ignatowsky, W. (1910). "Einige allgemeine Bemerkungen zum Relativitätsprinzip". *Verhandlungen der Deutschen Physikalischen Gesellschaft.* 12:788–96.

———(1911). "Das Relativitätsprinzip". Archiv für Mathematik und Physik. (3) 17:1–24; 18:17–41.

Jammer, M. (1961). Concepts of Mass in Classical and Modern Physics. Cambridge, MA: Harvard University Press.

———(1966). The Conceptual Development of Quantum Mechanics. New York: McGraw-Hill.

———(1974). The Philosophy of Quantum Mechanics: The Interpretations of Quantum Mechanics in Historical Perspective. New York: Wiley.

Jaśkowski, S. (1934). "On the rules of suppositions in formal logic". Studia Logica. 1:5–32 (reprinted in S. McCall, ed., Polish Logic 1920–1939, Oxford: Clarendon Press, 1967, pp. 231–58).

Jauch, J. and C. Piron (1963). "Can hidden variables be excluded in Quantum Mechanics?". Helvetica Physica Acta. 38:827–37.

Jordan, P. (1926). "Über kanonische Transformationen in der Quantenmechanik". Zeitschrift für Physik. 37:383–86; 38:513–17.

———(1927). "Über eine neue Begründung der Quantenmechanik". Zeitschrift für Physik. 40:809–38.

Joule, J. P. (1843). "On the heat evolved during the electrolysis of water". In J. P. Joule 1884, pp. 109–23 (read to the Manchester Literary and Philosophical Society on 24 January 1843).

———(1844). "On the changes of temperature produced by the rarefaction and condensation of air". In J. P. Joule 1884, pp. 171–89 (communicated on 20 June 1844 to the Royal Society of London, which only published an abstract).

———(1849). "On the mechanical equivalent of heat". In J. P. Joule 1884, pp. 298–328 (read to the Royal Society of London on 21 June 1849; published in Royal Society of London Philosophical Transactions, 140:61–82 (1850)).

———(1884). The Scientific Papers of James Prescott Joule. London: The Physical Society (an additional volume appeared in 1886).

Kant, I. (Ak.). Gesammelte Schriften. Herausgegeben von der K. Preussischen, bzw. Deutschen Akademie der Wissenschaften. Berlin, 1902ff.

———(1746). Gedanken von der wahren Schätzung der lebendigen Kräfte und Beurtheilung der Beweise derer sich Herr von Leibniz und andere Mechaniker in dieser Streitsache bedienet haben, nebst einigen vorhergehenden Betrachtungen welche die Kraft der Körper überhaupt betreffen. Königsberg: Martin Eberhard Dorn (reprinted in Kant, Ak. I, 1–181).

———(1755). Allgemeine Naturgeschichte und Theorie des Himmels, oder Versuch von der Verfassung und dem mechanischen Ursprunge des ganzen Weltgebäudes nach Newtonischen Grundsätzen abgehandelt. Königsberg und Leipzig: Johann Friedrich Petersen (reprinted in Kant, Ak. I, 215–368).

———(1756). Metaphysica cum Geometria Iunctae Usus in Philosophia Naturalis, cuius Specimen I. continet Monadologiam Physicam. Königsberg (reprinted in Kant, Ak. I, 473–87).

———(1766). Träume eines Geistersehers, erläutert durch Träume der Metaphysik. Königsberg: Kanter (reprinted in Kant, Ak. II, 315–73).

———(1768). "Von dem ersten Grunde des Unterschiedes der Gegenden im Raume". Wochentliche Königsbergsche Frag- und Anzeigungsnachrichten. N° 6–8, 13. und 20. Februar 1768 (reprinted in Kant, Ak. II, 375–83).

————(1770). *De mundi sensibilis atque intelligibilis forma et principiis.* Regiomonti: Io. Iac. Kanter (reprinted in Kant, Ak. II, 384–419).

————(1781). *Critik der reinen Vernunft.* Riga: Johann Friedrich Hartknoch (the first half, which was substantially modified in the second edition, is reprinted in Kant, Ak. IV, 1–252).

————(1783). *Prolegomena zu einer jeden künftigen Metaphysik, die als Wissenschaft wird auftreten können.* Riga: Johann Friedrich Hartknoch (reprinted in Kant, Ak. IV, 252–383).

————(1786). *Metaphysische Anfangsgründe der Naturwissenschaft.* Riga: Johann Friedrich Hartknoch (reprinted in Kant, Ak. IV, 465–565).

————(1787). *Critik der reinen Vernunft.* Zweyte hin und wieder verbesserte Auflage. Riga: Johann Friedrich Hartknoch (reprinted in Kant, Ak. III, pp. 1–552).

————(1933). *Critique of Pure Reason.* Second edition with corrections. Translated by N. Kemp Smith. London: Macmillan.

Kargon, R. and P. Achinstein, eds. (1987). *Kelvin's Baltimore Lectures and Modern Theoretical Physics: Historical and Philosophical Perspectives.* Cambridge, MA: MIT Press.

Kelvin, Lord. *See* Thomson, W.

Kepler, J. (GW). *Gesammelte Werke.* Herausgegeben von M. Caspar et al. München: C. H. Beck, 1939– .

Keynes, J. M. (1921). *A Treatise on Probability.* London: Macmillan.

King, H. R. (1956). "Aristotle without *Prima Materia*". *Journal for the History of Ideas.* 17:370–89.

Kirchhoff, G. (1860). "Über das Verhältnis zwischen dem Emissionsvermögen und dem Absorptionsvermögen der Körper für Wärme und Licht". *Annalen der Physik.* (2) 109:275–301.

Klein, F. (GA). *Gesammelte mathematische Abhandlungen.* Berlin: Springer, 1921–1923. 3 vols.

————(1871). "Über die sogenannte Nicht-Euklidische Geometrie". *Mathematische Annalen.* 4:573–625.

————(1872). *Vergleichende Betrachtungen über neuere geometrische Forschungen.* Erlangen: A. Duchert.

————(1873). "Über die sogenannte Nicht-Euklidische Geometrie (Zweiter Aufsatz)". *Mathematische Annalen.* 6:112–45.

————(1874). "Nachtrag zu dem 'zweiten Aufsatz' über Nicht-Euklidische Geometrie". *Mathematische Annalen.* 7:531–37.

————(1890). "Zur Nicht-Euklidischen Geometrie". *Mathematische Annalen.* 37:544–72.

————(1893). "Vergleichende Betrachtungen über neuere geometrische Forschungen". *Mathematische Annalen.* 43:63–100 (revised version of Klein 1872).

————(1911). "Über die geometrischen Grundlagen der Lorentz-Gruppe". *Physikalische Zeitschrift.* 12:17–27.

Kochen, S. and E. P. Specker (1967). "The problem of hidden variables in Quantum Mechanics". *Journal of Mathematics and Mechanics.* 17:59–87 (reprinted in C. A. Hooker 1975, pp. 293–328).

Koyré, A. (1923). *Descartes und die Scholastik*. Bonn: Bouvier.

——— (1965). *Newtonian Studies*. London: Chapman & Hall.

Kragh, H. (1996). *Cosmology and Controversy: The Historical Development of Two Theories of the Universe*. Princeton, NJ: Princeton University Press.

Kramers, H. (1924a). "The law of dispersion and Bohr's theory of spectra". *Nature*. 113:673–674 (reprinted in B. L. Van der Waerden 1967, pp. 177–80).

——— (1924b). "The quantum theory of dispersion". *Nature*. 114:310–11 (reprinted in B. L. Van der Waerden 1967, pp. 199–201).

Kramers, H. and W. Heisenberg (1925). "Über die Streuung von Strahlung durch Atome". *Zeitschrift für Physik*. 31:681–708 (English translation in B. L. Van der Waerden 1967, pp. 223–52).

Krönig, A. K. (1856). "Grundzüge einer Theorie der Gase". *Annalen der Physik*. (2) 99:315–22.

Kuhn, T. S. (1959). "Energy conservation as an example of simultaneous discovery". In M. Claggett 1959b, pp. 321–56 (reprinted in Kuhn 1977, pp. 66–104).

——— (1962). *The Structure of Scientific Revolutions*. Chicago: University of Chicago Press.

——— (1970). *The Structure of Scientific Revolutions*. 2nd Ed., enlarged. Chicago: University of Chicago Press.

——— (1977). *The Essential Tension: Selected Studies in Scientific Tradition and Change*. Chicago: University of Chicago Press.

——— (1978). *Black-Body Theory and the Quantum Discontinuity: 1894–1912*. Oxford: Clarendon Press.

Kuratowski, K. (1921). "Sur la notion d'ordre dans la théorie des ensembles". *Fundamenta Mathematicae*. 2:161–171.

Kyburg Jr., H. E. and H. E. Smokler, eds. (1980). *Studies in Subjective Probability*. Huntington, NY: Krieger.

Lagrange, J. L. (1788). *Mécanique analitique*. Paris: Chez la Veuve Desaint (facsimile reprint: Paris, Jacques Gabay, 1989).

Lambert, J. H. (1786). "Theorie der Parallellinien". *Magazin für die reine und angewandte Mathematik*. 2:137–64; 3:325–58 (written about 1766; posthumously published by J. Bernoulli).

Landau, L. D. and E. M. Lifshitz (1960). *Course of Theoretical Physics*. Vol. I, *Mechanics*. Translated from the Russian by J. B. Sykes and J. S. Bell. Oxford: Pergamon Press.

Lange, L. (1885). "Über das Beharrungsgesetz". *K. Sächsische Gesellschaft der Wissenschaften zu Leipzig; math.-phys. Cl. Berichte*. 37:333–51.

Laplace, P. S. (OC). *Œuvres complètes*. Paris: Gauthier-Villars, 1878–1912. 14 vols.

——— (1795). *Essai philosophique sur les probabilités*. Paris (later republished as the Introduction to Laplace, *Traité analytique des probabilités*, Paris 1814, reprinted in Laplace, OC, VIII).

Larmor, J. (1900). *Aether and Matter: A Development of the Dynamical Relations of the Aether to Material Systems on the basis of the Atomic Constitution of Matter, including a discussion of the influence of the Earth's motion on optical phenomena*. Being an Adams Prize Essay in the University of Cambridge. Cambridge: Cambridge University Press.

———(1904). "Lord Kelvin on optical and molecular dynamics". *Nature.* 70(Supplement):iii–v.

Leibniz, G. W. (GP). *Die philosophischen Schriften.* Herausgegeben von C. J. Gerhardt. Hildesheim: Olms, 1965. 7 vols. (unaltered reprint of the original edition published in Berlin, 1875–1890).

———(OFI). *Opuscules et fragments inédits.* Extraits des manuscrits de la Bibliothèque royale de Hanovre par L. Couturat. Hildesheim: Georg Olms, 1961 (facsimile reprint of the first edition: Paris, Alcan, 1903).

Lemaître, G. (1927). "Un univers homogène de masse constante et de rayon croissant, rendant compte de la vitesse radiale des nébuleuses extragalactiques". *Annales de la Société Scientifique de Bruxelles.* A 47:49–59.

Linde, A. (1983). "Chaotic inflation". *Physical Letters.* B 129:177–81.

———(1986). "Eternal chaotic inflation". *Modern Physics Letters.* A 1:81.

———(1987). "Inflation and quantum cosmology". In S. Hawking and W. Israel 1987, pp. 604–30.

Lindsay, R. B. and H. Margenau (1957). *Foundations of Physics.* New York: Dover (corrected reprint of first edition, published in 1936).

Lobachevsky, N. I. (ZGA). *Zwei geometrische Abhandlungen.* Aus dem Russischen übersetzt von F. Engel. Leipzig: Teubner, 1898–99. 2 vols.

[Locke, J.] (1690). *An Essay concerning Humane Understanding.* In four books. London: Printed for Thomas Basset and sold by Edward Mory. (Published anonymously; the author's name was added in the second edition.)

Locke, J. (1699). *Mr. Locke's Reply to the Right Reverend Lord Bishop of Worcester's Answer to his Second Letter.* London: Printed for A. and J. Churchil; and E. Castle.

London, F. (1926a). "Über die Jacobischen Transformationen der Quantenmechanik". *Zeitschrift für Physik.* 37:915–25.

———(1926b). "Winkelvariable und kanonische Transformationen in der Undulationsmechanik". *Zeitschrift für Physik.* 40:193–219.

London, F. and E. Bauer (1939). *La théorie de l'observation en mécanique quantique.* Paris: Hermann. Actualités scientifiques e industrielles, N° 775 (English translation, including a new paragraph by Fritz London, in J. A. Wheeler and W. H. Zurek 1983, pp. 217–59).

Lorentz, H. A. (CP). *Collected Papers.* Edited by P. Zeeman and A. D. Fokker. The Hague: Nijhoff, 1935–1939. 9 vols.

———(1892). "The relative motion of the earth and the ether". In Lorentz CP, vol. 4, pp. 219–23 (originally published in Dutch in *Versl. K. Nederlandse Akademie van Wetenschappen* 1:74–79 (1892)).

———(1895). *Versuch einer Theorie der electrischen und optischen Erscheinungen in bewegten Körpern.* Leiden: Brill (reprinted in Lorentz CP, vol. 5, pp. 1–137).

———(1897). "Über den Einfluss magnetischer Kräfte auf die Emission des Lichtes". *Annalen der Physik.* (3) 63:278–84.

———(1899). "Theorie simplifiée des phénomènes électriques et optiques dans les corps en mouvement". In Lorentz CP, vol. 5, pp. 139–55 (originally published in Dutch in *Versl. K. Nederlandse Akademie van Wetenschappen*, 7:507–22 (1899)).

———(1904). "Electromagnetic phenomena in a system moving with any velocity smaller than that of light". *K. Nederlandse Akademie van Wetenschappen. Proceedings.* 6:809–31 (reprinted in Lorentz, CP, vol. 5, pp. 172–97).

Lorenzen, P. and K. Lorenz (1978). *Dialogische Logik.* Darmstadt: Wissenschaftliche Buchgesellschaft.

Loschmidt, J. (1876). "Über den Zustand des Wärmegleichgewichtes eines Systemes von Körpern mit Rücksicht auf die Schwerkraft". *Sitzungsberichte der Akademie der Wissenschaften zu Wien, math.-naturwiss. Klasse.* 73:128–42.

Lovelock, D. and H. Rund (1975). *Tensors, Differential Forms, and Variational Principles.* New York: Wiley.

Lucas, J. R. (1984). *Space, Time and Causality: An Essay in Natural Philosophy.* Oxford: Clarendon Press.

Lüders, G. (1951). "Über die Zustandsänderung durch den Messprozess". *Annalen der Physik.* (5) 8:322–28.

Ludwig, G. (1953). "Der Messprozess". *Zeitschrift für Physik.* 135:483–511.

———(1954). *Grundlagen der Quantenmechanik.* Berlin: Springer.

———(1979). *Einführung in die Grundlagen der theoretischen Physik.* Band 4: Makrosysteme, Physik und Mensch. Braunschweig: Vieweg.

———(1984). *Einführung in die Grundlagen der theoretischen Physik.* Band 3: Quantentheorie. 2. Auflage. Braunschweig: Vieweg.

———(1984/85). *Foundations of Quantum Mechanics.* Translated by C. A. Hain. New York: Springer. 2 vols. (much revised and enlarged English edition of Ludwig 1954).

———(1985/87). *An Axiomatic Basis for Quantum Mechanics. Volume I: Derivation of Hilbert Space Structure. Volume II: Quantum Mechanics and Macrosystems.* Berlin: Springer.

Łukasiewickz, J. (1920). "On three-valued logic". In J. Łukasiewickz, *Selected Works.* Edited by L. Borkowski. Amsterdam: North-Holland, 1970; pp. 87–88 (first published in Polish in *Ruch Filosoficzny,* 5:169–71).

Mach, E. (1868). "Über die Definition der Masse". *Carl's Repertorium der Experimentalphysik.* 4:355–59 (reprinted in Mach 1909, pp. 50–54).

———(1872). *Die Geschichte und die Wurzel des Satzes von der Erhaltung der Arbeit.* Vortrag gehalten in der K. Böhm. Gesellschaft der Wissenschaften am 15. Nov. 1871. Prag: Calve.

———(1875). *Grundlinien der Lehre von den Bewegungsempfindungen.* Leipzig: Engelmann.

———(1883). *Die Mechanik in ihrer Entwicklung historisch-kritisch dargestellt.* Leipzig: Brockhaus.

———(1886). *Beiträge zur Analyse der Empfindungen.* Jena: Fischer.

———(1896). *Die Principien der Wärmelehre: Historisch-kritisch entwickelt.* Leipzig: Barth.

———(1905). *Erkenntnis und Irrtum: Skizzen zur Psychologie der Forschung.* Leipzig: Barth.

———(1906). *Die Analyse der Empfindungen und das Verhältnis des physischen zum psychischen*. Jena: Fischer (fifth edition, substantially enlarged, of Mach 1886).

———(1909). *Die Geschichte und die Wurzel des Satzes von der Erhaltung der Arbeit*. Vortrag gehalten in der K. Böhm. Gesellschaft der Wissenschaften am 15. Nov. 1871. Leipzig: Barth (unmodified reprint of Mach 1872).

———(1921). *Die Prinzipien der physikalischen Optik: Historisch und erkenntnispsychologisch entwickelt*. Leipzig: Barth.

———(AS). *The Analysis of Sensations and the Relation of the Physical to the Psychical*. Translated from the first German edition by C. M. Williams. Revised and supplemented from the fifth German edition by S. Waterlow. New York: Dover (English translation of Mach 1886/1906).

———(EI). *Erkenntnis und Irrtum: Skizzen zur Psychologie der Forschung*. Darmstadt: Wissenschaftliche Buchgesellschaft, 1968 (unmodified reprint of the fifth edition, Leipzig, 1926).

———(M). *Die Mechanik: historisch-kritisch dargestellt*. Darmstadt: Wissenschaftliche Buchgesellschaft, 1963 (unmodified reprint of the ninth edition, Leipzig, 1933).

Mackey, G. W. (1963). *The Mathematical Foundations of Quantum Mechanics: A lecture-note volume*. Reading, MA: Benjamin/Cummings.

Mainzer, K. (1995). "Quantentheorie". In J. Mittelstrass, ed., *Enzyklopädie Philosophie und Wissenschaftstheorie*. Band 3. Stuttgart: Metzler, 1995; pp. 429–36.

Maupertuis, P. L. M. (1732). *Discours sur les différentes figures des astres avec une exposition des systèmes de MM. Descartes et Newton*. Paris: Imprimerie Royale.

Maxwell, J. C. (1855/56). "On Faraday's lines of force". In Maxwell 1890, vol. I, pp. 155–229 (read on 10 December 1855 and 11 February 1856 at the Cambridge Philosophical Society; published in the Society's *Transactions*, 10 1 (1857)).

———(1860). "Illustrations of the dynamical theory of gases". In Maxwell 1890, vol. I, pp. 377–409. (Originally published in the *Philosophical Magazine*, (4) 19:19–32; 20:21–37 (1860).)

———(1861–62). "On physical lines of force". In Maxwell 1890, vol. I, pp. 451–513. (Originally published in the *Philosophical Magazine*.)

———(1864). "A dynamical theory of the electromagnetic field". In Maxwell 1890, pp. 526–97 (received on 27 October and read on 8 December 1864 at the Royal Society of London; published in the *Philosophical Transactions*, 155).

———(1866). "On the dynamical theory of gases". In Maxwell 1890, vol. II, pp. 26–78 (read at the Royal Society of London on 31 May 31 1866; originally published in the *Philosophical Transactions*, 157:49–88 (1867)).

———(1873). *A Treatise on Electricity and Magnetism*. Oxford: Clarendon Press.

———(1883). *Theory of Heat*. Seventh edition. London: Longmans, Green, and Co.

———(1890). *The Scientific Papers of James Clerk Maxwell*. Edited by W. D. Niven. Cambridge: Cambridge University Press. 2 vols. (unaltered reprint: New York, Dover, 1965).

———(1891). *A Treatise on Electricity and Magnetism.* Third edition. Oxford: Clarendon Press. 2 vols. (unabridged, slightly altered reprint: New York, Dover, 1954).

Mayo, D. G. (1996). *Error and the Growth of Experimental Knowledge.* Chicago: University of Chicago Press.

McMullin, E., ed. (1965). *The Concept of Matter in Greek and Medieval Philosophy.* Notre Dame: University of Notre Dame Press (reprint of the first part of *The Concept of Matter*, ed. by E. McMullin, Notre Dame: University of Notre Dame Press, 1963).

Mehra, J. and H. Rechenberg (1982–88). *The Historical Development of Quantum Theory.* New York: Springer. Five volumes bound as six.

Messiah, A. (1961). *Quantum Mechanics.* Translated from the French by G. M. Temmer and J. Potter. Amsterdam: North-Holland Publishing Co. 2 vols.

Michelson, A. A. (1881). "The relative motion of the earth and the luminiferous ether". *American Journal of Science.* 22:120–29.

———(1891). "On the application of interference methods to spectroscopic measurements". *Philosophical Magazine.* 31:338–46 (continued in *Philosophical Magazine*, 34:280–99 (1892)).

Michelson, A. A. and E. W. Morley (1887). "On the relative motion of the earth and the luminiferous ether". *American Journal of Science.* 34:333–45.

Migne, J. P., ed. (PG). *Patrologia graeca.* Paris, 1857–1866.

Mill, J. S. (1843). *A System of Logic Ratiocinative and Inductive.* London: John W. Parker.

———(1874). *A System of Logic Ratiocinative and Inductive, Being a Connected View of the Principles of Evidence and the Method of Scientific Investigation.* 8th Ed. New York: Harper & Brothers. (This edition, which appeared shortly after Mill's death, was the last one he corrected for publication. The University of Toronto Press published in 1973 a critical edition with all variant readings. Unfortunately, I no longer had access to it when I wrote this book.)

Miller, A. I., ed. (1990). *Sixty-Two Years of Uncertainty: Historical, Philosophical and Physical Inquiries into the Foundations of Quantum Mechanics.* New York: Plenum.

———(1994). *Early Quantum Electrodynamics: A Source Book.* Translations from the German by W. Grant. Cambridge: Cambridge University Press.

Minkowski, H. (1909). "Raum und Zeit". *Physikalische Zeitschrift.* 10:104–11.

———(1915). "Das Relativitätsprinzip". *Jahresbericht der Deutschen Mathematiker-Vereinigung.* 24:372–82. (Lecture delivered at the Göttingen Mathematical Society on 5 November 1907; printed also in *Annalen der Physik* (4) 47:927ff. (1915).)

Mittelstaedt, P. (1978). *Quantum Logic.* Dordrecht: Reidel.

———(1979). "Quantum Logic". In G. Toraldo di Francia 1979, pp. 264–99.

Mormann, T. (1996). "Categorical structuralism". In W. Balzer and C. U. Moulines, eds., *Structuralist Theory of Science: Focal Issues, New Results.* Berlin: de Gruyter, pp. 265–86.

Muller, F. A. (1997). "The Equivalence Myth of Quantum Mechanics". *Studies in the History and Philosophy of Modern Physics.* 28:35–61, 219–47.

Nersessian, N. J. (1984). *Faraday to Einstein: Constructing Meaning in Scientific Theories.* Dordrecht: Nijhoff.

482   References

Neumann, C. (1870). *Über die Principien der Galilei-Newton'schen Theorie.* Leipzig: Teubner.

Neurath, O. (1932/33). "Protokollsätze". *Erkenntnis.* 3:204–14. (Translated by G. Schick as "Protocol sentences" in Ayer 1959, pp. 199–208.)

Newton, I. (1687). *Philosophiae naturalis principia mathematica.* Londini: Jussu Societatis Regiae ac Typis Josephi Streater.

——(1713). *Philosophiae naturalis principia mathematica.* Editio secunda auctior et emendatior. Cantabrigiæ.

——(1726). *Philosophiae naturalis principia mathematica.* Editio tertia aucta & emendata. Londini: Apud Guil. & Joh. Innys, Regiæ Societatis typographos.

——(1934). *Mathematical Principles of Natural Philosophy and his System of the World.* Translated into English by A. Motte in 1729. Translations revised, and supplied with a historical and explanatory appendix, by F. Cajori. Berkeley: University of California Press. 2 vols.

——(1972). *Philosophiae naturalis principia mathematica,* 3rd Ed. (1726) with variant readings. Assembled and edited by A. Koyré and I. B. Cohen. Cambridge, MA: Harvard University Press. 2 vols.

——(1974). *Newton's Philosophy of Nature: Selections from his Writings.* Edited and arranged with notes by H. S. Thayer. New York: Hafner Press (fourth printing; originally published by Hafner Publishing Co. in 1953).

——(*Opticks*). *Opticks, or A Treatise on the Reflections, Refractions, Inflections & Colours of Light.* Based on the Fourth Edition, London, 1730. New York: Dover, 1979.

Noether, E. (1918). "Invariante Variationsprobleme". *Göttinger Nachrichten,* pp. 235–57.

Omnès, R. (1994). *The Interpretation of Quantum Mechanics.* Princeton, NJ: Princeton University Press.

O'Raifeartaigh, L., ed. (1972). *General Relativity: Papers in honour of J. L. Synge.* Oxford: Clarendon Press.

The Oxford English Dictionary (OED). Second Edition on Compact Disc for the Apple Macintosh®. Oxford: Oxford University Press, 1994.

Pais, A. (1982). *'Subtle is the Lord...': The Science and the Life of Albert Einstein.* Oxford: Clarendon Press.

——(1986). *Inward Bound: Of Matter and Forces in the Physical World.* Oxford: Clarendon Press.

——(1991). *Niels Bohr's Times, in Physics, Philosophy, and Polity.* Oxford: Clarendon Press.

Pars, L. A. (1965). *A Treatise on Analytical Dynamics.* New York: Wiley.

Pascal, B. *L'Œuvre de Pascal.* Texte établi et annoté par J. Chevalier. Paris: Gallimard.

Pasch, M. (1882). *Vorlesungen über neuere Geometrie.* Leipzig: Teubner.

Pauli, W. (1925a). "Über den Einfluss der Geschwindigkeitsabhängigkeit der Elektronenmasse auf den Zeemaneffekt". *Zeitschrift für Physik.* 31:373–85.

——(1925b). "Über den Zusammenhang des Abschlusses der Elektronengruppen im Atom mit der Komplexstruktur der Spektren". *Zeitschrift für Physik.* 31:765–83.

———(1926a). "Quantentheorie". In H. Geiger and K. Scheel, eds., *Handbuch der Physik*. Berlin: Springer. Vol. 23, pp. 1–278.

———(1926b). "Über das Wasserstoffspektrum vom Standpunkt der neuen Quantenmechanik". *Zeitschrift für Physik*. 36:336–63 (English translation in B. L. Van der Waerden 1967, pp. 387–415).

———(1940). "The connexion between spin and statistics". *Physical Review*. 58:716–22.

Peano, G. (1895–1908). *Formulaire des Mathématiques*. Torino: Bocca. 5 vols.

Peirce, C. S. (CP). *Collected Papers*. Edited by C. Hartshorne, P. Weiss, and A. W. Burks. Cambridge, MA: The Belknap Press of Harvard University Press. 1931–1958. 8 vols.

———(1892). "The doctrine of necessity examined". *The Monist*. 2:321–37 (reprinted in Peirce CP, 6.28–35).

———(1982–). *Writings of Charles S. Peirce: A Chronological Edition*. Max H. Fisch, General Editor. Bloomington: Indiana University Press.

———(RLT). *Reasoning and the Logic of Things*. The Cambridge Conferences: Lectures of 1898. Edited by K. L. Ketner. Cambridge, MA: Harvard University Press, 1992.

Penzias, A. A. and R. W. Wilson (1965). "A measurement of excess antenna temperature at 4080 MHz". *Astrophysical Journal*. 142:419–21.

Perrin, J. (1913). *Les Atomes*. Paris: Félix Alcan.

———*Atoms*. Translated by D. L. Hammick. Woodbridge, CT: Ox Bow Press.

Planck, M. (PAV). *Physikalische Abhandlungen und Vorträge*. Braunschweig: Vieweg, 1958. 3 vols.

———(1900). "Zur Theorie des Gesetzes der Energieverteilung im Normalspectrum". In Planck PAV, vol. I, pp. 698–716 (originally published in *Verhandlungen der Deutschen Physikalischen Gesellschaft*, 2:237–45 (1900)).

———(1906). "Das Prinzip der Relativität und die Grundgleichungen der Mechanik". In Planck PAV, vol. II, pp. 115–20 (originally published in *Verhandlungen der Deutschen Physikalischen Gesellschaft*, 4:136–41 (1906)).

Platon. *Œuvres Complètes*. Paris: Société d'Édition Les Belles Lettres, 1920ff. 13 vols. (I quote Plato by this edition. As is customary, reference is made to the pages of Henri Estienne's edition, Paris, 1578.)

Poincaré, H. (1891). "Les géométries non-euclidiennes". *Revue générale des sciences pures et appliquées*. 2:769–74 (reprinted with additions as Chapter III of Poincaré 1902).

———(1898). "La mésure du temps". *Revue de métaphysique et de morale*. 6:1–13 (reprinted in Poincaré, *La valeur de la science*, Paris: Flammarion, 1905).

———(1902). *La science et l'hypothèse*. Paris: Flammarion.

———(1905). "Sur la dynamique de l'électron". *Comptes rendus de l'Académie des Sciences*. 140:1504–08.

———(1906). "Sur la dynamique de l'électron". *Rendiconti del Circolo matematico di Palermo*. 21:129–75.

Pollard, H. (1976). *Celestial Mechanics*. The Mathematical Association of America. Carus Mathematical Monographs, No. 18.

Pólya, G. (1977). *Mathematical Methods in Science*. The Mathematical Association of America (revised edition; the book originally appeared in 1963 as Vol. 11 of the Studies in Mathematics series sponsored by the School Mathematics Studies Group).

Price, H. (1996). *Time's Arrow and Archimedes' Point: New Directions for the Physics of Time*. Oxford: Oxford University Press.

Putnam, H. (1957). "Three-valued logic". *Philosophical Studies*. 8:73–80 (reprinted in Putnam 1979, pp. 166–73).

——— (1962). "What theories are not". In Putnam 1979, pp. 215–27 (first published in E. Nagel, P. Suppes, and A. Tarski, eds., *Logic, Methodology and Philosophy of Science*, Stanford, CA: Stanford University Press, 1962).

——— (1965). "A philosopher looks at quantum mechanics". In Putnam 1979, pp. 130–58 (first published in R. G. Colodny 1965, pp. 75–101).

——— (1967). "Time and physical geometry". *Journal of Philosophy*. 64:240–47 (reprinted in Putnam 1979, pp. 198–205).

——— (1969). "The logic of quantum mechanics". In Putnam 1979, pp. 174–97. (First published as "Is logic empirical?" in R. Cohen and M. Wartofsky, eds., *Boston Studies in the Philosophy of Science*, Dordrecht: D. Reidel, Vol. 5, pp. 216–41).

——— (1970). "Is semantics possible?". In H. Putnam 1975, vol. 2, pp. 139–52 (first published in H. Kiefer and M. Munitz, eds. *Language, Belief and Metaphysics*, Albany: State University of New York Press, 1970).

——— (1973). "Explanation and reference". In H. Putnam 1975, vol. 2, pp. 196–214 (first published in Pearce and Maynard, *Conceptual Change*, Dordrecht: D. Reidel, 1973, pp. 199–221).

——— (1974). "How to think quantum-logically". *Synthese*. 29:55–61.

——— (1975a). *Mind, Language and Reality. Philosophical Papers, Volume 2*. Cambridge: Cambridge University Press.

——— (1975b). "The meaning of 'meaning'". In Putnam 1975, pp. 215–71 (first published in *Minnesota Studies in the Philosophy of Science*, 7).

——— (1979). *Mathematics, Matter and Method. Philosophical Papers, Volume 1* (2nd Ed.). Cambridge: Cambridge University Press.

——— (1981). "Quantum mechanics and the observer". *Erkenntnis*. 16:193–219 (reprinted in Putnam 1983, pp. 248–70).

——— (1983). *Realism and Reason. Philosophical Papers, Volume 3*. Cambridge: Cambridge University Press.

Ramsey, F. P. (1926). "Truth and probability". In F. P. Ramsey 1931, pp. 156–98.

——— (1931). *The Foundations of Mathematics and other Logical Essays*. London: Routledge & Kegan Paul.

Rédei, L. (1968). *Foundation of Euclidean and Non-Euclidean Geometries According to F. Klein*. Oxford: Pergamon Press.

Redhead, M. (1987). *Incompleteness, Nonlocality, and Realism: A Prolegomenon to the Philosophy of Quantum Mechanics*. Oxford: Clarendon Press.

——— (1988). "A philosopher looks at quantum field theory". In H. R. Brown and R. Harré 1988, pp. 9–23.

——— (1993). "The conventionality of simultaneity". In J. Earman, A. I. Janis, G. J. Massey, and N. Rescher 1993, pp. 103–28.

Reich, K. (1932). *Die Vollständigkeit der Kantischen Urteilstafel*. Berlin: Richard Schoetz.

Reichenbach, H. (1924). *Axiomatik der relativistischen Raum-Zeit-Lehre.* Braunschweig: Vieweg.

———(1928). *Philosophie der Raum-Zeit-Lehre.* Berlin: de Gruyter.

———(1944). *Philosophical Foundations of Quantum Mechanics.* Berkeley: University of California Press.

Riemann, B. (1867). "Über die Hypothesen, welche der Geometrie zugrunde liegen". *Göttinger Abhandlungen.* 13:133–52.

Rietdijk, C. W. (1966). "A rigorous proof of determinism derived from the special theory of relativity". *Philosophy of Science.* 33:341–44.

———(1976). "Special relativity and determinism". *Philosophy of Science.* 43:598–609.

Robertson, H. P. (1929). "On the foundations of relativistic cosmology". *National Academy of Sciences Proceedings.* 15:822–29.

———(1935/36). "Kinematics and world-structure". *Astrophysical Journal.* 82:284–301; 83:187–201, 257–71.

Robertson, H. P. and T. W. Noonan (1969). *Relativity and Cosmology.* Philadelphia: Saunders.

Robinson, H. M. (1974). "Prime matter in Aristotle". *Phronesis.* 19:168–88.

Roqué, X. (1997). "The manufacture of the positron". *Studies in History and Philosophy of Modern Physics.* 28B:73–129.

Rossi, B. and D. B. Hall (1941). "Variation of the rate of decay of mesotrons with momentum". *Physical Review.* 59:223–28.

Ruhla, C. (1992). *The Physics of Chance from Blaise Pascal to Niels Bohr.* Translated from the French by G. Baron. Oxford: Oxford University Press.

Rumford, Benjamin Count (CW). *Collected Works. Volume I, The Nature of Heat.* Edited by S. C. Brown. Cambridge, MA: The Belknap Press of Harvard University Press, 1968.

———(1798). "An experimental inquiry concerning the source of the heat which is excited by friction". *Royal Society of London Philosophical Transactions.* 88:80–102 (reprinted in Rumford CW, I, 1–26).

Russell, B. (1897). *An Essay on the Foundations of Geometry.* Cambridge: Cambridge University Press (unaltered reprint: New York, Dover, 1956).

———(1917). *Mysticism and Logic, and Other Essays.* London: Allen & Unwin.

———(1959). *My Philosophical Development.* London: Allen & Unwin.

Saccheri, G. (1733). *Euclides ab omni naevo vindicatus, sive conatus geometricus quo stabiliuntur prima ipsa universae geometriae principia.* Mediolani: Ex Typographia Pauli Antonii Montani (reprinted as Girolamo Saccheri's *Euclides vindicatus,* with facing English translation by G. B. Halsted: New York, Chelsea, 1986).

Schilpp, P. A., ed. (1949). *Albert Einstein: Philosopher-Scientist.* Evanston, IL: Library of Living Philosophers.

Schneider, I., ed. (1988). *Die Entwicklung der Wahrscheinlichkeitstheorie von den Anfängen bis 1933: Einführungen und Texte.* Darmstadt: Wissenschaftliche Buchgesellschaft.

Schrödinger, E. (WM). *Collected Papers on Wave Mechanics.* Together with Four Lectures on Wave Mechanics. Translated by J. F. Shearer and W. M. Deans. New York: Chelsea, 1982.

———(1926a). "Quantisierung als Eigenwertproblem. (Erste Mitteilung)".

Annalen der Physik. (4) 79:361–76 (English translation in Schrödinger WM, pp. 1–12).
———(1926b). "Quantisierung als Eigenwertproblem. (Zweite Mitteilung)". *Annalen der Physik.* (4) 79:489–527 (English translation in Schrödinger WM, pp. 13–40).
———(1926c). "Über das Verhältnis der Heisenberg-Born-Jordanschen Quantenmechanik zu der meinen". *Annalen der Physik.* (4) 79:734–56 (English translation in Schrödinger WM, pp. 45–61).
———(1926d). "Quantisierung als Eigenwertproblem. (Dritte Mitteilung: Störungstheorie, mit Anwendung auf den Starkeffekt der Balmerlinien)". *Annalen der Physik.* (4) 80:437–90 (English translation in Schrödinger WM, pp. 62–101).
———(1926e). "Quantisierung als Eigenwertproblem. (Vierte Mitteilung)". *Annalen der Physik.* (4) 81:109–39 (English translation in Schrödinger WM, pp. 102–23).
———(1928). *Four Lectures on Wave Mechanics.* Glasgow: Blackie & Son (reprinted in Schrödinger WM).
———(1935). "Die gegenwärtige Situation in der Quantenmechanik". *Naturwissenschaften.* 23:807–12, 823–28, 844–49 (English translation in J. A. Wheeler and W. H. Zurek 1983, pp. 152–67).
———(1995). *The Interpretation of Quantum Mechanics.* Dublin seminars (1949–1955) and other unpublished essays. Edited and with Introduction by M. Bitbol. Woodbridge, CN: Ox Bow Press.
Schwartz, E., ed. (ACO). *Acta conciliorum oecumenicorum.* Vol. 1.1.6. Berlin: de Gruyter, 1928.
Schwartz, L. (1957/1959). *Théorie des distributions.* Paris: Hermann. 2 vols.
Schwarzschild, K. (1916a). "Über das Gravitationsfeld eines Massenpunktes nach der Einsteinsche Theorie". *K. Preussische Akademie der Wissenschaften Sitzungsberichte,* pp. 189–96.
———(1916b). "Über das Gravitationsfeld einer Kugel aus inkompressibler Flüssigkeit nach der Einsteinsche Theorie". *K. Preussische Akademie der Wissenschaften Sitzungsberichte,* pp. 424–34.
Schweber, S. S. (1994). *QED and the Men Who Made It: Dyson, Feynman, Schwinger, and Tomonaga.* Princeton, NJ: Princeton University Press.
Schwinger, J. (1958). *Selected Papers on Quantum Electrodynamics.* New York: Dover.
Seelig, C. (1957). *Albert Einstein: A Documentary Biography.* Translated by M. Savill. London: Staples Press (English translation of Seelig, *Albert Einstein: eine dokumentarische Biographie,* Zürich: Europa Verlag, 1954).
Shamos, M. H., ed. (1959). *Great Experiments in Physics: First Hand Accounts from Galileo to Einstein.* New York: Holt, Rinehart and Winston.
Shapere, D. (1964). "The structure of scientific revolutions". *Philosophical Review.* 73:382–94 (reprinted in Shapere 1984, pp. 37–48).
———(1966). "Meaning and scientific change". In R. Colodny 1966, pp. 41–85 (reprinted in Shapere 1984, pp. 58–101).
———(1984). *Reason and the Search for Knowledge: Investigations in the Philosophy of Science.* Dordrecht: D. Reidel.

Shapiro, S. (1997). *Philosophy of Mathematics: Structure and Ontology.* New York: Oxford University Press.

Shimony, A. (1963). "Role of the observer in quantum theory". In Shimony 1993, pp. 3–33 (originally published in *American Journal of Physics*, 31:755–73 (1963)).

———(1990). "An exposition of Bell's Theorem". In A. I. Miller 1990, pp. 33–43 (reprinted, with some changes, in Shimony 1993, pp. 90–103).

———(1993). *Search for a Naturalistic World View. Volume II, Natural Science and Metaphysics.* Cambridge: Cambridge University Press.

Sklar, L. (1993). *Physics and Chance: Philosophical Issues in the Foundations of Statistical Mechanics.* Cambridge: Cambridge University Press.

Skyrms, B. (1987). "Dynamic coherence and probability kinematics". *Philosophy of Science.* 54:1–20.

Sneed, J. D. (1971). *The Logical Structure of Mathematical Physics.* Dordrecht: Reidel.

Solmsen, F. (1958). "Aristotle and Prime Matter: A reply to Hugh R. King". *Journal for the History of Ideas.* 19:243–52.

Sommerfeld, A. (1910a). "Zur Relativitätstheorie. I. Vierdimensionale Vektoralgebra.". *Annalen der Physik.* (4) 32:749–76.

———(1910b). "Zur Relativitätstheorie. II. Vierdimensionales Vektoranalysis". *Annalen der Physik.* (4) 33:649–89.

———(1915a). "Zur Theorie der Balmerschen Serie". *Sitzungsberichte der Bayerischen Akademie der Wissenschaften,* pp. 425–58.

———(1915b). "Die Feinstruktur der Wasserstoff und wasserstoffähnlichen Linien". *Sitzungsberichte der Bayerischen Akademie der Wissenschaften,* pp. 459–500.

———(1916). "Zur Quantentheorie der Spektrallinien". *Annalen der Physik.* (4) 51:1–94, 125–67.

———(1919). *Atombau und Spektrallinien.* Braunschweig: Vieweg.

———(1924). *Atombau und Spektrallinien.* Vierte umgearbeitete Auflage. Braunschweig: Vieweg.

Spinoza. *Opera.* Im Auftrag der Heidelberger Akademie der Wissenschaften herausgegeben von C. Gebhardt. Heidelberg: Carl Winters Universitätsbuchhandlung, n.d. 4 vols.

Stachel, J. (1986). "Do quanta need a new logic?". In R. G. Colodny 1986, pp. 229–347.

Stachel, J. and R. Torretti (1982). "Einstein's first derivation of mass-energy equivalence". *American Journal of Physics.* 50:760–63.

Stein, H. (1970). "On the notion of field in Newton, Maxwell, and beyond". *Minnesota Studies in the Philosophy of Science.* 5:264–310 (pp. 287ff. contain comments by Gerd Buchdahl and Mary Hesse, followed by Stein's reply).

———(1991). "On Relativity Theory and the openness of the future". *Philosophy of Science.* 58:147–67.

Stein, H. and A. Shimony (1971). "Limitations on measurement". In B. d'Espagnat 1971, pp. 56–76.

Strawson, P. F. (1959). *Individuals: An Essay in Descriptive Metaphysics.* London: Methuen.

Sudbery, A. (1986). *Quantum Mechanics and the Particles of Nature: An Outline for Mathematicians.* Cambridge: Cambridge University Press.

Suppe, F. (1967). "The Meaning and Use of Models in Mathematics and the Exact Sciences". Ph.D. dissertation. University of Michigan.

———, ed. (1977). *The Structure of Scientific Theories.* Second Edition. Urbana, IL: University of Illinois Press.

———(1989). *The Semantic Conception of Theories and Scientific Realism.* Urbana, IL: University of Illinois Press.

Suppes, P. (1960). "A comparison of the meaning and uses of models in mathematics and the empirical sciences". *Synthese.* 12:287–301.

———(1962). "Models of data". In E. Nagel, P. Suppes, and A. Tarski, eds., *Logic, Methodology and Philosophy of Science:* Stanford, CA: Stanford University Press, pp. 252–61.

———(1967). "What is a scientific theory". In S. Morgenbesser, ed., *Philosophy of Science Today.* New York: Basic Books, pp. 55–67.

———(1969). "The structure of theories and the analysis of data". In F. Suppe 1977, pp. 266–307.

Teller, P. (1973). "Conditionalization and observation". *Synthese.* 26:218–58.

———(1988). "Three problems of renormalization". In H. R. Brown and R. Harré 1988, pp. 73–89.

———(1995). *An Interpretive Introduction to Quantum Field Theory.* Princeton, NJ: Princeton University Press.

Thomson, J. (1884). "On the Law of Inertia, the Principle of Chronometry and the Principle of Absolute Clinural Rest, and of Absolute Rotation". *Royal Society of Edinburgh Proceedings.* 12:568–78.

Thomson, W. (1874). "The kinetic theory of the dissipation of energy". *Royal Society of Edinburgh Proceedings.* 8:325–34.

Toraldo di Francia, G., ed. (1979). *Problems in the Foundations of Physics. Proceedings of the International School of Physics "Enrico Fermi", Course LXXII.* Amsterdam: North-Holland.

Torretti, R. (1967). *Manuel Kant: Estudio sobre los fundamentos de la filosofía crítica.* Santiago de Chile: Ediciones de la Universidad de Chile.

———(1978). *Philosophy of Geometry from Riemann to Poincaré.* Dordrecht: Reidel (corrected reprint: Dordrecht, Reidel, 1984).

———(1983). *Relativity and Geometry.* Oxford: Pergamon (corrected reprint: New York, Dover, 1996).

———(1990). *Creative Understanding: Philosophical Reflections on Physics.* Chicago: University of Chicago Press.

Truesdell, C. (1968). *Essays in the History of Mechanics.* New York: Springer.

Uhlenbeck, G. E. and S. Goudsmit (1925). "Ersetzung der Hypothese vom unmechanischen Zwang durch eine Forderung bezüglich des inneren Verhaltens jedes einzelnen Elektrons". *Naturwissenschaften.* 13:953–54.

Van der Waerden, B. L., ed. (1967). *Sources of Quantum Mechanics.* Amsterdam: North-Holland (unaltered reprint: New York, Dover, 1968).

van Fraassen, B. C. (1972). "A formal approach to the philosophy of science". In R. G. Colodny 1972, pp. 303–66.

Voigt, W. (1887). "Über das Doppler'sche Prinzip". *Göttinger Nachrichten*, pp. 41–51.

von Arnim, J. (SVF). *Stoicorum veterum fragmenta*. Leipzig: Teubner, 1903–1905. 4 vols.

von Laue, M. (1911). "Zur Dynamik der Relativitätstheorie". *Annalen der Physik*. (4) 35:524–42.

———(1920). "Theoretisches über neuere optische Beobachtungen zur Relativitätstheorie". *Physikalische Zeitschrift*. 21:659–62.

———(1955). *Die Relativitätstheorie. 1. Band. Die spezielle Relativitätstheorie*. 6. durchgesehene Auflage. Braunschweig: Vieweg.

von Mises, R. (1964). *Mathematical Theory of Probability and Statistics*. New York: Academic Press.

von Neumann, J. (1927). "Mathematische Begründung der Quantenmechanik". *Nachrichten der Gesellschaft der Wissenschaften zu Göttingen*. pp. 1–57.

———(1932). *Mathematische Grundlagen der Quantenmechanik*. Berlin: Springer.

———(1955). *Mathematical Foundations of Quantum Physics*. Translated by E. T. Beyer. Princeton, NJ: Princeton University Press (English translation of von Neumann 1932).

Wald, R. M. (1984). *General Relativity*. Chicago: University of Chicago Press.

Walker. A. G. (1935). "On the formal comparison of Milne's kinematical system with the systems of general relativity". *Monthly Notices of the Royal Astronomical Society*. 95:263–69.

———(1937). "On Milne's theory of world-structure". *London Mathematical Society Proceedings*. 42:90–127.

Warwick, A. (1995). "The sturdy Protestants of science: Larmor, Trouton, and the Earth's motion through the ether". In J. Z. Buchwald 1995, pp. 300–43.

Waterston, J. J. (1846). "On the physics of media that are composed of free and perfectly elastic molecules in a state of motion". *Royal Society of London Philosophical Transactions*. 183A:5–79 (only an abstract appeared in the year of submission: Royal Society of London Proceedings 5:604 (1846)).

Watts, I. (1725). *Logick: or, the right use of reason in the enquiry after truth, with a variety of rules to guard against error, in the affairs of religion and human life, as well as in the sciences*. London. (The 20th Ed. of this influential textbook was printed in Glasgow by William Smith for John Bryce and Robert Farie in 1779.)

Weinberg, S. (1972). *Gravitation and Cosmology: Principles and Applications of the General Theory of Relativity*. New York: Wiley.

———(1995/96). *The Quantum Theory of Fields*. Cambridge: Cambridge University Press. 2 vols.

Westfall, R. S. (1971). *Force in Newton's Physics: The Science of Dynamics in the Seventeenth Century*. London: Macdonald.

———(1980). *Never at Rest: A Biography of Isaac Newton*. Cambridge: Cambridge University Press.

Wheeler, J. A. and W. H. Zurek, eds. (1983). *Quantum Theory and Measurement*. Princeton, NJ: Princeton University Press.

Whewell, W. (1837). *History of the Inductive Sciences, from the Earliest to the Present Times*. London: J. W. Parker. 3 vols.

———(1840). *The Philosophy of the Inductive Sciences, founded upon their History*. London: J. W. Parker.

———(1844). "On the fundamental antithesis of philosophy". *Cambridge Philosophical Society Proceedings*. 8:170–81 (reprinted in Whewell 1847, II, 647–68, and in R. E. Butts 1968, pp. 54–75).

———(1847). *The Philosophy of the Inductive Sciences, founded upon their History*. A new edition, with corrections and additions, and an appendix, containing philosophical essays previously unpublished. London: J. W. Parker. 2 vols. (reprint: London, Frank Cass, 1967).

———(1858). *Novum Organum Renovatum*. Being the second part of the Philosophy of the Inductive Sciences. 3rd Ed., with large additions. London: John W. Parker & Son.

———(1860). *On the Philosophy of Discovery*. Chapters Historical and Critical, including the completion of the third edition of the Philosophy of the Inductive Sciences. London: John W. Parker & Son (reprint: New York, Burt Franklin, 1971).

Whitehead, A. N. and B. Russell (1910–1913). *Principia Mathematica*. Cambridge: Cambridge University Press. 3 vols.

Whittaker, E. T. (1928). "Note on the law that light-rays are the null-geodesics of a gravitational field". *Cambridge Philosophical Society Proceedings*. 24:32–34.

———(1951/53). *A History of the Theories of Aether and Electricity*. London: Thomas Nelson & Sons (unaltered reprint: New York, Dover, 1989).

Wigner, E. P. (1939). "On unitary representations of the inhomogeneous Lorentz group". *Annals of Mathematics*. 40:149–204.

———(1952). "Die Messung quantenmechanischer Operatoren". *Zeitschrift für Physik*. 133:101–08.

———(1970). "On hidden variables and quantum mechanical probabilities". *American Journal of Physics*. 38:1005–1009.

———(1983). "Interpretation of Quantum Mechanics". In J. A. Wheeler and W. H. Zurek 1983, pp. 260–314 (lectures originally given in the Physics Department of Princeton University during 1976; revised for publication in 1981).

Will, C. M. (1981). *Theory and Experiment in Gravitational Physics*. Cambridge: Cambridge University Press.

Williams, C. J. W. (1982). *Aristotle's De Generatione et Corruptione*. Translated with notes. Oxford: Clarendon Press.

Wilson, W. (1915). "The quantum theory of radiation and line spectra". *Philosophical Magazine*. 29:795–802.

Wise, M. N. and C. Smith. "The practical imperative: Kelvin challenges the Maxwellians". In R. Kargon and P. Achinstein 1987, pp. 323–48.

Wittgenstein, L. (1922). *Tractatus Logico-Philosophicus*. London: Routledge and Kegan Paul.

———(1953). *Philosophische Untersuchungen*. Edited by R. Rhees and G. E. M. Anscombe. Oxford: Blackwell.

Wolf, A. (1939). *A History of Science, Technology, and Philosophy in the Eighteenth Century*. New York: Macmillan.

Wolter, A. B. (1963). "The Ockamist critique". In E. McMullin 1965, pp. 124–46.

Wróblewski, A. (1985). "De mora luminis: A spectacle in two acts with a prologue and an epilogue". *American Journal of Physics*. 53:620–30.

Young, T. (1807). *A Course of Lectures on Natural Philosophy and the Mechanical Arts*. London: J. Johnson.

Zawirski, Z. (1932). "Les logiques nouvelles et le champ de leur application". *Revue de métaphysique et de morale*. 39:503–19 (Zawirski's proposal for a three-valued quantum logic was first put forward in Polish, in a Poznan journal, in 1931).

Zeeman, P. (1897). "On the influence of magnetism on the nature of the light emitted by a substance". *Philosophical Magazine*. 43:226–39 (English translation of two articles published in Dutch in 1896).

Zermelo, E. (1896). "Über einen Satz der Dynamik und die mechanische Wärmetheorie". *Annalen der Physik*. (3) 57:485–94.

Index

α-particles, 309, 311
abduction, 228–30
aberration of starlight, 251n
absolute space idle in Newtonian
    mechanics, 53–54
accident, 54
action: by contact, 75, 80; faster-
    than-light, 354
Aeschylus, 130n
aether, 181n, 251–52; Aristotle's, 11,
    16, 175n; Descartes's, 175n;
    Lorentz's, 180, 252, 427;
    Maxwell's, 173–80; motion of
    earth in, 251; partial drag, 251,
    260n
aim of physical theory: Berkeley,
    102, 103; Duhem, 243;
    Helmholtz, 184; Mach, 237, 238;
    Maxwell, 199; Newton, 41, 77
aitema, 148
aitia, 130n
Alexander, H. G., 56n, 99n
Alpher, Ralph A., 304n, 305
Ampère, André-Marie, 79n, 82–83,
    171, 178
analogies of experience (Kant), 120–
    38, 232n, 369
analogy: in mathematics and
    philosophy, 134; in physics, 173,
    198

analytic statements, 98, 110n
analytical mechanics, 84–96, 176
and/or, 438n
Anderson, Carl, 394
angle between two vectors, 445
angular momentum of earth, 51n
Anscombe, G. E. M., 128n
anthropic principle, 400–401
anticipations of perception (Kant),
    120, 369
antinomies (Kantian), 107, 109,
    140–45, 434
antiperistasis, 11
Antiphon, 405
antisymmetric, 133; tensor, 428
Apollonius of Perga, 12n, 27, 432
Appell, Paul, 239n
approximation, 74, 231, 246
Aquinas, Thomas, 8
Arabatzis, Theodore, 309n
Arana, Juan, 80n
Archimedes, 7, 26
Aristotle, 1, 2, 3, 5, 6, 8–13, 14, 36,
    41, 79, 130n, 140n, 175n, 188n,
    310, 379, 416, 422, 431;
    experience and experiment, 3;
    four simple bodies, 10; two-tiered
    universe, 8
arithmetic and intuition of time,
    118–19

493

Darrigol, Olivier, 308n, 316n, 317n, 318
Darwin, Charles, 423, 435
Davidson, Donald, 423
Davisson, Clinton Joseph, 326
Davy, Humphry, 181n, 182n
de Broglie, Louis, 128, 325–26, 327, 332, 335, 374n
de Finetti, Bruno, 438
de Gandt, François, 43n, 61n, 67n
de Maria, M., 395n
de Sitter, Willem, 301, 302n, 303, 305
de Volder, Burcher, 99
Debs, Talal A., 280n
decoherence, 364–67
deduction from phenomena, 61, 78; see also Newton's rules of philosophy
deduction, transcendental, 111
deductive inference, 226–27
deferent, 12n
degeneracy, degree of, 447
Democritus, 15n, 310
dense (in a topological space), 451
density operator, see statistical operator
density, 42n, 124
Descartes, René, 1, 2, 6, 8, 10, 14, 17–20, 21, 30, 31, 32, 33, 34, 35, 36, 42, 43, 46, 65n, 71, 76, 79, 81, 113, 126, 175n, 184, 274n, 400, 402, 406, 432
determinacy, thoroughgoing, 109, 141, 434
determinant, 449
determinism, 94, 132, 142–44, 230–34, 334, 337–38, 401; kinematical, 280–83; logical, 280
DeWitt, Bryce S., 368, 392
Dicke, R. H., 305
Dieudonné, Jean, 413n
differentiable function, 159

differentiable manifold, see manifold, smooth
differential equations, 59n, 75n, 131, 132, 133–34; and determinism, 93–94, 131–32
Dirac, Paul Adrien Maurice, 321, 330n, 336, 339, 388, 393, 394, 395, 396, 430n, 452
Dirac notation for inner product of vectors, 330n, 388, 446–47
Dirac's δ function, 452
distance: symmetric function, 275
distributive laws, 454
Doppler effect, 40, 214; see also redshift
Drake, Stillman, 24n, 398
du Fay, Charles, 81
Dugas, René, 84n
Duhem, Pierre, 186n, 216, 242–48, 430
Duhem–Quine thesis, 430
Duns Scotus, 2
dynamical system, 84–94; conservative, 88, 206; holonomic, 88, 90
dynamis, Greek for 'force', 84
Dyson, Freeman, 395

Earman, John, 297n, 305n, 306n, 429n, 440n, 441
effective field theories, 395
efficiency of heat engine, 188
Ehlers, Jürgen, 271
Ehrenfest, Paul, 210n, 297
Ehrenfest, Tatiana, 210n
eidos, 9
eigenvalue, see proper value
eigenvector, see proper vector
Einstein, Albert, 35n, 37, 42, 73n, 74, 75n, 105, 137, 147, 167, 180, 204, 205, 211, 234, 241, 242, 246n, 248, 249–306, 307, 308, 310, 313–16, 318, 321,

498

Index

322n, 326, 349–55, 373, 385,
417, 418, 424, 429, 435–36,
442n
Einstein field equations, 297–98,
426; modified in 1917, 300
Einstein static world model, 167,
299–300
Einstein tensor, 298
Einstein time, 256, 273, 283n;
consistency conditions, 256–57
electromagnetic field, 172–73, 175–
79, 250, 427, 428–29; energy,
176, 184; momentum, 176;
tensor, 428
electromagnetic waves, 174, 178,
179, 180
electromotive force, 176
electron, 309; Greek for 'amber',
81
electrotonic state, 177
element of reality, 349, 350
elliptic geometry, 156
Ellis, George F. R., 56n, 143, 401
Ellis, R. L., 200
Empedocles, 36
energetics, 186n, 242n
energy, 35n, 175n, 182, 183–86,
188n, 286; kinetic, 186; of
electromagnetic field, 176, 184;
potential, 186; psychic, 186n; see
also conservation of energy
energy–momentum tensor, 295
ensemble, 204, 347–48
entailment, 379n
entropy, 188, 191–95, 207n, 208,
213
epicycle, 12n
EPR paradox, 308, 349–55, 356,
373, 378; fallacy in EPR
argument, 353
equations of motion: Hamilton, 93,
325, 326, 338; Lagrange, 90, 91,
94, 95; Newton, 84
equivalence principle: Newton's, 60;

Einstein's, 290, 292; strong and
weak, 290n
Erlangen Program, 152–55, 264
error statistics (Mayo), 228n, 441n
Euclid, 7, 147–49, 406, 408, 429
Euclid's fifth postulate, 148, 151
Euclidian geometry, 114–15, 116–
117, 162, 164, 169n, 378;
difference with Lobachevskian
geometry, 149
Eudoxus, 14, 61n
Euler, Leonhard, 84, 105, 106, 168,
239
event (meaning 'spacetime point'),
261
Everett (III), Hugh, 368, 387–93
exclusion principle, 319, 321
exp(x), 196n
expectation value, 344
experiment, 3, 17, 144, 245–46,
371, 401, 425n
explanation, scientific, 102, 242–43,
244n
explication of conceptions, 219

facts and concepts, 431, 434, 437
Falkenburg, Brigitte, 328n, 355n,
395n, 417n, 430n
false assumptions: use of in physics,
26
Fano, U., 347
Faraday, Michael, 83, 171, 172,
173, 177, 178, 180n, 309
Fermat, Pierre, 20
Fermi, Enrico, 187, 188, 321
Février, Paulette, 379
Feyerabend, Paul K., 220, 229,
379n, 404, 421
Feynman, Richard P., 177n, 178n,
179n, 428
field (algebraic), 168n, 444
field (physical), 168–80; effective,
395; electric, 170;
electromagnetic, 175–80;

metaphysica specialis, 110, 113, 138, 140
meter, 262, 274
metric, Minkowski, 267, 292, 299, 303, 429; Riemannian, 165, 267; semi-Riemannian, 269, 426, 429
Michell, John, 82n
Michelson, Albert Abraham, 52n, 251, 252, 312, 424
microwave background radiation, *see* thermal radiation, cosmic background
Migne, J. P., 406n
Mill, John Stuart, 216, 232n, 407
Miller, A. I., 150n, 394
Milne, E. A., 304
mind and body, 20, 234, 400
Minkowski, Hermann, 109n, 257, 260–71, 292n, 295, 299, 412, 419, 427, 428, 429
Mittelstaedt, Peter, 356n, 358n
mixture of quantum states, 345, 346
Möbius strip, 158
model, 415, 416–17; two senses of word, 415n
momentum: classical, 18, 35–36, 42; Cartesian, 18; relativistic, 287; term meaning 'increment', 21; *see also* conservation of momentum, generalized momentum, quantity of motion
monad, 81, 101n, 104
Moore, Henry, 158
Morley, Edward W., 251, 252, 424
Mormann, Thomas, 414
morphe, 9
Mosterín, Jesús, 306n
motion: absolute and relative distinguished by Newton, 53; Aristotle's concept, 9, 19; Descartes's concept, 17, 30–31; forced, 11, 16; natural, 10, 16; of missiles, 11, 20–21, 27–30; two

coexisting in one body, 28; uniform, 7, 21, 51; uniformly accelerated, 21–26; *see also* equations of motion, kinesis, perpetual motion, *phora*, quantity of motion,
Motte, Andrew, 48n
Moulines, C. Ulises, 402, 414, 426n
Muller, F. A., 329n
multilinear, *see* n-linear
Murphey, M. G., 223n

n-linear function, 447
n-manifold, 158
n-tuple, ordered, 443n
nabla operator, 95
nature: Aristotelian definition, 9; book written in mathematical language, 15, 432; economy of, 72, 406; frame of, 41; origin of idea, 8
necessity, 121–22, 129–31, 134
neighborhood, 456
Nernst, Walther, 194, 316n
Nersessian, Nancy, 180n
Neumann, Carl, 51, 53
Neurath, Otto, 402–403, 404, 421
neutral element, 153, 443
Newton, Isaac, 1, 14, 20n, 21, 39n, 41–79, 98, 100, 105, 106, 113, 162, 170, 222, 234, 239–41, 246n, 249, 250, 282n, 283, 288, 297, 322, 373, 377, 378, 385, 406, 418, 429, 433, 436, 442n
Newton's Laws of Motion, 33, 45, 87, 175, 180, 259–60, 283; First, 45–46, 51, 53–54, 239, 240–41, 273; Second, 47–48, 284; Third, 48–49, 184
Newton's rules of philosophy, 61, 69–74
Neyman, Jerzy, 228
Nicol, William, 347n

proper vector, 339, 358n, 447
Prosperi, G. M., 356n, 364
psi (ψ) function, 327–28, 332;
collapse of, 361, 362; defined in
configuration space, 332; of the
universe, 337, 387; probabilistic
interpretation, 333–36, 373
Ptolemy, 12n
Putnam, Hilary, 281, 288, 379n,
383–86, 407, 422
Pythagoras's theorem, 108n, 109n,
118; for arbitrarily many
dimensions, 450

QED (quantum electrodynamics),
394–96
QM (quantum mechanics), 128,
131, 308, 396, 419; background,
308–21; Copenhagen
interpretation, 368–73, 392;
genesis, 321–36; groundlines,
336–48; HV-extensions, 374, 375;
incomplete? 349–55;
indeterminacy relations, 348–49,
369; many-worlds interpretation,
387–93; paradoxes, 349–67;
philosophies, 367–93
quadrivium, 4
qualities: occult, 77, 78; primary
and secondary, 15–16, 76, 398
quantifier, 227n
quantity of matter, 18, 42, 124, 288;
see also mass
quantity of motion, 18, 34, 35, 42
quantum, 254n, 307–308;
electrodynamics, see QED;
mechanics, see QM
quantum condition, 312; sharpened,
325
quantum evolution, two laws? 341–
42, 360, 387
quantum logic, 378–86
quantum numbers, 312

quantum theory: old, 309–13, 319,
321, 355; relativistic, 393–97, 419
Quêtelet, Lambert Adolphe Jacques,
180
Quine, Willard van Orman, 430

radioactive decay, 309, 314, 333
radioactivity, 309
Ramsey, Frank Plumpton, 438
random sequence, 200n
Rankine, William John, 35n, 185,
188n
ratios, universal calculus of, 6–7
Rayleigh–Jeans "law", 205
reason as guide, 138, 139
Rechenberg, Helmut, 308n
reciprocity principle, 258
Redhead, Michael, 273n, 280n,
336n, 353n, 357n, 375n, 396,
452n
redshift, cosmological, 249;
gravitational, 291n
reference without sense, 288, 422
reflection, 56
Reich, Klaus, 111n
Reichenbach, Hans, 273, 274, 275–
77, 379, 421
Reichenbach time, 275–76;
modified, 276–77
relativistic quantum theories, 393–
97
relativity of motion: Descartes, 31;
Kant, 146
relativity principle (Einstein), 253,
256, 259, 287n; (Newton), 53,
250, 259
relativity theory, 42, 249–306, 308;
alleged conflict with quantum
nonlocality, 355; see also GR
(general relativity), SR (special
relativity)
Remond, Nicolas, 101n
renormalization, 395